Berries and Berry Bioactive Compounds in Promoting Health

Food Chemistry, Function and Analysis

Series editors:
Gary Williamson, *Monash University, Australia*
Alejandro G. Marangoni, *University of Guelph, Canada*
Graham A. Bonwick, *AgriFoodX Limited, UK*
Catherine S. Birch, *AgriFoodX Limited, UK*

Titles in the series:

1: Food Biosensors
2: Sensing Techniques for Food Safety and Quality Control
3: Edible Oil Structuring: Concepts, Methods and Applications
4: Food Irradiation Technologies: Concepts, Applications and Outcomes
5: Non-extractable Polyphenols and Carotenoids: Importance in Human Nutrition and Health
6: Cereal Grain-based Functional Foods: Carbohydrate and Phytochemical Components
7: Steviol Glycosides: Cultivation, Processing, Analysis and Applications in Food
8: Legumes: Nutritional Quality, Processing and Potential Health Benefits
9: Tomato Chemistry, Industrial Processing and Product Development
10: Food Contact Materials Analysis: Mass Spectrometry Techniques
11: Vitamin E: Chemistry and Nutritional Benefits
12: Anthocyanins from Natural Sources: Exploiting Targeted Delivery for Improved Health
13: Carotenoid Esters in Foods: Physical, Chemical and Biological Properties
14: Eggs as Functional Foods and Nutraceuticals for Human Health
15: Rapid Antibody-based Technologies in Food Analysis
16: DNA Techniques to Verify Food Authenticity: Applications in Food Fraud
17: Advanced Gas Chromatography in Food Analysis
18: Handbook of Food Structure Development
19: Mitigating Contamination from Food Processing
20: Biogenic Amines in Food: Analysis, Occurrence and Toxicity
21: Nutrition and Cancer Prevention: From Molecular Mechanisms to Dietary Recommendations
22: Health Claims and Food Labelling
23: Nutraceuticals and Human Health: The Food-to-supplement Paradigm
24: Nutritional Signalling Pathway Activities in Obesity and Diabetes
25: The Chemistry and Bioactive Components of Turmeric
26: Foodomics
27: Food Proteins and Peptides: Emerging Biofunctions, Food and Biomaterial Applications
28: Handbook of Antioxidant Methodology: Approaches to Activity Determination
29: Fats and Associated Compounds: Consumption and Human Health

30: Oral Processing and Consumer Perception: Biophysics, Food Microstructures and Health
31: Development of Trans-free Lipid Systems and their Use in Food Products
32: Advanced Spectroscopic Techniques for Food Quality
33: Berries and Berry Bioactive Compounds in Promoting Health

How to obtain future titles on publication:
A standing order plan is available for this series. A standing order will bring delivery of each new volume immediately on publication.

For further information please contact:
Book Sales Department, Royal Society of Chemistry, Thomas Graham House, Science Park, Milton Road, Cambridge, CB4 0WF, UK
Telephone: +44 (0)1223 420066, Fax: +44 (0)1223 420247,
Email: booksales@rsc.org
Visit our website at www.rsc.org/books

Berries and Berry Bioactive Compounds in Promoting Health

Edited by

Dorothy Klimis-Zacas
The University of Maine, USA
Email: dorothea@maine.edu

and

Ana Rodriguez-Mateos
King's College London, UK
Email: ana.rodriguez-mateos@kcl.ac.uk

ROYAL SOCIETY
OF **CHEMISTRY**

Food Chemistry, Function and Analysis No. 33

Print ISBN: 978-1-83916-216-9
PDF ISBN: 978-1-83916-545-0
EPUB ISBN: 978-1-83916-546-7
Print ISSN: 2398-0656
Electronic ISSN: 2398-0664

A catalogue record for this book is available from the British Library

The Royal Society of Chemistry is a charity, registered in England and Wales, Number 207890, and a company incorporated in England by Royal Charter (Registered No. RC000524), registered office: Burlington House, Piccadilly, London W1J 0BA, UK, Telephone: +44 (0) 20 7437 8656.

Visit our website at www.rsc.org/books

Printed in the United Kingdom by CPI Group (UK) Ltd, Croydon, CR0 4YY, UK

Preface

The last 20 years have evidenced a dramatic increase in research related to the health benefits of berries and berry bioactive compounds. Novel insights into the diverse composition and structure of berry phytochemicals, their bioavailability, metabolism and impact on the gut microbiome are among the key areas that have been avidly investigated. Additionally, research on the health benefits of edible berries, their bioactive compounds and related metabolites has also exploded, resulting in new knowledge regarding their role in chronic diseases, their molecular mode of action and their clinical significance in promoting health and preventing chronic disease. Their beneficial impact on chronic diseases such as cardiovascular disease, diabetes mellitus and cancer and their role in attenuating inflammation and impacting the gut microbiome has been documented in *in vitro, in vivo* and clinical studies. "Berries and Berry Bioactive Compounds in Promoting Health" reviews state-of-the-art research findings in this area. This comprehensive book, comprised of 13 chapters, brings together international experts in their corresponding fields to review the current literature, identify knowledge gaps and limitations of the current research and propose future directions. Its uniqueness lies in its approach to presenting recent and comprehensive science-based evidence on areas that have been largely and robustly researched, focusing on human intervention studies, molecular mode of action and intraindividual responses to dietary interventions.

Unique topics covered include the chemistry and analytical methods of detection of berry bioactive compounds, bioavailability, metabolism and factors that affect them, including individual variability, interaction with the gut microbiome and their role and mechanism(s) of action on chronic diseases such as cardiovascular disease, diabetes mellitus, obesity-induced inflammation, neurodegenerative diseases, cancer and wound healing.

Food Chemistry, Function and Analysis No. 33
Berries and Berry Bioactive Compounds in Promoting Health
Edited by Dorothy Klimis-Zacas and Ana Rodriguez-Mateos

Published by the Royal Society of Chemistry, www.rsc.org

To advance research on the health benefits of berries and their bioactive-containing compounds, the complexity of chemical structures and diversity of the bioactive compounds contained in berries necessitate sophisticated and advanced analytical methodologies. Chapter 1 presents a summary of current practices, focusing on modern analytical techniques used in the analysis of berry bioactives in foods and biological samples, with tips for quality assurance and reporting practices.

The bioavailability and metabolism of berry (poly)phenols is discussed in Chapter 2, which explores recent evidence demonstrating that contrary to previous beliefs, berry (poly)phenols are highly bioavailable as a result of the action of the gut microbiota that enable phenolic catabolites to be absorbed in substantial amounts. As a result, the complexity of the composition of the colonic microflora has been associated with inter-individual variability and the classification of subjects as non-, low- and high-producers of phenolic catabolites. Inter-individual variability is also responsible for the conflicting evidence we observe in human studies after dietary berry interventions. Chapter 3 explores and addresses factors of inter-individual variability in cardiometabolic health following berry (poly)phenol consumption from evidence from human clinical trials, reviews and meta-analyses of randomized controlled trials.

Currently, berries are being investigated as a new potential arsenal of antimicrobials and prebiotics because of their ability to selectively inhibit enteric pathogens while promoting the growth of beneficial microorganisms. The critical role of gut microbiota in berry bioactive compound metabolism and their resultant role in health, as well as the specific constituents of berries and microorganisms that support this beneficial relationship, are presented in Chapter 4.

Evidence from recent advances of the role of dietary berries in cardiovascular disease and their molecular mechanisms involved has been documented from epidemiological, clinical and animal studies. Specifically, berries have increasingly been demonstrated to exert cardiovascular-protective effects, including the beneficial modulation of vascular function (Chapter 5) by acting on multiple targets such as activating the endothelial nitric oxide signaling pathway, suppressing endothelial inflammation, modulating redox signaling, altering lipid metabolism and modifying vascular glycocalyx (Chapter 6).

Upon examining the role of berries in cognition and cognitive function with age from epidemiological as well as animal and human intervention studies, it has been documented that diets rich in berries are associated with a reduced risk of cognitive decline with age and improvements in cognitive function. This area is increasingly being explored recently, considering neurodegenerative diseases such as Alzheimer's, Parkinson's, amyotrophic lateral sclerosis and others (Chapter 7). A review of the mechanisms of several neurodegenerative diseases and the potential of berry flavonoids, in particular anthocyanins, and their metabolites in tackling the mechanisms behind the onset and the progression of neurodegenerative diseases to elucidate potential targets and new trends in the future of neurodegenerative research is also reviewed (Chapter 8).

Evidence from *in vitro* and *in vivo* studies indicates an array of beneficial effect for berries against cancer and highlights some mechanisms including reactive species scavenging and inflammasome suppression (Chapter 9). Since there is limited clinical research corroborating these findings, it is challenging to identify dietary doses and establish dietary recommendations at this time.

A review of the last 20 years of observational and clinical studies on the relationship between intake of dietary berries and type 2 diabetes mellitus (T2DM) indicate that long-term dietary berry intake is associated with a lower risk of the disease in adults, supporting the consumption of dietary berries to prevent, delay, and treat T2DM (Chapter 10).

Data on the relationship between the intake of berries and capacity to counteract obesity-induced inflammation in human intervention trials indicate that berries may help mitigate the inflammatory state in overweight/obese subjects without having a direct effect on obesity (Chapter 11). The main molecular pathways involved in the observed beneficial effects on chronic inflammation from *in vitro* and animal studies are explored (Chapter 12).

A recent area of research has focused on wound healing and the role of berries and berry (poly)phenols. The wound healing potential of various bioactive molecules is presented and discussed (Chapter 13).

This book is directed at researchers and scientists working in biomedical sciences and conducting research in the area of berries and their bioactive compounds; health care professionals, *i.e.*, nutritionists, dietitians and clinicians; nutraceutical and natural food products developers (industry) and food regulators.

We would like to thank the staff at the Royal Society of Chemistry, Liv Towers, Katie Morrey and Nicki Dennis, for their professionalism and support, and the production team for their attention to detail and timely production of the book. Additional thanks go to the contributing authors for their professional collaboration, scholarly reviews and timely submission of their chapters. We also thank our families for their constant understanding and support in realizing this book.

Dorothy Klimis-Zacas PhD, FACN and Anna Rodriquez-Matteos PhD

Contents

Food Chemistry, Function and Analysis No. 33
Berries and Berry Bioactive Compounds in Promoting Health
Edited by Dorothy Klimis-Zacas and Ana Rodriguez-Mateos

Published by the Royal Society of Chemistry, www.rsc.org

CHAPTER 1

Bioactive Compounds of Berries: Chemistry and Analytical Methods of Detection

C. D. KAY,* A. NEILSON AND M. FERRUZZI

North Carolina State University, Food Bioprocessing and Nutrition Sciences Department, Plants for Human Health Institute, Kannapolis, NC, USA
*Email: cdkay@ncsu.edu

1.1 Introduction

Due to the diversity of (poly)phenol structures and their localization within a fruit morphology, and those modified as a result of human or microbial enzymes, multiple extraction and separation techniques have been developed and reported,[1–3] including spectrophotometry, gas chromatography, liquid chromatography, thin-layer chromatography and capillary electrophoresis. Similarly, there are a diverse array of techniques utilized for identification and quantitation, including flame ionization, thermal conductivity, near-infrared, mass, ultraviolet–visible (UV–Vis) and fluorescence spectrometry, light scattering, refractive index, as well as nuclear magnetic resonance (NMR) spectroscopy. No single technique is ideal for resolving, identifying and quantitating all subclasses of berry (poly)phenols, which comprise low-molecular-weight phenols (*i.e.*, benzoic acids) to large polymers (*i.e.*, procyanidins, hydrolyzable tannins, *etc.*). Dozens of textbooks, book chapters and review articles over the years have been devoted to flavonoid/(poly)phenol characterization in plants and animals, notably

Food Chemistry, Function and Analysis No. 33
Berries and Berry Bioactive Compounds in Promoting Health
Edited by Dorothy Klimis-Zacas and Ana Rodriguez-Mateos

Published by the Royal Society of Chemistry, www.rsc.org

The Systematic Identification of Flavonoids,[4] *The Flavonoids: Advances in Research since* 1980,[5] a series of articles in *Current Protocols in Food and Analytical Chemistry*,[6–8] and *Methods in Polyphenol Analysis*.[9] The textbook, *Methods in Polyphenol Analysis*[9] describes in considerable detail methods for polyphenol extraction, identification and quantification across food and biological matrices. The text covers (poly)phenol extraction from foods, biological fluids and tissues, liquid chromatography–mass spectrometry (LC–MS) identification practices, high pressure/high-performance liquid chromatography (HPLC) with coulometric detection, application of liquid chromatography–nuclear magnetic resonance spectroscopy (LC–NMR) in structure elucidation and purification and quantitation of phenolic, (poly)-phenols and procyanidins. More modern, notable texts and reviews include *Photocatalysis – Fundamentals, Materials and Potential*;[10] *Analytical Methods of Phenolic Compounds*,[11] and *Phenolics in Foods*.[12] As (poly)phenols have been covered extensively in a number of previous and recent reviews and textbooks, the present chapter aims to provide a brief summary of current practices, with a focus on modern techniques, tips for best reporting practices, and assuring qualitative and quantitative accuracy, which are not covered in detail in previous texts. Practical methodological considerations (*i.e.*, advice and tips) are provided in detail as supplementary material (Appendix).

1.2 Chemistry of Bioactive Berry Compounds

The most consumed berries (fresh-raw) in the USA are strawberry, blueberry and raspberry (44%, 29% and 15%, respectively), followed by blackberry and other, which include cranberry (9% and 3.4%, respectively) (https://www.statista.com/statistics/191512/fresh-berry-category-share-in-2011/).[13] The primary (*i.e.*, found in highest abundance) biologically active compounds reported in these berries include sugars, fiber, anthocyanins, hydroxycinnamic acids, vitamin C, flavonols, hydroxybenzoic acids, flavan-3-ols, ellagitannins, procyanidins and stilbenes (*listed in order of relative abundance*; Table 1.1). Sugar and fiber are by far the most prominent nutritive components and are generally reported between 12–15 g per 100 g and 2–5 g per 100 g (respectively), while the remaining non-nutrient (poly)-phenols are reported between 0.1 and 170 mg per 100 g, or collectively between 95 and 300 mg per 100 g of fresh weight (http://phenol-explorer.eu; http://www.foodb.ca; https://fdc.nal.usda.gov/, 10/10/2020).[14,15] Some berries, such as strawberry and raspberry, contain relatively high amounts of vitamin C (60 and 25 mg per 100 g, respectively). Even though the focus of this chapter is on non-nutrient berry (poly)phenols, sugar, fiber and vitamin C should not be discounted when considering the broader array of constituents within a berry or when developing methods of extraction, as they provide considerable matrix components which can impact extraction efficiency, chromatographic performance, compatibility with selected detection systems and detection limits. In the present chapter, we focus on analytical considerations for the most abundant (poly)phenols (Table 1.1) within the

Table 1.1 Bioactive components in berries.[a]

Bioactive components	Polyphenol subclasses
Anthocyanins	Flavonoids
Hydroxycinnamic acids	Phenolic acids
Flavonols	Flavonoids
Hydroxybenzoic acid	Phenolic acids
Flavan-3-ols	Flavonoid
Ellagitannins, gallotannins, ellagic and coumaroyl-glycosides	Hydrolyzable tannins
Procyanidins (flavan-3-ol polymers)	Condensed tannins
Resveratrol	Stilbenes

[a]Phytochemicals arranged in order of highest average relative abundance to lowest (top to bottom; https://fdc.nal.usda.gov/; http://phenol-explorer.eu, 10102020),[14,15] based on the most consumed berries (fresh-raw) in the USA (Share of berries sales in the US 2020, by type www.statista.com, 10102020).

most consumed berries, including anthocyanins, hydroxycinnamic acids, flavonols, hydroxybenzoic acids, flavan-3-ols and their polymers, procyanidins and ellagitannins; and their known biological (human, animal and microbial) metabolites (Table 1.2).

In these most highly consumed berries, anthocyanins are generally reported highest in blackberries and blueberries, while hydroxycinnamic acids are highest in blueberries, and flavonols in blueberries and cranberries (http://phenol-explorer.eu; http://www.foodb.ca; https://fdc.nal.usda.gov/, 10/10/2020).[14,15] Further, hydroxybenzoic acids are found in highest abundance in blackberries and cranberries, flavan-3-ols in blackberries, ellagitannins in raspberries and procyanidins in strawberries. Although numerous reported bioactivities of berries and their phytochemical constituents exist, including cardio-metabolic, anti-diabetic, anti-inflammatory, cognitive performance and anti-cancer activities,[16] there is currently no direct evidence to suggest any single phytochemical is responsible for the health effects observed following berry consumption.

1.2.1 Anthocyanins, Flavonols and Flavan-3-ols

Anthocyanins, flavonols and flavan-3-ols are subclasses of flavonoids, which share the same basic structural characteristics, each having a base structure containing two linked aromatic rings (A-ring and B-ring) bridged by a heterocycle containing three carbons and one oxygen (C-ring), with each aromatic ring containing at least one aromatic hydroxyl (see Scheme 1.1). Anthocyanins, flavonols and flavan-3-ols are further divided into subclasses based on functional groups on the C-ring and hydroxylation and methoxylation patterns of the B-ring. The primary distinguishing feature separating these subclasses is C-ring unsaturation, where anthocyanins have double bonds between carbon 1–2 and 3–4, while flavonols contain a 2–3 double bond and flavan-3-ols contain no C-ring unsaturation. Other distinguishing characteristics include anthocyanins having a positively charged oxygen (at acidic pH) in their heterocyclic C-ring (C-1), while flavonols contain a C-4 ketone.

Table 1.2 Polyphenol metabolites.

Precursor	Microbial product/Phase II precursor[a]
Flavonoids	
Anthocyanins	Hydroxycinnamic acids
	3-(Hydroxyphenyl)propanoic acids
	Hydroxyphenylacetic acids
	Hydroxybenzoic acids
	Benzene-di/tri-ols
	Hippuric acids
Flavonols	Hydroxyphenylacetic acids
	Hydroxybenzoic acids
Flavan-3-ols	Hydroxyphenyl-gamma-valerolactones
	5-(Hydroxyphenyl)-gamma-valeric acids
	3-(3-Hydroxyphenyl)propanoic acids
	3-(Hydroxyphenyl)hydracrylic acid
	Hydroxyphenylacetic acids
	Hydroxybenzoic acids
	Hippuric acids
Phenolic acids	
Hydroxycinnamic acids	
Hydroxybenzoic acids	Benzene-di/tri-ols
Hydrolyzable tannins	
Ellagitannins	Ellagic acid
	Urolithins
Gallotannins	Ellagic acid
	Gallic acid
	Urolithins
	Benzoic acids
Coumaroyl-glycosides	Coumaric acids
	Flavonoids
Procyanidins	
	Flavan-3-ols
	Hydroxyphenyl-gamma-valerolactone
	5-(Hydroxyphenyl)-gamma-valeric acid

[a]Structures present in native unconjugated form but most often found conjugated with sulfate, glucuronide, methyl and glycine as a result of human phase II biotransformation. Hippuric acid is a commonly reported terminal metabolite of (poly)phenols and phenolic catabolites.[17,74]

Further, anthocyanins and flavonols are found in nature glycosylated, while flavan-3-ols are most commonly non-glycosylated, remaining distributed in aglycone form.[17] These unique structural differences, *i.e.*, unsaturation, charge state, glycosylation and hydroxylation, afford differential analytical consideration as they impact stability, recovery and detection characteristics.

1.2.2 Hydroxybenzoic and Hydroxycinnamic Acids

Phenolic acids, including hydroxybenzoic and hydroxycinnamic acids (see Scheme 1.2), contain a six-carbon atom (C_6) phenyl ring and differ by the

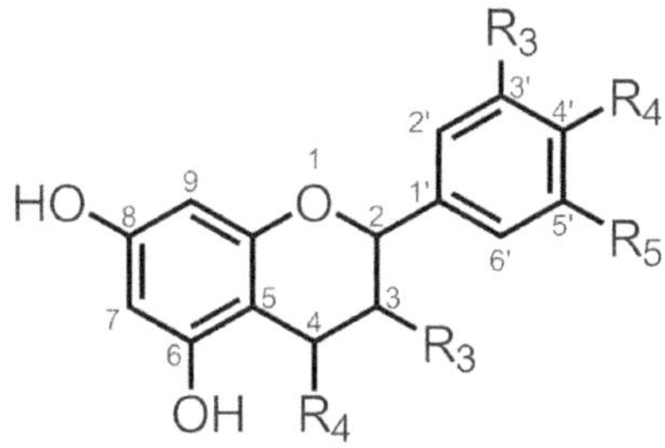

Scheme 1.1 Flavonoid structure (anthocyanins, flavonols and flavan-3-ols). R3: H, OH, O-glycoside, O-glucuronide, sulfate; R4: H, O; R3′, 4′, 5′: H, OH, OCH_3.

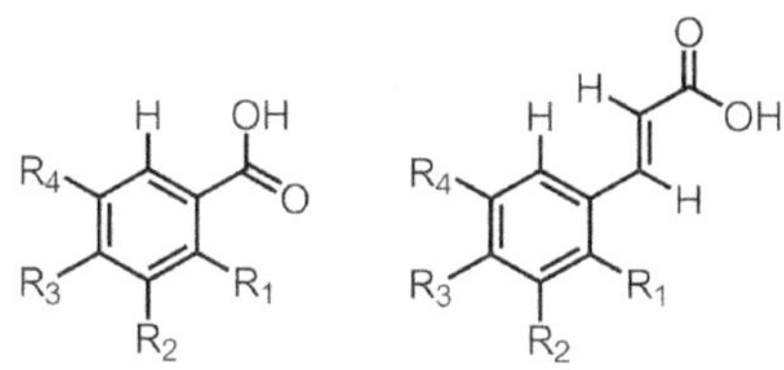

Scheme 1.2 Hydroxybenzoic and hydroxycinnamic acid structure. R1, 2, 3, 4: H, OH, OCH_3.

number of carbons on their side chain. Hydroxybenzoic acids are denoted C_1, having only a single carbon side chain, while cinnamic acids are denoted C_3 phenolics, having a three-carbon side chain. Both being acids, their side-chain terminates with a carboxyl group. Cinnamic acids have a future distinguishing feature, as their side-chain has a C1–C2 double bond (unsaturation). Together these features provide the descriptive nomenclature C_6–C_1 or C_6–C_3 (hydroxybenzoic and hydroxycinnamic acids, respectively).[17,18]

1.2.3 Hydrolyzable Tannins

Gallotannins are composed of a central polyol, typically quinic acid or glucose, with phenolic acids (typically gallic acid) esterified at various positions and in various chain lengths (see Scheme 1.3). Hydrolyzable tannins are referred to as such due to their ability to be hydrolyzed by moderate levels of acid or base, whereas condensed tannins (see below) require stronger nucleophilic reagents for hydrolysis.[19]

1.2.4 Procyanidins

Procyanidins are non-hydrolyzable (*i.e.*, condensed) tannins composed of flavan-3-ol monomer subunits linked by C–C inter-flavan bonds (see Scheme 1.4). In berries, the flavan-3-ol monomers are primarily (+)-catechin and (−)-epicatechin. Procyanidins are further subdivided into A and B types. B-type procyanidins are the simplest, with C4–C6 inter-flavan

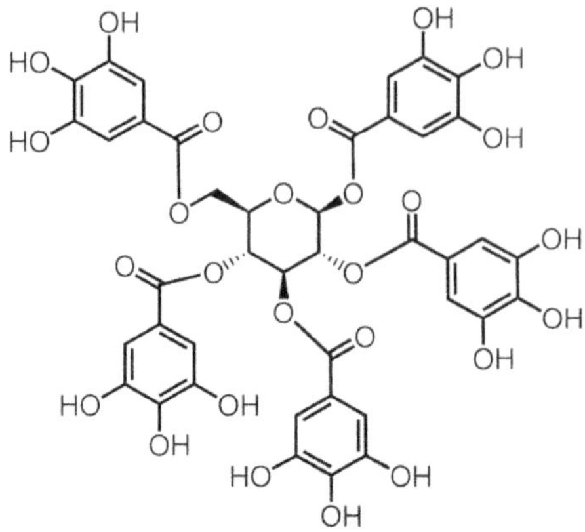

Scheme 1.3 Hydrolyzable tannin structure.

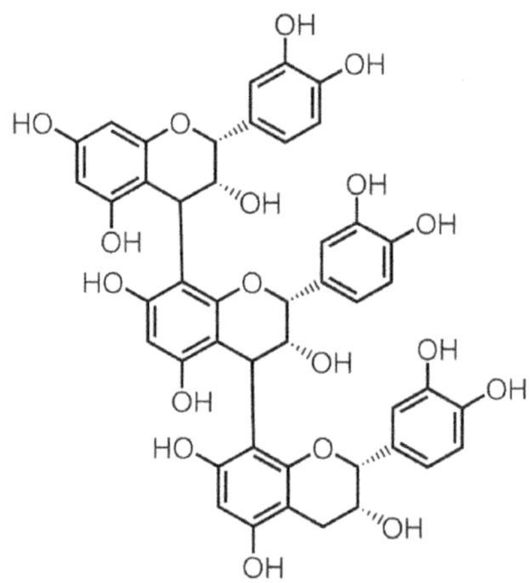

Scheme 1.4 Procyanidin structure.

bonds. A-type procyanidins are more complex, with C4–C8 and C2–*O*–C7 linkages. B-type procyanidins are widely distributed in berries, with A-type procyanidins primarily found in some *Vaccinium* species, particularly cranberries.[20,21]

1.2.5 Phenolic Microbial Metabolites

Phenolic metabolites, products of microbial catabolism of larger polyphenol and/or polyphenol polymers, are generally annotated as C6 structures, signifying they have a six-carbon ring (see Scheme 1.5). These benzene derivative C6 rings, in their simplest forms, are bound with hydroxyl (OH) or methoxy (OCH_3) subgroups forming structures like dihydroxybenzene (*catechol*) or 1,2-dihydroxy-4-methylbenzene (4-*methylcatechol or homo-catechol*). Microbes can cleave these rings forming even smaller structures

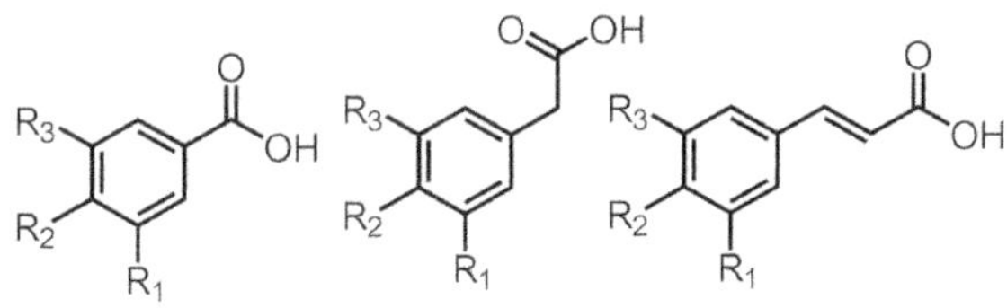

Scheme 1.5 Phenolic metabolite structure (C1, C2, C3). R1, 2, 3 H, OH, OCH_3, O-glucuronide, sulfate.

such as muconates, oxalocrotonates and short-chain fatty acids; however, these low-molecular-weight metabolites often escape detection methods such as electrospray ionization-mass spectrometry (ESI-MS) and UV–Vis. Most frequently, phenolic metabolites of berry (poly)phenols are identified as phenolic acids, C6 rings containing carbon side chains of varying lengths, terminating with a carboxyl (COOH) group. Side chains are generally annotated C1, C2, C3, depending on the number of carbons in the chain. Together this nomenclature is denoted C6–C1, C6–C2, C6–C3, *etc.*[22] These side chains contain double bonds and hydroxyl groups and provide chromophores or ions providing detection using common ESI-MS and UV–Vis-detectors. The nomenclature of microbial metabolites differs by positioning of the double bonds (or site of unsaturation) and oxygen in the side chain. For example, an unsaturated C3 side chain with a double bond between C2 and C3 is referred to as a cinnamic acid.[17,18] Phenolic acids can be derived from most polyphenol forms; however, some microbial metabolites are unique to a specific subclass. For example, hydroxyphenylacetic, (hydroxyphenyl)propanoic and benzoic acid can be derived broadly from polyphenols, while urolithins are specifically derived from the catabolism of ellagitannins. Similarly, gamma-valerolactones are unique to flavan-3-ols. Urolithins differ from many berry-derived microbial metabolites as they possess two phenyl rings and lack a side chain. Phenolic metabolites derived from the microbial catabolism of berry (poly)phenols can be absorbed and found in the circulation as simple phenolics or as conjugated host metabolites, undergoing phase II conjugation in the intestine, liver and tissues, forming glucuronide, sulfate, methyl and/or glycine conjugates.[16] As with their precursor (poly)phenols, differences in conjugation, unsaturation, hydroxylation and methoxylation afford differential analytical considerations.[16–21]

1.3 Instrumentation

Traditionally, polyphenols in berry research were characterized and quantified using HPLC. As polyphenols are chromophores absorbing at wavelengths as high as 450–500 nM (and above), multi-wavelength detectors such as diode-array (DAD) or UV–Vis spectrophotometry were commonly used. Many reviews focusing on anthocyanin characterization (detailing packing materials, acid modifiers, mobile phase compositions, fragmentation patterns, *etc.*)

have been previously published.[23–25] Originally, techniques focused on HPLC using single or multi-visual wavelength or UV–Vis–diode array detectors capable of collecting in-line spectra. These techniques are still useful today for the tentative identification and quantitation of anthocyanins and, in many cases, are more sensitive than MS due to the fact that anthocyanins possess a high-intensity chromophore at acidic pH. However, the most utilized modern instruments for (poly)phenol analysis is ultra-performance liquid chromatography/ultra-high-pressure liquid chromatography (UPLC/UHPLC) coupled with mass spectrometry (MS). Gas chromatography–mass spectrometry (GC–MS) is less commonly used for polyphenols due to their polarity and poor volatility but is likely to gain (or re-gain) popularity for characterizing low-molecular-weight microbial metabolites, which often lack a significant chromophore for spectrophotometric characterization or ionize poorly using ESI-MS. To date, any number of MS platforms have proven effective at characterizing polyphenols from berries and their metabolites, including triple quadrupole mass spectrometer (QqQ), time of flight (TOF) and ion trap; each having its own strengths and weaknesses (as discussed below). While fluorescence is generally less suitable for monomeric (poly)phenol analysis, larger polymers such as procyanidins are most commonly detected using fluorescence spectrometry.[26,27]

Procyanidins present unique challenges for analysis due to their immense (almost limitless) structural variation (*i.e.*, numerous combinations of possible monomer subunits, linkages and degrees of polymerization), which leads to poor resolution performance of common separation methods, poor analytical sensitivity (with sensitivity inversely proportional to molecular weight) and dearth of available analytical standards.[26,28] Due to these factors, most research has focused on the smaller, more easily characterized procyanidins (dimers, trimers, *etc.*). Larger procyanidin oligomers and polymers may be identified by more complex analytical techniques such as matrix-assisted laser desorption/ionization-time of flight mass spectrometry (MALDI-TOF-MS).[29] However, even these methods have significant limitations, and thus the vast majority of complex procyanidins are characterized by somewhat non-specific techniques. These include normal-phase liquid chromatography (LC), hydrophilic–lipophilic interaction chromatography (HILIC),[30] or methods that degrade the polymers and quantify the resulting monomers to identify end-units and monomer ratios (such as thiolysis or phluoroglucinolysis).[27]

1.3.1 HPLC *vs.* UPLC

LC remains the most commonly used separation technique for the characterization of anthocyanins and broader (poly)phenolics, including microbial and host metabolites in fruits, vegetables and biological fluids.[31] Extensive lists of columns, LC conditions and MS platforms have previously been published.[31] In the past 15 years, ultra-performance liquid chromatography (UPLC) has emerged as an eventual replacement for HPLC. UPLC does not differ theoretically from HPLC but rather capitalizes on column and LC platforms with higher optimal operational pressures and sub 2 μm particle size for stationary phases that enable higher levels of resolution. By affording column pressures in excess of 15 000 psi, higher flow rates can be achieved even with reductions in particle sizes without compromising column resolution. This results in improved efficiency and overall resolution due to reduced diffusion and shorter run times.

1.3.2 QqQ *vs.* TOF *vs.* Ion Trap MS

Many modes and configurations of mass spectrometers are reported for (poly)phenol analysis from berries. These include ionization modes such as ESI and MALDI coupled to detectors such as QqQ, TOF, quadrupole time of flight (QTOF) and ion trap (such as Orbitrap); while ion mobility (IM) MS is becoming more popular in recent years.[31] As (poly)phenols have adequate ionization in negative mode ESI, the negative mode is often used for their detection due to reduced background noise to achieve higher S/N ratios, leading to lower limits of detection and better sensitivity compared to similar approaches in positive mode ionization. Anthocyanins, due to their flavylium cations at acid pH, lend themselves more to positive mode ESI; however, most flavonoids will ionize effectively in either positive or negative mode. Some have even reported negative mode ESI can be useful for anthocyanin characterization, as this often leads to higher sensitivity and more characteristic ions [M − [2H](−); [M − 2H + H(2)O](−)], adducts (resulting from acid modifiers) and doubly charged ions, which may be useful in characterizing anthocyanins relative to other (poly)phenols in complex mixtures.[32] To date, the most common instrument configuration for anthocyanin analysis is ESI with QqQ or TOF MS platforms.[31]

1.3.3 QqQ

Triple quadrupole mass spectrometry (QqQ MS) is a highly versatile platform, excelling in situations with broad-spectrum analytical objectives, but is not ideal for qualitative characterization as they provide only nominal mass (nominal mass resolution is insufficient for definitive exact mass conformation required for qualitative identification). The optimal (and most common) operation mode is using multiple reaction monitoring (MRM) or scheduled multiple reaction monitoring (sMRM) MS detection for its selectivity and sensitivity. The main draw of QqQ MS is that it can analyze several hundred mass transitions in a run, therefore proving ideal for the analysis of complex mixtures such as plant or animal tissue extracts. QqQ MS can provide full scan acquisition but is more commonly used for product ion scanning, *i.e.*, parent, daughter or neutral loss scanning functions, which are useful in the characterization of plant and human metabolite conjugates. The QqQ MS has a wide range of applications due to its scan speed and various fragmentation profiling capabilities. QqQ-linear ion trap

(Q-Trap) is an enhanced QqQ, where the final quadrupole can function as a virtual ion trap (or is replaced with a linear ion trap), and is capable of collision-induced dissociation (CID) fragmentation profiling, providing MS(*n*) scanning capabilities and often high mass resolution or accurate mass. Q-Trap data-dependent acquisition modes allow for selected/multiple reaction monitoring (SRM/MRM) and triggered enhanced product ion (EPI) MS/MS scanning for structural elucidation, making it ideal for screening applications.[33]

1.3.4 TOF

The TOF platforms offer theoretically unlimited mass range scanning, which is useful for characterizing large compounds such as hydrolyzable tannins and procyanidins that exceed the typical mass ranges of QqQ platforms. TOF instruments also provide high-resolution mass accuracy required for molecular formula determination/confirmation, allowing discrimination of compounds with similar but different molecular masses (isobaric ions having the same unit mass with different exact masses). Hybrid quadrupole time-of-flight MS (Q-TOF) is often the chosen instrument for metabolite screening (plant or animal) and is preferred for their full scan sensitivity, mass accuracy and high data-acquisition rates. These instruments are also commonly used in metabolomics as they provide detection of a broad range of analytes (known and unknown) in complex mixtures. The advantage of Q-TOF over TOF instruments is the ability to obtain more detailed structural elucidation information, providing MS(E) fragmentation data.[33]

1.3.5 Trap

Ion traps such as Orbitrap are high-resolution ion trap mass analyzers often used in proteomics and untargeted metabolomics, due to their ability to scan tens of thousands of MS features with ultra-high resolution and reliable mass accuracy. LTQ Orbitrap is a hybrid high-resolution mass analyzer combined with a linear ion trap, thus providing accurate mass and having the capacity for MS(*n*) scanning functionality, providing added functionality and versatility for structural elucidation. Traps and TOF have similar functionality but have one drawback relative to QqQ MS, as they are less compatible with high-throughput rapid chromatography applications, such as high-throughput quantitative metabolite screening in complex mixtures.[33]

Practical Instrumentation considerations (i.e., advice and tips) are provided in detail as supplementary material (Appendix).

1.4 Extraction

Accurate quantitation requires sound sample preparation techniques that maximize recovery of the analytes in question while optimizing the reproducibility and robustness of the method. When separating (poly)-phenols from fruits such as berries, or biological matrices, there are many matrix components to consider, including sugars, proteins, carbohydrates, lipids and fibers, which can negatively impact extraction efficiency and quantitative accuracy. In a systematic review published by Singh *et al.* (2020)[34] summarizing LC–MS methods for the identification and quantification of anthocyanins in fruits and vegetables, the authors found that extraction and sample preparation techniques were the most critical factors in establishing quantitative accuracy and consistency. Many methods of extraction can be applied to berries or berry-derived products, including refluxing, heating, supercritical extraction, *etc.*; however, the most frequently utilized techniques in laboratory environments are liquid–liquid extraction (LLE), solid–liquid extraction (SLE) and solid-phase extraction (SPE). Each method has its advantages and disadvantages. Struck *et al.* (2016)[35] reviewed common processing and chemical analysis techniques (Table 1.3) for extracting and assessing (poly)phenols in berry pomace and containing details on a wide variety of berries, including blueberry, strawberry, raspberry, blackberry, chokeberry, black currant. Multiple processing variables were explored across published works; the most common extraction conditions for low-molecular-weight (poly)phenols in berries involved the use of organic solvents such as methanol, ethanol or acetonitrile, water and an acid modifier (often formic acid or citric acid if using MS).[31,35]

Many reviews on methods for the extraction of berry anthocyanins in plants, juices, wines and biological fluids have been published over the years. Originally, techniques focused on extraction using LLE, SLE or large open columns packed with sorbent materials such as Amberlite XAD resin or

Table 1.3 Basic techniques used in sample pre-processing and extractions.

Pre-processing	Processing
Disruption	**Extraction**
Pressing	Soaking
Extrusion	Alcohol extraction
Milling/grinding	Acetone extraction
Homogenization/blending	Water extraction
Heating	CO_2 extraction
Sonication	**Separation**
Stirring/shaking/vortexing	Solid phase
Cleanup	Liquid–liquid
Physical separation	Solid–liquid (adsorbent; *i.e.*, amberlite)
Sieving	**Concentration**
Filtration	Freeze drying
Dialyses	Oven drying
Precipitation	
Centrifugation	
Dilution	
pH adjusting (acidic)	
Enzyme hydrolysis	
Acid hydrolysis	

prepacked SPE single cartridges.[2,25,31] More recently, Kafkas *et al.*[1] provided a review of advanced methods for the analysis of phenolics in fruits and reported low-molecular-weight phenolics and higher-molecular-weight polyphenols (such as flavonoids) were generally extracted using the same extraction media (*i.e.*, packing materials and sorbents) and conditions, while larger polymers required alternative methods. For example, hydrolyzable tannins (widely distributed in berries),[36] due to their structural complexity, require techniques such as methanolysis, which is used to break the hydrolyzable tannins down into their polyol and phenolic acid subunits, for analytical extraction and characterization of subunit ratios.[37–39]

1.4.1 Disruption Techniques

Primary sample preparation methods are critical to establishing the best conditions for extraction and subsequent analysis from berry components. Primary extraction from fresh tissues, frozen tissues and freeze-dried powders are widely reported. In all cases, adequate homogenization of samples is critical to ensure a representative sample for the cellular disruption that will facilitate the release of phenolics for subsequent solvent extraction. Some common techniques include, but are not limited to, physical homogenization, heat treatment, ultrasonication and microwave-assisted extraction.[40,41]

1.4.2 Cleanup Techniques

For cells, tissues and organs, it is critical to lyse cells, ensuring the release of intracellular compounds. This is typically performed using a lysis buffer (such as RIPA: Radioimmunoprecipitation Assay Buffer), in combination with fine homogenization, a French press, ultrasonication, or magnetic bead disruption. Enzymatic digestion may be used for particularly difficult matrices, such as bone or connective tissues. Subsequently, protein–polyphenol bonds are disrupted using reagents such as phosphoric acid, and proteins and phospholipids are precipitated using organic solvents (acetonitrile, methanol) and zinc sulfate. Centrifugation or filtration is then used to remove the insoluble material, followed by re-extraction/washing with the solvent of choice. Extracellular components from urine or cell culture media can often be treated using "dilute-and-shoot" procedures; however, some form of sample "cleanup" is recommended (*i.e.*, SPE, LLE) and often followed by concentration. For biological fluids, the process is similar, although cell lysis and disruption are typically omitted due to the absence of cells in these materials.

1.4.3 Liquid–Liquid, Solid–Liquid

LLE and SLE have a long history of use in polyphenol extraction from plants due to their often simple and rapid nature, in addition to their ability to be scaled up for use on large masses of materials. Such methods use alcohols, acetone, hexane, ethyl acetate, diethyl ether, chloroform and often require the addition of varying proportions of water in order to capture more polar phenolic compounds such as highly hydroxylated flavonoids, hydroxybenzoic and hydroxycinnamic acids and larger procyanidins. Also, some of the solvents used in LLE and SLE are inherently dangerous to work with and require consideration. Although these techniques are useful for bulk extraction, they often have higher variation in quantitative analysis when using low sample volumes. Additional techniques are often required to remove solvent residues from final mixtures prior to analysis or consumption as extracts. One benefit of these procedures is that they rapidly separate polar polyphenols from other unwanted matrix components such as oils and waxes. More modern techniques include supercritical fluid extraction, which benefits from being solvent free and requires relatively no additional sample cleanup.[2,42]

Polarity is the number one consideration when establishing an extraction solvent or suitable chromatography column and solvent system. Polyphenols can be challenging to universally extract with any single solvent due to their varying polarities. For example, procyanidins are considerably more polar than their monomer subunits and these co-exist in a single botanical matrix-like cocoa, cinnamon and fruits, including grape. Catechin and epicatechin can be extracted effectively with methanol, ethyl acetate and diethyl ether, whereas procyanidins composed of these monomers are most efficiently extracted with mixtures of acetone, water and acetic acid. As such, complex extraction schemes are often reported in the literature, with fractions often requiring combination prior to final preparation for analyses. The introduction of water to extraction solvents is a common practice but presents challenges for subsequent concentration, often eliminating nitrogen evaporation or vacuum evaporation as practical options (where large volumes are at play), necessitating the use of a freeze dryer. Thus, care should be taken to select the extraction solvent that provides the best combination of efficiency while facilitating the desired post-extraction processes.[43]

Practical liquid–liquid and solid–liquid extraction considerations (i.e., advice and tips) are provided in detail as supplementary material (Appendix).

1.4.4 SPE

A common sample preparation technique used in berry research is SPE, which is useful in removing complex matrix components from plant and animal tissues that may not be fractionated during LLE while retaining aqueous polar (poly)phenols. SPE as a "cleanup" technique is preferred in quantitative workflows as its easily scalable, uses relatively low amounts of solvent, comes in a large number of sorbent bed materials, is relatively rapid, highly reproducible, can be used to concentrate samples, and results in final samples which are free of instrument interfering matrix components. Ninety-six well micro-elution SPE has proven particularly useful in biological matrix cleanup as it is amenable to low volumes of biological fluids and facilitates high-throughput sample processing. SPE can also be used for solid tissue, generally following dilution, homogenization and centrifugation steps. If non-compatible solvents are required for tissue homogenization and LLE, samples can be dried and re-dissolved in appropriate dilute loading solvents after centrifugation and concentration. Polyphenol stability can often be achieved by adjusting pH using acids such as formic acid or hydrochloric acid (HCl).[44]

Practical solid-phase extraction (SPE) considerations (i.e., advice and tips) are provided in detail as supplementary material (Appendix).

1.5 Separation (Columns, Packing Materials and Mobile Phases)

There are numerous columns and stationary phases on the market supplied by any number of manufactures, which provide appropriate qualities for polyphenol and phenolic metabolite separation. Most suppliers offer column selection guides, application notes or apps, which are useful in identifying an appropriate stationary phase. In 2010, Ignat *et al.* published a review on methods for characterizing (poly)phenolic compounds in fruits and vegetables,[2] providing summaries of reported extraction and LC conditions. Here the most commonly reported methods of separation included the use of standard C18 or modified C18 columns and mixtures of acetonitrile, water and formic acid. More recently, Singh *et al.*[34] published a systematic review of LC–MS methods for the identification and quantification of anthocyanins in fruits and vegetables. The review summarized many types of column stationary phase, mobile phase, acid modifiers and detection techniques, citing reverse-phase C18 as the most common stationary phase and acetonitrile or methanol, and water with formic acid as mobile phase components. Other common stationary phases used in polyphenol research are reverse-phase octadecyl modified C18 (*i.e.*, polar-embedded/end-capped), phenyl-hexyl modified, pentafluorophenyl propyl modified (fluorinated silica/PFP) and ammonium sulfonic acid modified (HILIC) columns. PFP columns are particularly useful for retaining low-molecular-weight phenolics, which may otherwise not be sufficiently retained using C18 columns. For establishing simple methods involving a few analytes, application notes provided by the manufacturers are generally useful; however, when choosing a specialized column/stationary phase for complex mixtures and requiring separation of a broad-spectrum of structures of varying size and polarities, compromises in resolution (peak shape *vs.* retention) may be required. HILIC (typically diol) columns are becoming more common, particularly where required to resolve highly polar polymers such as procyanidins. HILIC and associated solvent systems are also more compatible with ESI than traditional normal-phase solvent systems, which are optimal for atmospheric pressure chemical ionization (APCI). UPLC-HILIC columns facilitate rapid separations compatible with ESI and reduce run time considerably from traditional reverse phase columns (*e.g.*, from 1 h to a few minutes).[30]

While samples should be filtered or otherwise clarified of particulate matter through centrifugation prior to loading into the LC system, the use of guard columns is standard practice and generally paired to the primary column as recommended by the manufacturer. A guard column is used to protect the primary column and prolong its lifespan, which is dependent on the stationary phase, mobile phase, pore size and flow/pressure utilized. Guard column integrity is generally monitored by a % increase in pressure from standard operating pressures and/or retention time drift. Loss of peak shape can also be an indication of failure of the guard column and should be checked prior to the complete replacement of the analytical column. It is generally accepted that peak retention time drift should not exceed a 1.5% coefficient of variation (CV) across multiple injections.[45]

The decision of what column length, internal diameter (ID) and pore size to utilize for a given method is largely made based on acceptable run-time and separation or resolution requirements. C18 phases are generally superior at retaining larger, more polar polyphenols while maintaining optimal peak shape; however, retention of smaller and less polar analytes, such as microbial metabolites, can be compromised. Alternatively, a modified silica phase such as PFP is capable of retaining low-molecular-weight phenolic metabolites with higher effectiveness; however, peak shapes of larger, more polar structures are often compromised. The impact of peak width or resolution on qualitative or quantitative data is dependent on the detection method utilized and may not be an important consideration for mass spectrometers with high duty cycle-time. In such cases, having a unique analyte mass, fragmentation profile and an appropriate number of data points captured is often more important than resolution. Generally, for reproducible quantitative MS data, 10–20 data points per peak are recommended.[46] This value generally refers to peaks of appropriate shape and size; for peaks larger than 10 s and/or abnormally shaped peaks, a proportionately greater number of data points should be captured.

1.5.1 Reverse-phase Liquid Chromatography

The most common technique for the separation of low-molecular-weight (poly)phenols is reverse-phase LC. Most (poly)phenol chromatography is reported using a mixture of formic acid and acetonitrile or methanol, although other acids such as acetic acid, trifluoroacetic acid (TFA) or HCl are not uncommon when using colorimetric detection, TFA or HCl are generally avoided when using MS detectors, do to their ion suppressing qualities. Formic acid is typically reported in the mobile phase between 1 and 5% for colorimetric detection or 0.1–1.5% for MS. Optimizing the acid modifier (type and concentration) is important for polyphenols and phenolics, as it affects both chromatographic separation (stationary phase-specific) and ionization. Generally, a compromise is required between peak width or shape and peak intensity when using MS, as acids are known to influence ionization efficiency and can serve as ion suppressors. This is less important for UV–Vis detectors, which remain less impacted and often benefit from improvements in peak width and height. This requires significant method development when transitioning older methodology based on LC–UV to newer LC–MS and UPLC–MS platforms. When optimizing for complex mixtures of precursor (poly)phenols, their microbial metabolites and biological conjugates, the compromise between peak width and intensity should be made in consideration with the limit of detection, limit of quantitation and integration thresholds, which are relative to the signal-to-noise within a given matrix. Higher acid mobile phase concentrations are generally required to resolve berry anthocyanins but will often reduce the detection limit of low-molecular-weight phenolic metabolites when using MS as a detection method.[47]

1.5.2 Normal Phase, HILIC/Diol/Aqueous Reverse Phase

Reverse-phase separation is most commonly used for (poly)phenol monomers. However, normal-phase separation and HILIC ("aqueous normal-phase") separations are particularly well-suited for the resolution of highly polar polymers such as procyanidins.[26,30,48] Normal-phase and HILIC solvent systems are not optimal for ESI interfaces (HILIC is better, but still not ideal), and thus post-column ionization enhancers such as ammonium formate are often used to improve ionization.[30,37] There are some negative consequences with the introduction of post-column buffers, such as source contamination/fouling and solubility with organic solvents, which can often lead to blockages in LC lines, injector ports, columns, *etc.*

1.6 Method Development and Optimization

The principles for method development remain constant regardless of (poly)phenol or berry bioactive targets, whether developing a method for the extraction of components from foods, tissues or liquids. The key is formulating a method development strategy with predefined endpoints, constraints and expectations, with consideration for the continuum from method development, to optimization and validation. In defining the goals of any analysis, it is important to understand, there is no "one-size-fits-all" strategy, particularly when developing a method for the characterization of a diversity of analytes from a complex matrix. Prioritizing the most important deliverable/outcomes, and acknowledging that sacrifices will likely have to be made on other parameters, is a fundamental principle of method development. Method optimization is generally a process used to fine-tune a method, either improving precision and accuracy or assay efficiency (time, materials use, *etc.*). Method optimization generally follows the same principles of method development: establish defined endpoints with predefined constraints and expectations and execute the strategy. A list of common tips for method development and optimization is provided below.

Practical method development and optimization considerations (i.e., advice and tips) are provided in detail as supplementary material (Appendix).

1.7 Precision and Performance

Precision is the assessment of how close measurement results are to each other. Precision can be established by providing evidence of variability in assay sensitivity and specificity, such as providing CV of calculated concentrations of internal standards or quality control (QC) samples, regression slopes, limits of quantitation, retention time, ion-pairing ratios, peak quality parameters, *etc.* Acceptable % CV for quantitation is generally considered below 15% within the expected range for a given analyte, and as much as 20% CV is considered acceptable at the lower limit of quantitation (LLOQ) and HLQ.[49–51] Assay and instrument performance are generally monitored by including blanks (and/or double blanks) and QC samples inserted randomly in a run at 10–20% of injection frequency.[49–51]

The Horwitz ratio (HorRat) is a normalized performance parameter often used as a measure of inter-laboratory precision as a function of analyte concentration. It can, however, be applied to assess the relative reproducibility of a method within a laboratory and allows for comparison across a range of metabolites. It is calculated as a ratio of the Relative Standard Deviation for an analyte response, RSDr, to that calculated from by the equation ($PRSDr = 2C^{-0.15}$) from the mean concentration, C, found or added, expressed as a mass fraction.[52] HorRat values can be calculated for individual phenolic species to assess the relative precision of the broader method. Ranges of 0.5–2.0 typically suggest high reproducibility for individual metabolites. However, it is not uncommon for HorRat values to be below 0.5 for species at the lower end of the calibration range of the method.

Practical considerations (i.e., advice and tips) for improving precision and accuracy are provided in detail as supplementary material (Appendix).

1.8 Accuracy

Accuracy is the assessment of how close a test result is to the expected value and has relevance to both qualitative and quantitative analyses. Qualitative accuracy can be established by providing evidence of variation in retention time, accurate mass, ion painting ratios, MS/MS fragmentation, *etc.* Quantitative accuracy can be established by providing evidence of regression linearity, actual *vs.* returned concentrations, limits of detection and quantitation, *etc.* Where available, stable isotope dilution can be employed for optimal reproducibility and accuracy. In this quantification technique, stable (^{15}N, ^{13}C and/or ^{2}H) isotopes of each analyte are used as the internal standard in each sample.[53] These isotopes typically have identical elution and ionization properties and are thus optimal internal standards. Using stable isotopes as internal standards is problematic for many polyphenols, as only a small number of these compounds are available in the labeled form. Labeled isotopes are much more common for nutrients, endogenous and drug metabolites, and thus may be of limited utility for many berry compounds. However, when possible, the use of stable isotopes is considered a "gold standard" for quantitation.

The accuracy of a method can also be established through the application of extraction methods to standard reference materials (SRMs) for which the expected value is known. These are available from the National Institute of Standards and Technology (NIST) for a limited array of polyphenol-rich foods, including baking chocolate, cocoa extract, green tea leaves and extracts where a subset of phenolic metabolites have been quantified and can serve as the gold standard. However, the availability of validated SRM for fruits, including berry fruits, remains limited to non-existent.

1.8.1 Qualitative Accuracy

Qualitative accuracy is important when interpreting studies reporting the identification of analytes (compounds, metabolites, *etc.*) using untargeted analysis techniques (*e.g.*, untargeted metabolomics) or in situations where reference standards are not available. The difficulty in interpreting qualitative accuracy is that a significant proportion of studies in the literature do not sufficiently describe how analytes were identified. This is particularly the case for conjugates of secondary plant metabolites (*i.e.*, sugar conjugation or acetylation) or conjugates of human phase-2 metabolism (*i.e.*, glucuronide and sulfate), which are most often not commercially available as reference standards. In these situations, where reference standards are not available to assign structural identity (*i.e.*, provide retention time, fragmentation profiling, *etc.*), identification should be reported as "putative". Assigning an identification based solely on mass, UV–Vis or infrared (IR) spectrometry features is of particular concern with larger molecules and/or compounds having significant numbers of isomers. When it is necessary, investigators reporting qualitative data should describe the identification as an "annotation" or "putative identification", and ideally report other possible isomers matching the molecular formula.

The qualitative accuracy of many studies appears inherently low, particularly as there are no standardized practices for reporting MS methodologies, making it difficult to compare findings reported in the literature.[16] Consequently, it is likely that misclassified and falsely identified metabolites are abundant in the literature and propagated in online metabolomics databases, as highlighted in many publications.[16,54,55] Unfortunately, most researchers in the (poly)polyphenolic field report positive identification based on a single assignment (one precursor mass or precursor to product ion (MS^2) match), following low-resolution nominal mass scanning (*i.e.*, one decimal place accuracy). This is problematic due to the fact that numerous compounds can have similar, or the same, integer/unit masses (isobaric ions). Significant mass resolution (down to 5–6 decimal places) and mass accuracy (>5 ppm difference between measured mass and assigned molecular formula mass) are required for confidence in a molecular formula determination. Furthermore, many investigators neglect to factor in the differences between monoisotopic exact mass (measured by MS) *vs.* average molecular weight (factoring in average isotope abundances) when interpreting mass spectra. This can be extremely problematic for large compounds with numerous carbons or other elements presenting significant isotopic variability. Likewise, compounds with identical molecular formulae but distinct structures will have the same exact mass. Moreover, as there are few reference standards available for confirmation or MS optimization of biological conjugates, false-positive identification is likely, particularly as signal quality (*e.g.*, peak width, number of data points captured, signal to noise, *etc.*) is not reported in the majority of studies. Structures such as hydroxylated cyclic and polycyclic structures (*i.e.*, polyphenols) are also prone to multiple gas phase artifacts, including water elimination (loss −17), hydrogen elimination (−1), radical fragmentation (loss −15, −14), Retro-Diels–Alder (RDA) reactions (C-ring electron rearrangement) (+ or −2), gas-phase dimers, or combinations thereof (such as loss −13 or −12).[56–58] Finally, in-source fragmentation following soft-ionization is often observed for sugars and phase II conjugates, particularly sulfate and glucuronide conjugates.

1.8.2 Quantitative Accuracy

Developing a quantitatively accurate assay generally begins with establishing a linear regression derived from a serial dilution of reference standards in a background matrix representative of the samples requiring analysis. Assay linearity and its precision are established by performing a number of calibration or regression curve analyses (*i.e.*, number of replicates), with accuracy established based on acceptable % CV. Assay linearity essentially defines assay quality relative to quantitative accuracy and is particularly important in establishing reproducibility at the lowest and highest accurate concentrations for the assay. Assay sensitivity is the ability to accurately detect small changes in concentration. Sensitivity is often related to the ability to detect changes at the LLOQ and is influenced by signal/noise and limit of detection but technically applies to any concentration reflected by the standard curve regression. Sensitivity can be negatively impacted by many factors, including carryover and matrix effects.[59]

1.8.3 Semi-quantitative or Relative Quantitative Accuracy

In berry research, semi-quantitative or relative quantitative assays are readily used and require discussion with respect to establishing assay performance.

Berries are frequently consumed as juices (including cranberry, blackcurrant, *etc.*) and many rapid/simple analytical techniques have been established for characterizing total (poly)phenols in these often relatively "clean" fluid matrices. These rapid and cost-effective techniques have the greatest utility in industry for monitoring within-process or batch variability; however, they generally lack the analytical specificity required for academic laboratory research. For example, the Folin–Ciocalteu (F–C), Lowenthal permanganate (L-P), 4-dimethylaminocinnamaldehyde (DMAC) and the bovine serum albumin (BSA) precipitation assays are useful in establishing relative differences between treatments, but when evaluated for absolute qualitative and quantitative accuracy, they lack specificity and sensitivity. As exemplified in a recent publication by Ma *et al.* recently (2017),[60] the authors explored the accuracy of such methods and found the methods provided differential accuracy for quantifying various subclasses of polyphenols and absolute quantitative values for total (poly)phenols were not comparable across the assays. Consistency within a given assay was suitable. However, each assay had a considerably different linear range, indicating the assays have the greatest utility for studying changes to total (poly)phenol content within a common matrix (*i.e.*, juice) and assay, but limited capacity to characterize and quantify individual (poly)phenol contents within and across assays.

1.8.4 Surrogate Regression

When interpreting evidence from existing studies, it is often not apparent how metabolites are quantified. In most situations, reference standards do not exist. In such cases, quantitation should be described as "putative", as surrogate reference standard curves are used for quantitation, which is highly inaccurate in most instances. This practice of using surrogate analytes assumes the response factor of one analyte is similar to another. A response factor is typically defined in spectrometry as the ratio between a signal intensity or area (or area: intensity ratio) and the known quantity of the analyte. For chromatographic techniques which use colorimetric detectors, this is less of an issue, particularly for compounds having similar chromophores; for MS detection, which uses voltage to create a mass-to-charge ratio, response factors for seemingly similarly structured compounds can be extremely dissimilar. Most evidence indicates MS response-factors are highly variable between chemical species, and the use of surrogate regression curves leads to significant over and under-estimation.[61] Investigators reporting quantitative data for compounds that do not have commercially available references standards should identify these instances as "putative quantitation". This is particularly true for conjugated host metabolites of phenolic species. In validating methodology for flavan-3-ol metabolites in plasma, authors have reported greater than 100-fold difference in quantitative concentrations established using response factors derived from the regression lines of authentic catechin, epicatechin and their corresponding biosynthetically generated and authenticated glucuronides and O-methyl-glucuronide conjugates.[61,62] Considering the lack of commercial availability for most (poly)phenol metabolites, estimations are commonly established using aglycone forms, which is likely to lead to even greater inaccuracies. A list of common tips for achieving accuracy is provided below.

Practical considerations (i.e., advice and tips) for improving accuracy are provided in detail as supplementary material (Appendix).

1.9 Method Validation

Method validation is the process of establishing the suitability of an assay for its intended use. Methods for quantification of berry phenolics and metabolites in food and biological matrices should follow standardized validation approaches. One approach, adopted by the AOAC Association of Official Analytical Chemists in collaboration with NIH-Office of dietary supplements and the Food and Drug Administration, is outlined in the Guidelines for Single-Laboratory Validation of Chemical Methods for Dietary Supplements and Botanicals.[63,64] In the simplest sense, this guidance document provides a framework from which to establish the performance characteristics of any method within a lab. Performance, in this case, includes the parameters already discussed above, including selectivity, repeatability, precision and accuracy. It may not be necessary for every laboratory assay to satisfy all of the validation criteria outlined in the guidelines for dietary supplements and botanicals protocols, and many other guidelines exist;[50,51,63,65–67] however, quality assurance requires some level of validation for every assay. Below summarizes some of the more commonplace validation processes used in single user, single lab assay validation.

1.9.1 Robustness/Ruggedness

The robustness or ruggedness of an overall method is defined by the extent to which select factors, such as environment, storage, extraction condition or otherwise, impact the accuracy and precision of a method. It is recommended to establish robustness during method development and refinement or optimization, as well as assessment of final assay validation. Importantly, robustness should be re-established when scaling a method or translating one method for use with different sample matrices, such as applying a method used for fruit on biological tissue. Traditional approaches commonly explored one critical variable at a time (*e.g.*, sample weight, extraction time or solvent volumes/temperature). However, it is possible to determine the impact of multiple factors at once using methods defined by Youden & Steiner (1975).[68]

1.9.2 Recovery/Extraction Efficiency

Recovery/extraction efficiency is the fraction of an analyte (or analytes) that can be recovered within a given process. Often presented as % recovery, there are many factors that lead to poor or inconsistent recovery efficiency. These most often relate to matrix interference and lack of solvent or solute compatibility with process stages. Sample matrix effects can affect both analyte recovery and potentially analytical response. Therefore, it is appropriate to conduct calibrations using analyte-free matrices or "background blanks" as well as to rely on internal standards to define matrix interactions/effects. This is most feasible for some sample matrices such as blood plasma/serum and urine, where synthetic or blank matrices are commercially available or can be generated within one's laboratory. However, for many food matrices, especially whole foods such as fruits, juices and derived products, it is not always possible to establish comparable matrix blanks. Also, as discussed previously, the lack of common SRMs for berries, derived food composites and metabolites continues to limit the ability to adjust calibrations and improve the accuracy of new and existing methods. In the absence of such materials, assessment of (poly)phenolic extraction recovery and matrix effects is often difficult. Some laboratories or companies produce their own SRMs as a composite of multiple batches and store them under conditions that will prevent sample degradation. On rare occasions, companies or laboratories may register their SRMs with NIST and provide them as NIST reference materials for public and commercial access. Establishing NIST reference materials and protocols require multi-lab ring trials to establish performance, as recently established for production and analysis of RM 8403 cocoa flavanol extract (NIST SP #260-207).[69]

Practical considerations (i.e., advice and tips) for method validation are provided in detail as supplementary material (Appendix).

1.10 Method Transfer

Method transfer is the process of applying a method to a new process for which it was not originally intended when conceived, developed, optimized and/or validated. Methods transfer does not always require complete re-development but should include tests for performance and often requires re-optimization when assay quality parameters fail to meet acceptable precision and accuracy criteria. The most common method transfers reported in (poly)phenol analysis are scaling of extraction to accommodate higher throughput and lower sample size and transfer of instrumentation, such as moving from HPLC to UPLC method or photon (wavelength/frequency) detector to mass detector. Aside from re-establishing assay performance as outlined above, there are other practical considerations that should be considered when transferring methods for (poly)phenol analysis of plant or biological matrices. A list of common tips for more efficient methods of transfer is provided below. Here we use the example of transferring an HPLC-photon detection method (*e.g.*, wavelength/frequency detector) to a UPLC–MS detection method.

Practical considerations (i.e., advice and tips) for method transfer are provided in detail as supplementary material (Appendix).

1.11 Metabolomic Practices (Untargeted *vs.* Broad-spectrum)

Traditionally, approaches to characterize phenolic compounds in foods and biological samples were based on targeted methodology; however, untargeted metabolomic-based approaches are becoming more commonplace. Targeted approaches, as the name suggests, are methods/workflows focused on expected or known analytes, contain limited numbers of spectral features relative to untargeted metabolomic assays, and are more amenable to quantitative or semi-quantitative data analysis.[70] Metabolomic approaches can be completely untargeted, which is essentially a qualitative technique, or broad-spectrum, which can be qualitative, relative-quantitative or even quantitative, depending on the study design, use of reference/calibration standards and data analysis strategy. Targeted approaches tend to use QqQ platforms, while untargeted approaches most often use high-resolution MS systems (HRMS) such as TOF and ion traps (such as Orbitrap) or hybrids (*e.g.*, Triple-TOF, LIT Orbitrap). Both targeted and untargeted approaches can be conducted using similar and varied analytical platforms, although UPLC-coupled HRMS is most commonly used for (poly)phenolic metabolomic analysis. There is an advantage of using HRMS over low-resolution QqQ MS, as it provides accurate/exact mass, which can be used to differentiate compounds with the same nominal mass, thus greatly reducing possible numbers of isomers. This is important for untargeted analyses, where the presence of multiple compounds with similar nominal masses is much more likely than in targeted analyses. Further, accurate mass fragmentation provides valuable information for structural conformation. HRMS platforms can still be used for targeted quantitative approaches and are often as or more sensitive; however, as they devote considerable cycle time for mass resolution, they are limited by the number of analytes that can be analyzed with quantitative accuracy within a given run. The major advantage of HRMS is it can scan for tens of thousands of signals within a given run, and in many instances, identification can be confirmed without the use of reference standards, particularly if MS(2) or MS(3) CID data is available.[71]

1.11.1 Quantitation

Strategies exist for drawing quantitative conclusions for non-targeted techniques; however, these should be considered relative-quantitation over absolute quantitation and should be interpreted with care, as ionization efficiency is not consistent across compounds or concentrations, and the use of proportional abundance or surrogate reference standard regression equations can lead to significant under- or overestimation (as discussed above). Techniques such as isotope dilution and radio-labeling have attempted to improve quantitative accuracy; however, this cannot be applied to the abundance of spectral features in an untargeted dataset and generally focus on a few targeted features such as in the study of specific metabolic pathways.[72]

1.11.2 Sample Preparation

Sample preparation methods also differ in untargeted metabolomics relative to targeted approaches. Targeted and quantitative approaches optimize sample preparation procedures to maximize recovery, whereas untargeted qualitative approaches optimize for a maximum number of features (and are thus more "generic"). Untargeted sample preparation method development procedures are generally less laborious than targeted approaches and focus on protein precipitation (solvent and centrifugation) and optimization of concentration (often dilution) to maximize the number of spectral features while minimizing matrix effects (background noise and ion suppression). The advantage of this approach is that it limits the loss of many compound classes, which may not be compatible with extraction process/matrices as used in SPE, LLE or SLE. A limitation of untargeted sample preparation techniques (Table 1.4) is it may result in increased contamination of instruments with non-compatible analytes and leads to increased abundance of adducts and artifacts. The use of standardized protocols, internal standards and quality control and pooled samples are often used to validate assay reproducibility and account for batch effects.[71]

1.11.3 Identification

Most commonly used metabolomic approaches to characterize (poly)-phenolic compounds in foods have used a profiling approach which essentially matches precursor ions in captured datasets to databases or spectral libraries, using monoisotopic accurate mass and molecular formula. One drawback of this approach is that the exact masses of precursors can match many isomeric compounds. Furthermore, in-source fragmentation of conjugates and gas-phase artifacts can increase the potential for false-positive identification of precursor structures. Additional techniques such as product-ion scanning [MS(2) or MS(3)] provide further structural information and are often required for positive identification.[71] Although online databases capture significant spectral data (including accurate mass, MS/MS spectra, adducts, *etc.*) for the characterization of plant-based compounds, data for phenolic compounds from processed foods and human metabolism is limited. Compound discovery platforms and workflows aim to

Table 1.4 Targeted and untargeted benefits and limitations.

Untargeted approaches		Targeted approaches	
Benefits	Limitations	Benefits	Limitations
Analytes not *a priori* selected	Non-quantitative	Quantitative	Prior knowledge of sample composition required
No need for reference standards	Expensive instrumentation and software	Cheaper instrumentation	Reference standards required (considerable lead time for Serial dilution production)
Screen wide range of analytes	Reliant on spectral databases	High specificity	Number of analytes restricted
Limited method development required	Longer acquisition time for MS(2) or MS(*n*) fragmentation profiling	Rapid scanning	Considerable method development required (compound optimization, MRM development, *etc.*)
Isotope deconvolution possible	Confidence reporting required to confirm identification over annotation (*i.e.*, when retention time match was established based on reference standard over spectral feature match in database)	High quantitative accuracy	Often use low-resolution QqQ preventing positive identification if reference standards are not available
Simple sample preparation techniques	Prone to matrix interference, spectral artifacts and source contamination	Involved sample preparation techniques, limited matrix effects and source contamination	Often need for extensive sample cleanup to reduce matrix effects

reduce the number of unannotated signals in untargeted datasets and are important for discovery and hypothesis generation, but annotated structures still require conformation *via* structural elucidation experimentation or referee standards. Confirmatory analysis using MS(2), MS(3), MS(*n*) or MS(E) techniques has some limitations. Drawbacks of MS(*n*) scanning includes the design of suitable MS(*n*) experiments requires advanced knowledge; voltage ramping is required for fragmentation of sugars or biological phase II conjugates (glucuronides and sulfate) relative to base structure fragmentation; datasets acquired have greater complexity and structural identification often requires manual interpretation for conformation. Newly emerging techniques in HRMS, including developments in IM MS, software, machine learning and online databases, will undoubtedly reduce some of the limitations of untargeted techniques (Table 1.4), leading to increased use of untargeted and broad-spectrum metabolomics in (poly)phenol research.

1.12 Data Interpretation and Reporting

Without prior knowledge or evidence of the voltages required to produce predicted or expected fragments from published works (or found in online databases such as HMDB, PubChem, *etc.*), structural elucidation from low-resolution techniques such as MS/MS is difficult. Further, the use of single mass in selected ion monitoring (SIM) or full scan analysis lends a high probability of false-positive identification. Further, as polyphenol conjugates suffer from in-source fragmentation, misidentification of precursor structures is likely in situations where reference standards are not available. In such circumstances, compounds should be reported as putatively identified or annotated, and interpretation should be considered uncertain until further confirmatory analysis using MS(*n*) and high-resolution techniques is performed.

Structural elucidation requires understanding the instrumentation environment, such as voltage and chemical ionization, fragmentation, and impact of the matrix within the source and mass detector, all of which complicate data interpretation. Without awareness of these potential limitations, there is a high likelihood of spectral data misinterpretation.[73] For example, not all peaks in an MS spectrum originate from the compounds of interest and are as likely to result from the chemical and physical background environment (*i.e.*, artifacts). Also, as the MS signal is spread over different isotopes, fragments and adducts, it is inherently difficult to adequately characterize product ions that can originate from the loss of part of a molecule or molecular rearrangement. Therefore, structure elucidation with MS often leads to the partial assignment of a structure rather than fully identified structures.

The level of confidence in qualitative or quantitative data is dependent on the details captured and reported. Therefore, it is important to capture multiple chromatographic (*i.e.*, retention time, peak width, retention time peak drift, and resolution) and spectroscopic (signal stability, accurate mass, fragmentation pattern) features as many of these can be verified by others or through use of online data repositories. This is particularly important when reporting novel structures or in situations where no reference standards are available, as identity cannot be established/confirmed based on a single spectral feature.[73]

1.13 Recommended Minimal Analytical Method and Data Reporting Criteria

Recommendations on reporting requirements for flavonoids in research have been previously published[55] and focused primarily on the design and reporting of nutritional intervention studies but also touched on research challenges, limitations and inappropriate analytical methodologies. The themes described in this chapter are in-line with the 2015 recommendations; however, the more modern focus on the metabolome and microbiome in berry and (poly)phenol research adds further complexities which need to be considered, as discussed herein. Building on this past initiative, a 2020 special article[18] addressed issues with required reporting of (poly)phenol microbial and human metabolite nomenclature. Appropriate and consistent chemical description brings us one step closer to best reporting practices. However, best practice guidelines for reporting analytical methodology should be the focus of future position/policy papers, particularly considering the widespread use of MS for identification and quantitation. Without adopting consistent practices and reporting criteria, the replication of experimental findings will be difficult and incorporation of nutrition intervention findings into databases and metabolomic initiatives will be hindered.

1.13.1 Ideal Minimal Reporting Criteria

- Platform: Make and model of both flow-system (*i.e.*, U/HPLC) and detector (*i.e.*, DAD or MS), and in the case of MS, source ionization interface (ESI API, *etc.*), ionization mode (positive or negative) and scanning type, for example, SIM, SRM/MRM, sMRM, *etc.*
- Voltage: Reporting collision energy is useful when replicating a method for the same instrument and is a common element captured in online MS and metabolomic data repositories and databases.
- Gradient: Mobile phase assay conditions, mobile phase composition, flow rate and gradient program.
- Column: Column manufacturer, make/packing material, particle size, pore size, ID and length and column oven temperature.
- HPLC/UPLC-auto-sampler: Temperature, injection volume and method run-time.
- Materials/reference standards: Chemical name and manufacturer of all reference standards (ideally product number or CAS numbers).
- Qualitative data analysis: Qualitative reference standard details, retention time, accurate or MS(n) m/z, or indicate "putative identification/annotation" where no reference standard was available.

- Quantitative data analysis: Include a linear range of reference standard curves and reference standard matrix composition. Reference standard curve linearity and integration methodology is useful for replication of experiments or verification of assay results.
- Assay Precision and accuracy (*i.e.*, quality): % CV for internal standards either in a blank or matched matrix, frequency of assay blanks and quality control samples. Extraction efficiency details provide confidence in assay quality. LOD, LLOQ, HLQ and % CV of regression slopes provide further confirmation of data quality for studies comprising multiple batches of samples run over long periods of time.

1.14 General Considerations and Conclusions

Due to the diversity of (poly)phenol structures, their distribution and pre- and post-absorptive metabolism, substantial analytical considerations are required when developing or utilizing assays for berry phytochemicals. Multiple extractions and analytical techniques suitable for characterization and quantitation are presented herein. Sound method development, optimization, validation and reporting practices are required to capture the impact of environment, processing and metabolism on berry phytochemicals. Many texts have been devoted to providing guidance for such an initiative. Understanding best analytical practices, and most importantly, reporting practices will allow for the effective advancement of berry research. However, findings from studies can only be verified when they report details required to establish study or experiment quality and identify limitations. Identifying or realizing methodological limitations should be considered a strength, not a weakness, as all science is limited by the capacities of the present technology/instrumentation and available funding. Knowing the limitations of the analytical instrumentation used for identification and quantitation is required to prevent over-interpretation of study findings. Reporting key methodological details and process limitations will allow for efficient/effective progression of the berry health research field.

Abbreviations

IM	Ion mobility
LC–NMR	liquid chromatography–NMR
MALDI-TOF	matrix-assisted laser desorption/ionization-time of flight
CV	coefficient of variation
ESI-MS	electrospray ionization mass spectrometry
HPLC	high pressure/high-performance liquid chromatography
HILIC	hydrophilic–lipophilic interaction chromatography
LC–MS	liquid chromatography–mass spectrometry
LLE	liquid–liquid extraction
MRM	multiple reaction monitoring

NMR	nuclear magnetic resonance spectroscopy
QTOF	quadrupole time of flight
sMRM	scheduled multiple reaction monitoring
SPE	solid-phase extraction
SLE	solid–liquid extraction
TFA	trifluoroacetic acid
QqQ	triple quadrupole mass spectrometer
UPLC/UHPLC	ultra-performance liquid chromatography/ultra-high-pressure liquid chromatography

Appendix

Items for Consideration: Instrumentation

- Counterintuitively, lower injection volumes typically provide more consistent data, particularly when transferring methods from HPLC to UPLC or to columns of lower ID and pore size. Less volume on-column lends to reduced peak width, less impact of the sample diluent, higher chromatographic resolution, more consistent peak integration and greater peak height and intensity.
- For modern UPLC columns, optimizing methodology for relatively high and consistent pressure leads to more reliable chromatographic resolution, better peak shape (less tailing and fronting), reduced peak width, greater MS peak intensity and reduced retention time drift (*i.e.*, greater precision/reproducibility).
- Caution should be taken when characterizing structures using any detection technique, particularly in instances where chemical reference standards are unavailable (aside from perhaps NMR) or where compounds cannot be appropriately chromatographically resolved. In such cases, a single spectral feature (*e.g.*, exact mass, wavelength or mass transition) cannot be used to definitively identify a structure; multiple scanning experiments or platforms are required for structural conformation in these instances.

Items for Consideration: Extraction

- Solubility is influenced by extraction time and temperature.
- Extraction efficiency is dependent on solute-to-solvent ratio, particle size, bonding properties of matrix components, pH and temperature.
- Sequential extractions with apolar solvents (defatting) should be used prior to full LLE or SLE to minimize potential interferences from lipid-soluble components.
- During any extraction process, one should be aware that many phenolic compounds are prone to oxidation and even condensation. Shorter extraction times at lower temperatures (<50 °C) are recommended, as is the use of antioxidants such as ascorbic acid.

- Extraction of low-molecular-weight (poly)phenols is often more efficient in solvents such as methanol, while higher-molecular-weight polymers generally require less polar solvents such as acetone.[74]

Items for Consideration: SPE

The use of:

- reference standards spiked into SPE samples and solutions at loading and post-elution allows for rapid method development and validation.
- nitrogen or argon positive pressure manifolds can provide faster methods over the use of vacuum manifolds, in addition to improving sample stability, recovery and reducing variability (*i.e.*, from evaporation, sample loss, column obstruction, *etc.*).
- 96-well SPE along with loading plates and multi-channel pipettes reduces variability over single column/cartridge methods.
- high-quality 96-well septum/sealing mats reduce evaporation and allow for direct transfer to instrument auto-samplers for rapid analysis.
- LLE, in combination with SPE as a final cleanup, has the advantage of limiting the abundance of known MS ion suppressors in sample eluents, thereby enhancing the quality of the final LC–MS analysis of samples.

Items for Consideration: Method Development and Optimization

- A review of published literature methods to establish variables that differ helps establish typical ranges in variables utilized, allowing for testing of parameters across the reported ranges, such as low, medium and high concentrations, times, temperatures, pH, *etc.*
- Test variables in a predefined order and stick to a predetermined plan. A second method development experiment can refine variables that were less than effective.
- Have an *a priori* plan for judging findings, including a hierarchy of what variables will be compared and what outcomes are primary *vs.* secondary (*i.e.*, absolute recovery, precision, accuracy, *etc.*).
- The use at least three replicates of every variation at minimum to establish % CV (5–7 replicates are commonly recommended).
- The use of 96 well plates allows for making multiple comparisons across method parameters in a single method development experiment.
- Retain solvent phases or SPE eluent which would normally be discarded until results are analyzed as these can be used for troubleshooting if findings are ambiguous.
- Overcompensate for extraction solvent volume, number of replicate extractions or column washes or incubation time (temperature dependent), *etc.*, as these can be refined at the method optimization stage.

- It is recommended to establish method validation after a method has been developed and optimized. If replicates and internal standards are used and collected at every stage, along with sufficient matrix blanks, data will be available to establish a number of method validation parameters from the development and validation experiments, thus streamlining further validation efforts if required.
- Having defined quality control/assurance (QC/QA) samples (standards, extracts, *etc.*) and blanks to run before, during, and after critical runs can facilitate early detection of performance issues (*i.e.*, loss of sensitivity, retention time shifts, carryover, *etc.*) when running unknown samples, or large continuous runs of samples.
- When attempting to characterize a biological conjugate using MS, and no reference standard exists, for example, cyanidin-3-glucuronide, optimize fragmentation voltages for the aglycone (*i.e.*, cyanidin), and a comparable glucoside (*i.e.*, cyanidin-3-glucoside) as voltages required to fragment the precursor and major product (*i.e.*, aglycone) are often similar.
- Compounds putatively identified or compounds which do not have an available reference standard can still be optimized for voltage and sMRM parameters (*i.e.*, transitions, time windows, integration thresholds, *etc.*). Voltage ramping experiments can be conducted to establish optimized collision energies and automated or manual flow injection analysis experiments using concentrated and dilute samples allow for establishing sMRM parameters, linearity and matrix effects.
- Relative quantitation of compounds putatively identified or compounds which do not have available reference standards can still be established by assigning regression equations from compounds of high structural similarity or by using published response factors if available.
- Internal and external standards are highly recommended during all method development, optimization and validation procedures. Ideally, unique internal standards would be added at various stages of the process. Internal standards are added at a constant concentration to all samples, including blanks and calibration standards for correction of analyte loss or performance drift. Internal standards are also important to prove or improve the precision of an assay and aid in method development and method/instrument troubleshooting. Internal standards are particularly useful in establishing losses during assay processes, such as extraction, evaporation, centrifugation, *etc.* An external standard is similar to an internal standard but is run alone to account for instrument performance, matrix effects or to generate standard reference regression curves. Often, internal and external standard responses are compared to establish method and instrument compatibility, drift and for troubleshooting and establishing inter- and intra-user and -laboratory validation.

Items for Consideration: Precision and Performance

Optimize:

- for minimal injection volume to prevent column "overload", and provide improved peak shape, reduced carryover, and prevent ion suppression
- matrix "cleanup" procedures to remove ion suppressors from the background matrix
- sample composition (aqueous-to-organic ratio) to improve atomization and volatility
- source parameters such as gas, temperature, and *x*- and *y*-axis to prevent excessive ion entrance into Q0
- peak shape by altering pressure trace (*i.e.*, flow) to prevent large dips in system pressure known to broaden peak shape and cause fronting and tailing
- integration thresholds suitably above background signal/noise to avoid integration of background signals and irregular fronting and tailing peaks
- if using low injection volumes, sample organic solvent composition has a limited impact on chromatography
- clean source cone regularly to prevent ion contamination
- change capillary frequently to prevent arching or spray pattern degradation
- avoid excessive reference standard curve ranges by establishing regressions that only capture expected concentrations to improve reproducibility at limits of detection (LLOQ or HLOQ)
- Most parameters of precision and accuracy required for assay validation can be established in a single 96-well plate experiment, by running multiple blanks, reference standards and calibrations. It is recommended to separate reference standards by blanks, along with replicate injections of analytes at the assay lower quality control (LQC), medium quality control (MQC) and high quality control (HQC) limits.

Items for Consideration: Accuracy

- How a reference standard is dissolved can impact its assay performance. Therefore, it is recommended to always look up known solubility prior to dissolving a reference standard, particularly if it is of low quantity and high expense.
- When dissolving reference standards for MS use, it is important to use the highest quality MS grade solvents (including water) to prevent instrument and sample contamination.
- Primary stock solutions of reference standards can often be dissolved in dimethyl sulfoxide (DMSO) to prolong stability.

- If solubility dictates DMSO or water be used in high abundance in primary stock solutions, secondary stocks should dilute using more volatile solvents, such as acetonitrile or methanol, as water and DMSO can negatively impact declustering and ionization in ESI-MS; particularly DMSO, where it is recommended that final working concentrations remain below 5% (ideally <1–2% for quantitative analysis). High water abundance is less problematic, particularly if not injected directly into the source. If syringe infusion is being used for compound optimization, increasing the source temperature and declustering potential will improve atomization and ionization.
- Matrix-matched blanks and reference standard calibration curves (*i.e.*, standard curves or serial dilutions) are required for quantitative mass spectrometry due to matrix suppression of ionization. Furthermore, and similar to colorimetric detection, the background matrix of a sample negatively impacts background noise, peak integration and, therefore, accurate quantitation.
- The gold standard method for establishing a suitable assay matrix involves the use of a pool of all samples to be analyzed for background matrix effects. In situations where this is not possible, a pooled subset of baseline samples is often utilized or a commercial matrix (*i.e.*, pooled multi-donor serum, plasma, urine, tissue, *etc.*) or universally accepted synthetic matrix.[50,75,76]
- Quantitative accuracy in mass spectroscopy cannot be established if a matched-matrix is not utilized or without significant matrix validation, such as first establishing no matrix effects exist and the sample matrix has dilution integrity.
- Investigators reporting quantitative data for compounds that do not have commercially available references standards should identify these instances as "putative quantitation" (or semi-quantitative or relative quantitation).
- In the absence of commercially available SRM, investigators should consider establishing an "in-house" reference food or biological materials by pooling of materials, which can be used routinely as control samples to monitor changes in extraction efficiencies and/or LC/LC–MS performance.

Items for Consideration: Method Validation

- Percent recovery and matrix effects (described above) should be determined for each phenolic species assayed. However, the absence of analytical standards for some phenolic metabolites and many microbial and host metabolites precludes this. In these cases, it is recommended that one use recovery and matrix effects from the most structurally similar phenolic species where authenticated standards are available.

- Certified (poly)phenolic standards should be used to establish recovery and may be added to the analyte-free matrix (when available) at 1–2 times the concentration range anticipated in the test products.
- If a (poly)phenolic-free matrix is not available, such as often in the case of berries, a standard addition method can be used to establish accurate recovery.
- Recovery and matrix effects can be determined by plotting the response against actual concentrations of the reference standards, using pre- and post-extraction spiking with internal standards, and comparing peak areas, heights or area-to-intensity ratios of analytes within a blank and true sample matrix.

Items for Consideration: Method Transfer

- It is important to consider acid/pH modifiers when transitioning from HPLC-UV–Vis to MS detection. pH modifiers are often used in the HPLC-UV–Vis mobile phase as an organic modifier to improve peak shape and chromatographic separation. This often affects chemical interactions between the stationary and mobile phase, which is particularly apparent for anthocyanin analysis.
- Mobile phase pH modifiers can have a significant impact on ionization efficiency (providing a charged state) and result in adduct formation, hampering identification and quantitation.
- It is most common to use formic acid as a volatile organic modifier in (poly)phenol research as it improves peak shape while having minimal impact on ionization efficiency relative to other acids such as TFA and HCl, which are well-known ion suppressors, especially in positive mode.
- 1–5% formic acid by volume in the mobile phase will generally provide narrow chromatography peak widths of 0.3–0.5 min (concentration dependent) and lends to optimal chromatography of polyphenols using UV–Vis detection methods but generally leads to ion suppression and loss of sensitivity using MS detectors.
- Formic acid levels of 0.1–1% will lead to increased sensitivity in MS but often at the detriment of peak width, particularly for anthocyanins. This can often be overcome by moving to a higher flow/pressure system or narrow bore column, which can narrow peak width (injection volume dependent). This dichotomy between impacts of acid on resolution and sensitivity is often referred to as the "acid effect" and is particularly prevalent for anthocyanins.
- It is important to note the concentrations of acid in the mobile phase in publications, as it can impact LOD and LLOQ, and potentially quantitatively accuracy.
- When transitioning from detection methods such as HPLC-UV–Vis to MS, it is often assumed that MS will provide a higher level of qualitative and quantitative certainty. However, MS comes with an added level of uncertainty, particularly with respect to polyphenols, which undergo

many reactions (gas-phase artifacts) in the high-energy gas-phase of the MS, such as molecular rearrangement, Retro-Diels–Alder reactions, quasimolecular ion formation, radical formation, *etc.* Because of these confounders, misidentification of precursor polyphenols and their phase II metabolites is assumed to be highly prevalent when using a single transition or spectral feature for identification.

Acknowledgements

CDK was supported by the USDA National Institute of Food and Agriculture (Hatch/Kay-Colin; 1011757).

References

1. N. E. Kafkas, M. Kosar, A. T. Öz and A. E. Mitchell, Advanced Analytical Methods for Phenolics in Fruits, *J. Food Qual.*, 2018, **2018**, 3836064.
2. I. Ignat, I. Volf and V. I. Popa, A critical review of methods for characterisation of polyphenolic compounds in fruits and vegetables, *Food Chem.*, 2011, **126**(4), 1821–1835.
3. G. Domínguez-Rodríguez, M. L. Marina and M. Plaza, Strategies for the extraction and analysis of non-extractable polyphenols from plants, *J. Chromatogr. A*, 2017, **1514**, 1–15.
4. T. J. Mabry, K. R. Markham and M. B. Thomas. *Systematic Identification of Flavonoïds*, 1970, DOI: 10.1007/978-3-642-88458-0.
5. J. B. Harborne, *The Flavonoids: Advances in Research Since 1980*, Springer, 2013.
6. R. W. Durst and R. E. Wrolstad, Separation and Characterization of Anthocyanins by HPLC, *Curr. Protoc. Food Anal. Chem.*, 2001, **00**(1), F1.3.1–F1.3.13.
7. L. E. Rodriguez-Saona and R. E. Wrolstad, Extraction, Isolation, and Purification of Anthocyanins, *Curr. Protoc. Food Anal. Chem.*, 2001, **00**(1), F1.1.1–F1.1.11.
8. M. M. Giusti and R. E. Wrolstad, Characterization and Measurement of Anthocyanins by UV-Visible Spectroscopy, *Curr. Protoc. Food Anal. Chem.*, 2001, **00**(1), F1.2.1–F1.2.13.
9. C. Santos-Buelga and G. Williamson, *Methods in Polyphenol Analysis*, Royal Society of chemistry, 2003.
10. P. Pichat, *Photocatalysis: Fundamentals, Materials and Potential*, MDPI AG-Multidisciplinary Digital Publishing Institute, 2016.
11. I. Ignat, I. Volf and V. I. Popa Analytical Methods of Phenolic Compounds, in *Natural Products: Phytochemistry, Botany and Metabolism of Alkaloids, Phenolics and Terpenes*, ed. Ramawat K. G. and Mérillon J.-M., Springer Berlin Heidelberg, 2013, pp. 2061–2092.
12. A. Aires, Phenolics in foods: Extraction, analysis and measurements, *Phenolic Compd.*, 2017, 61–88.

13. Statista. https://www.statista.com/statistics/191512/fresh-berry-category-share-in-2011/. Accessed 8 Feb 2020.
14. USDA. USDA Database for the Flavonoid Content of Selected Foods (Release 3.0). http://www.ars.usda.gov/Services/docs.htm?docid=6231.
15. J. A. Rothwell, J. Perez-Jimenez and V. Neveu, *et al.*, Phenol-Explorer 3.0: a major update of the Phenol-Explorer database to incorporate data on the effects of food processing on polyphenol content, *Database*, 2013, bat070.
16. G. Williamson, C. D. Kay and A. Crozier, The Bioavailability, Transport, and Bioactivity of Dietary Flavonoids: A Review from a Historical Perspective, *Compr. Rev. Food Sci. Food Saf.*, 2018, **17**(5), 1054–1112.
17. S. Dwivedi, C. Malik and V. Chhokar, Molecular structure, biological functions, and metabolic regulation of flavonoids, *Plant Biotechnology: Recent Advancements and Developments*, Springer, 2017, 171–188.
18. C. D. Kay, M. N. Clifford and P. Mena, *et al.*, Recommendations for standardizing nomenclature for dietary (poly) phenol catabolites, *Am. J. Clin. Nutr.*, 2020, **112**(4), 1051–1068.
19. T. Okuda, T. Yoshida and T. Hatano, Hydrolyzable tannins and related polyphenols, *Fortschritte der Chemie organischer Naturstoffe/Progress in the Chemistry of Organic Natural Products*, Springer, 1995, 1–117.
20. K. R. Määttä-Riihinen, M. P. Kähkönen, A. R. Törrönen and I. M. Heinonen, Catechins and procyanidins in berries of vaccinium species and their antioxidant activity, *J. Agric. Food Chem.*, 2005, **53**(22), 8485–8491.
21. R. L. Prior, S. A. Lazarus, G. Cao, H. Muccitelli and J. F. Hammerstone, Identification of procyanidins and anthocyanins in blueberries and cranberries (Vaccinium spp.) using high-performance liquid chromatography/mass spectrometry, *J. Agric. Food Chem.*, 2001, **49**(3), 1270–1276.
22. J. Frank, N. K. Fukagawa and A. R. Bilia, *et al.*, Terms and nomenclature used for plant-derived components in nutrition and related research: efforts toward harmonization, *Nutr. Rev.*, 2020, **78**(6), 451–458.
23. C. R. Welch, Q. Wu and J. E. Simon, Recent Advances in Anthocyanin Analysis and Characterization, *Curr. Anal. Chem.*, 2008, **4**(2), 75–101.
24. J. Martín, M. J. Navas, A. M. Jiménez-Moreno and A. G. Asuero, Anthocyanin pigments: Importance, sample preparation and extraction, *Phenolic Ccompounds–Natural Ssources, Iimportance and Aapplications*, InTech, 2017, pp. 117–152.
25. G. Mazza, J. E. Cacace and C. D. Kay, Methods of analysis for anthocyanins in plants and biological fluids, *J. AOAC Int.*, 2004, **87**(1), 129–145.
26. R. J. Robbins, J. Leonczak and J. C. Johnson, *et al.*, Method performance and multi-laboratory assessment of a normal phase high pressure liquid chromatography–fluorescence detection method for the quantitation of flavanols and procyanidins in cocoa and chocolate containing samples, *J. Chromatogr. A*, 2009, **1216**(24), 4831–4840.
27. M. R. Dorenkott, L. E. Griffin and K. M. Goodrich, *et al.*, Oligomeric Cocoa Procyanidins Possess Enhanced Bioactivity Compared to

Monomeric and Polymeric Cocoa Procyanidins for Preventing the Development of Obesity, Insulin Resistance, and Impaired Glucose Tolerance during High-Fat Feeding, *J. Agric. Food Chem.*, 2014, **62**(10), 2216–2227.
28. A. P. Neilson, S. F. O'Keefe and B. W. Bolling, High-Molecular-Weight Proanthocyanidins in Foods: Overcoming Analytical Challenges in Pursuit of Novel Dietary Bioactive Components, *Annu. Rev. Food Sci. Technol.*, 2016, **7**(1), 43–64.
29. L. Ye and A. Neilson, Comparison of A-type Proanthocyanidins in cranberry and Peanut skin extracts using matrix assisted laser desorption ionization-time of flight mass spectrometry, *J. Mol. Genet. Med.*, 2016, **10**, 1–7.
30. K. C. Racine, A. H. Lee, A. C. Stewart, K. W. Blakeslee and A. P. Neilson, Development of a rapid ultra performance hydrophilic interaction liquid chromatography tandem mass spectrometry method for procyanidins with enhanced ionization efficiency, *J. Chromatogr. A*, 2019, **1594**, 54–64.
31. P. Ongkowijoyo, D. A. Luna-Vital and E. Gonzalez de Mejia, Extraction techniques and analysis of anthocyanins from food sources by mass spectrometry: An update, *Food Chem.*, 2018, **250**, 113–126.
32. J. Sun, L. Lin and P. Chen, Study of the mass spectrometric behaviors of anthocyanins in negative ionization mode and its applications for characterization of anthocyanins and non-anthocyanin polyphenols, *Rapid Commun. Mass Spectrom.*, 2012, **26**(9), 1123–1133.
33. T. Rousu, J. Herttuainen and A. Tolonen, Comparison of triple quadrupole, hybrid linear ion trap triple quadrupole, time-of-flight and LTQ-Orbitrap mass spectrometers in drug discovery phase metabolite screening and identification in vitro – amitriptyline and verapamil as model compounds, *Rapid Commun. Mass Spectrom.*, 2010, **24**(7), 939–957.
34. M. Chandra Singh, C. Kelso, W. E. Price and Y. Probst, Validated liquid chromatography separation methods for identification and quantification of anthocyanins in fruit and vegetables: A systematic review, *Food Res. Int.*, 2020, **138**, 109754.
35. S. Struck, M. Plaza, C. Turner and H. Rohm, Berry pomace–a review of processing and chemical analysis of its polyphenols, *Int. J. Food Sci. Technol.*, 2016, **51**(6), 1305–1318.
36. P. Arapitsas, Hydrolyzable tannin analysis in food, *Food Chem.*, 2012, **135**(3), 1708–1717.
37. Z. Lei, J. Jervis and R. F. Helm, Use of Methanolysis for the Determination of Total Ellagic and Gallic Acid Contents of Wood and Food Products, *J. Agric. Food Chem.*, 2001, **49**(3), 1165–1168.
38. I. Mueller-Harvey, Analysis of hydrolysable tannins, *Anim. Feed Sci. Technol.*, 2001, **91**(1), 3–20.
39. A. G. Newsome, Y. Li and R. B. van Breemen, Improved Quantification of Free and Ester-Bound Gallic Acid in Foods and Beverages by UHPLC–MS/MS, *J. Agric. Food Chem.*, 2016, **64**(6), 1326–1334.

40. T. Belwal, S. M. Ezzat and L. Rastrelli, *et al.*, A critical analysis of extraction techniques used for botanicals: Trends, priorities, industrial uses and optimization strategies, *TrAC, Trends Anal. Chem.*, 2018, **100**, 82–102.
41. C. S. Dzah, Y. Duan and H. Zhang, *et al.*, The effects of ultrasound assisted extraction on yield, antioxidant, anticancer and antimicrobial activity of polyphenol extracts: A review, *Food Biosci.*, 2020, **35**, 100547.
42. E. Brglez Mojzer, M. Knez Hrnčič, M. Škerget, Ž. Knez and U. Bren, Polyphenols: Extraction Methods, Antioxidative Action, Bioavailability and Anticarcinogenic Effects, *Molecules*, 2016, **21**(7), 901.
43. K. M. Goodrich and A. P. Neilson, Simultaneous UPLC–MS/MS analysis of native catechins and procyanidins and their microbial metabolites in intestinal contents and tissues of male Wistar Furth inbred rats, *J. Chromatogr. B*, 2014, **958**, 63–74.
44. S. Sang, M.-J. Lee, Z. Hou, C.-T. Ho and C. S. Yang, Stability of Tea Polyphenol (−)-Epigallocatechin-3-gallate and Formation of Dimers and Epimers under Common Experimental Conditions, *J. Agric. Food Chem.*, 2005, **53**(24), 9478–9484.
45. M. M. Deshpande, V. S. Kasture, M. Mohan and M. J. Chavan, Bioanalytical Method Development and Validation: A, *Recent Adv. Anal. Chem.*, 2019, 97.
46. A.-C. Dubbelman, F. Cuyckens, L. Dillen, G. Gross, R. J. Vreeken and T. Hankemeier, Mass spectrometric recommendations for Quan/Qual analysis using liquid-chromatography coupled to quadrupole time-of-flight mass spectrometry, *Anal. Chim. Acta*, 2018, **1020**, 62–75.
47. S. Gao, Z. P. Zhang and H. T. Karnes, Sensitivity enhancement in liquid chromatography/atmospheric pressure ionization mass spectrometry using derivatization and mobile phase additives, *J. Chromatogr. B: Anal. Technol. Biomed. Life Sci.*, 2005, **825**(2), 98–110.
48. M. A. Kelm, J. C. Johnson, R. J. Robbins, J. F. Hammerstone and H. H. Schmitz, High-Performance Liquid Chromatography Separation and Purification of Cacao (Theobroma cacao L.) Procyanidins According to Degree of Polymerization Using a Diol Stationary Phase, *J. Agric. Food Chem.*, 2006, **54**(5), 1571–1576.
49. D. Zimmer, New US FDA draft guidance on bioanalytical method validation versus current FDA and EMA guidelines: chromatographic methods and ISR, *Bioanalysis*, 2014, **6**(1), 13–19.
50. European Medicines Agency, Guideline on bioanalytical method validation, 2011, EMEA/CHMP/EWP/192217/2009.
51. D. Tsikas, Bioanalytical method validation of endogenous substances according to guidelines by the FDA and other organizations: Basic need to specify concentration ranges, *J. Chromatogr. B*, 2018, **1093–1094**, 80–81.
52. W. Horwitz and R. Albert, The Horwitz ratio (HorRat): A useful index of method performance with respect to precision, *J. AOAC Int.*, 2006, **89**(4), 1095–1109.

53. E. Ciccimaro and I. A. Blair, Stable-isotope dilution LC–MS for quantitative biomarker analysis, *Bioanalysis*, 2010, **2**(2), 311–341.
54. C. D. Kay, G. Pereira-Caro, I. A. Ludwig, M. N. Clifford and A. Crozier, Anthocyanins and flavanones are more bioavailable than previously perceived: a review of recent evidence, *Annu. Rev. Food Sci. Technol.*, 2017, **8**, 155–180.
55. D. A. Balentine, J. T. Dwyer and J. W. Erdman, *et al.*, Recommendations on reporting requirements for flavonoids in research, *Am. J. Clin. Nutr.*, 2015, **101**(6), 1113–1125.
56. R. W. Binkley, G. W. Dillow, T. W. Flechtner, W. Winnik and M. J. Tevesz, Fragmentation of collisionally activated hydroxyphenoxide anions in the gas phase, *Org. Mass Spectrom*, 1994, **29**(9), 491–495.
57. A. S. Abdelhameed, M. W. Attwa, H. A. Abdel-Aziz and A. A. Kadi, Induced in-source fragmentation pattern of certain novel (1Z,2E)-N-(aryl)propanehydrazonoyl chlorides by electrospray mass spectrometry (ESI-MS/MS), *Chem. Cent. J.*, 2013, **7**, 16.
58. D. P. Demarque, A. E. Crotti, R. Vessecchi, J. L. Lopes and N. P. Lopes, Fragmentation reactions using electrospray ionization mass spectrometry: an important tool for the structural elucidation and characterization of synthetic and natural products, *Nat. Prod. Rep.*, 2016, **33**(3), 432–455.
59. M. Saeed, A. Sage and C. Wolff, *et al.*, *Guide to Achieving Reliable Quantitative LC–MS Measurements*, 2013.
60. S. Ma, C. Kim and A. P. Neilson, *et al.*, Comparison of Common Analytical Methods for the Quantification of Total Polyphenols and Flavanols in Fruit Juices and Ciders, *J. Food Sci.*, 2019, **84**(8), 2147–2158.
61. J. I. Ottaviani, R. Y. Fong, G. Borges, H. Schroeter and A. Crozier, Use of LC–MS for the quantitative analysis of (poly)phenol metabolites does not necessarily yield accurate results: Implications for assessing existing data and conducting future research, *Free Radical Biol. Med.*, 2018, **124**, 97–103.
62. J. W. Blount, B. W. Redan and M. G. Ferruzzi, *et al.*, Synthesis and Quantitative Analysis of Plasma-Targeted Metabolites of Catechin and Epicatechin, *J. Agric. Food Chem.*, 2015, **63**(8), 2233–2240.
63. A. International and Guideline Working G, AOAC INTERNATIONAL guidelines for validation of botanical identification methods, *J. AOAC Int.*, 2012, **95**(1), 268–272.
64. W. Horwitz, AOAC guidelines for single laboratory validation of chemical methods for dietary supplements and botanicals, USA: AOAC International, Gaithersburg, MD, 2002, 12–19.
65. Health UDo, Services H. Bioanalytical method validation, guidance for industry. http://www.fda.gov/cder/guidance/4252fnl.htm. 2001.
66. Food and Drug Administration, U. S. "Bioanalytical method validation guidance for industry." US Department of Health and Human Services, 2018, pp. 1–41.
67. D. ZimmerNew US FDA draft guidance on bioanalytical method validation versus current FDA and EMA guidelines: chromatographic methods and ISR*Bioanalysis*2013, 6(1), 13–19.

68. W. J. Youden and E. H. Steiner *Statistical Manual of The Association of Official Analytical Chemists*, Aoac International, 1975.
69. C. A. Rimmer, K. A. Lippa and J. Yen, *et al.*, Production and Analysis of RM 8403 Cocoa Flavanol Extract, *NIST Spec. Publ.*, 2020, **260**, 207.
70. V. Lelli, A. Belardo and A. M. Timperio, From Targeted Quantification to Untargeted Metabolomics, *Metabolomics*, IntechOpen, 2021.
71. S. Piovesana, C. Cavaliere, A. Cerrato, C. M. Montone, A. Laganà and A. L. Capriotti, Developments and pitfalls in the characterization of phenolic compounds in food: From targeted analysis to metabolomics-based approaches, *TrAC, Trends Anal. Chem.*, 2020, **133**, 116083.
72. A. Kruve, *Strategies for Drawing Quantitative Conclusions from Nontargeted Liquid Chromatography–High-Resolution Mass Spectrometry Analysis*, ACS Publications, 2020.
73. T. De Vijlder, D. Valkenborg, F. Lemière, E. P. Romijn, K. Laukens and F. Cuyckens, A tutorial in small molecule identification via electrospray ionization-mass spectrometry: The practical art of structural elucidation, *Mass Spectrom. Rev.*, 2018, **37**(5), 607–629.
74. E. Brglez Mojzer, M. Knez Hrnčič, M. Škerget, Ž. Knez and U. Bren, Polyphenols: Extraction Methods, Antioxidative Action, Bioavailability and Anticarcinogenic Effects, *Molecules*, 2016, **21**(7), 901.
75. P. Panuwet, R. E. Hunter Jr and P. E. D'Souza, *et al.*, Biological matrix effects in quantitative tandem mass spectrometry-based analytical methods: advancing biomonitoring, *Crit. Rev. Anal. Chem.*, 2016, **46**(2), 93–105.
76. D. Tsikas, Bioanalytical method validation of endogenous substances according to guidelines by the FDA and other organizations: Basic need to specify concentration ranges, *J. Chromatogr. B*, 2018, **1093–1094**, 80–81.

CHAPTER 2

Occurrence, Bioavailability and Metabolism of Berry (Poly)phenols

GIUSEPPE DI PEDE,[a] CLAUDIA FAVARI,[a] LETIZIA BRESCIANI,[b] TAHANI MAZYAD ALMUTAIRI,[c] DANIELE DEL RIO[a] AND ALAN CROZIER*[c,d]

[a] Human Nutrition Unit, Department of Food and Drug, University of Parma, Via Volturno 39, 43125 Parma, Italy; [b] Human Nutrition Unit, Department of Food and Drug, University of Parma, Via Volturno 39, 43125 Parma, Italy; [c] Department of Chemistry, King Saud University, Riyadh 11451, Saudi Arabia; [d] School of Medicine, Dentistry and Nursing, University of Glasgow, Glasgow, G12 8QQ, UK
*Email: alan.crozier44@gmail.com

2.1 (Poly)phenolic Compounds in Berries

Berries are a rich source of micronutrients and phytochemicals, and the most notable of which are (poly)phenolic compounds.[1,2] (Poly)phenols found in berries include phenolic acids, flavonoids and polymeric tannins, which comprise ellagitannins (ETs) and proanthocyanidins (PACs), respectively also known as hydrolyzable and condensed tannins. The levels of these constituents are influenced by plant age and environmental and agronomic factors.[3]

The vibrant red, purple and dark blue colors of berries are due to anthocyanins; the aglycones, namely cyanidin, delphinidin, malvidin,

Food Chemistry, Function and Analysis No. 33
Berries and Berry Bioactive Compounds in Promoting Health
Edited by Dorothy Klimis-Zacas and Ana Rodriguez-Mateos

Published by the Royal Society of Chemistry, www.rsc.org

pelargonidin, peonidin and petunidin, referred to as anthocyanidins, differ by virtue of the presence and absence of hydroxy or methoxy groups at the C-3′ and C-5′ positions (Table 2.1). Because of their instability, anthocyanidins *per se* are not found *in planta* but exist as anthocyanins in which the anthocyanidin moiety is *O*-linked to a range of mono-, di-, and trisaccharides which can be acetylated to a range of aromatic and aliphatic acids (Figure 2.1).

This provides structural stability[4–6] and results in a diverse array of anthocyanin structures being found in berries and other plant sources.[3,7] Anthocyanin contents are typically high in blueberry (*Vaccinium corymbosum*),

Table 2.1 Structures of major anthocyanins.

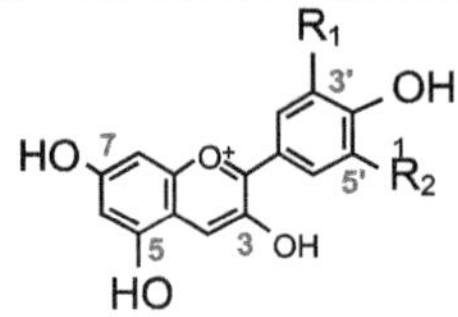

Anthocyanidin	R_1	R_2	MW
Pelargonidin	H	H	271
Cyanidin	OH	H	287
Delphinidin	OH	OH	303
Peonidin	OCH_3	H	301
Petunidin	OCH_3	OH	317
Malvidin	OCH_3	OCH_3	331

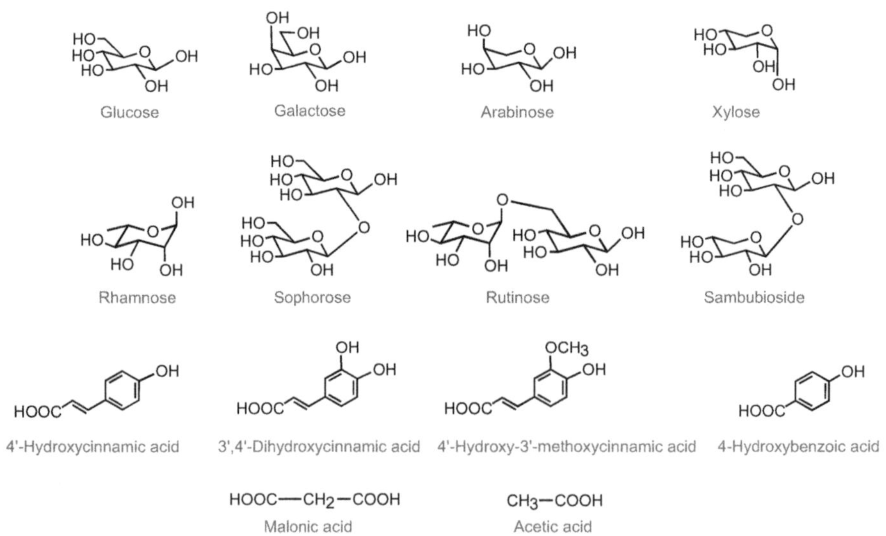

Figure 2.1 Selected sugars and aromatic and aliphatic acids that commonly occur in anthocyanin structures.

blackberry (*Rubus fruticosus* L.) and bilberry (*Vaccinium myrtillus*), which can contain up to 495 mg kg^{-1} fresh weight (FW).[8] Levels can be influenced by agronomic conditions and genotype.[3,9,10] Delphinidin-3-*O*-rutinoside is the main anthocyanin in blackcurrants (*Ribes nigrum*). Cyanidin glycosides are found in blackberry, elderberry (*Sambucus nigra*), American cranberry (*Vaccinium macrocarpon*), raspberry (*Rubus idaeus*) and redcurrant (*Ribes rubrum*).[9] The predominant anthocyanin in strawberry (*Fragaria* × *ananassa*) is pelargonidin-3-*O*-glucoside. Blueberries have a complex anthocyanin profile, which includes malvidin-3-*O*-arabinoside and the 3-*O*-galactosides of delphinidin, cyanidin, petunidin and malvidin, with many other minor anthocyanins also being present.[7,9] Cranberries, like blueberries, are a member of the *Ericaceae* but have a very different spectrum of anthocyanins with cyanidin and peonidin-based compounds predominating.[7,9]

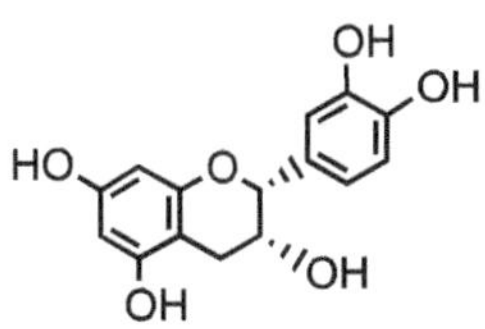

(−)-Epicatechin **1**

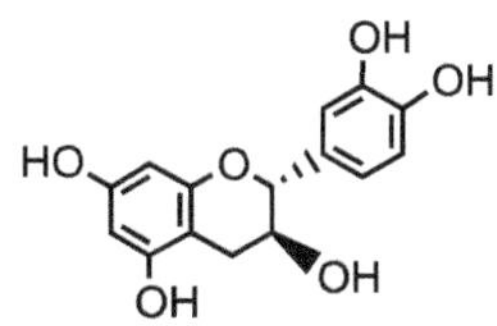

(+)-Catechin **2**

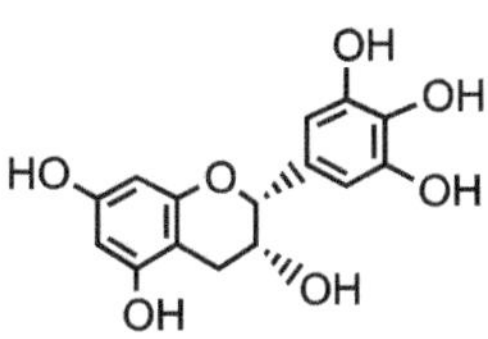

(−)-Epigallocatechin **3**

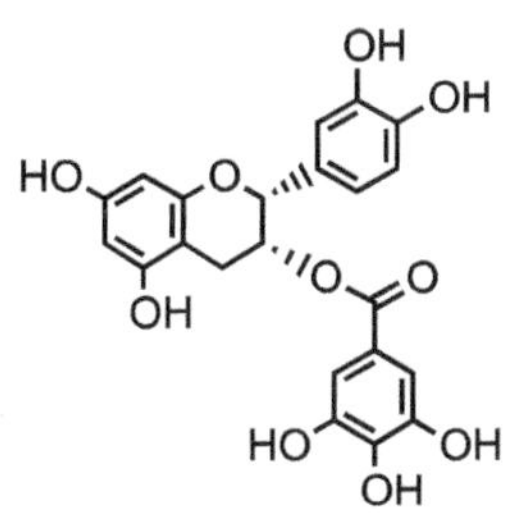

(−)-Epicatechin-3-*O*-gallate **4**

Other flavonoids of dietary relevance occurring in berries include flavan-3-ols.[1] Although less abundant than in fresh cocoa (*Theobroma cacao*) beans and young leaves of green tea (*Camellia sinensis*),[11] American cranberry (73 mg kg^{-1} FW), red raspberry (44 mg kg^{-1} FW) and strawberry (42 mg kg^{-1} FW) contain the monomeric flavan-3-ols (−)-epicatechin **1** and its enantiomer (+)-catechin 2. (−)-Epigallocatechin **3** and (−)-epicatechin-3-*O*-gallate **4** occur less widely and in lower amounts.[1,9,12] Among ten native varieties of Canadian berries, Alpine blueberry (*Vaccinium uliginosum* L.) contains the highest level of flavan-3-ols (1020 mg kg^{-1} FW).[12] PAC dimers and trimers have been quantified in genotypes of Saskatoon berry (*Amelanchier alnifolia* Nutt.) and American cranberry (259 and 189 mg kg^{-1} FW for dimer and trimer contents, respectively).[9,13] Chokeberry (*Aronia melanocarpa*) and American cranberry are rich sources of PACs (~690 mg kg^{-1} FW).[12,14] American cranberry represents a source of both A- and B-type

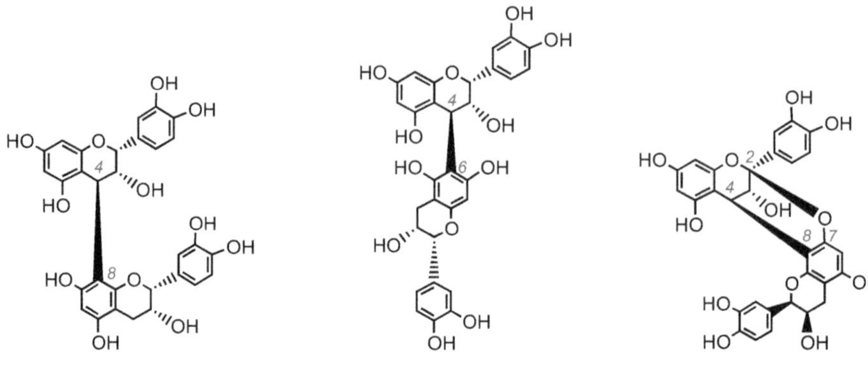

Figure 2.2 Selected structures of type A and B procyanidins.

PACs.[9,14] B-type PACs are formed by oxidative coupling between the C-4 of the upper monomer and the C-6 or C-8 of the adjacent lower unit to create oligomers or polymers. A-type PACs have an additional ether bond between C-2 in the C-ring of one monomer and C-7 in the A-ring of the other monomer (Figure 2.2). PACs can occur as polymers up to 50 units. Most quantitative estimates of flavan-3-ols in berries provide a simplified assessment of what exists *in planta*. Firstly, the monomers (−)-epicatechin and (+)-catechin also exist as (+)-epicatechin **5** and (−)-catechin **6**. The stereoisomers can be separated by chiral chromatography but not reversed-phase HPLC. A- and B-type PAC dimers and trimers can be separated by reversed-phase HPLC but PACs with a higher degree of polymerization are retained on the column. As a consequence, analyses of many berry PACs are limited to dimers and trimers.

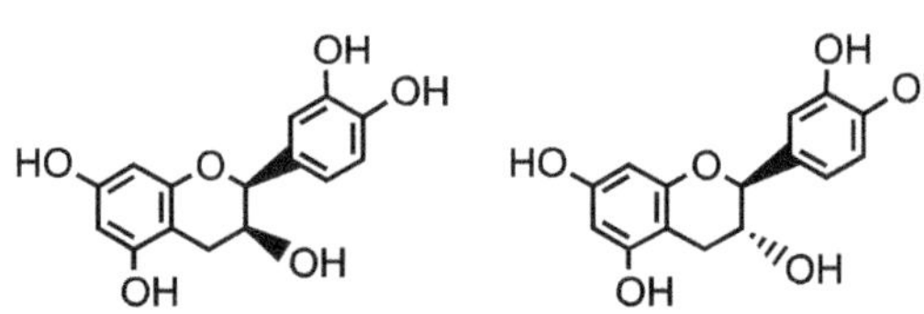

(+)-Epicatechin **5** (−)-Catechin **6**

Separation of PACs with a degree of polymerization (DP) of 2–7 subunits can be achieved using hydrophobic interaction chromatography (HILIC).[15] Details of a robust HILIC method with a Torus diol HPLC column have recently been reported.[16] Wang and colleagues have recently identified 304 individual A-type and B-type PACs from cranberry through reversed-phase HPLC coupled with ion-mobility quadrupole time-of-flight mass spectrometry.[17,18] Analysis of higher molecular weight PACs, which can exist in substantial amounts, is not straightforward but has been accomplished with grape seed PACs using acid-catalyzed depolymerization in the presence of a nucleophile and gel permeation chromatography based on the procedures of

Kennedy and Taylor and Bindon and Kennedy.[19–21] Estimation of the overall levels of type A cranberry PACs using the colorimetric 4-(dimethylamino)-cinnamaldehyde (DMAC) assay has been evaluated by Feliciano *et al.*[22] For the analysis of polymeric PACs, various techniques can be employed making use of their structural properties, *i.e.* binding for proteins and/or polysaccharides,[23,24] or their conversion to cyanidin aglycones upon thermal depolymerization in acid.[25,26]

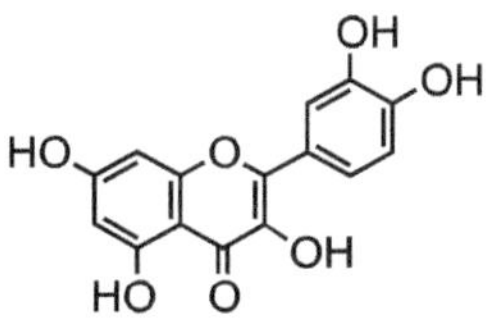

Quercetin **7**

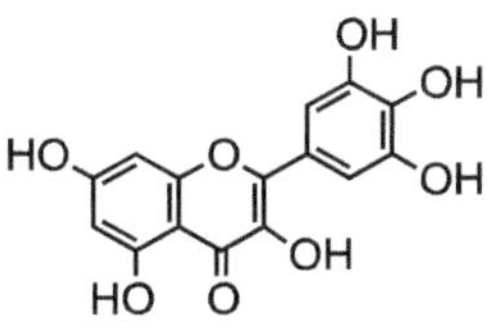

Myricetin **8**

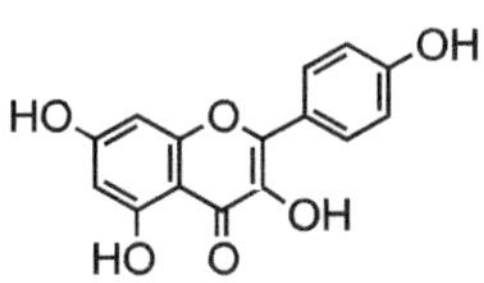

Kaempferol **9**

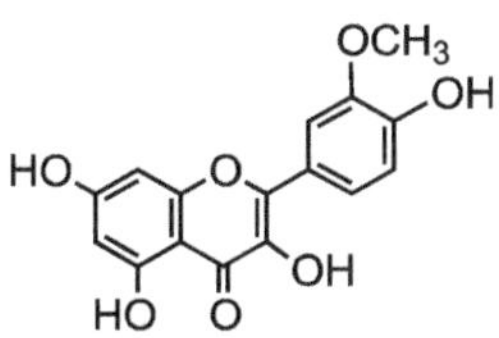

Isorhamnetin **10**

Flavonols are mostly found in berries as glycosides of quercetin **7**, myricetin **8** and, to a lesser extent, kaempferol **9** and isorhamnetin **10**.[1] Eight glycosides of quercetin were quantified in Polish and Canadian berry cultivars, among which quercetin-3-*O*-arabinoglucoside, quercetin-3-*O*-rutinoside and quercetin-3-*O*-galactoside were the most abundant.[13] American cranberries contain approximately 1000 mg total flavonols per kilogram of FW, with quercetin-*O*-glycosides being the main representatives.[2] Myricetin-*O*-glycosides (3-*O*-glucoside, -galactoside and -rutinoside derivatives) are the most abundant flavonols in blackcurrant cultivars (89–203 mg kg^{-1} FW),[9] while kaempferol glycosides accumulate in cherry (*Prunus avium*) (39–50 mg kg^{-1} FW) and blueberry (23–30 mg kg^{-1} FW).[27–29]

The ETs are derivatives of ellagic acid **11**, made up of one or more hexahydroxydiphenoyl (HHDP) moieties esterified to a sugar, typically glucose.[30,31] Due to the many possibilities for the HHDP moieties to link with the sugar and to their marked tendency to form ET oligomers and polymers, a great structural variability and complexity exist.[30] They are found in many berries, including raspberry, strawberry, blackberry, blueberry, American cranberry and cloudberry (*Rubus chamaemorus*).[30] Ellagic acid can account for more than 50% of the total phenolic content of berries, principally in the form of ETs,[3] which are considered to be the main contributors to astringency in strawberries and raspberries.[3] Red raspberries contain approximately 1 mg kg^{-1} of ellagic acid compared to ~300 mg kg^{-1} of ETs, mainly sanguiin H-6 **12** and lambertianin C **13**.[9]

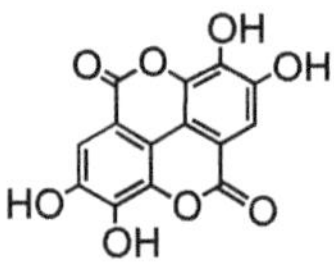

Ellagic acid **11**

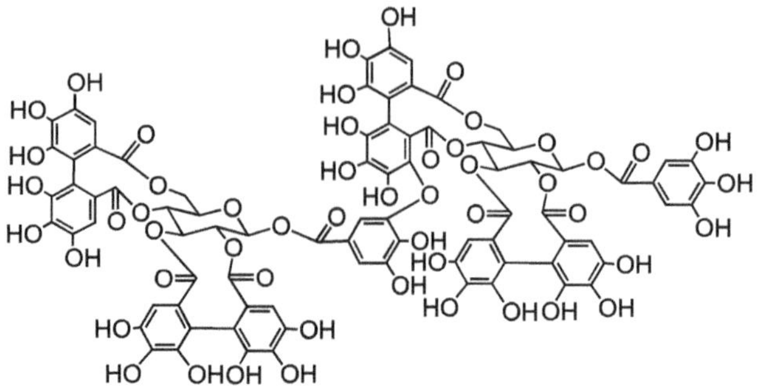

Sanguiin H-6 **12**

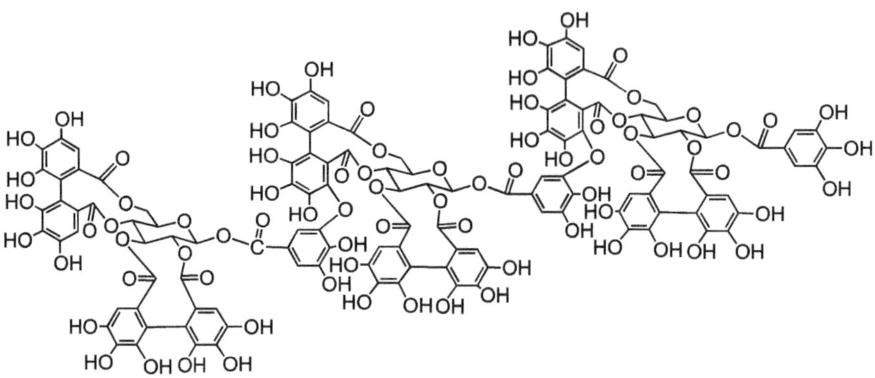

Lambertianin C **13**

Hydroxycinnamic acid and hydroxybenzoic acid derivatives are the most abundant phenolic acids in berries[1] and have been detected in amounts up to 1390 mg kg^{-1} FW.[12] Chokeberry and chokecherry (*Prunus virginiana*) fruits have a similar phenolic acid content (630 mg kg^{-1} FW), followed by wild blueberry (*aka* low-bush blueberry) (*Vaccinium angustifolium*) (510 mg kg^{-1} of FW), containing mainly 5-*O*-caffeoylquinic acid **14** (chlorogenic acid) and 4-*O*-caffeoylquinic acid **15** (neochlorogenic acid).[12] Among the hydroxycinnamic acid derivatives, 4′-hydroxycinnamic acid **16** (*p*-coumaric acid) has been found in the highest concentrations in Swedish bilberries (510 mg kg^{-1} dry weight [DW]). The 4′-hydroxy-3′-methoxycinnamic acid **17** (ferulic acid) content of cloudberries is similar to that of bilberries (390 mg kg^{-1} DW), while a 3′,4′-dihydroxycinnamic acid **18** (caffeic acid) content of 690 and 830 mg kg^{-1} DW has been reported, respectively, in bilberries and lingonberries. 4-Hydroxybenzoic acid **19**, 3,4,5-trihydroxybenzoic acid **20** (gallic acid), 2-hydroxybenzoic acid **21** (salicylic acid) and 4-hydroxy-3-methoxybenzoic acid **22** (vanillic acid) are the main hydroxybenzoic acid derivatives in berry fruits. 3,4,5-Trihydroxybenzoic acid and 4-hydroxy-3-methoxybenzoic acid are found in lingonberries and cloudberries, at respective concentrations of 120 and 2730 mg kg^{-1} DW.[32] The (poly)phenol content of berries can be affected by post-harvest UV light, organic farming and location.[3,33]

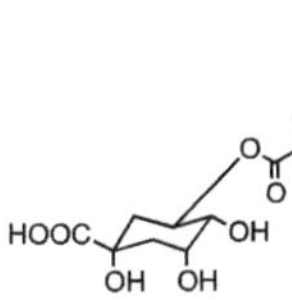
5-*O*-Caffeoylquinic acid **14**

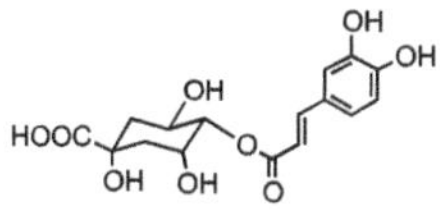
4-*O*-Caffeoylquinic acid **15**

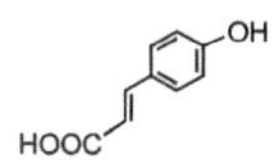
4'-Hydroxycinnamic acid **16**

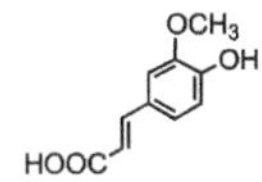
4'-Hydroxy-3'-methoxycinnamic acid **17**

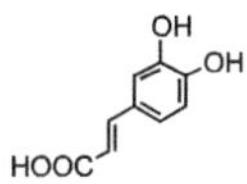
3',4'-Dihydroxycinnamic acid **18**

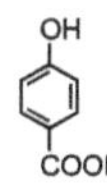
4-Hydroxybenzoic acid **19**

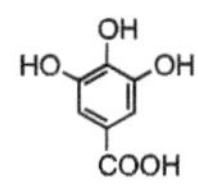
3,4,5-Trihydroxybenzoic acid **20**

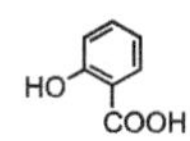
2-Hydroxybenzoic acid **21**

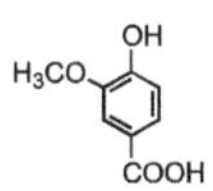
4-Hydroxy-3-methoxybenzoic acid **22**

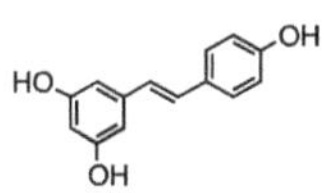
trans-Resveratrol **23**

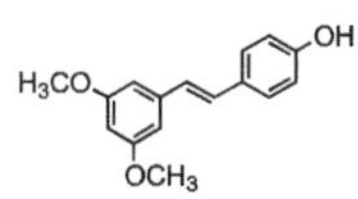
trans-Pterostilbene **24**

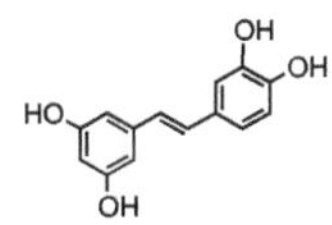
trans-Piceatannol **25**

Stilbenes accumulate in berries to a much more limited extent than other (poly)phenols, with *trans*-resveratrol **23**, *trans*-pterostilbene **24** and *trans*-piceatannol **25** being found mainly as glycosides in deerberry (*Vaccinium stamineum*), blueberry, cowberry and lingonberry (*Vaccinium vitis-idaea*).[34]

Although berries are usually consumed as fresh fruit which are perishable and, thus, have limited season availability, at least from local growers, a range of frozen, freeze-dried fruit, juices, capsules, purees, jams and canned products have been developed to facilitate "all year round" consumption. The technology employed for the manufacture of these products has not been investigated in detail, but processing may well affect the natural (poly)phenol content.[14,35,36]

2.2 Absorption, Disposition, Metabolism and Excretion of Berry (Poly)phenols

Typically, absorption of (poly)phenol glycosides is associated with hydrolysis releasing aglycones as a result of the action of lactase phlorizin hydrolase (LPH) in the brush border of the small intestine epithelial cells. LPH exhibits broad substrate specificity for flavonoid-*O*-β-D-glucosides, and the released aglycone can enter the epithelial cells by passive diffusion, as a result of its increased lipophilicity and its proximity to the enterocyte membrane.[37] An alternative hydrolytic step is mediated by cytosolic β-glucosidase (CBG) within the epithelial cells. In order for CBG-catalyzed hydrolysis to occur, the polar glucosides must be transported into the epithelial cells, possibly with the involvement of the active sodium-dependent glucose transporter SGLT1.[38] Thus, there are two possible routes by which the glucoside conjugates are hydrolyzed, namely LPH-diffusion and transport-CBG, with the resultant aglycones appearing in the epithelial cells.[39]

Prior to passage into the bloodstream, (poly)phenol aglycones undergo phase II metabolism in the enterocytes, forming sulfate, glucuronide and/or methylated metabolites through the respective action of sulfotransferases, uridine-5′-diphosphate glucuronosyltransferases and catechol-*O*-methyltransferases. There is an efflux of at least some of the metabolites back into the lumen of the small intestine,[40,41] and this is thought to involve members of the ATP-binding cassette family of transporters, including multidrug resistance protein and P-glycoprotein. Once in the portal bloodstream, metabolites pass to the liver, where they may be subjected to further phase II metabolism.[39]

The considerable amounts of (poly)phenols not absorbed from the small intestine pass to the colon,[42–44] where the microbiota can cleave conjugating moieties and the resultant aglycones undergo ring fission, leading to the production of phenolic catabolites and aromatic compounds. These low-molecular-weight catabolites can be absorbed, and a portion is subjected to phase II metabolism before being excreted in urine in substantial quantities, which in most instances is well in excess of that of metabolites that enter the circulatory system *via* the small intestine.[45,46] When these metabolites and catabolites are included in the overall estimation of absorption, disposition, metabolism and excretion (ADME), which is synonymous with bioavailability in nutrition research (European Food Information Council 2010),[47] it is evident that many (poly)phenols are highly bioavailable. Although there is speculation,[48] as yet there is no compelling evidence that enterohepatic recirculation in humans results in substantial recycling of (poly)phenol metabolites back into the small intestine *via* biliary excretion.[49]

In bioavailability studies, it is of note that while plasma profiles can supply useful information, they are not an accurate quantitative guide of absorption because the presence of metabolites and catabolites in the circulatory system is transient as they are rapidly removed from the bloodstream *via* renal excretion. More appropriate estimates of absorption are obtained by determining cumulative urinary excretion, typically over a 24–48 h period, post-intake.[46,50,51] Analysis of (poly)phenol phase II glucuronide, sulfate and methylated metabolites in biofluids has evolved markedly over the last 20 years. Initially, in the absence of reference compounds, samples were subjected to acid or enzyme hydrolysis and the released aglycones were analyzed by HPLC with absorbance, fluorescence or electrochemical detection. This provided, at best, very limited information on the identity of the phase II metabolites. Gradually this methodology was replaced by HPLC–MS, which, even in the absence of reference compounds, facilitated the partial identification of metabolites as methyl, sulfate or glucuronide derivatives,[39] although it now realized that there could be major inaccuracies with quantitative estimates.[52] In recent years, a number of methods have been devised for the synthesis of (poly)phenol metabolites, a number of which can now be obtained from commercial sources. This has also been accompanied by the increasing use of ultra-high-performance liquid chromatography (uHPLC) and high-resolution mass spectrometry (HRMS). These developments are discussed in Chapter 1.[53]

2.2.1 Anthocyanins

The majority of studies on the post-absorption fate of anthocyanins have focussed on berries. The pioneering studies were carried out at the end of the 1990s.[54] For many years, anthocyanins were considered to be poorly bioavailable in humans,[55] with the trace quantities of unchanged glycosylated structures being observed in plasma within 30–90 min of intake and in urine collected 0–24 h after ingestion.[56–58] However, as noted above, the current perceived wisdom is that (poly)phenols, including anthocyanins, are more bioavailable than previously envisaged when ADME takes into deglycosylation, formation of phase II glucuronide, sulfate and methoxy metabolites, and to a much greater extent, ring fission and the formation of phenolic catabolites during passage along the gastrointestinal (GI) tract, as well as post-absorption metabolism following entry into the circulatory system.[6,39,59]

The low-level absorption of parent anthocyanin glycosides compared to their phase II metabolites has been reviewed by Cassidy & Minihane and is exemplified in a study by McKay *et al.*[60,61] After the consumption of cranberry juice, peonidin derivatives, including the 3-*O*-galactoside, arabinoside and glucoside conjugates, were detected in plasma by HPLC with absorbance detection at 520 nm, together with cyanidin-3-*O*-arabinoside, cyanidin-3-*O*-galactoside and cyanidin-3-*O*-glucoside. They reached their maximum concentration (C_{max}) 0.9–4.7 h after juice intake (T_{max}). The C_{max} of these native anthocyanins, expressed as cyanidin-3-*O*-glucoside equivalents, ranged between 77 and 981 $pg\,mL^{-1}$, which was at least 1000-fold lower than the C_{max} of anthocyanin glucuronide metabolites. In parallel, urinary excretion of parent anthocyanins was >30-fold lower than that of their phase II glucuronide metabolites.[61]

Similar findings emerged from a study in which volunteers ingested strawberries where the predominant anthocyanin was pelargonidin-3-*O*-glucoside. The only anthocyanin to be detected in plasma was a pelargonidin glucuronide with a T_{max} of 1.1 h and a C_{max} of 274 $nmol\,L^{-1}$, quantified as pelargonidin-3-*O*-glucoside equivalents. The amount of the glucuronide excreted in urine over a 24 h period after strawberry intake was 1.5 μmol compared to 55 nmol of the ingested pelargonidin-3-*O*-glucoside, a 27-fold difference.[62]

Hydroxylation of the anthocyanidin B-ring impacts absorption with the excretion of metabolites of strawberry pelargonidin-3-*O*-glucoside

(Figure 2.1) being equivalent to 1.8% of intake,[63] whereas metabolites of blackberry cyanidin-3-*O*-glucoside are reported at 10-fold lower concentrations.[64] Ileal fluid collected from ileostomists following the ingestion of bilberries showed that malvidin and petunidin glycosides, characterized by methoxy-groups on the B-ring, were recovered in greater quantities than cyanidin and delphinidin glycosides which have only B-ring hydroxyls. This would suggest that *O*-methylation of anthocyanins confers increased stability and/or reduced absorption.[65]

Anthocyanin phase II metabolites derived from monosaccharides, most notably, glucosides, appear to be absorbed more readily in the upper GI tract than those originating from di- and trisaccharide conjugates. To some degree, this could be a consequence of the monosaccharides being more readily deglycosylated by LPH and/or CBG. The enhanced susceptibility of monosaccharides to hydrolysis by these enzymes is supported by the data of González-Barrio *et al.*, which showed that after the consumption of raspberries by ileostomists, there was a 5.9% recovery of cyanidin-3-*O*-glucoside and a 40–93% recovery of cyanidin di- and trisaccharides in ileal fluid (Table 2.2).[66]

A subsequent study, in which urine and plasma were collected after volunteers with a functioning colon had consumed raspberries, indicated that cyanidin released in the small intestine undergoes only level conversion to a glucuronide metabolite which appeared in plasma 1 h after berry intake with a C_{max} of 0.1 nmol L^{-1}.[67] Most of the anthocyanidin probably underwent a pH-dependent transformation to a retro-chalcone structure that was converted to 3′,4′-dihydroxycinnamic acid **18**.[68] The cinnamic acid did not accumulate in detectable amounts in either plasma or urine, probably as a result of rapid phase II metabolism to 3′-methoxycinnamic acid-4′-sulfate **26** (ferulic acid-4′-sulfate), 3′-methoxycinnamic acid-4′-glucuronide **27** (ferulic acid-4′-glucuronide) and 4′-methoxycinnamic acid-3′-glucuronide **28** (isoferulic acid-3′-glucuronide) all of which appeared in plasma 1.0–1.5 h after

Table 2.2 Recovery of the cyanidin-derived anthocyanins in ileal fluid collected 0–4 h, 4–7 h and 7–24 h after the consumption of 300 g raspberries. After González-Barrio *et al.*[66,a]

Anthocyanins	0–4 h	4–7 h	7–24 h	Total
Cyanidin-3-*O*-glucoside	1.5 ± 0.2	0.1 ± 0.0	nd	1.6 ± 0.2 (5.9%)
Cyanidin-3-*O*-sophoroside	41 ± 1	3.3 ± 1.0	0.4 ± 0.2	45 ± 2 (40%)
Cyanidin-3-*O*-rutinoside	5.6 ± 0.2	0.2 ± 0.1	0.1 ± 0.1	5.9 ± 0.4 (54%)
Cyanidin-3-*O*-(2″-*O*-glucosyl)rutinoside	22 ± 1	1.5 ± 0.4	0.2 ± 0.1	24 ± 1 (53%)
Cyanidin-3-O-(2″-*O*-xylosyl)rutinoside	1.2 ± 0.1	0.1 ± 0.4	nd	1.3 ± 0.9 (93%)

[a]Data presented as means values in μmol ± standard error ($n = 4$). Italicized figures in parentheses represent the amounts recovered as a percentage of the quantity ingested. nd – not detected. Anthocyanins were not detected in ileal fluid samples collected >24 h after raspberry intake.

raspberry intake with C_{max} values ranging from 14 to 47 nmol L^{-1} before undergoing a series of conversions potentially by the pathways illustrated in Figure 2.3.[67] Other cyanidin breakdown products arising, arguably, mainly from the A-ring of the anthocyanin, namely 1,2-dihydroxybenzene **29** (catechol), 1,3-dihydroxybenzene **30** (resorcinol) and 1,2,3- trihydroxybenzene **31**, were identified by González-Barrio *et al.*[69] following *ex vivo* anaerobic fecal incubations with raspberry extracts. These products are included in Figure 2.3.

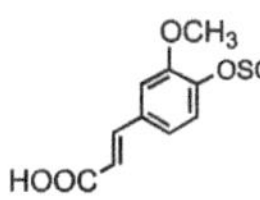

3'-Methoxycinnamic acid-4'-sulfate **26**

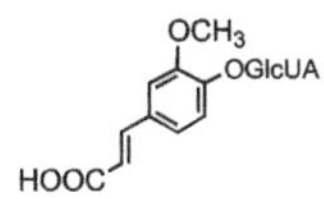

3'-Methoxycinnamic acid-4'-glucuronide **27**

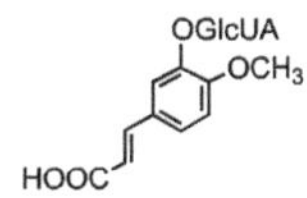

4'-Methoxycinnamic acid-3'-glucuronide **28**

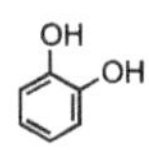

1,2-Dihydroxybenzene **29**

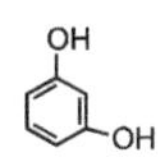

1,3-Dihydroxybenzene **30**

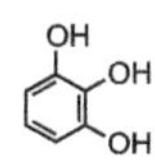

1,2,3-Trihydroxybenzene**31**

In the preparation of the proposed pathways illustrated in Figure 2.3, the following points were taken into consideration. The conversion of cyanidin to 3′,4′-dihydroxycinnamic acid could involve pH-mediated degradation as well as microbial and mammalian enzymes. Subsequent methylation and glucuronide, sulfate and glycine conjugation are mammalian in origin. Dehydroxylation and demethoxylation are almost certainly mediated by the gut microflora, whereas demethylation and hydrogenation steps can be mediated by both microbial and mammalian enzymes. For convenience, the pathways in Figure 2.3 show C_6–C_3 catabolites being converted by two α-oxidations to C_6–C_1 compounds by microflora and/or mammalian enzymes. However, it is plausible that the C_6–C_3 catabolites progress directly to C_6–C_1 structures *via* β-oxidation and that C_6–C_3 catabolites arise independently, possibly by α-oxidation. Alternatively, C_6–C_2 structures could arise from the intact anthocyanidin by an independent gut flora-mediated pathway. In reality, further complexity is introduced as there are multiple points at which catabolites might be absorbed. For example, a percentage of some C_6–C_3 catabolites could be absorbed and undergo β-oxidation and/or mammalian phase II conjugation while the balance is subjected to microbial hydrogenation and/or β-oxidation prior to absorption and mammalian conjugation. Also, for some catabolites, mammalian conjugation either does not occur or is incomplete. The production of benzene derivatives almost certainly involves microbial enzymes.

An intervention study by de Ferrars *et al.* provided further evidence that, following ingestion, anthocyanins were converted predominantly to phenolic catabolites.[70] Post-menopausal women ingested an elderberry extract containing 500 mg of anthocyanins, mainly in the form of cyanidin-3-*O*-glucoside and cyanidin-3-*O*-sambubioside. In total, 21 metabolites/catabolites were detected. The phenolic catabolites, which included

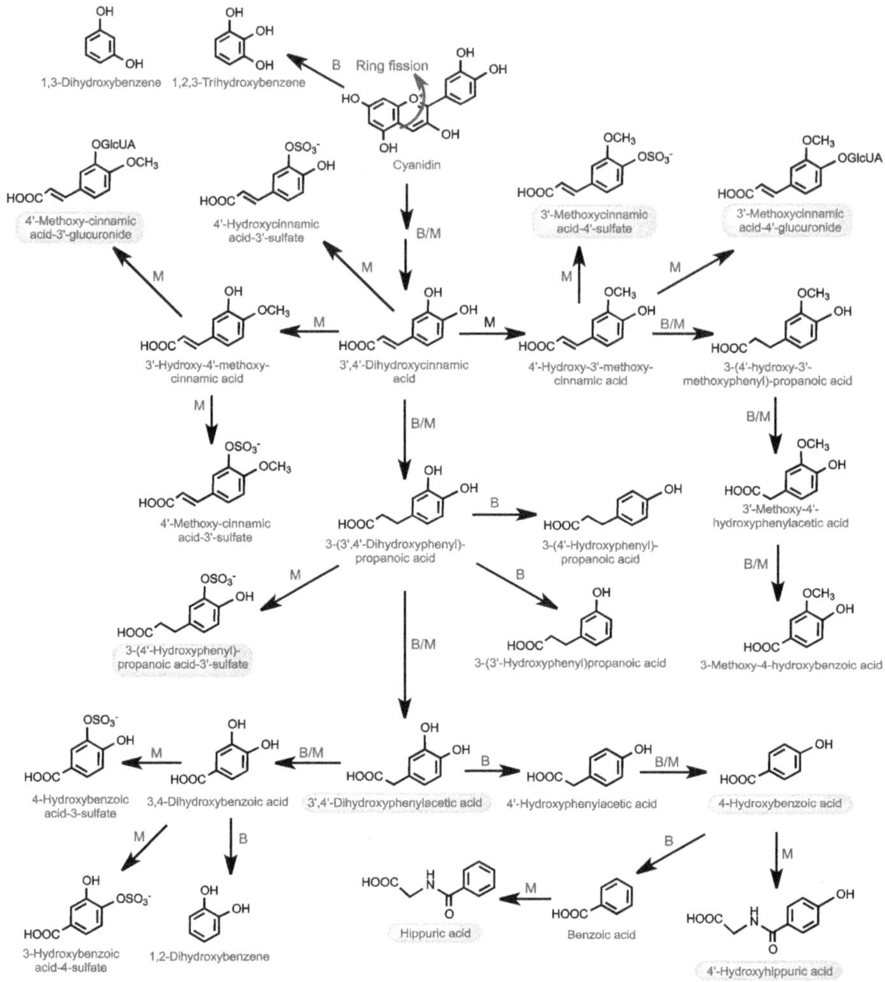

Figure 2.3 Proposed pathways for the conversion of cyanidin-based anthocyanins to phenolic acids and related compounds based on urinary excretion in feeds with red raspberries and *in vitro* fecal incubations.[67,69] M – steps catalyzed by mammalian enzymes. B – steps probably catalyzed by bacterial enzymes. M/B – conversions that could involve mammalian and/or bacterial enzymes. Abbreviation: GlcUA, glucuronic acid. Compounds with boxed names indicate major components in 0–24 h urine.

sulfates of 4-hydroxybenzoic acid **32**, 3,4-dihydroxybenzoic acid **33** and 4-hydroxy-3-methoxybenzoic acid **22**, attained a C_{max} of 1237 nmol L^{-1}, well in excess of the 34 nmol L^{-1} of the parent anthocyanin glycosides, while excretion of the phenolic catabolites in urine was 60-fold higher than that of anthocyanin conjugates.

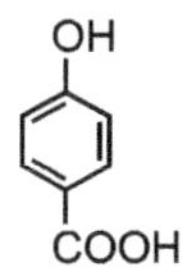

4-Hydroxybenzoic acid **32**

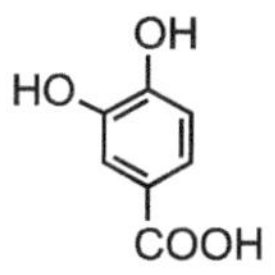

3,4-Dihydroxybenzoic acid **33**

The most notable anthocyanin bioavailability investigation to date is that of Czank *et al.* and de Ferrars *et al.*, who fed $^{13}C_5$-labeled cyanidin-3-*O*-glucoside (Figure 2.4) to human participants.[71,72] After 48 h, recovery was established as 44% of the ^{13}C dose, with 6.9% in breath, 32% in feces and 5.4% urine. The particular advantage of using [6,8,10,3′,5′-$^{13}C_5$]cyanidin-3-*O*-glucoside was that it had three ^{13}C molecules incorporated into the A-ring and two into the B-ring. Numerous phenolic products were detected in plasma and urine, and because of the ^{13}C label, it was possible to ascertain whether they were derived from the A- or B-ring of cyanidin. The capacity to determine relatively low levels of incorporation of the ^{13}C label into the pools of phenolic compounds enabled cyanidin-derived catabolites to be identified and quantified in much more detail than the raspberry feeding study of Ludwig *et al.*,[67] where catabolites derived from cyanidin were reported only where there was a statistically significant increase in the urinary pool after supplementation. Nonetheless, the structural conversions illustrated in Figure 2.3, based primarily on outputs from raspberry interventions, are broadly in line with the array of ^{13}C-labeled products detected by Czank *et al.* and de Ferrars *et al.*, who in their studies observed conversions to additional benzoic acid and benzene derivatives.[71,72]

There are reports that anthocyanins exist as hundreds of derivatives of phase II conjugates that persist for prolonged periods in tissues and the circulation.[73–75] These inferences are based on an untargeted metabolomics approach using single-ion monitoring and low-resolution

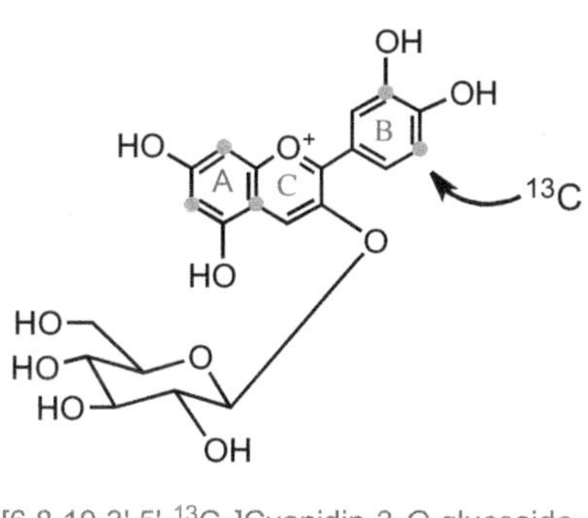

[6,8,10,3′,5′-$^{13}C_5$]Cyanidin-3-*O*-glucoside

Figure 2.4 Structure of [6,8,10,3′,5′-$^{13}C_5$]cyanidin-3-*O*-glucoside. The five highlighted carbons show the positions of the ^{13}C-label.

triple-quadrupole mass spectrometry and are not supported by more appropriate analyses based on high-resolution time-of-flight or orbitrap mass spectrometry.[76–79] Until more convincing evidence is forthcoming, proposals that following ingestion anthocyanins exist as hundreds of long-lived phase II metabolites, none of which have been structurally identified, lack credibility.

2.2.2 Flavan-3-ols

As noted in Section 2.1, flavan-3-ols are found *in planta* in a variety of forms. Most bioavailability research with berries has focused on the monomer (−)-epicatechin, which occurs in comparatively low amounts and this has hindered the identification and quantification of specific metabolites, a task that has been further compounded by difficulties in obtaining reference compounds of glucuronide, sulfate and methyl phase II metabolites.[80,81] Synthesis procedures have been devised, but they are complex and as a result, obtaining metabolites is a specialized and expensive process.[82–84]

There is one report on the ADME of [2-^{14}C](−)-epicatechin in humans, which provides useful insights into the likely fate of the monomer in berry-based studies.[85] Measurement of radioactivity in plasma indicated a biphasic profile with maxima at 1 h and 6 h (Figure 2.5). The total radioactivity in plasma never exceeded 2% of intake. The radioactivity with a T_{max} of 1 h consisted of 12 structurally related (−)-epicatechin metabolites (SREMs), principally (−)-epicatechin-3′-glucuronide **34**,

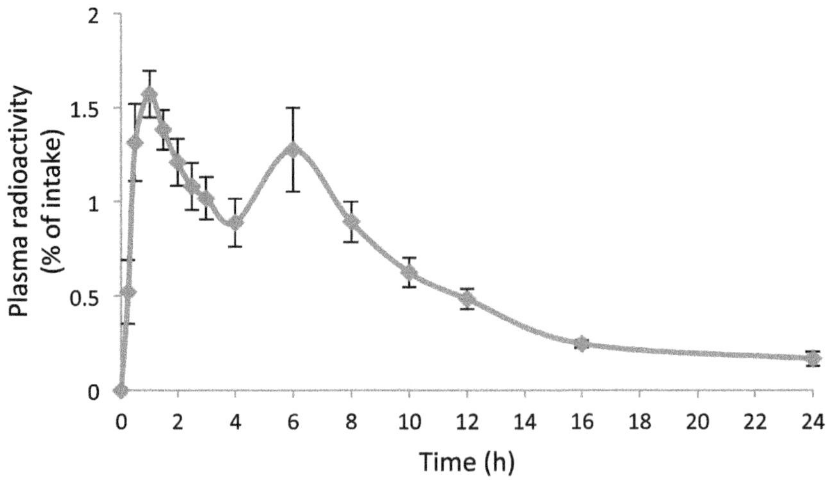

Figure 2.5 Radioactivity detected in plasma collected 0–24 h after the ingestion of 300 μCi (207 μmoles) of [2-^{14}C](−)-epicatechin by volunteers.[85] Data presented as mean values ± standard error ($n = 8$) and expressed as a percentage of the ingested radioactivity.

(−)-epicatechin-7-glucuronide **35**, (−)-epicatechin-3′-sulfate **36** and 3′-methoxy-(−)-epicatechin-5-sulfate **37**, which, the T_{max} indicates, were absorbed in the upper GI tract.

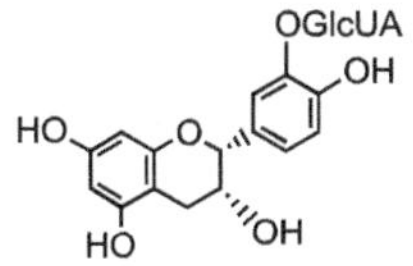

(−)-Epicatechin-3'-glucuronide **34**

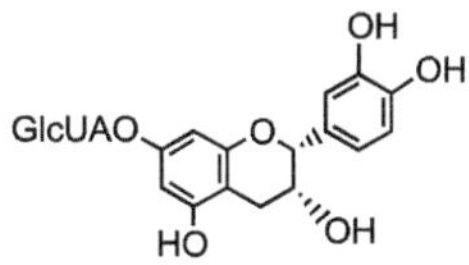

(−)-Epicatechin-7-glucuronide **35**

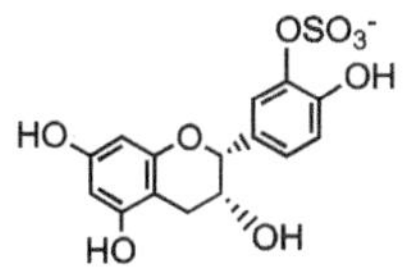

(−)-Epicatechin-3'-sulfate **36**

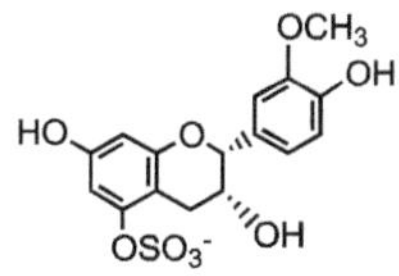

3'-Methoxy-(−)-epicatechin-5-sulfate **37**

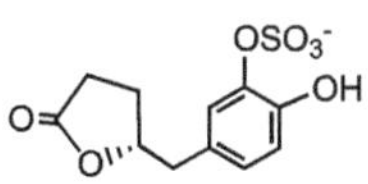

5-(4'-Hydroxyphenyl)-γ-valerolactone-3'-sulfate **38**

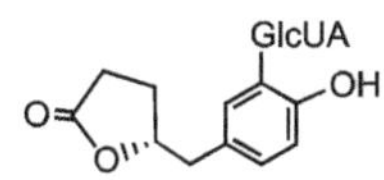

5-(4'-Hydroxyphenyl)-γ-valerolactone-3'-glucuronide **39**

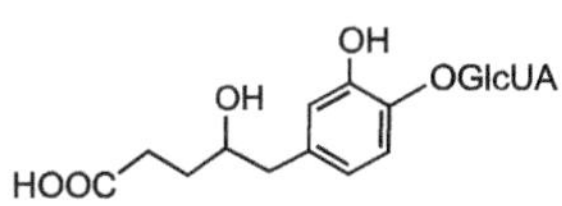

4-Hydroxy-5-(3'-Hydroxyphenyl)valeric acid-4'-glucuronide **40**

(−)-Epicatechin that was not absorbed in the small intestine, and its metabolites effluxed back in the lumen of the GI tract, passed to the colon where they were subjected to the action of the microbiota. Hence, radioactivity appearing in the circulatory system with a T_{max} of 6 h (Figure 2.5) was associated with the resultant 5C-ring fission metabolites (5C-RFMs), principally in the form of 5-(4′-hydroxyphenyl)-γ-valerolactone-3′-sulfate **38** and 5-(4′-hydroxyphenyl)-γ-valerolactone-3′-glucuronide **39**. Other 5C-RFMs detected in plasma included 4-hydroxy-5-(3′-hydroxyphenyl)valeric acid-4′-glucuronide **40**. Urinary recovery 0–48 h after intake indicated excretion of SREMs and 5C-RFMs that was equivalent, respectively, to 20% and 42% of (−)-epicatechin intake, while there was also a 7% excretion of 2C- and/or 3C-ring fission metabolites (2C/3C-RFMs) and 21% excretion of hippuric acids. Thus, the recovery of metabolites absorbed in the colon was 70%. Voided feces contained 10% of the ingested radioactivity as a diversity of ring fission metabolites.[86] Potential routes for the degradation of (−)-epicatechin by microbiota in the colon and subsequent phase II metabolism in colonocytes and/or hepatocytes are summarized in Figure 2.6. Additional details can be found in papers by Monagas *et al.*,[87] Rodriguez-Mateos *et al.*,[50] Borges *et al.*,[86] Mena *et al.*[88] and Liu *et al.*[89]

A further study with (−)-epicatechin by Actis-Goretta *et al.* made use of a Loc-I-gut intestinal technique to introduce (−)-epicatechin into an isolated section of the jejunum.[40] This revealed that there was an efflux of absorbed metabolites back into the intestinal lumen and that the main metabolite effluxing back into the jejunum was (−)-epicatechin-3′-sulfate **36**, while (−)-epicatechin-3′-glucuronide **34** was absorbed in the circulatory system and excreted in urine to a greater extent. The data obtained in this study have also indicated that enterohepatic recirculation of flavan-3-ol metabolites was, at most, a very minor event.

Bioavailability studies on flavan-3-ols using berries include that of Pimpão *et al.*,[79] who investigated the 24 h metabolic fate of (−)-epicatechin **1** and (+)-catechin **2**, following acute intake by healthy volunteers of a (poly)phenol-rich fruit puree consisting of blueberries, blackberries, raspberries, strawberry tree (*Arbutus unedo*) and crowberries (*Corema album* L.). Structurally related (−)-epicatechin metabolites in the form of two (epi)catechin sulfate and two glucuronide conjugates were excreted in urine at high levels after 2 h, suggesting their absorption in the upper GI tract. Post-ingestion of a grape and wild blueberry mix, the circulating levels of SREMs was investigated by Philip *et al.*[90] SREMs, including an (epi)catechin-glucuronide, two (epi)catechin-sulfate isomers and a methoxy(epi)catechin-sulfate, had a C_{max} in plasma 3 h post-intake, arguably indicative of flavan-3-ol monomer absorption in the proximal GI tract. Neither of these studies had access to SREMs as reference compounds, so the metabolites could only be partially identified, nor could they be accurately quantified.[91]

Following the intake of cranberry juice containing 1741 mg of total flavan-3-ols, 22 free and phase-II derivatives of the colon-derived 5C-RFMs were identified in plasma and urine samples, of which 13 were quantified.[92] 5-(4′-Hydroxyphenyl)-γ-valerolactone-3′-glucuronide **39**, 5-(3′-hydroxyphenyl)-γ-valerolactone-4′-glucuronide **41** and two 5-(hydroxyphenyl)-γ-valerolactone-sulfate isomers were the most abundant plasma 5C-RFMs.[92] 5-(4′-Hydroxyphenyl)-γ-valerolactone-3′-glucuronide was reported to be the main metabolite to be excreted in urine post-cranberry intake followed by 5-(3′-hydroxyphenyl)-γ-valerolactone-4′-glucuronide and a 5-(hydroxyphenyl)-γ-valerolactone-sulfate.[92]

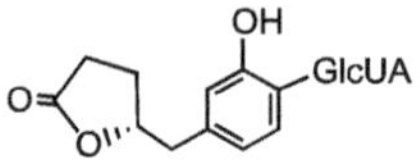

5-(3'-Hydroxyphenyl)-γ-valerolactone-4'-glucuronide **41**

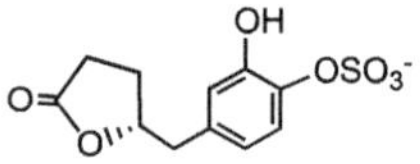

5-(3'-Hydroxyphenyl)-γ-valerolactone-4'-sulfate **42**

After strawberry tree ingestion, containing principally (epi)catechin and dimeric PACs, neither native flavan-3-ols nor SREMs were detected in plasma,[93] whereas an (epi)catechin-sulfate was the most abundant metabolite in urine along with lower levels of several sulfated and glucuronidated derivatives.[93] Similarly, (epi)catechin and PACs were not detected in plasma

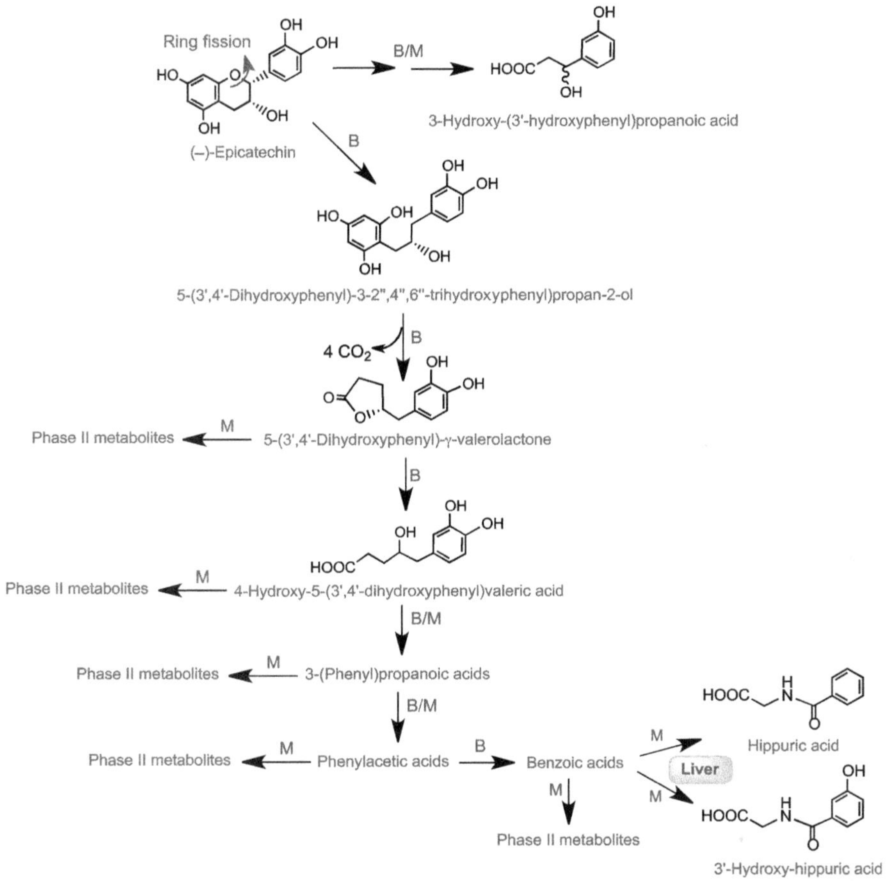

Figure 2.6 Proposed routes for the metabolism of (−)-epicatechin passing from the small to the large intestine. M – steps catalyzed by mammalian enzymes. B – steps probably catalyzed by bacterial enzymes. M/B – conversions that could involve mammalian and/or bacterial enzymes. There is evidence that tail units of proanthocyanidins, like (−)-epicatechin, are converted to 5-(3′,4′-dihydroxyphenyl)-3-(2″,4″,6″-trihydroxyphenyl)propan-2-ol.

or urine collected after consumption of cranberry juice, although PACs were the most abundant flavan-3-ols in the drink.[94] Generally, the amount of native cranberry flavan-3-ols detected in the circulatory system and urine is negligible compared to that of colonic metabolites principally in the form of 5C-RFMs. Following the ingestion of cranberry PACs, or wild blueberry PACs with DPs of 3–10, only one 5C-RFM derivative, namely 5-(3′-hydroxyphenyl)-γ-valerolactone-4′-sulfate **42**, was quantified in plasma and urine.[95,96]

PACs have not been detected in plasma or urine in most human bioavailability investigations. Among the exceptions, the study by McKay *et al.* detected the accumulation of a variety of metabolites, including trace amounts of PAC A2 dimer (see Figure 2.1), in urine after the ingestion of a cranberry juice.[61] Low levels of PAC dimers in biofluids have also been reported by other investigators.[97–101]

It is evident that high-molecular-weight PACs are not absorbed and that there is extremely limited absorption of dimers and trimers into the circulatory systems. The substantial amounts of PACs that occur in many berries, therefore, pass into the lower bowel. The available evidence obtained by Appeldoorn *et al.*[102] and Stoupi *et al.*[103] indicates they are degraded by the colonic microbiota, with scission of the interflavan bond releasing tail units which, like monomers, undergo ring fission, being converted *via* 5-(3′,4′-dihydroxyphenyl)-3-(2′′,4′′,6′′-trihydroxyphenyl)propan-2-ol **43** to 5C-RFMs.[89] The 5C-RFMs, as shown in Figure 2.5, can be further degraded to lower-molecular-weight phenolic acids and aromatic compounds, which are also derived from the catabolism of the upper and lower PAC extension units. These catabolites, unlike the parent PACs, are readily absorbed *in situ*.[50,88] Voided feces are likely to contain unabsorbed PACs and, although as yet they have not been analyzed, partially degraded PAC chains.[19]

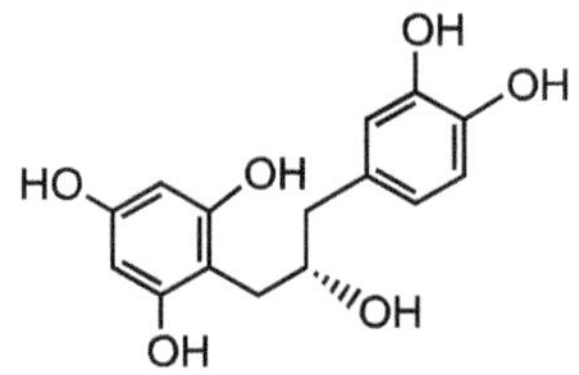

5-(3',4'-Dihydroxyphenyl)-3-(2",4",6"-trihydroxyphenyl)propan-2-ol **43**

2.2.3 Ellagitannins

Studies on the ADME of **11** acid and ETs with berries have focussed principally on strawberries and raspberries.[31,66,67,69,93,104,105] Both types of berries contain sanguiin H-6 **12** and ellagic acid, both as the aglycone and conjugated with a sugar. Raspberries also contain sanguiin H-10 **44** and lambertianin C **10**,[66] and strawberries strictinin **45**,[93] casuarictin **46** and agrimoniin **47**.[106] Much of the non-berry research on the bioavailability of ETs has involved pomegranate juice, which is rich in ETs, including punicalagin **48**.[107,108]

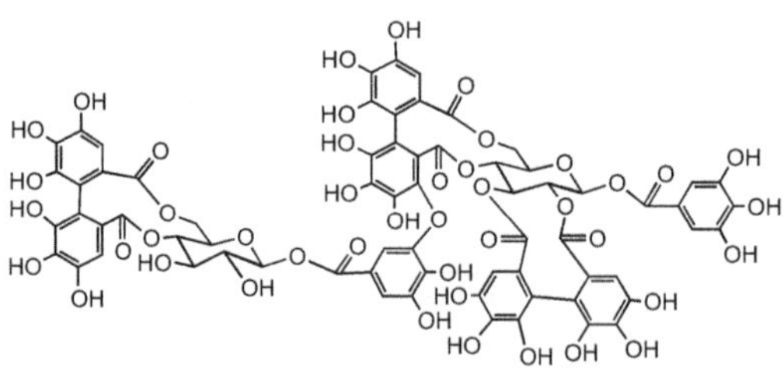

Sanguiin H-10 **44**

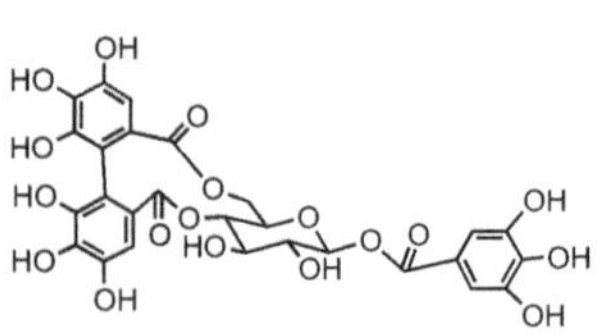

Strictinin **45**

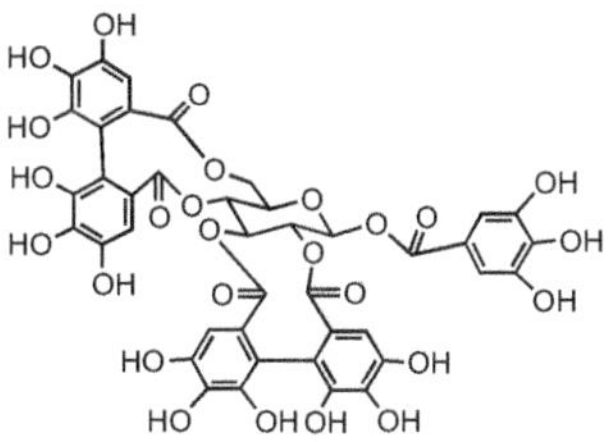

Casuarictin **46**

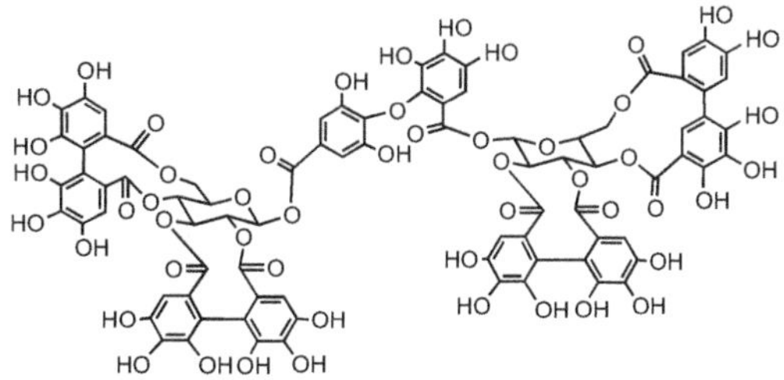

Agrimoniin **47**

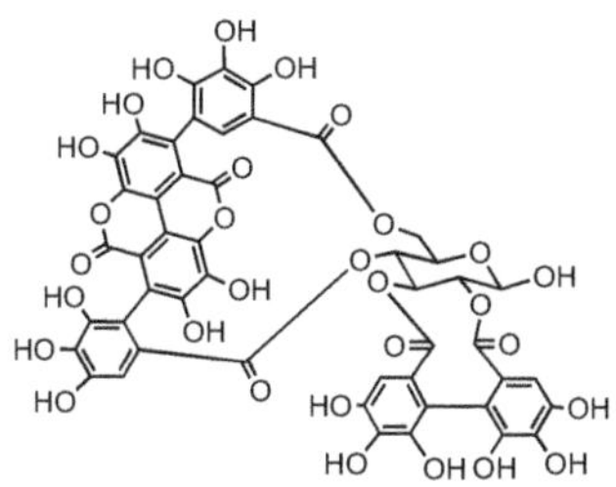

Punicalagin **48**

When ingested, ETs remain stable in the stomach and are either hydrolyzed in the small intestine due to pH changes or remain in their intact form. ETs and ellagic acid are poorly absorbed in the small intestine and reach the colon, as observed in a parallel intervention study with healthy participants and ileostomists who ingested 300 g of raspberries.[66] The 0–48 h ileal fluids contained only sanguiin H-6 **12** of the three ETs identified in raspberries, with a recovery corresponding to the 23% of intake. Ellagic acid recovery was 241% of intake, indicating ET hydrolysis in the upper GI tract.[66] No ETs and ellagic acid or ellagic acid glycosides were detected in plasma samples of healthy subjects,[66] an observation confirmed by Mosele *et al.*[93] However, in pomegranate juice intervention studies with healthy volunteers with an intact colon, ETs did not accumulate in plasma, although nmol L^{-1} levels of free ellagic acid were detected 1 h post-ingestion.[107,108] Similar findings were recently obtained in a trial that correlated ET metabolites from raspberries with improvements in endothelial function.[109] These results suggest low-level absorption of ellagic acid, either from the dietary source *per se* or released from ETs or ellagic acid glycosides, in the upper bowel. Ellagic acid appears in the circulatory system mainly as the free acid,[108] but it can also undergo phase II metabolism: ellagic acid-glucuronide, dimethylellagic acid (DMEA) and DMEA-glucuronide have been detected in plasma in nmol L^{-1} concentrations and in urine in μmol amount.[66,67,93,107,108]

Ellagic acid and ETs reaching the large intestine are substrates for the colonic microbiota-mediated conversion into urolithins (6*H*-dibenzo[*b*,*d*]pyran-6-one). Firstly, ETs are hydrolyzed into ellagic acid by tannin-hydrolase.[110] *Ex vivo* fecal incubations suggest that the conversion of

ETs to ellagic acid may be the rate-limiting step in the formation of urolithins.[69] Ellagic acid undergoes lactone-ring cleavage and decarboxylation, yielding an intermediate 3,4,8,9,10-pentahydroxy-urolithin (urolithin M5). Subsequent dehydroxylation reactions occur, generating tetrahydroxy-urolithins, namely 3,4,8,10-tetrahydroxy-urolithin (urolithin E), 3,8,9,10-tetrahydroxy-urolithin (urolithin M6) and 3,4,8,9-tetrahydroxy-urolithin (urolithin D), trihydroxy-urolithins, namely 3,8,10-trihydroxy-urolithin (urolithin M7) and 3,8,9-trihydroxy-urolithin (urolithin C) and finally dihydroxy-urolithins, namely 3,8-dihydroxy-urolithin (urolithin A) and its isomer 3,9-dihydroxy-urolithin (isourolithin A) and 3-hydroxy-urolithin (urolithin B) (Figure 2.7).[110]

Urolithin production from ETs and ellagic acid appears to be linked to individual gut microbiota composition, although there is limited information on the bacterial species that are involved. Currently, bacteria species from human feces, belonging to the *Coriobacteriaceae*, *Gordonibacter urolithinfaciens* and *Gordonibacter pamelaeae*, have been shown to carry out the first catabolic steps resulting in 3,8,9-trihydroxy-urolithin formation.[110] Other microbial species and environmental conditions leading to di- and monohydroxy-urolithins remain to be unraveled.

The progressive dehydroxylation steps from penta- to di- and monohydroxy-urolithins lead to an increase in lipophilicity and consequently an increase in absorption. Typically, 3,8-dihydroxy-urolithin, 3,9-dihydroxy-urolithin and 3-hydroxy-urolithin are the main urolithins found in human plasma and urine, as phase II glucuronide metabolites.[110] These metabolites first appear in human plasma ~8 h after ET and ellagic acid ingestion and reach C_{max} after 48–72 h.[48,67] Urinary excretion starts ~16 h post-ingestion of intake and can continue for a further 4 days.[31,104]

In a recent review, Tomás-Barberán and colleagues estimated that the main urolithins attain C_{max} in human plasma in the μmol L^{-1} range (0.024–35 μmol L^{-1} and 0.0045–0.745 μmol L^{-1} for 3,8-dihydroxy-urolithin and for 3,9-dihydroxy-urolithin glucuronides, respectively, and 0.012–7.3 μmol L^{-1} for urolithin-3-glucuronide).[110] Lower levels, around 0.005 μmol L^{-1}, of free urolithins have also been detected in plasma usually when analysis has made use of uHPLC–HRMS techniques with low limits of detection.[104,110] Other minor metabolites identified and quantified in plasma and urine include a hydroxy-urolithin-sulfate, a urolithin-sulfoglucuronide and a urolithin-sulfate.[67,109,110]

The highest urinary recovery after berry consumption was described by Truchado and colleagues,[104] who reported a mean 92-h urinary excretion of urolithin conjugates of 58 ± 48% of the administered 62 mg of ellagic acid in a strawberry supplement. In a previous study, Cerdá and colleagues observed lower-mean recovery of ETs in 0–56 h urine, equal to 2.8 ± 4.4% for strawberry, 3.4 ± 4.4% for raspberry, 6.5 ± 5.7% for oak-aged red wine and 16.6 ± 28% for walnuts.[31] In agreement with this study, recoveries of urolithins in 0–48 h urine after raspberry consumption by ten individual volunteers ranged from 0 to 8.6% of ellagic acid and ET intake.[66]

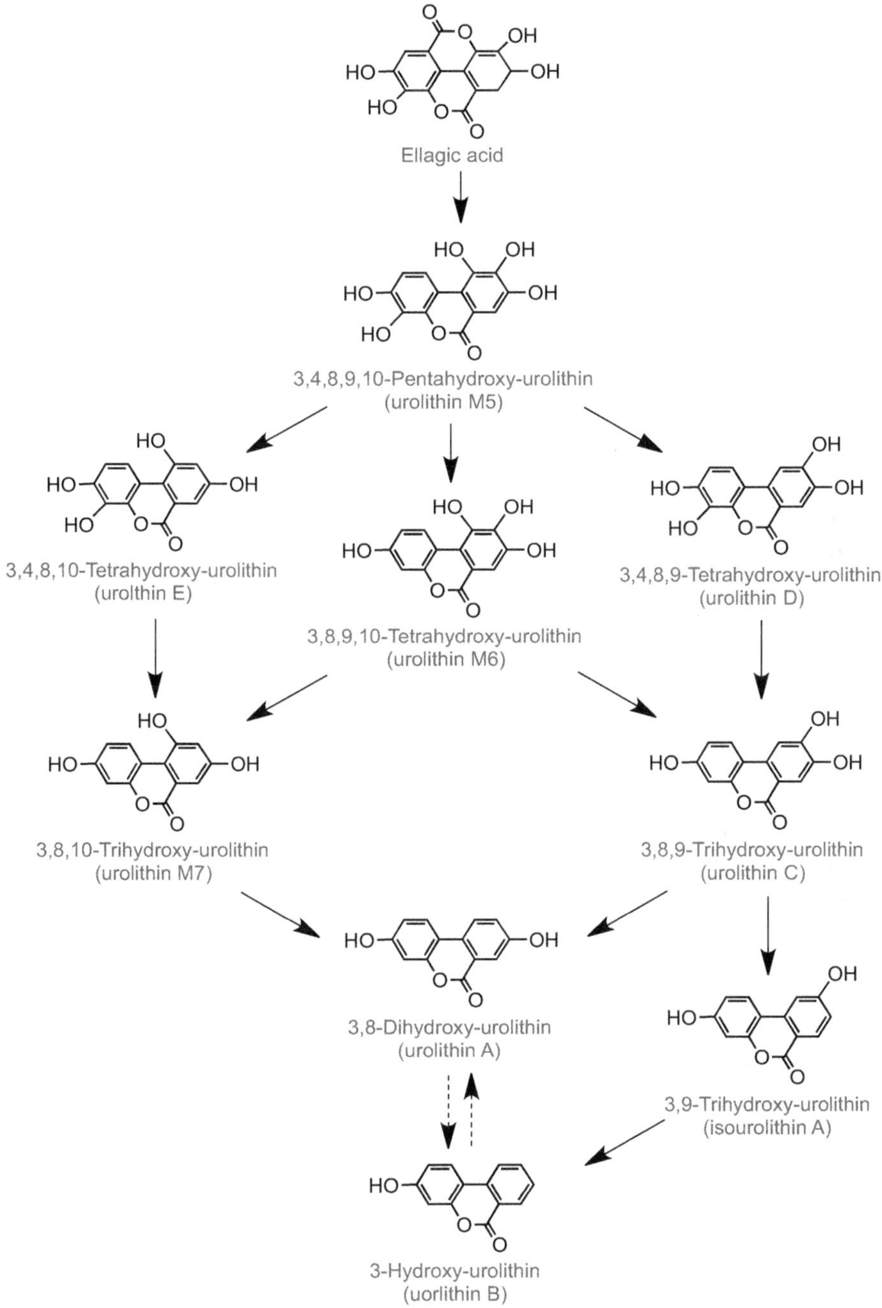

Figure 2.7 In the colon, ellagitannins (ETs) are converted to ellagic acid, which the resident microbiota can convert to urolithins by the routes illustrated.

The effect of ellagic acid/ET ratio on urolithin production is controversial. González-Sarrías *et al.* showed that higher free ellagic acid promotes higher urolithin production after the consumption of two pomegranate extracts containing different ellagic acid/ETs ratios.[108] This is in keeping with the fecal incubations of González-Barrio *et al.*, which showed that ellagic acid was converted to urolithins more readily than ETs.[69] In contrast, Truchado and colleagues reported that urolithin production was not affected by an increase in ellagic acid/ETs ratio promoted by thermal processing.[104] This, however, could be a matrix effect as urolithin excretion was compared after the ingestion of fresh strawberries and a strawberry purée. A further complicating factor could be the release of ETs from achenes in the processed purée.

The analysis of human feces has unveiled relatively low but variable concentrations of ETs, ellagic acid and urolithin aglycones.[48,110] Tomás-Barberán and colleagues reported fecal concentrations of unconjugated urolithins in the range of 7–7180 $\mu g\,g^{-1}$ for 3,8-dihydroxy-urolithin, 13–1894 $\mu g\,g^{-1}$ for 3-hydroxy-urolithin and 36–207 $\mu g\,g^{-1}$ for 3,9-dihydroxy-urolithin.[110] This is in line with the variable conversions of ellagic acid and pomegranate punicalagin **48** to urolithins by fecal samples from different subjects reported by González-Barrio *et al.*[69]

It is clear that there is a large inter-individual variability (IIV) with urolithin production, and this has been associated with varying microbiota compositions. This variability is not only quantitative with low *vs.* high producers[31] but also qualitative with more subjects excreting 3,8-dihydroxy-urolithin than 3-hydroxy-urolithin derivatives,[66,69] giving birth to the concept of metabolic phenotypes, which is discussed further in Section 2.3.[111]

2.2.4 Flavonols

Flavonol glycosides of quercetin **7**, myricetin **8**, kaempferol **9** and isorhamnetin **10** occur widely throughout the plant kingdom.[7,34] Feeding with onions (where about 200 g of fried onions were consumed) has shown that quercetin-*O*-glucosides are bioavailable with the aglycone being absorbed in the small intestine and appearing in the bloodstream as glucuronide, sulfate and methyl metabolites, the main ones being quercetin-3′-sulfate **49** and quercetin-3-glucuronide **50.**[112,113] Urinary excretion 0–24 h after intake was equivalent to 4.7% of the consumed flavonol glucosides. In contrast, the disaccharide, quercetin-3-*O*-rutinoside (rutin) in tomatoes, passes to the distal GI tract where it is deglycosylated by microbial enzymes and appears in plasma in ~500-fold lower concentrations than after quercetin glucoside intake, as quercetin 3-glucuronide and isorhamnetin-3-glucuronide **51.**[45] Most of the quercetin released in the colon undergoes microbiota-mediated ring fission giving rise to phenylpropanoic, phenylacetic and hippuric acids, which were excreted in urine in amounts equivalent to 23% of flavonol intake.[45] A subsequent study with fecal incubations showed that the quercetin catabolites were further degraded to benzoic acids and hydroxybenzene derivatives.[114]

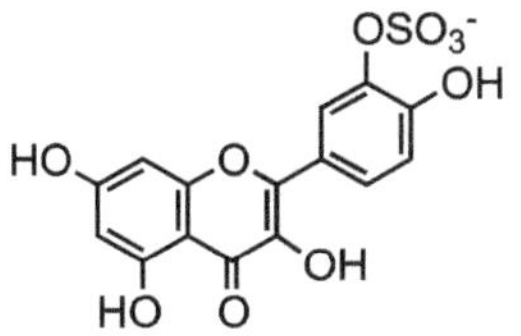

Quercetin-3'-sulfate **49**

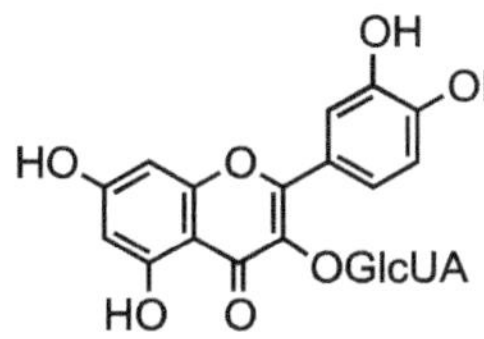

Quercetin-3-glucuronide **50**

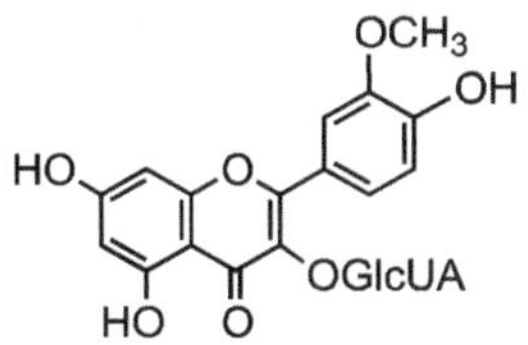

Isorhamnetin-3-glucuronide **51**

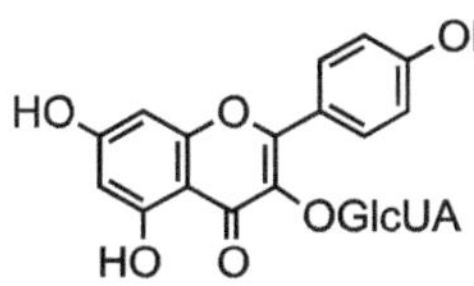

Kaempferol-3-glucuronide **52**

While flavonols do occur in berries, they are not major constituents and are present along with high concentrations of other (poly)phenols.[115,116] For instance, a wild blueberry drink used in an acute bioavailability study contained 302 mg of total (poly)phenols, of which 31 mg were flavonols,[95] consisting of a mixture of myricetin and quercetin glycosides.[116] Also present were 150 mg anthocyanins, 64 mg caffeoylquinic esters, 52 mg of PACs, 4 mg of flavan-3-ol monomers and 1 mg of benzoic and cinnamic acids.[95] At baseline, prior to intake, plasma contained 3 ± 1 nmol L^{-1} of quercetin-3-glucuronide, which after 2 h increased significantly to 5 ± 1 nmol L^{-1}. Substantial amounts of an assortment of (poly)phenolic compounds were also detected in plasma 2 h after blueberry intake, but most were present prior to supplementation and likewise with the profile of phenolics excreted in urine. In the circumstances, elucidating which of the diversity of phenolic catabolites originate from what constituents in the drink is not feasible in the majority of instances. The one exception was 5-(3′-hydroxyphenyl)-γ-valerolactone-4′-sulfate **42**, which can originate from flavan-3-ol monomers and the tail unit of PACs (Section 2.2.2).

In a bioavailability study with cranberry juice, volunteers ingested 787 mg of (poly)phenols comprising 710 mg of PACs, 16 mg anthocyanins, 5 mg flavan-3-ol monomers, 31 mg flavonols, quercetin and myricetin aglycones and glycosides, 5 mg 5-*O*-caffeoylquinic acid and 19 mg phenolic acids.[94] Along with low levels of kaempferol **9**, kaempferol-3-glucuronide **52** and quercetin-3-glucuronide, 5-(3′-hydroxyphenyl)-γ-valerolactone-4′-sulfate and numerous phenolics and their phase II metabolites were detected in plasma and urine and a detailed metabolic map was presented summarizing their potential origins.[94]

Along with many other (poly)phenols, chokeberries (*A. melanocarpa*) contain flavonol aglycones and quercetin-, kaempferol- and myricetin-*O*-glycosides. After feeding a chokeberry extract to healthy men, analysis of plasma and urine detected low levels of kaempferol-3-glucuronide, quercetin-3′-sulfate **49** and quercetin-3-glucuronide **50**.[117]

2.2.5 Other (Poly)phenols

Flavanones, most notably the 7-*O*-rutinosides of hesperetin **53** and naringenin **54** occur predominately in citrus fruits.[7] They are, however, found in very small amounts in chokeberries (*Aronia* spp.).[117,118] The bioavailability of these flavanones, which are absorbed in the distal GI tract as phase II metabolites after microbiota-mediated aglycone release, has been investigated principally with orange juice.[51,119–121] Unsurprisingly, no specific flavanone phase II metabolites were detected in either plasma or urine after ingestion of a 500 mg *A. melanocarpa* extract that contained only 5 nmol of hesperetin and 20 nmol of naringenin.[117]

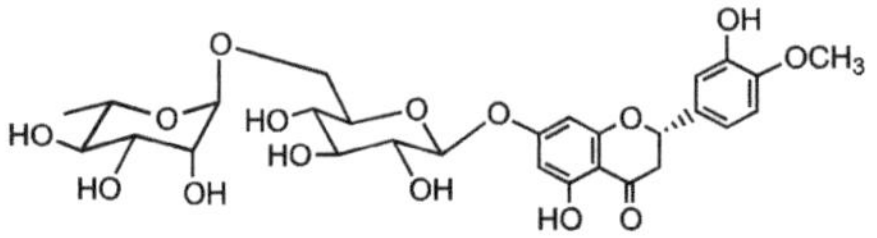

Hesperetin-7-*O*-rutinoside **53**

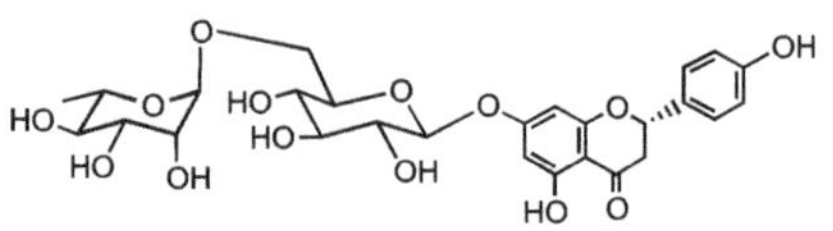

Naringenin-7-*O*-rutinoside **54**

The majority of studies on the bioavailability of 5-*O*-caffeoylquinic acid **14** have used coffee beverages that contain especially high levels of chlorogenic acid together with smaller amounts of other acyl quinic acid esters.[7] The 3C-RFMs that accumulate transiently in plasma after ingestion of 412 μmol of chlorogenic acids include 3′-methoxycinnamic acid-4′-sulfate **26**, 4′-hydroxycinnamic acid-3′-sulfate **55** (caffeic acid-3′-sulfate), 3-(4′-hydroxy-3′-methoxyphenyl)propanoic acid **56** (dihydroferulic acid) and 3-(3′-methoxyphenyl)propanoic acid-4′-sulfate **57** (dihydroferulic acid-4′-sulfate).[122]

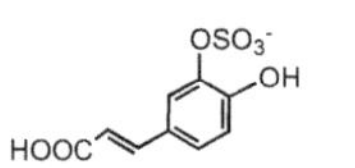

4′-Hydroxycinnamic acid-3′-sulfate **55**

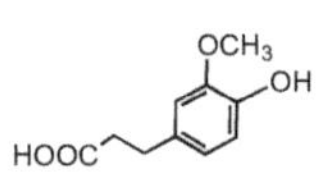

3-(4′-Hydroxy-3′-methoxyphenyl)propanoic acid **56**

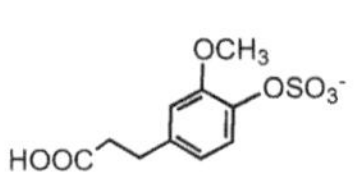

3-(3′-Methoxyphenyl)propanoic acid-4′-sulfate **57**

5-*O*-Caffeoylquinic acid and related acyl-quinic acids have been detected in low amounts in a number of berry species.[34] The exceptions, which contain substantial amounts, are wild blueberry and highbush blueberry (*V. corymbosum*). In the study where a wild blueberry drink was consumed, as mentioned in Section 2.2.4, the supplement contained 64 mg (181 μmol) of chlorogenic acids per 100 g FW of berries.[95] Post-intake, numerous phenolic compounds were detected and quantified in plasma and urine, including 3C-RFMs, that could be catabolites of the ingested chlorogenic acids. However, many of these compounds were also detected in plasma and urine after ingestion of cranberries which contained much lower amounts of chlorogenic acids.[94] Ascertaining their origins is, therefore, not feasible, especially as they could also be degradation products of other (poly)phenols, including anthocyanins, flavan-3-ols and flavonols.

There have been a few dietary studies on the bioavailability of the stilbene *trans*-resveratrol **23**, principally because it is a very minor component in the human diet and is found, at best, in trace levels in berries. The findings of ADME studies that have been carried out with stilbenes, none of which involved berries, have been reviewed by Vitaglione *et al.*[123]

2.3 Inter-individual Variability in the Metabolism of Berry (Poly)phenolics

The beneficial effects of (poly)phenol-rich foods on the promotion of cardiometabolic health have been widely demonstrated on some key risk factors.[124–128] Nevertheless, despite this large body of evidence, the role of (poly)phenols in cardiometabolic protection has, as yet, not been demonstrated consistently.[129,130] This could be due to the heterogeneity in individuals' response to the consumption of dietary (poly)phenols.[130,131] However, this IIV and its impact on physiological outcomes have, as yet, not been investigated in depth.[132]

IIV in the production of metabolites has been observed with many bioactives, including ETs, flavanones, (−)-epicatechin and ETs.[83,108,117] Some factors, including sex, age and dietary habits, may influence (poly)phenol ADME. However, due to the colonic metabolism of a large fraction of ingested (poly)phenols, the microbial population in the lower bowel is arguably the most important factor modulating IIV.[131] Particularly well-established examples of IIV are the formation of urolithins from ETs, equol **58** from the isoflavone daidzein **59** and 8-prenylnaringenin **60**, a hop prenylflavonoid, for which metabolic phenotypes (*aka* metabotypes) with different microbial functionalities have been proposed.[111,133,134]

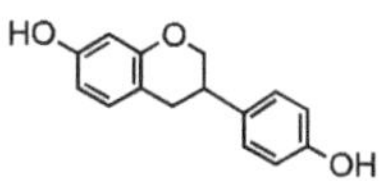

Equol **58**

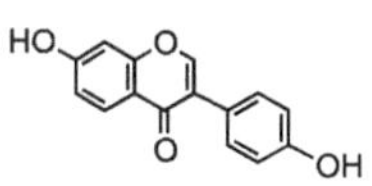

Daidzein **59**

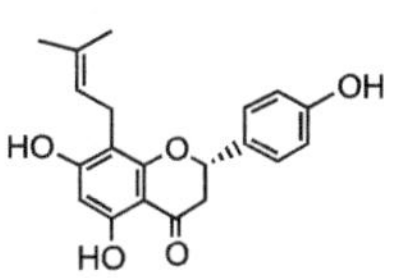

8-Prenylnaringenin **60**

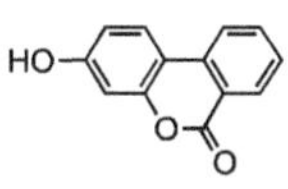

3-Hydroxy-urolithin **61**

Metabotypes have been proposed for the conversion of flavan-3-ols to 5C-RFMs.[135] Three phenotypes have been observed with urolithins independent of volunteers' health status and demographic characteristics, including age, gender and body mass index. A higher percentage of 3-hydroxy-urolithin **61** producers was observed in volunteers with chronic illness associated with gut microbial imbalance (dysbiosis).[111] This is in keeping with the proposed role of microbial composition in inter-individual differences and highlights the need for a new generation of intervention studies designed to capture events underlying this and other IIVs.[136]

IIV in the metabolism of berry (poly)phenols began with studies on bioavailability with raspberry and strawberry, which identified high and low 3-hydroxy-urolithin producers from ET and ellagic acid.[31,66,118,137] After strawberry consumption, the excretion of 3-hydroxy-urolithin ranged from 0.05 to 6.3% and from 0.21 to 7.6% after raspberry ingestion.[31] Feliciano *et al.* assessed (poly)phenol IIV after blueberry consumption.[95] 1,2-Dihydroxybenzene **29** derivatives and benzoic acids showed the lowest IIV in plasma (36%–58% and 50%–63%, respectively), whereas benzaldehydes, phenylacetic acids and benzoic acids were excreted with more uniformity (47%–69%, 50%–54% and 40%–71%, respectively). Phenylpropanoic acids, 1,2,3-trihydroxybenzene **31** and 5C-RFMs in plasma and urine exhibited IIV. The IIV for the sum of (poly)phenol catabolites was marginally lower in plasma (40%–48%) than urine (47%–54%).

In general, bioavailability pharmacokinetic parameters are expressed as mean or median values accompanied by figures for standard deviation or standard error. Recently, the coefficient of variation (CV or CV%) has begun to be used in relation to IIV. The CV of the plasma area under the curve (AUC) of total (poly)phenol metabolites quantified after consumption of cranberry juice was 53% and the CV for C_{max} was 51%. The CV for the AUC varied between 48% for 4-hydroxybenzaldehyde **62** and 163% for a 1-methoxy-hydroxybenzene-sulfate. The CV for C_{max} varied between 43% for 3-(3′-methoxyphenyl)propanoic acid-4′-sulfate **57** and 216% for 4-hydroxy-3-methoxybenzoic acid **22**.[138]

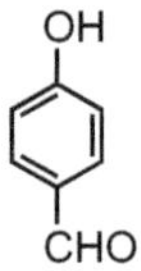

4-Hydroxybenzaldehyde **62**

The targeted analysis of 5C-RFMs after consumption of increasing doses of flavan-3-ols in cranberry juice highlighted a high variability among subjects. CV for plasma C_{max} ranged from 32% to 297%, similar to those reported for the 0–24 h AUC (32% to 277%). CV calculated for 24 h-cumulative urinary excretion for the sum of 5C-RFMs varied from 138% to 289% depending on the treatment.[92]

2.4 Biomarkers of Berry Intake and Future Research

In recent years, biomarkers of food intake have attracted interest as a promising tool to objectively assess the consumption of certain food items or food groups. In order to be validated and defined as such, candidate biomarkers of food intake must meet certain criteria.[139] In particular, a biomarker of intake has to be specific to a particular food and be detectable in plasma or urine in a manner reflective of intake.

Berries contain high levels of micronutrients and fibers and are among the richest sources of (poly)phenolic compounds, none of which are exclusive to berries, whether considering individual berry species or berries as a group.[140,141] Urolithins are produced specifically from ETs and ellagic acid (Section 2.2.2), which unfortunately are not exclusive to berries. The same applies to flavan-3-ol-derived 5C-RFMs, which, while biomarkers of flavan-3-ol intake, are not specific biomarkers of berry consumption. In this more restrictive role, urinary SREMs and 5C-RFMs were used as biomarkers of flavan-3-ol intake in a study involving the analysis of urine from about 25 000 participants in the EPIC Norfolk study.[142] This revealed that high flavan-3-ol intake from tea, pome fruits, berries, cocoa-derived products and nuts was associated with significantly lower systolic blood pressure in men and women. Furthermore, hypertensive participants had a stronger inverse association between the biomarkers and systolic blood pressure compared to normotensive participants. Flavan-3-ol intake could, therefore, have a role in the maintenance of cardiovascular health on a population scale.

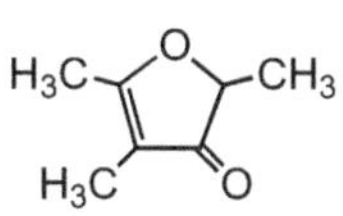

Furaneol **63**

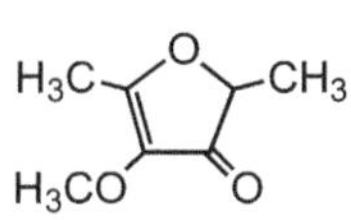

mesifurane **64**

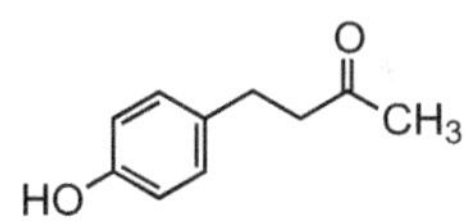

4-(4'-Hydroxyphenyl)-2-butanone **65**

In a recent review, aroma compound metabolites have been proposed as potential biomarkers of intake of specific berries, including furaneol **63** and mesifurane **64** phase II metabolites of strawberry and 4-(4′-hydroxyphenyl)-2-butanone **65** metabolites of raspberry.[141] However, furaneol is found in tropical fruits such as kiwi and pineapple and is also a product of Maillard reactions produced during cooking. To further compound matters, 4-(4′-hydroxyphenyl)-2-butanone is being used increasingly as a food additive and so is found in many aromatized foods and, up to now, its bioavailability and metabolism in humans have not been investigated.[141] It has been suggested that these compounds, in combination with other metabolites, such as pelargonidin metabolites and/or urolithins, could be used as part of a multi-metabolite biomarker panel.[143,144] This approach to predict intake would appear to be more appropriate than focussing on a single biomarker.[141] However, much research remains to be carried out to validate potential biomarkers and evaluate their applicability. More untargeted metabolomics studies aimed at discovering specific biomarkers of berry intake are required and efforts need to be devoted to identifying possible discriminating metabolites, since this step remains the major bottleneck of biomarker research.

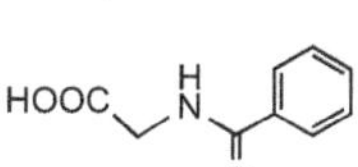

Hippuric acid **66**

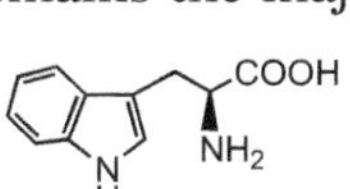

Tryptophan **67**

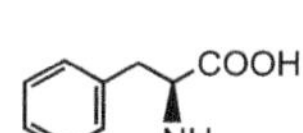

Phenylalanine **68**

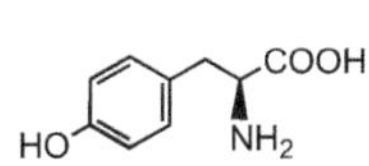

Tyrosine **69**

As outlined in Section 2.2, with the exception of the 5C-RFMs and urolithins, most 2/3C-RFMs, benzoic acid and benzene derivatives are colonic degradation products that can be derived from many (poly)phenols. To further complicate matters, some of these catabolites, such as hippuric acid **66**, are produced by the body independent of dietary intake, from benzoic acid **32**, tryptophan **67**, phenylalanine **68** and tyrosine **69**.[145–148] Determining to what degree individual catabolites are derived from the (poly)phenol intake is, therefore, less than straightforward. Clarification could be achieved using ^{13}C-labeled (poly)phenols in ADME studies. Indeed, as noted in Section 2.2.1, this approach has been used by Czank *et al.* and de Ferrars *et al.*, who fed [6,8,10,3′,5′-$^{13}C_5$] cyanidin-3-*O*-glucoside (Figure 2.5) to human volunteers.[71,72] The particular advantage of using the $^{13}C_5$-labeled cyanidin-3-*O*-glucoside was that it was possible to determine what proportion of individual metabolites/catabolites were endogenous as opposed to originating from the labeled anthocyanin. Furthermore, as the cyanidin-3-*O*-glucoside had three ^{13}C molecules incorporated into the A-ring and two into the B-ring, it was also possible to ascertain whether metabolites/catabolites detected in plasma and urine were derived from the A- or B-ring of cyanidin.

2.5 Concluding Remarks

In conclusion, berries represent important dietary sources of many types of (poly)phenols, which following ingestions initially undergo a variety of patterns metabolism in enterocytes of the small intestine prior to absorption into the circulatory system. (Poly)phenols not absorbed in the upper GI tract enter the colon where microbiota-mediated breakdown yields a variety of potentially bioactive phenolics that can undergo phase II metabolism in colonocytes prior to absorption into the bloodstream, which can be followed by further modification by hepatocytes. For many years, it was believed that dietary (poly)phenols were poorly absorbed; however, it is now realized that when their diverse metabolites and catabolites are taken into account, they are highly bioavailable and yield a diversity of potentially bioactive phenolic compounds. Elucidating the pathways involved in these events represents a major analytical challenge, and progress is being made on a number of fronts with acute and chronic feeding studies and *in vitro* and *ex vivo* model test systems coupled with major refinements in analytical methodology. Key biomarkers of intake, including phase II metabolites of 5-(phenyl)-γ-valerolactones and urolithins, are beginning to be identified. Despite some gaps, due to inaccurate quantification of (poly)phenol metabolites, new insights into the colonic fate of ingested berry (poly)-phenols and metabolomic approaches may eventually be useful in accurately determining levels of consumption. However, IIV may represent a challenge that needs to be taken into account when evaluating the beneficial effects of (poly)phenol-rich berries, and other plant-based foods, on the promotion of cardiometabolic health.

Abbreviations

5C-RFMs	5C-ring fission metabolites
3C-RFMs	3C-ring fission metabolites
2C-RFMs	2C-ring fission metabolites
ADME	absorption, disposition, metabolism and excretion
AUC	area under the curve
CBG	cytosolic β-glucosidase
C_{max}	maximum concentration
CV	coefficient of variation
DMAC	4-(dimethylamino)cinnamaldehyde
DMEA	dimethylellagic acid
DP	degree of polymerization
DW	dry weight
ETs	ellagitannins
FW	fresh weight
GI	gastrointestinal
HHDP	hexahydroxydiphenoyl
HILIC	hydrophobic interaction chromatography
HRMS	high-resolution mass spectrometry
IIV	Inter-individual variability
LPH	lactase phlorizin hydrolase
PACs	proanthocyanidins
SREMs	structurally related (−)-epicatechin metabolites
T_{max}	time to reach C_{max}
uHPLC	ultra-high-performance liquid chromatography

Acknowledgements

T.M.A. and A.C. are grateful for support from the Researchers Supporting Project of King Saud University, Riyadh, Saudi Arabia [grant number RSP-2020/273].

References

1. L. Lavefve, L. R. Howard and F. Carbonero, Berry polyphenols metabolism and impact on human gut microbiota and health, *Food Funct.*, 2020, **11**, 45.
2. C. Tuberoso, Berries: Properties, Consumption and Nutrition, Nova Biomedical Books, New York, 2012.
3. L. Di Vittori, L. Mazzoni, M. Battino and B. Mezzetti, Pre-harvest factors influencing the quality of berries, *Sci. Hortic.*, 2018, **233**, 310.
4. J. Fleschhut, F. Kratzer, G. Rechkemmer and S. E. Kulling, Stability and biotransformation of various dietary anthocyanins in vitro, *Eur. J. Nutr.*, 2006, **45**, 7.

5. T. K. McGhie and M. C. Walton, The bioavailability and absorption of anthocyanins: Towards a better understanding, *Mol. Nutr. Food Res.*, 2007, **51**, 702.
6. C. D. Kay and A. Crozier, Pharmacokinetics, in *Anthocyanins from Natural Sources: Exploiting Targeted Delivery for Improved Health*, ed. M. S.-L. Brooks and G. B. Celli, Royal Society of Chemistry, 2019, vol. 2019, p. 159.
7. A. Crozier, T. Yokota, I. B. Jaganath, S. Marks, M. Saltmarsh and M. N. Clifford, Secondary metabolites in fruits, vegetables, beverages and other plant-based dietary components, in *Plant Secondary Metabolites: Occurrence, Structure and Role in the Human Diet*, ed. A. Crozier, M. N. Clifford and H. Ashihara, Blackwell Publishing Ltd, Oxford, UK, 2006, p. 208.
8. S. H. Nile and S. W. Park, Edible berries: Bioactive components and their effect on human health, *Nutrition*, 2014, **30**, 134.
9. D. Del Rio, G. Borges and A. Crozier, Berry flavonoids and phenolics: bioavailability and evidence of protective effects, *Br. J. Nutr.*, 2010, **104**, S67.
10. R. Veberic, A. Slatnar, J. Bizjak, F. Stampar and M. Mikulic-Petkovsek, Anthocyanin composition of different wild and cultivated berry species, *LWT - Food Sci. Technol.*, 2015, **60**, 509.
11. A. Crozier, I. B. Jaganath and M. N. Clifford, Dietary phenolics: Chemistry, bioavailability and effects on health, *Nat. Prod. Rep.*, 2009, **26**, 1001.
12. S. Dudonné, P. Dubé, F. F. Anhê, G. Pilon, A. Marette, M. Lemire, C. Harris, E. Dewailly and Y. Desjardins, Comprehensive analysis of phenolic compounds and abscisic acid profiles of twelve native Canadian berries, *J. Food Compos. Anal.*, 2015, **44**, 214.
13. S. Lachowicz, J. Oszmiański and S. Pluta, The composition of bioactive compounds and antioxidant activity of Saskatoon berry (*Amelanchier alnifolia* Nutt.) genotypes grown in central Poland, *Food Chem.*, 2017, **235**, 234.
14. E. Pappas and K. M. Schaich, Phytochemicals of cranberries and cranberry products: Characterization, potential health effects, and processing stability, *Crit. Rev. Food Sci. Nutr.*, 2009, **49**, 741.
15. R. J. Robbins, J. Leonczak, J. C. Johnson, J. Li, C. Kwik-Uribe, R. L. Prior and L. Gu, Method performance and multi-laboratory assessment of a normal phase high pressure liquid chromatography-fluorescence detection method for the quantitation of flavanols and procyanidins in cocoa and chocolate containing samples, *J. Chromatogr. A*, 2009, **1216**, 4831.
16. U. Bussy, B. R. May, Y. Olanrewaju, G. Hewitt, N. Anderson, A. Crozier, J. I. Ottaviani and C. Kwik-Uribe, Reliable, accessible and transferable method for the quantification of flavanols and procyanidins in foodstuffs and dietary supplements, *Food Funct.*, 2020, **11**, 131.

17. Y. Wang, N. Vorsa, P. D. B. Harrington and P. Chen, Nontargeted metabolomic study on variation of phenolics in different cranberry cultivars using UPLC-IM-HRMS, *J. Agric. Food Chem.*, 2018, **66**, 12206.
18. Y. Wang, P. de, B. Harrington, T. Chang, X. Wu and P. Chen, Analysis of cranberry proanthocyanidins using UPLC-ion mobility-high-resolution mass spectrometry, *Anal. Bioanal. Chem.*, 2020, **412**, 3653.
19. G. Pereira-Caro, S. Gaillet, J. L. Ordóñez, P. Mena, L. Bresciani, K. A. Bindon, D. Del Rio, J.-M. Rouanet, J. M. Moreno-Rojas and A. Crozier, Bioavailability of red wine and grape seed proanthocyanidins in rats, *Food Funct.*, 2020, **11**, 3986.
20. J. A. Kennedy and A. W. Taylor, Analysis of proanthocyanidins by high-performance gel permeation chromatography, *J. Chromatogr. A*, 2003, **995**, 99.
21. K. A. Bindon and J. A. Kennedy, Ripening-induced changes in grape skin proanthocyanidins modify their interaction with cell walls, *J. Agric. Food Chem.*, 2011, **59**, 2696.
22. R. P. Feliciano, M. P. Shea, D. Shanmuganayagam, C. G. Krueger, A. B. Howell and J. D. Reed, Comparison of isolated cranberry (*Vaccinium macrocarpon* Ait.) proanthocyanidins to catechin and procyanidins A2 and B2 for use as standards in the 4-(dimethylamino)cinnamaldehyde Assay, *J. Agric. Food Chem.*, 2012, **60**, 4578.
23. J. F. Harbertson, J. A. Kennedy and D. O. Adams, Tannin in skins and seeds of Cabernet Sauvignon, Syrah, and Pinot Noir berries during ripening, *Am. J. Enol. Vitic.*, 2002, **53**, 54.
24. C. J. Sarneckis, R. G. Dambergs, P. Jones, M. Mercurio, M. J. Herderich and P. A. Smith, Quantification of condensed tannins by precipitation with methyl cellulose: Development and validation of an optimised tool for grape and wine analysis, *Aust. J. Grape Wine Res.*, 2006, **12**, 39.
25. J. H. Grabber, W. E. Zeller and I. Mueller-Harvey, Acetone enhances the direct analysis of procyanidin- and prodelphinidin-based condensed tannins in lotus species by the butanol-HCl-iron assay, *J. Agric. Food Chem.*, 2013, **61**, 2669.
26. P. E. Shay, J. A. Trofymow and C. P. Constabel, An improved butanol-HCl assay for quantification of water-soluble, acetone: Methanol-soluble, and insoluble proanthocyanidins (condensed tannins), *Plant Methods*, 2017, **13**, 63.
27. W. M. Dabeek and M. V. Marra, Dietary quercetin and kaempferol: bioavailability and potential cardiovascular-related bioactivity in humans, *Nutrients*, 2019, **11**, 2288.
28. L. Iglesias-Carres, A. Mas-Capdevila, F. I. Bravo, M. Mulero, B. Muguerza and A. Arola-Arnal, Optimization and characterization of Royal Dawn cherry (*Prunus avium*) phenolics extraction, *Sci. Rep.*, 2019, **9**, 1.
29. V. Neveu, J. Perez-Jimenez, F. Vos, V. Crespy, L. du Chaffaut, L. Mennen, C. Knox, R. Eisner, J. Cruz, D. Wishart and A. Scalbert,

Phenol-Explorer: an online comprehensive database on polyphenol contents in foods, *Database*, 2010, **2010**, bap024.
30. J. M. Landete, Ellagitannins, ellagic acid and their derived metabolites: A review about source, metabolism, functions and health, *Food Res. Int.*, 2011, **44**, 1150.
31. B. Cerdá, F. A. Tomás-Barberán and J. C. Espín, Metabolism of antioxidant and chemopreventive ellagitannins from strawberries, raspberries, walnuts, and oak-aged wine in humans: Identification of biomarkers and individual variability, *J. Agric. Food Chem.*, 2005, **53**, 227.
32. E. Hajazimi, R. Landberg and G. Zamaratskaia, Simultaneous determination of flavonols and phenolic acids by HPLC-CoulArray in berries common in the Nordic diet, *LWT - Food Sci. Technol.*, 2016, **74**, 128.
33. K. Karppinen, L. Zoratti, N. Nguyenquynh, H. Häggman and L. Jaakola, On the Developmental and Environmental Regulation of Secondary Metabolism in *Vaccinium* spp. Berries, *Front. Plant Sci.*, 2016, **7**, 655.
34. PhytoHub, http://phytohub.eu/ (last accessed Janaury 2021).
35. L. R. Howard, R. L. Prior, R. Liyanage and J. O. Lay, Processing and storage effect on berry polyphenols: challenges and implications for bioactive properties, *J. Agric. Food Chem.*, 2012, **60**, 6678.
36. A. Michalska and G. Łysiak, Bioactive compounds of blueberries: post-harvest factors influencing the nutritional value of products, *Int. J. Mol. Sci.*, 2015, **16**, 18642.
37. A. J. Day, F. J. Cañada, J. C. Díaz, P. A. Kroon, R. Mclauchlan, C. B. Faulds, G. W. Plumb, M. R. A. Morgan and G. Williamson, Dietary flavonoid and isoflavone glycosides are hydrolysed by the lactase site of lactase phlorizin hydrolase, *FEBS Lett.*, 2000, **468**, 166.
38. J. M. Gee, M. S. DuPont, A. J. Day, G. W. Plumb, G. Williamson and I. T. Johnson, Intestinal transport of quercetin glycosides in rats involves both deglycosylation and interaction with the hexose transport pathway, *J. Nutr.*, 2000, **130**, 2765.
39. G. Williamson, C. D. Kay and A. Crozier, The bioavailability, transport, and bioactivity of dietary flavonoids: A review from a historical perspective, *Compr. Rev. Food Sci. Food Saf.*, 2018, **17**, 1054.
40. L. Actis-Goretta, A. Lévèques, M. Rein, A. Teml, C. Schäfer, U. Hofmann, H. Li, M. Schwab, M. Eichelbaum and G. Williamson, Intestinal absorption, metabolism, and excretion of (–)-epicatechin in healthy humans assessed by using an intestinal perfusion technique, *Am. J. Clin. Nutr.*, 2013, **98**, 924.
41. L. Actis-Goretta, T. P. Dew, A. Lévèques, G. Pereira-Caro, M. Rein, A. Teml, C. Schäfer, U. Hofmann, M. Schwab, M. Eichelbaum, A. Crozier and G. Williamson, Gastrointestinal absorption and metabolism of hesperetin-7-*O*-rutinoside and hesperetin-7-*O*-glucoside in healthy humans, *Mol. Nutr. Food Res.*, 2015, **59**, 1651.
42. A. Stalmach, H. Steiling, G. Williamson and A. Crozier, Bioavailability of chlorogenic acids following acute ingestion of coffee by humans with an ileostomy, *Arch. Biochem. Biophys.*, 2010, **501**, 98.

43. A. Stalmach, W. Mullen, H. Steiling, G. Williamson, M. E. J. Lean and A. Crozier, Absorption, metabolism, and excretion of green tea flavan-3-ols in humans with an ileostomy, *Mol. Nutr. Food Res.*, 2010, **54**, 323.
44. G. Borges, M. E. J. Lean, S. A. Roberts and A. Crozier, Bioavailability of dietary (poly)phenols: a study with ileostomists to discriminate between absorption in small and large intestine, *Food Funct.*, 2013, **4**, 754.
45. I. B. Jaganath, W. Mullen, C. A. Edwards and A. Crozier, The relative contribution of the small and large intestine to the absorption and metabolism of rutin in man, *Free Radical Res.*, 2006, **40**, 1035.
46. D. Del Rio, A. Rodriguez-Mateos, J. P. E. Spencer, M. Tognolini, G. Borges and A. Crozier, Dietary (poly)phenolics in human health: structures, bioavailability, and evidence of protective effects against chronic diseases, *Antioxid. Redox Signal.*, 2013, **18**, 1818.
47. Nutrient bioavailability: Getting the most out of food (EUFIC), https://www.eufic.org/en/food-today/article/nutrient-bioavailability-getting-the-most-out-of-food, (last accessed 23 December 2020).
48. C. Garcia-Muñoz and F. Vaillant, Metabolic fate of ellagitannins: implications for health, and research perspectives for innovative functional foods, *Crit. Rev. Food Sci. Nutr.*, 2014, **54**, 1584.
49. A. Crozier, Absorption, metabolism, and excretion of (−)-epicatechin in humans: an evaluation of recent findings, *Am. J. Clin. Nutr.*, 2013, **98**, 861.
50. A. Rodriguez-Mateos, D. Vauzour, C. G. Krueger, D. Shanmuganayagam, J. Reed, L. Calani, P. Mena, D. Del Rio and A. Crozier, Bioavailability, bioactivity and impact on health of dietary flavonoids and related compounds: an update, *Arch. Toxicol.*, 2014, **88**, 1803.
51. G. Pereira-Caro, G. Borges, J. van der Hooft, M. N. Clifford, D. Del Rio, M. E. J. Lean, S. A. Roberts, M. B. Kellerhals and A. Crozier, Orange juice (poly)phenols are highly bioavailable in humans, *Am. J. Clin. Nutr.*, 2014, **100**, 1378.
52. J. I. Ottaviani, R. Y. Fong, G. Borges, H. Schroeter and A. Crozier, Use of LC-MS for the quantitative analysis of (poly)phenol metabolites does not necessarily yield accurate results: Implications for assessing existing data and conducting future research, *Free Radical Biol. Med.*, 2018, **124**, 97.
53. C. D. Kay, in *Berry Bioactive Compounds and their Health Benefits*, ed. D. J. Klimis-Zacas and A. Rodriguez-Mateos, Royal Society of Chemistry, 2021, (update when page numbers are available).
54. T. Miyazawa, K. Nakagawa, M. Kudo, K. Muraishi and K. Someya, Direct intestinal absorption of red fruit anthocyanins, cyanidin-3-glucoside and cyanidin-3,5-diglucoside, into rats and humans, *J. Agric. Food Chem.*, 1999, **47**, 1083.
55. X. Wu, G. Cao and R. L. Prior, Absorption and metabolism of anthocyanins in elderly women after consumption of elderberry or blueberry, *J. Nutr.*, 2002, **132**, 1865.

56. G. Cao and R. L. Prior, Anthocyanins are detected in human plasma after oral administration of an elderberry extract, *Clin. Chem.*, 1999, **45**, 574.
57. G. Cao, H. U. Muccitelli, C. Sánchez-Moreno and R. L. Prior, Anthocyanins are absorbed in glycated forms in elderly women: a pharmacokinetic study, *Am. J. Clin. Nutr.*, 2001, **73**, 920.
58. P. E. Milbury, G. Cao, R. L. Prior and J. Blumberg, Bioavailablility of elderberry anthocyanins, *Mech. Ageing Dev.*, 2002, **123**, 997.
59. C. D. Kay, G. Pereira-Caro, I. A. Ludwig, M. N. Clifford and A. Crozier, Anthocyanins and flavanones are more bioavailable than previously perceived: a review of recent evidence, *Annu. Rev. Food Sci. Technol.*, 2017, **8**, 155.
60. A. Cassidy and A.-M. Minihane, The role of metabolism (and the microbiome) in defining the clinical efficacy of dietary flavonoids, *Am. J. Clin. Nutr.*, 2017, **105**, 10.
61. D. L. McKay, C. Y. O. Chen, C. A. Zampariello and J. B. Blumberg, Flavonoids and phenolic acids from cranberry juice are bioavailable and bioactive in healthy older adults, *Food Chem.*, 2015, **168**, 233.
62. W. Mullen, C. A. Edwards, M. Serafini and A. Crozier, Bioavailability of pelargonidin-3-*O*-glucoside and its metabolites in humans following the ingestion of strawberries with and without cream, *J. Agric. Food Chem.*, 2008, **56**, 713.
63. C. Felgines, S. Talavéra, M.-P. Gonthier, O. Texier, A. Scalbert, J.-L. Lamaison and C. Rémésy, Strawberry anthocyanins are recovered in urine as glucuro- and sulfoconjugates in humans, *J. Nutr.*, 2003, **133**, 1296.
64. C. Felgines, S. Talavera, O. Texier, A. Gil-Izquierdo, J.-L. Lamaison and C. Remesy, Blackberry anthocyanins are mainly recovered from urine as methylated and glucuronidated conjugates in humans, *J. Agric. Food Chem.*, 2005, **53**, 7721.
65. M. Kraus, K. Kahle, F. Ridder, M. Schantz, W. Scheppach, P. Schreier and E. Richling, Colonic availability of bilberry anthocyanins in humans, in *Flavor and Health Benefits of Small Fruits*, ed. M. C. Qian and A. M. Rimando, American Chemical Society, 2010, p. 159.
66. R. González-Barrio, G. Borges, W. Mullen and A. Crozier, Bioavailability of anthocyanins and ellagitannins following consumption of raspberries by healthy humans and subjects with an ileostomy, *J. Agric. Food Chem.*, 2010, **58**, 3933.
67. I. A. Ludwig, P. Mena, L. Calani, G. Borges, G. Pereira-Caro, L. Bresciani, D. Del Rio, M. E. J. Lean and A. Crozier, New insights into the bioavailability of red raspberry anthocyanins and ellagitannins, *Free Radical Biol. Med.*, 2015, **89**, 758.
68. M. N. Clifford, Anthocyanins - nature, occurrence and dietary burden, *J. Sci. Food Agric.*, 2000, **80**, 1063.
69. R. González-Barrio, C. A. Edwards and A. Crozier, Colonic catabolism of ellagitannins, ellagic acid, and raspberry anthocyanins: In vivo and in vitro studies, *Drug Metab. Dispos.*, 2011, **39**, 1680.

70. R. M. de Ferrars, A. Cassidy, P. Curtis and C. D. Kay, Phenolic metabolites of anthocyanins following a dietary intervention study in post-menopausal women, *Mol. Nutr. Food Res.*, 2014, **58**, 490.
71. C. Czank, A. Cassidy, Q. Zhang, D. J. Morrison, T. Preston, P. A. Kroon, N. P. Botting and C. D. Kay, Human metabolism and elimination of the anthocyanin, cyanidin-3-glucoside: A ^{13}C-tracer study, *Am. J. Clin. Nutr.*, 2013, **97**, 995.
72. R. M. de Ferrars, C. Czank, Q. Zhang, N. P. Botting, P. A. Kroon, A. Cassidy and C. D. Kay, The pharmacokinetics of anthocyanins and their metabolites in humans, *Br. J. Pharmacol.*, 2014, **171**, 3268.
73. W. Kalt, Y. Liu, J. E. McDonald, M. R. Vinqvist-Tymchuk and S. A. E. Fillmore, Anthocyanin metabolites are abundant and persistent in human urine, *J. Agric. Food Chem.*, 2014, **62**, 3926.
74. W. Kalt, J. E. McDonald, Y. Liu and S. A. E. Fillmore, Flavonoid metabolites in human urine during blueberry anthocyanin intake, *J. Agric. Food Chem.*, 2017, **65**, 1582–1591.
75. W. Kalt, J. E. McDonald, M. R. Vinqvist-Tymchuk, Y. Liu and S. A. E. Fillmore, Human anthocyanin bioavailability: effect of intake duration and dosing, *Food Funct.*, 2017, **8**, 4563.
76. S. Bajad and V. Shulaev, Highly-parallel metabolomics approaches using LC-MS^2 for pharmaceutical and environmental analysis, *TrAC, Trends Anal. Chem.*, 2007, **26**, 625.
77. J. J. J. van der Hooft, R. C. H. de Vos, V. Mihaleva, R. J. Bino, L. Ridder, N. de Roo, D. M. Jacobs, J. P. M. van Duynhoven and J. Vervoort, Structural elucidation and quantification of phenolic conjugates present in human urine after tea intake, *Anal. Chem.*, 2012, **84**, 7263.
78. B. Zhou, J. F. Xiao, L. Tuli and H. W. Ressom, LC-MS-based metabolomics, *Mol. BioSyst.*, 2012, **8**, 470.
79. R. C. Pimpão, T. Dew, M. E. Figueira, G. J. McDougall, D. Stewart, R. B. Ferreira, C. N. Santos and G. Williamson, Urinary metabolite profiling identifies novel colonic metabolites and conjugates of phenolics in healthy volunteers, *Mol. Nutr. Food Res.*, 2014, **58**, 1414.
80. A. Serra, A. Macià, M.-P. Romero, M.-J. Salvadó, M. Bustos, J. Fernández-Larrea and M.-J. Motilva, Determination of procyanidins and their metabolites in plasma samples by improved liquid chromatography–tandem mass spectrometry, *J. Chromatogr. B*, 2009, **877**, 1169.
81. A. Serra, A. Macià, M.-P. Romero, J. Valls, C. Bladé, L. Arola and M.-J. Motilva, Bioavailability of procyanidin dimers and trimers and matrix food effects in in vitro and in vivo models, *Br. J. Nutr.*, 2010, **103**, 944.
82. E. S. Mull, M. Van Zandt, A. Golebiowski, R. P. Beckett, P. K. Sharma and H. Schroeter, A versatile approach to the regioselective synthesis of diverse (−)-epicatechin-β-d-glucuronides, *Tetrahedron Lett.*, 2012, **53**, 1501.
83. M. Zhang, G. E. Jagdmann, M. Van Zandt, R. Sheeler, P. Beckett and H. Schroeter, Chemical synthesis and characterization of epicatechin

glucuronides and sulfates: Bioanalytical standards for epicatechin metabolite identification, *J. Nat. Prod.*, 2013, **76**, 157.

84. M. Zhang, G. Erik Jagdmann, M. Van Zandt, P. Beckett and H. Schroeter, Enantioselective synthesis of orthogonally protected (2R,3R)-(−)- epicatechin derivatives, key intermediates in the de novo chemical synthesis of (−)-epicatechin glucuronides and sulfates, *Tetrahedron: Asymmetry*, 2013, **24**, 362.
85. J. I. Ottaviani, G. Borges, T. Y. Momma, J. P. E. Spencer, C. L. Keen, A. Crozier and H. Schroeter, The metabolome of [2-^{14}C](−)-epicatechin in humans: Implications for the assessment of efficacy, safety, and mechanisms of action of polyphenolic bioactives, *Sci. Rep.*, 2016, **6**, 1.
86. G. Borges, J. I. Ottaviani, J. J. J. van der Hooft, H. Schroeter and A. Crozier, Absorption, metabolism, distribution and excretion of (−)-epicatechin: A review of recent findings, *Mol. Aspects Med.*, 2018, **61**, 18.
87. M. Monagas, M. Urpi-Sarda, F. Sánchez-Patán, R. Llorach, I. Garrido, C. Gómez-Cordovés, C. Andres-Lacueva and B. Bartolomé, Insights into the metabolism and microbial biotransformation of dietary flavan-3-ols and the bioactivity of their metabolites, *Food Funct.*, 2010, **1**, 233.
88. P. Mena, L. Bresciani, N. Brindani, I. A. Ludwig, G. Pereira-Caro, D. Angelino, R. Llorach, L. Calani, F. Brighenti, M. N. Clifford, C. I. R. Gill, A. Crozier, C. Curti and D. Del Rio, Phenyl-γ-valerolactones and phenylvaleric acids, the main colonic metabolites of flavan-3-ols: synthesis, analysis, bioavailability, and bioactivity, *Nat. Prod. Rep.*, 2019, **36**, 714.
89. C. Liu, J. Vervoort, K. Beekmann, M. Baccaro, L. Kamelia, S. Wesseling and I. M. C. M. Rietjens, Interindividual differences in human intestinal microbial conversion of (−)-epicatechin to bioactive phenolic compounds, *J. Agric. Food Chem.*, 2020, **68**, 14168.
90. P. Philip, P. Sagaspe, J. Taillard, C. Mandon, J. Constans, L. Pourtau, C. Pouchieu, D. Angelino, P. Mena, D. Martini, D. Del Rio and D. Vauzour, Acute intake of a grape and blueberry polyphenol-rich extract ameliorates cognitive performance in healthy young adults during a sustained cognitive effort, *Antioxidants*, 2019, **8**, 650.
91. J. I. Ottaviani, R. Y. Fong, G. Borges, H. Schroeter and A. Crozier, Use of LC-MS for the quantitative analysis of (poly)phenol metabolites does not necessarily yield accurate results: Implications for assessing existing data and conducting future research, *Free Radical Biol. Med.*, 2018, **124**, 97.
92. C. Favari, P. Mena, C. Curti, G. Istas, C. Heiss, D. Del Rio and A. Rodriguez-Mateos, Kinetic profile and urinary excretion of phenyl-γ-valerolactones upon consumption of cranberry: a dose-response relationship, *Food Funct.*, 2020, **11**, 3975.
93. J. I. Mosele, A. Macià and M.-J. Motilva, Understanding of human metabolic pathways of different sub-classes of phenols from *Arbutus unedo* fruit after an acute intake, *Food Funct.*, 2016, **7**, 1700.

94. R. P. Feliciano, A. Boeres, L. Massacessi, G. Istas, M. R. Ventura, C. Nunes dos Santos, C. Heiss and A. Rodriguez-Mateos, Identification and quantification of novel cranberry-derived plasma and urinary (poly)phenols, *Arch. Biochem. Biophys.*, 2016, **599**, 31.
95. R. Feliciano, G. Istas, C. Heiss and A. Rodriguez-Mateos, Plasma and urinary phenolic profiles after acute and repetitive intake of wild blueberry, *Molecules*, 2016, **21**, 1120.
96. R. P. Feliciano, A. Boeres, L. Massacessi, G. Istas, M. R. Ventura, C. Nunes Dos Santos, C. Heiss and A. Rodriguez-Mateos, Identification and quantification of novel cranberry-derived plasma and urinary (poly)phenols, *Arch. Biochem. Biophys.*, 2016, **599**, 31.
97. R. R. Holt, S. A. Lazarus, M. Cameron Sullards, Q. Y. Zhu, D. D. Schramm, J. F. Hammerstone, C. G. Fraga, H. H. Schmitz and C. L. Keen, Procyanidin dimer B2 [epicatechin-(4β-8)-epicatechin] in human plasma after the consumption of a flavanol-rich cocoa, *Am. J. Clin. Nutr.*, 2002, **76**, 798.
98. A. Sano, J. Yamakoshi, S. Tokutake, K. Tobe, Y. Kubota and M. Kikuchi, Procyanidin B1 is detected in human serum after intake of proanthocyanidin-rich grape seed extract, *Biosci., Biotechnol., Biochem.*, 2003, **67**, 1140.
99. M. Urpi-Sarda, M. Monagas, N. Khan, R. M. Lamuela-Raventos, C. Santos-Buelga, E. Sacanella, M. Castell, J. Permanyer and C. Andres-Lacueva, Epicatechin, procyanidins, and phenolic microbial metabolites after cocoa intake in humans and rats, *Anal. Bioanal. Chem.*, 2009, **394**, 1545.
100. R. P. Feliciano, E. Mecha, M. R. Bronze and A. Rodriguez-Mateos, Development and validation of a high-throughput micro solid-phase extraction method coupled with ultra-high-performance liquid chromatography-quadrupole time-of-flight mass spectrometry for rapid identification and quantification of phenolic metabolites in human plasma and urine, *J. Chromatogr. A*, 2016, **1464**, 21.
101. G. Sasot, M. Martínez-Huélamo, A. Vallverdú-Queralt, M. Mercader-Martí, R. Estruch and R. M. Lamuela-Raventós, Identification of phenolic metabolites in human urine after the intake of a functional food made from grape extract by a high resolution LTQ-Orbitrap-MS approach, *Food Res. Int.*, 2017, **100**, 435.
102. M. M. Appeldoorn, J.-P. Vincken, A.-M. Aura, P. C. H. Hollman and H. Gruppen, Procyanidin dimers are metabolized by human microbiota with 2-(3,4-dihydroxyphenyl)acetic acid and 5-(3,4-dihydroxyphenyl)-γ-valerolactone as the major metabolites, *J. Agric. Food Chem.*, 2009, **57**, 1084.
103. S. Stoupi, G. Williamson, J. W. Drynan, D. Barron and M. N. Clifford, A comparison of the in vitro biotransformation of (-)-epicatechin and procyanidin B2 by human faecal microbiota, *Mol. Nutr. Food Res.*, 2010, **54**, 747.
104. P. Truchado, M. Larrosa, M. T. García-Conesa, B. Cerdá, M. L. Vidal-Guevara, F. A. Tomás-Barberán and J. C. Espín, Strawberry

processing does not affect the production and urinary excretion of urolithins, ellagic acid metabolites, in humans, *J. Agric. Food Chem.*, 2012, **60**, 5749.
105. A. K. Sandhu, M. G. Miller, N. Thangthaeng, T. M. Scott, B. Shukitt-Hale, I. Edirisinghe and B. Burton-Freeman, Metabolic fate of strawberry polyphenols after chronic intake in healthy older adults, *Food Funct.*, 2018, **9**, 96.
106. M. Gasperotti, D. Masuero, F. Mattivi and U. Vrhovsek, Overall dietary polyphenol intake in a bowl of strawberries: The influence of *Fragaria* spp. in nutritional studies, *J. Funct. Foods*, 2015, **18**, 1057.
107. N. P. Seeram, S. M. Henning, Y. Zhang, M. Suchard, Z. Li and D. Heber, Pomegranate juice ellagitannin metabolites are present in human plasma and some persist in urine for up to 48 hours, *J. Nutr.*, 2006, **136**, 2481.
108. A. González-Sarrías, R. García-Villalba, M. Á. Núñez-Sánchez, J. Tomé-Carneiro, P. Zafrilla, J. Mulero, F. A. Tomás-Barberán and J. C. Espín, Identifying the limits for ellagic acid bioavailability: A crossover pharmacokinetic study in healthy volunteers after consumption of pomegranate extracts, *J. Funct. Foods*, 2015, **19**, 225.
109. G. Istas, R. P. Feliciano, T. Weber, R. Garcia-Villalba, F. Tomas-Barberan, C. Heiss and A. Rodriguez-Mateos, Plasma urolithin metabolites correlate with improvements in endothelial function after red raspberry consumption: A double-blind randomized controlled trial, *Arch. Biochem. Biophys.*, 2018, **651**, 43.
110. F. A. Tomás-Barberán, A. González-Sarrías, R. García-Villalba, M. A. Núñez-Sánchez, M. V. Selma, M. T. García-Conesa and J. C. Espín, Urolithins, the rescue of "old" metabolites to understand a "new" concept: Metabotypes as a nexus among phenolic metabolism, microbiota dysbiosis, and host health status, *Mol. Nutr. Food Res.*, 2017, **61**, 1500901.
111. F. A. Tomás-Barberán, R. García-Villalba, A. González-Sarrías, M. V. Selma and J. C. Espín, Ellagic acid metabolism by human gut microbiota: consistent observation of three urolithin phenotypes in intervention trials, independent of food source, age, and health status, *J. Agric. Food Chem.*, 2014, **62**, 6535.
112. A. J. Day and G. Williamson, Biomarkers for exposure to dietary flavonoids: a review of the current evidence for identification of quercetin glycosides in plasma, *Br. J. Nutr.*, 2001, **86**, S105.
113. W. Mullen, C. A. Edwards and A. Crozier, Absorption, excretion and metabolite profiling of methyl-, glucuronyl-, glucosyl- and sulpho-conjugates of quercetin in human plasma and urine after ingestion of onions, *Br. J. Nutr.*, 2006, **96**, 107.
114. I. B. Jaganath, W. Mullen, M. E. J. Lean, C. A. Edwards and A. Crozier, In vitro catabolism of rutin by human fecal bacteria and the antioxidant capacity of its catabolites, *Free Radical Biol. Med.*, 2009, **47**, 1180.

115. W. Zheng and S. Y. Wang, Oxygen radical absorbing capacity of phenolics in blueberries, cranberries, chokeberries, and lingonberries, *J. Agric. Food Chem.*, 2003, **51**, 502.
116. G. Borges, A. Degeneve, W. Mullen and A. Crozier, Identification of flavonoid and phenolic antioxidants in black currants, blueberries, raspberries, red currants, and cranberries, *J. Agric. Food Chem.*, 2010, **58**, 3901.
117. G. Istas, E. Wood, M. Le Sayec, C. Rawlings, J. Yoon, V. Dandavate, D. Cera, S. Rampelli, A. Costabile, E. Fromentin and A. Rodriguez-Mateos, Effects of aronia berry (poly)phenols on vascular function and gut microbiota: A double-blind randomized controlled trial in adult men, *Am. J. Clin. Nutr.*, 2019, **110**, 316.
118. R. Koli, I. Erlund, A. Jula, J. Marniemi, P. Mattila and G. Alfthan, Bioavailability of various polyphenols from a diet containing moderate amounts of berries, *J. Agric. Food Chem.*, 2010, **58**, 3927.
119. G. Pereira-Caro, I. A. Ludwig, T. Polyviou, D. Malkova, A. García, J. M. Moreno-Rojas and A. Crozier, Identification of plasma and urinary metabolites and catabolites derived from orange juice (poly)phenols: analysis by high-performance liquid chromatography-high-resolution mass spectrometry, *J. Agric. Food Chem.*, 2016, **64**, 5724.
120. G. Pereira-Caro, T. Polyviou, I. A. Ludwig, A. M. Nastase, J. M. Moreno-Rojas, A. L. Garcia, D. Malkova and A. Crozier, Bioavailability of orange juice (poly)phenols: The impact of short-term cessation of training by male endurance athletes, *Am. J. Clin. Nutr.*, 2017, **106**, 791.
121. G. Pereira-Caro, M. N. Clifford, T. Polyviou, I. A. Ludwig, H. Alfheeaid, J. M. Moreno-Rojas, A. L. Garcia, D. Malkova and A. Crozier, Plasma pharmacokinetics of (poly)phenol metabolites and catabolites after ingestion of orange juice by endurance trained men, *Free Radical Biol. Med.*, 2020, **160**, 784.
122. A. Stalmach, W. Mullen, D. Barron, K. Uchida, T. Yokota, C. Cavin, H. Steiling, G. Williamson and A. Crozier, Metabolite profiling of hydroxycinnamate derivatives in plasma and urine after the ingestion of coffee by humans: Identification of biomarkers of coffee consumption, *Drug Metab. Dispos.*, 2009, **37**, 1749.
123. P. Vitaglione, S. Sforza and D. Del Rio, Occurrence, bioavailability and metabolism of reseveratrol, in *Flavonoids and Related Compounds: Bioavailability and Function*, ed. J. P. E. Spencer and A. Crozier, CRC Press, 2012, p. 167.
124. C. Del, Bo', V. Deon, J. Campolo, C. Lanti, M. Parolini, M. Porrini, D. Klimis-Zacas and P. Riso, A serving of blueberry (*V. corymbosum*) acutely improves peripheral arterial dysfunction in young smokers and non-smokers: Two randomized, controlled, crossover pilot studies, *Food Funct.*, 2017, **8**, 4108.
125. R. Menezes, A. Rodriguez-Mateos, A. Kaltsatou, A. González-Sarrías, A. Greyling, C. Giannaki, C. Andres-Lacueva, D. Milenkovic,

E. R. Gibney, J. Dumont, M. Schär, M. Garcia-Aloy, S. A. Palma-Duran, T. Ruskovska, V. Maksimova, E. Combet and P. Pinto, Impact of flavonols on cardiometabolic biomarkers: a meta-analysis of randomized controlled human trials to explore the role of inter-individual variability, *Nutrients*, 2017, **9**, 117.
126. A. González-Sarrías, E. Combet, P. Pinto, P. Mena, M. Dall'Asta, M. Garcia-Aloy, A. Rodríguez-Mateos, E. R. Gibney, J. Dumont, M. Massaro, J. Sánchez-Meca, C. Morand and M. T. García-Conesa, A systematic review and meta-analysis of the effects of flavanol-containing tea, cocoa and apple products on body composition and blood lipids: exploring the factors responsible for variability in their efficacy, *Nutrients*, 2017, **9**, 746.
127. M. T. García-Conesa, K. Chambers, E. Combet, P. Pinto, M. Garcia-Aloy, C. Andrés-Lacueva, S. De Pascual-Teresa, P. Mena, A. K. Ristic, W. J. Hollands, P. A. Kroon, A. Rodríguez-Mateos, G. Istas, C. A. Kontogiorgis, D. K. Rai, E. R. Gibney, C. Morand, J. C. Espín and A. González-Sarrías, *Int. J. Mol. Sci.*, 2018, **19**, 694.
128. D. Martini, L. Chiavaroli, A. González-Sarrías, L. Bresciani, S. A. Palma-Duran, M. Dall'asta, G. E. Deligiannidou, M. Massaro, E. Scoditti, E. Combet, V. Maksimova, M. Urpi-Sarda, C. A. Kontogiorgis, C. Andrés-Lacueva, E. R. Gibney, D. Del Rio, C. Morand, M. Garcia-Aloy, A. Rodriguez-Mateos and P. Mena, Impact of foods and dietary supplements containing hydroxycinnamic acids on cardiometabolic biomarkers: a systematic review to explore inter-individual variability, *Nutrients*, 2019, **11**, 1805.
129. M. T. García-Conesa, Dietary polyphenols against metabolic disorders: How far have we progressed in the understanding of the molecular mechanisms of action of these compounds?, *Crit. Rev. Food Sci. Nutr.*, 2017, **57**, 1769.
130. E. R. Gibney, D. Milenkovic, E. Combet, T. Ruskovska, A. Greyling, A. González-Sarrías, B. de Roos, F. Tomás-Barberán, C. Morand and A. Rodriguez-Mateos, Factors influencing the cardiometabolic response to (poly)phenols and phytosterols: a review of the COST Action POSITIVe activities, *Eur. J. Nutr.*, 2019, **58**, 37.
131. C. Manach, D. Milenkovic, T. Van de Wiele, A. Rodriguez-Mateos, B. de Roos, M. T. Garcia-Conesa, R. Landberg, E. R. Gibney, M. Heinonen, F. Tomás-Barberán and C. Morand, Addressing the inter-individual variation in response to consumption of plant food bioactives: Towards a better understanding of their role in healthy aging and cardiometabolic risk reduction, *Mol. Nutr. Food Res.*, 2017, **61**, 1600557.
132. D. Milenkovic, C. Morand, A. Cassidy, A. Konic-Ristic, F. Tomás-Barberán, J. M. Ordovas, P. Kroon, R. De Caterina and A. Rodriguez-Mateos, Interindividual variability in biomarkers of cardiometabolic health after consumption of major plant-food bioactive compounds and the determinants involved, *Adv. Nutr.*, 2017, **8**, 558.

133. S. Bolca, S. Possemiers, V. Maervoet, I. Huybrechts, A. Heyerick, S. Vervarcke, H. Depypere, D. De Keukeleire, M. Bracke, S. De Henauw, W. Verstraete and T. Van de Wiele, Microbial and dietary factors associated with the 8-prenylnaringenin producer phenotype: a dietary intervention trial with fifty healthy post-menopausal Caucasian women, *Br. J. Nutr.*, 2007, **98**, 950.
134. K. D. R. Setchell, N. M. Brown, S. Summer, E. C. King, J. E. Heubi, S. Cole, T. Guy and B. Hokin, Dietary factors influence production of the soy isoflavone metabolite *S*-(−)equol in healthy adults, *J. Nutr.*, 2013, **143**, 1950.
135. P. Mena, I. A. Ludwig, V. B. Tomatis, A. Acharjee, L. Calani, A. Rosi, F. Brighenti, S. Ray, J. L. Griffin, L. J. Bluck and D. Del Rio, Inter-individual variability in the production of flavan-3-ol colonic metabolites: preliminary elucidation of urinary metabotypes, *Eur. J. Nutr.*, 2019, **58**, 1529.
136. C. Morand, B. De Roos, M. T. Garcia-Conesa, E. R. Gibney, R. Landberg, C. Manach, D. Milenkovic, A. Rodriguez-Mateos, T. Van de Wiele and F. Tomas-Barberan, Why interindividual variation in response to consumption of plant food bioactives matters for future personalised nutrition, *Proc. Nutr. Soc.*, 2020, **79**, 225.
137. C. I. R. Gill, G. J. Mcdougall, S. Glidewell, D. Stewart, Q. Shen, K. Tuohy, A. Dobbin, A. Boyd, E. Brown, S. Haldar and I. R. Rowland, Profiling of phenols in human fecal water after raspberry supplementation, *J. Agric. Food Chem.*, 2010, **58**, 10389.
138. R. Feliciano, C. Mills, G. Istas, C. Heiss and A. Rodriguez-Mateos, Absorption, metabolism and excretion of cranberry (poly)phenols in humans: A dose response study and assessment of inter-individual variability, *Nutrients*, 2017, **9**, 268.
139. L. O. Dragsted, Q. Gao, A. Scalbert, G. Vergères, M. Kolehmainen, C. Manach, L. Brennan, L. A. Afman, D. S. Wishart, C. Andres Lacueva, M. Garcia-Aloy, H. Verhagen, E. J. M. Feskens and G. Praticò, Validation of biomarkers of food intake - critical assessment of candidate biomarkers, *Genes Nutr.*, 2018, **13**, 14.
140. J. Beattie, A. Crozier and G. Duthie, Potential health benefits of berries, *Curr. Nutr. Food Sci.*, 2005, **1**, 71.
141. M. Ulaszewska, M. Garcia-Aloy, N. Vázquez-Manjarrez, M. T. Soria-Florido, R. Llorach, F. Mattivi and C. Manach, Food intake biomarkers for berries and grapes, *Genes Nutr.*, 2020, **15**, 17.
142. J. I. Ottaviani, A. Britten, D. Lucarelli, R. Luben, A. A. Mulligan, M. A. Lentjes, R. Fong, N. Gray, P. B. Grace, D. H. Mawson, A. Tym, A. Wierzbicki, N. G. Forouhi, K.-T. Khaw, H. Schroeter and G. G. C. Kuhnle, Biomarker-estimated flavan-3-ol intake is associated with lower blood pressure in cross-sectional analysis in EPIC Norfolk, *Sci. Rep.*, 2020, **10**, 17964.
143. M. Garcia-Aloy, M. Rabassa, P. Casas-Agustench, N. Hidalgo-Liberona, R. Llorach and C. Andres-Lacueva, Novel strategies for improving

dietary exposure assessment: Multiple-data fusion is a more accurate measure than the traditional single-biomarker approach, *Trends Food Sci. Technol.*, 2017, **69**, 220.
144. M. Garcia-Aloy, R. Llorach, M. Urpi-Sarda, S. Tulipani, R. Estruch, M. A. Martínez-González, D. Corella, M. Fitó, E. Ros, J. Salas-Salvadó and C. Andres-Lacueva, Novel multimetabolite prediction of walnut consumption by a urinary biomarker model in a free-living population: the predimed study, *J. Proteome Res.*, 2014, **13**, 3476.
145. M. N. Clifford, E. L. Copeland, J. P. Bloxsidge and L. A. Mitchell, Hippuric acid as a major excretion product associated with black tea consumption, *Xenobiotica*, 2000, **30**, 317.
146. H. L. Self, R. R. Brown and J. M. Price, Quantitative studies on the metabolites of tryptophan in the urine of swine, *J. Nutr.*, 1960, **70**, 21.
147. H.-D. Grümer, Formation of hippuric acid from phenylalanine labelled with carbon-14 in phenylketonuric subjects, *Nature*, 1961, **189**, 63.
148. J. W. Bridges, M. R. French, R. L. Smith and R. T. Williams, The fate of benzoic acid in various species, *Biochem. J.*, 1970, **118**, 47.

CHAPTER 3

Factors Affecting the Inter-individual Variability in the Cardiometabolic Response to Berry (Poly)phenols

PAUL YOUNG TIE YANG, MADELEINE S. WILLIAMS, RACHEL GIBSON AND ANA RODRIGUEZ-MATEOS*

Department of Nutritional Sciences, School of Life Course and Population Sciences, Faculty of Life Sciences and Medicine, King's College London, London, UK
*Email: ana.rodriguez-mateos@kcl.ac.uk

3.1 Introduction

Cardiometabolic diseases are one of the primary drivers of both disease and mortality as well as healthcare costs in many western countries. They include those affecting the heart or blood vessels, such as coronary artery disease, coronary heart disease, congestive heart failure, myocardial infarction, peripheral artery disease, stroke, and hypertension.[1–3] Together, these cardiometabolic diseases, including type 2 diabetes, hyperglycemia, dyslipidemia, obesity, and non-alcoholic fatty liver disease, are the number one cause of both global morbidity and deaths.[1] Intermediate markers of such diseases include dyslipidemia (elevated low-density lipoproteins [LDL], total cholesterol, and triglycerides, and low levels of high-density lipoproteins [HDL]),

Food Chemistry, Function and Analysis No. 33
Berries and Berry Bioactive Compounds in Promoting Health
Edited by Dorothy Klimis-Zacas and Ana Rodriguez-Mateos

Published by the Royal Society of Chemistry, www.rsc.org

hypertension (elevated systolic and/or diastolic blood pressure [SBP/DBP]), impaired glucose intolerance and insulin resistance (assessed by HbA1c, homeostasis model assessment of insulin resistance [HOMA-IR]). More recent biomarkers of cardiometabolic disease risk, and in particular related to vascular function, are endothelial function, assessed as flow-mediated dilation (FMD),[4] and arterial stiffness, indicated by pulse wave velocity (PWV) and augmentation index (AIx).[5,6] Studies indicate that cardiometabolic disease could be prevented by improving lifestyle factors, diet being one of the most important ones. Epidemiological studies have shown that consumption of fruit and vegetables is associated with a decrease in cardiometabolic disease risk.[7–9] Such food products are rich not only in fiber, minerals, and vitamins but also in plant bioactives known as (poly)phenols.[10,11] Associations between high intake of dietary (poly)phenols and reduction in cardiometabolic disease risk have also been shown in various epidemiological studies.[12,13] Berries are naturally rich in (poly)phenols, in particular anthocyanins, ellagitannins, flavan-3-ols, flavonols, and phenolic acids.[10] While many studies have found significant improvements in biomarkers of cardiometabolic disease risk following (poly)phenol supplementation, results from randomized-controlled trials (RCT) have shown large variations between individuals.[14] This inter-individual variability in the cardiometabolic benefits of (poly)phenol consumption suggests that some individuals may be more likely than others to benefit from their consumption. This chapter will summarize the current understanding of the inter-individual variability in cardiometabolic response to berry (poly)phenol consumption and the main factors involved and will consider how these findings may impact nutrition recommendations and clinical practice. We will discuss the evidence on the most abundant berry (poly)phenols (anthocyanins, ellagitannins, flavan-3-ols, flavonols, and phenolic acids) and will include all commonly defined culinary berries such as strawberry, blackberry, raspberry, blueberry, boysenberry, blackcurrants, sea buckthorn berry, cranberry, elderberry, aronia berry, bilberry, black raspberry, chokeberry, and mulberry.

3.2 Effects of Berry (Poly)phenols on Biomarkers of Cardiometabolic Disease Risk

Berries, in general, are considered excellent health-promoting foods; a number of berries have even been marketed as 'superfoods' due to their many health benefits, although often, those claims are not supported by robust scientific evidence. Fresh berries are inherently high in water, fiber, and antioxidants such as vitamins C and E, as well as micronutrients including folic acid, calcium, lutein, alpha- and beta-carotene, and selenium.[15] Notably, they have high content in non-nutrient bioactive compounds such as (poly)phenols, which are present in berries in high amounts, in particular as anthocyanins, ellagitannins, flavan-3-ols, flavonols, and phenolic acids.

3.2.1 Anthocyanins

Anthocyanins are one of the (poly)phenol subclasses most abundant in berries and responsible for their characteristic colors. Anthocyanins are particularly abundant in dark blue and red type berries such as chokeberries, raspberries, cranberries, bilberries, and blueberries.[16] The most commonly found anthocyanins in berries are pelargonidin, cyanidin, delphinidin, peonidin, petunidin, and malvidin derivatives.[10] They are usually found attached to a sugar moiety, such as galactose, glucose, arabinose, rutinose, or xyloside, and they can also be found acylated, creating a unique, wide, and diverse anthocyanin profile for each type of berry.[17] Anthocyanin content in berries can vary from a few milligrams to a few grams per 100 g in fruit (dry weight).[18]

Numerous reports of their health benefits exist, from preclinical models to human studies. A recent *meta*-analysis of 19 prospective studies reported that a higher intake of anthocyanins in the diet reduced the risk of chronic heart disease by 9% and the risk of cardiovascular disease (CVD) mortality by 8%.[19] A growing number of *meta*-analyses of well-conducted randomized controlled studies have also shown health benefits, including significant improvements in endothelial function measured as FMD, and arterial stiffness, measured as PWV.[20] Additionally, anthocyanins were reported to improve plasma biomarkers such as fasting glucose, 2 h post-prandial glucose, total cholesterol, and LDL.[21] The evidence regarding improvements in blood pressure is, however, more divided. Several *meta*-analyses were not able to establish a direct link between anthocyanin consumption and blood pressure reductions,[22,23] which was in agreement with several RCTs.[24,25] However, other RCTs conducted with berries rich in anthocyanins did show improvements in blood pressure, pointing out the need to conduct more studies and to investigate potential factors affecting the variability in response to anthocyanins.[26,27]

3.2.2 Ellagitannins

Ellagitannins belong to the hydrolyzable tannin class of (poly)phenols, and they are complex derivatives of ellagic acid. Ellagitannins are most abundant in walnuts and pomegranates; however, they can also be found in berries such as cloudberry, raspberry, rose hip, strawberry, and sea buckthorn, where their concentrations range from 1 to 330 mg per 100 g fresh weight.[16] Ellagitannins found in berries include sanguiin H6, lambertianin C, and sanguiin H5, and they usually coexist along with other (poly)phenols in much lower concentrations, compared to anthocyanins.[28] For these reasons, most of the current evidence has been obtained from trials using pomegranate and walnut sources as interventions, where ellagitannins are mostly in the form of punicalagin.[29] Very few studies have investigated the effect of raspberry ellagitannins. A pilot study using red raspberries correlated significant acute improvements in FMD with plasma urolithin A levels (a metabolite derived from ellagitannin metabolism).[30] *In vitro* work also showed that ellagic acid and some urolithins had antiatherogenic effects.[31] These cardioprotective properties were confirmed in a *meta*-analysis of RCT using walnut and pomegranate ellagitannins which documented significant reductions in waist circumference, total cholesterol, LDL, and triacylglycerol (TAG).[32] More trials should be conducted with berry ellagitannins to confirm these findings.

3.2.3 Flavan-3-ols

Flavan-3-ols are a broad class of (poly)phenols that include monomers, oligomers, and high-molecular-weight polymers. Green tea, black tea, and cocoa are the most important sources of flavan-3-ols but they can also be found in berries, which contain mostly proanthocyanidin oligomers.[33] The flavan-3-ol content varies widely amongst berries, ranging from as low as 1.4 up to 1170 mg per 100 g of dry weight (post-lyophilization).[18] Interestingly, different proanthocyanidin structures are found amongst berries and can be categorized into A-type and B-type depending on the stereoconfiguration and linkage between monomers. B-type procyanidins are characterized by a single interflavan bond between carbon-4 of the B-ring and either carbon-8 or carbon-6 of the C-ring. This B-type is the most abundant in berries. A-type procyanidins have an interflavan bond as well as a second ether linkage between the A-ring hydroxyl group and carbon-2 of the A-ring. This A-type is found in cranberries in particular,[34,35] which are one of the most concentrated sources of flavan-3-ols amongst berries.[36]

Beneficial properties of proanthocyanidins have been suggested in various studies. A prospective study found a significant decrease in the risk ratio of coronary heart disease (RR = 0.77, 95% CI = [0.65, 0.92]) and overall mortality (RR = 0.78, 95% CI = [0.72, 0.84]), with increased proanthocyanidin intake after adjustment of the age and energy intake on 34 492 CVD-free postmenopausal women (Iowa Women's Health Study).[13] Several RCTs have shown beneficial effects of berry intake on cardiovascular health where flavan-3-ol metabolites correlated with improvements in vascular function.[37,38] These improvements observed should be interpreted with caution as flavan-3-ols coexist with other (poly)phenols present in higher proportions, such as anthocyanins, phenolic acids, and flavonols. However, the evidence gathered from cocoa and tea *meta*-analysis suggests that flavan-3-ols can improve cardiovascular health as the flavan-3-ols they contain (though mostly monomeric *i.e.* (−)-epicatechin) have been shown to have a positive effect on blood pressure, endothelial function, and blood lipids.[39–41]

3.2.4 Flavonols

Flavonols are ubiquitous (poly)phenols in the plant kingdom. The most common of them are quercetin, kaempferol, myricetin, and isorhamnetin.[33] Spectrographic analysis of berries revealed great variations in flavonol content across berries, ranging from 6.3 to 790 mg per 100 g of dry weight, bog whortleberry being one of the most abundant sources.[18]

The benefits of flavonol consumption on vascular health have been reported in various epidemiological studies,[42,43] as well as in several RCTs.[44,45] Quercetin, one of the main berry flavonols, has shown blood pressure lowering effects in a *meta*-analysis of RCT, as well as capacities to improve lipids profile.[46,47] Flavonols, as a whole, were also reported to improve the lipid profile (total cholesterol, LDL cholesterol, TAG, and HDL cholesterol), and fasting plasma glucose, after chronic supplementation.[48]

In berries, flavonols coexist with other (poly)phenols, which make the direct contribution of flavonols still unclear. Several interventions using berries that are known to contain flavonols documented improvements in cardiometabolic disease risk markers,[49–51] and an aronia berry intervention identified significant associations between changes in FMD, flavonol metabolites, and specific gut microbial genera.[37]

3.2.5 Phenolic Acids

Phenolic acids can be divided into two groups: benzoic and hydroxycinnamic acid derivatives. Benzoic acids found in berries include gallic acid, ellagic acids, vanillic acid, coumaric acid, and hydroxybenzoic acids, and hydroxycinnamic acid derivatives found in berries include ferulic, caffeic, cinnamic, and caffeoylquinic acids, being chlorogenic acid the most representative one.[52] Gallic acid and chlorogenic acids are abundant in blueberry and blackberry. Ferulic acid is also present in blueberries, strawberries, raspberries, blackberries, and cranberries, while bilberries and blackcurrants are important sources of *p*-coumaric and caffeic acids. Chokeberries are rich in caffeic and chlorogenic acids.[52] On average, the content of phenolic acids found in berries is 220 mg per 100 g of blueberries, 27 mg per 100 g of blackberries, 10 mg per 100 g of raspberries, and 9 mg per 100 g of strawberries.[53]

A chronic chokeberry intervention containing hydroxycinnamic acid led to improvements in blood pressure[54] and another acute blueberry intervention containing phenolic acids studied their effects on FMD. Interestingly, effects observed after two hours were correlated with increases in plasma metabolites possibly derived from the metabolism of several (poly)phenols, including 5-*O*-caffeoylquinic acid. The variety of the (poly)phenol content in berries makes it challenging to identify the direct contribution of phenolic acids toward the observed effects.[51] Therefore, several studies have investigated specific phenolic acids and tried to identify their potential benefits on cardiometabolic disease risk biomarkers. Notably, hydroxycinnamic acids have been associated with a significant reduction in body mass index (BMI) and waist circumference, significant improvements in fasting glucose, significant reduction in homeostasis model assessment of insulin resistance (HOMA-IR), as well as significant reductions in one or more of TAG, LDL, total cholesterol, or increase in HDL.[55,56]

3.3 Inter-individual Variability in the Cardiometabolic Response to Berry (Poly)phenols

While the evidence is accumulating to show the potential health benefits of (poly)phenol consumption, a large heterogeneity in response and in the magnitude of the effects have been observed. Identified factors that may be important for explaining inter-individual variability include differences in bioavailability and, in particular, gut microbial metabolism and genetic polymorphisms.[57,58] Other factors of interest include age, sex, BMI, pathophysiological status, and medication, which could be responsible for the heterogeneity of physiological responses to (poly)phenol consumption.[14,59,60]

3.3.1 Variability in the Bioavailability of Berry (Poly)phenols

To establish a relationship between biological effects and berry (poly)phenol consumption, it is important to have a good understanding of their bioavailability and tissue distribution. Some (poly)phenols can be absorbed from the stomach or the small intestine and can undergo further conjugation transformations in the intestine enterocytes and liver to give methyl, glucuronide, and sulfate derivatives (phase II metabolites), while the majority are metabolized by the gut microbiota in the colon, where several ring fission and deconjugation reactions occur, leading to a large number of smaller phenolic metabolites in circulation, including phenolic acids and hydroxycinnamates, which can be further conjugated by mammalian enzymes in the enterocytes or liver prior to reaching the systemic circulation.[61,62]

Some evidence has shown that there is a high inter-individual variability in the metabolism and bioavailability of (poly)phenols.[60] This has been demonstrated through recovery studies in biological fluids. In a cranberry juice study (480 mL), the recovery of anthocyanins in urine showed important between-subject variability, ranging from less than 0.1% to 3.2%.[63] Important variations were also reported in another cranberry intervention containing several types of (poly)phenol, where the coefficient of variation ranged between 43% and 216% for the maximal plasmatic concentration observed between participants.[64] Similar important variations were observed with blueberries where volunteers showed a coefficient of variation of 44% in plasma fluids, 2 hours post-consumption. When looking specifically at the classes of (poly)phenols, catechols, benzoic acids, and flavonols showed the lowest inter-individual variability in plasma (36%–58%, 50%–63%, and 81%–91%, respectively), whereas in urine, benzaldehydes, phenylacetic, and benzoic acids were the compounds excreted with less variability (47%–69%, 50%–54%, and 40%–71%, respectively). The authors highlighted that the background diet could be a confounding factor. Their work, however, suggested that (poly)phenol structures could have an impact on the variability in bioavailability.[50]

It has been demonstrated in some pharmacogenomic studies that some drugs from the alkaloid family are metabolized by individuals in different patterns: poor, intermediate, or extensive and ultra-rapid metabolizers.[65] These differences are thought to be due to specific genes and enzymes of the xenobiotic metabolism pathways, which displayed different variants. Interestingly, most plant food bioactives are also metabolized by the same polymorphic carriers and enzymatic systems, leading to suggest that (poly)phenol metabolism could also be affected by these genetic polymorphisms.[60] Slow caffeine metabolizers are characterized by the CYP1A2*1F variant of the *CYP1A2* gene, while rapid metabolizers possess the wild-type allele.[66] Other plant food bioactives such as the glucosinolates found in broccoli are metabolized by phase II conjugation enzymes named glutathione *S*-transferases (GSTs). The bioavailability of isothiocyanates (metabolites of glucosinolates) has been observed to be up to 20% higher in subjects showing the GSTM1 variant of the *GSTM* gene coding for GSTs.[67] Catechol-*O*-methyltransferase (COMT) enzymes are responsible for polyphenolic methylation. An RCT using tea flavanols showed the interaction between COMT genotype and the flavanol-rich green tea extract as higher levels of methylated epigallocatechin were found in volunteers with the high activity (GG) COMT group.[68] No work has been conducted specifically with berry (poly)phenols. However, the current literature that explored plant food bioactives suggests that investigating genetic polymorphisms could help explain the inter-individual variations observed regarding (poly)phenol metabolism.

Finally, the variability observed in the effects could be attributed to the gastrointestinal tract variability following oral consumption and before absorption. Gastrointestinal tract variability following drug intake has been previously reported, with several potential causes occurring in several compartments of the gastrointestinal tract, including the stomach, small intestine, and colon.[69] In the stomach, gastric conditions can vary according to the feeding state of volunteers, with the pH being higher (around pH = 3.5) following absorption of a meal[70] compared to the fasting state (pH = 1.7–3.3).[71–73] These changes have previously been reported to influence the availability of certain oncological drugs[74] and could potentially affect the absorption of (poly)phenols in a similar way. Other reported parameters of interest are the viscosity and osmolarity of fluids, composition of gastric fluid, gastric emptying time, and to a lesser extent, volume of fluids.[69] Similar parameters were reported as a source of inter-individual variability in the small intestine, including intestinal fluid volume and composition, intestinal pH, viscosity, and osmolarity to which can be added transit time, epithelial permeability, absorption into blood and lymph, and enterohepatic circulation.[69] In the colon, reported factors of variability include colonic transit time, colonic luminal composition as well as the composition of the microbiome.[69] Sub-populations have been identified to be more prone to display high variability in drug bioavailability including people who underwent bariatric surgery or surgical resection, as well as people suffering from chronic constipation, small intestinal motility disorder, anorexia, and cachexia.[69] While the evidence available is still limited and related to some drugs rather than (poly)phenols, these factors related to the gastrointestinal tract may be worth investigating to understand the inter-individual variability observed post-consumption as some drugs possess similar structures and degradation mechanisms in the body.[61,62,65]

3.3.2 Variability in Gut Microbial Metabolism of Berry (Poly)phenols

As mentioned previously, certain berry (poly)phenols such as ellagitannins are metabolized by the gut microbiome. As it is known that there is high variability in gut microbiome composition and function amongst individuals, differences in gut microbial metabolism of (poly)phenols will likely have an important role in explaining the inter-individual variability in cardiometabolic response.[60]

Feeding studies with berries have shown high recoveries of intact (poly)-phenols in non-ileostomic patients, suggesting that the majority of berry (poly)phenols are not absorbed in the small intestine and reach the colon,[14] where they are metabolized by the gut microbiome. A large number of metabolites are formed due to the many reactions occurring, such as oxidation, demethylation, decarboxylation, or deconjugation, which break down (poly)phenols into smaller phenolic acids and aromatic catabolites.[75] High variability in gut microbial metabolism has been documented, in particular for certain polyphenols such as isoflavones and ellagitannins,[76,77] which is likely due to the high variability in gut microbiome composition. In fact, specific (poly)phenol converting enterotypes (gut microbial composition patterns) corresponding to low or high metabolite producers have been proposed, and a classification of individuals into their capacity to metabolize (poly)phenols (metabotypes) has been established so far for isoflavones and ellagitannins.[78] Although isoflavones are not relevant to berries, ellagitannins are abundant, especially in raspberries, strawberries, and blackberries and several studies have detected the presence of urolithins, the main ellagitannin gut microbial metabolites, in urine and plasma after berry consumption.[30,76,79] In recent years, the existence of three different types of urolithin metabotypes (UM) have been reported: UM-0 (non-producers of urolithins), UM-A (producers or urolithin A only) or UM-B (producing urolithin A, urolithin B, and isourolithin A).[76,78,80] These metabotypes are associated with particular enterotypes, as some very specific bacteria strains have been identified as being able to convert ellagic acid into urolithins. *Gordonibacter urolithinfaciens* and *Gordonibacter pamelaeae* convert ellagic acid into urolithin[81] and *Ellagibacter isourolithinifaciens* converts ellagic acid into isourolithin A.[82] The presence of metabotypes has been consistently reported in several cohorts with UM-0, UM-A, and UM-B representing around 10%, 70%, and 20% of the population, respectively, seemingly independently of the place where the study was conducted and the ellagitannin source used, as most of the work has been conducted in Spain and one study with a low number of participants in Brazil.[29,83] The abundance of metabotypes seemed to be determined mainly by age.[76] The non-producer metabotype UM-0 remained stable at 10%, while UM-A decreased from 80% in the 5–10 years old strata, to stabilize at 50% past the 31–40 years strata. UM-B showed an opposite increase. When authors considered other covariables, they found no association between metabotypes and sex, BMI, weight, health status, diet, and health status at different age ranges (from 5 to 90 years old).[76] Very few studies have investigated the link between UM and cardiovascular health, and none on berries. A study using ellagitannin from pomegranate extract showed that UM-B volunteers had the highest baseline vascular risk.[84] Interestingly, biomarkers of cardiometabolic disease improved only in UM-B volunteers with a dose-dependent relationship, while the effects were correlated with both *Gordonibacter* levels and urolithin production, thus confirming the relation between the gut microbiome, metabolite production, and cardiometabolic disease risk.[84] A small pilot study in 10 healthy men reported improvements in endothelial function (FMD) after red raspberry consumption, and a correlation could be established only between urolithin A metabolites and improvements in endothelial function. Although very speculative due to the low number of participants, a non-significant trend toward increased FMD response in metabotype A was observed in comparison with metabotype B volunteers.[30]

Background diet has also been proposed as an important factor as a gut microbiome modulator since it can vary importantly across population groups. Protein, fat carbohydrates, and (poly)phenols have all been reported to be able to influence the gut microbiome composition, sometimes in periods as short as 24 h.[85] As it has been reported before, the current evidence supports the idea that (poly)phenol metabolites could be involved in the cardiovascular health benefits of berry consumption.[30,38] Since many (poly)phenols were metabolized by the gut microbiome, it leads us to suggest that the gut microbiome plays a key role in the modulation of the cardiometabolic disease risk biomarkers by being responsible for the conversion of parent (poly)phenols into bioactive compounds.[86–88] This idea that the gut microbiome could be a source of inter-individual variability has been supported by Eker *et al.* In their review, they show that recoveries and concentrations of anthocyanin metabolites found in human samples (blood, urine, and ileal fluid) exhibited important variations across individuals and were related to gut microbial diversity.[89]

In return, it has been suggested that (poly)phenol intake can modulate the gut microbiome and diversity. *In vitro* experiments found that quercetin slowed down intestinal bacterial growth or even completely stopped it in a dose-dependent way. Catechin, on the contrary, did not influence the growth of most bacteria studied. Investigators concluded that flavonoid aglycones may inhibit the growth of some intestinal bacteria, meaning that they could interfere with gut microbial composition and therefore indirectly interfere with their own bioavailability.[90] *In vitro* experiments conducted in rodent models showed up-regulation of the *Actinobacteria* and *Akkermansia*, down-regulation of *Bifidobacteria* and changes in *Lactobacillus* with different directions depending on the study.[91–94] A human intervention with aronia berry documented significant increases in *Anaerostipes* as compared to the placebo group after 12 weeks.[37] All this evidence supports the idea of a bi-directional interaction between gut microbiome and (poly)phenols, though the mechanisms and direct links remain unclear. More experiments are needed as the literature also suggests relationships between cardiovascular health and gut microbial composition.

Patients with coronary heart disease were reported to have a decreased proportion of *Bacteroidetes* and an increased proportion of *Firmicutes*.[95] A lower abundance of butyrate producers was also observed. Butyrate, along with other short-chain fatty acids, is the end product of the fermentation of dietary fibers, and the main energy source for colonocytes maintaining the gut mucosal barrier.[96] Interestingly, this ratio is strongly affected by diet, as a high-fat diet reduces the abundance of *Bacteroidetes* and favors the growth of *Firmicutes*, which is also associated with obesity.[97,98] Based on animal experimentation, blackberry and raspberry consumption could not reverse this trend in rats; however, consumption of honeyberry was able to.[99] Blueberry and blackcurrant significantly increased the *Bacteroidetes* and *Actinobacteria* populations and it was postulated that the anthocyanins present could reduce the oxygen tension (partial pressure of oxygen in blood) in the gut, which may benefit the growth of oxygen-sensitive bacteria.[92] As *Bacteroidetes* and *Actinobacteria* are obligate anaerobes, this would lead to reducing populations of *Firmicutes* and consequently decreasing *Firmicutes*/*Bacteroidetes* ratio.[98] These findings suggest that (poly)phenol consumption may also have an indirect role in influencing the inter-individual variability in vascular health through its prebiotic effect and capacity to influence the gut microbial composition.

3.3.3 Age

Age is the single greatest risk factor for CVD and is associated with an increased chance of developing cardiometabolic risk factors, including arterial stiffness, hypertension, and endothelial dysfunction.[39] Aging is also a complex phenomenon hiding comorbidities and other confounding factors. For example, when getting older, not only elevated baseline blood pressure and degradation of endothelial function but also changes in the bioavailability of food and changes of the endocrine mechanisms are observed. These changes could be partly related to the changes in the gut microbiome. Duda-Chodak *et al.* report in their review that the microbiome changes over the life of an individual due to changes in diet, following the loss of taste/smell or the exclusion of specific foods increasing cardiovascular risk (high salt, cholesterol, or sugar).[100] The microbial diversity also changes throughout the life span due to the intake of drugs such as antibiotics or changes in the immune system.[100]

The authors also report that elderly individuals are characterized by a reduced abundance of *Ruminococcus* and *Blautia* spp. And a diminished abundance of several butyrate producers, while younger subjects have an increased amount of *Escherichia* in comparison.[100] Caution is therefore necessary when interpreting the effects of age in relation to cardiovascular health following (poly)phenol consumption.

Although no studies have specifically investigated the effects of age on the cardiometabolic response to berry consumption, a *meta*-analysis aiming to determine the effects of berry consumption on cardiovascular risk factors performed subgroup analysis based on age. Eighteen RCT were included and divided into a younger subgroup (under 50 years old, $N=8$) and an older subgroup (50 and above, $N=10$). The authors found no significant effect of age on the changes in plasma lipid profile after berry consumption.[101] Few studies have investigated the effect of age on the cardiovascular health benefits of (poly)phenols present in berries, although none of them investigated the effects using berry interventions; so, results should be considered with caution. Consumption of a flavan-3-ol-rich cocoa drink for 2 weeks improved SBP and AIx in older volunteers but not in younger subjects,[102] in agreement with a previous study by Fisher *et al.* who found that blood pressure reductions were higher in the elderly as compared to the younger group.[103] A plausible explanation for these effects would be that older individuals will have a lower absorption and bioavailability of (poly)phenols. However, in another study with healthy men, no significant age-dependent differences were found in the absorption, distribution, metabolism, and excretion of cocoa flavan-3-ols, except for a lower excretion of gut microbial metabolites in the older group.[104] A study conducted with flavonols showed a higher lowering of blood pressure in the younger individuals subgroup (25–50 years), as compared to the whole group (25–65 years).[45] Another study confirmed these findings by reporting a bigger decrease in SBP in younger subjects after the consumption of 150 mg per day of quercetin for 6 weeks.[44] These trials did not specifically aim at determining the influence of age factor; the evidence available thus remains limited.

Several *meta*-analyses tried to evaluate the impact of age on the vascular response with different (poly)phenol classes not always originating from berries. Alshammari *et al.* intended to quantify the contribution of the age factor to the vascular health benefits of flavan-3-ol consumption originating from various sources (not only berries). Defining the younger group as volunteers under 40 years of age, and the older group as over 40, they found that while the flavan-3-ol intervention significantly improved TAG, LDL, and HDL markers, such improvements were lost in the older group.[39] Similar observations were made in a flavonol *meta*-analysis, where plasma biomarkers significantly improved in the mixed group (studies recruiting populations younger and older than 40 years of age) but not in the older group.[48] In both *meta*-analyses, authors suggest that the loss of significance may be due to the low number of studies.

While age appears to be an important interindividual factor, it is difficult to quantify its contribution toward the health benefits of (poly)phenol consumption.

3.3.4 Sex

Another important interindividual factor to consider is sex. While CVDs are affecting both sexes, the prevalence of CVD is not the same. In 2007, women accounted for more than 60% of the USA stroke deaths while the prevalence of CHD was higher for any age subgroup below 75 years old.[105] It is also known that sex and particular sex hormones have an impact on the metabolism of food. The incidence of single nucleotides polymorphisms (SNP) has been shown to be different depending on sex. For example, allele 5 of the hormone-sensitive lipase was found predominantly in males, leading to a more marked decrease in the lipolytic rate of abdominal fat cells. Fat metabolism has also been shown to be different across sexes with different body fat repartition as an example.[106] Based on this information, researchers have investigated the effects of sex on inter-individual variability in cardiovascular response to (poly)phenol consumption. Several studies attempted to compare the changes observed following (poly)phenol consumption between men and women.[32,40,48,56]

Though no study has been specifically designed to determine the direct influence of sex on berry (poly)phenol consumption, improvements have been observed after berry anthocyanin intake in both men and women. A prospective study associated high anthocyanin intake with a reduced risk of myocardial infarction in young and middle-aged women[12] and RCTs have shown significant improvements in women.[26,107] Studies conducted in males with anthocyanin-rich berries also reported improvement in vascular function.[37,51] A *meta*-analysis of RCT investigating the effects of ellagitannin and anthocyanin-rich foods including berries showed significant improvements in DBP only in men with anthocyanin interventions,[32] while Menezes *et al.* found no significant sex effect when performing the subgroup analysis in their *meta*-analysis related to flavonol consumption,[48] possibly due to the low number of studies available. A larger *meta*-analysis conducted by Gonzalez-Sarrias *et al.* using flavan-3-ol rich foods documented a significant reduction of waist circumference in both men and women. They also observed that total cholesterol and LDL cholesterol were significantly reduced in females only, whereas BMI was significantly lowered only in males.[40]

While studies tend to indicate that sex may be an important factor of inter-individual variability, no studies have been conducted specifically to investigate this, so based on the available data, no conclusions can be made.

3.3.5 Body Mass Index

The World Health Organization defines overweight and obesity as abnormal or excessive fat that accumulates and presents a risk to health. BMI allows us to classify people amongst different categories: underweight (under 18.5 kg m^{-2}), normoweight (18.5 to 24.9 kg m^{-2}), and overweight (25 kg m^{-2} and above). People with obesity have a BMI over 30 kg m^{-2}. Being overweight or obese increases CVD risk.[108,109]

When looking more specifically at berry interventions, several RCTs reported significant improvements in cardiovascular risk factors, including post-prandial insulin, in overweight or obese volunteers using either berry mixes or blueberries.[107,110]

Similarly, an intervention using Montmorency cherry improved vascular dysfunction, assessed *via* FMD in overweight men.[111] Another study using a (poly)phenol-rich acai berry supplementation in overweight men showed that consumption was associated with improvements in vascular function.[49] An anthocyanin and ellagitannin *meta*-analysis from Garcia-Conesa *et al.* found that anthocyanins and ellagitannins had no significant effects in the normoweight-only group. When stratifying by BMI, they, however, found significant decreases in SBP, DBP, and total cholesterol in overweight/obese volunteers after intake of anthocyanins. A similar pattern was observed with ellagitannin interventions to lower SBP.[32]

González-Sarrías *et al.* documented that consumption of flavan-3-ol-rich products had a greater effect in reducing cardiometabolic risk, including decreasing BMI, waist circumference, total cholesterol, and LDL and increasing HDL in overweight volunteers.[40] When they compared both BMI subgroups in terms of changes in BMI, waist circumference, total cholesterol, HDL, and TAGs, they did not observe significant differences. However, 61% of the total between-study variance in BMI could be explained by the respective baseline BMI values indicating that baseline values should be considered when assessing changes in BMI.[40] This finding was replicated by Alshammari *et al.*, who reported that flavan-3-ol consumption had significantly greater effects on HOMA-IR in overweight compared to normoweight individuals, although baseline BMI did not impact the effect on insulin.[39] Subgroup analysis by BMI was inconclusive in one flavonol *meta*-analysis due to the low availability of studies reporting the necessary data.[48] One analysis found that flavonol supplementation did not improve total cholesterol, LDL, or HDL in overweight individuals in 5 studies, while most studies involving healthy weight or mixed healthy and overweight individuals found significant improvements in these lipid markers. Still, TAG appeared to improve significantly in the overweight subgroup.[48] When performing *meta*-analysis on studies using phenolic acids, Martini *et al.* found that improvements in SBP or DBP occurred more frequently in participants who were overweight. Similarly, studies reporting improvements in lipid profile had greater baseline LDL levels and higher prevalence of overweight than those that did not.[56]

The current evidence tends to suggest that improvements observed after (poly)phenol consumption are higher in overweight and obese volunteers, which supports the hypothesis that increasing the berry content of their diet would reduce their overall cardiovascular risk.

3.3.6 Health Status

Another important parameter to take into consideration when assessing the health benefits of (poly)phenol consumption is the health status of participants. While most studies use healthy volunteers, volunteers with a disease and at-risk volunteers are important groups to consider.

Baseline disease status, including the presence of overweight/obesity, hypertension, type 2 diabetes, for example, as well as lifestyle choices such as smoking and dietary habits additionally, may play a role in inter-individual variability. Improvements in vascular biomarkers have been reported in several clinical studies with berries as the intervention including healthy volunteers,[30,37] individuals at risk of CVD,[26,112,113] and volunteers with higher CVD risk, in particular those with metabolic syndrome.[24,114–116] A *meta*-analysis reported significant improvements in total cholesterol and LDL after berry consumption only in volunteers at risk after intake, while no significant effects were observed in the healthy group.[101] An RCT conducted with 31 pre-hypertensive men showed higher HDL cholesterol levels after consuming 640 mg of anthocyanins daily for 4 weeks. A systematic review[117] reported that LDL levels were also significantly lowered in several clinical trials conducted with dyslipidemic or hyperlipidemic volunteers following the consumption of anthocyanin-rich food products[118–121] while it was not the case for studies including healthy participants.[122,123] The *meta*-analysis performed by Garcia-Conesa *et al.* documented inconsistent results regarding the effects of anthocyanin intake on blood pressure that were independent of health status, including healthy individuals, those on medication, those labeled at-risk, and those with diagnosed CVD. For example, waist circumference improved only in the at-risk subgroup, and total cholesterol decreased only in the healthy and disease subgroups, but not in the group classified at-risk. Other cardiometabolic biomarkers including BMI, LDL, HDL, TAG, glucose, insulin, and HOMA-IR were unaffected by health or medication status.[32] When considering ellagitannin interventions, healthy participants experienced reductions in total cholesterol and glucose, while those at-risk benefitted from reduced LDL and SBP, as well as small but significant improvements in total cholesterol, TAG, and HDL. Those in the disease subgroup benefitted only from a small but significant reduction in waist circumference and an increase in FMD. Data further implied larger reductions in blood glucose in healthy participants compared to those with reported disease.[32]

When examining other (poly)phenols from diverse sources, it has been suggested that medication could interfere with (poly)phenols as *meta*-analysis using flavan-3-ol intervention showed that non-medicated individuals, those whose cardiometabolic conditions or risk factors were not controlled by medications had better improvements in blood lipids as compared to the medicated subgroup.[40] Another flavan-3-ol systematic review from Rangel-Huerta *et al.* showed that participants classified as high risk had the greatest reductions in BMI and waist circumference and those with hyperlipidemia saw a greater increase in HDL and those with diabetes or hypertension a greater decrease in LDL.[124] Flavonol consumption was linked by *meta*-analysis to greater impact on individuals with a diagnosed disease or dyslipidemia compared to healthy individuals, suggesting a possible correlation between disease state or overall risk and cardioprotective effects of flavonol intake. Furthermore, stratifying by baseline, lipid levels showed improvements following flavonol intake only in TAG and HDL in individuals in the dyslipidemic subgroup following flavonol intake. Researchers concluded that flavonol supplementation may be more beneficial in improving HDL for healthy individuals, LDL in diseased individuals, and more specifically, TAG and HDLs in dyslipidemic individuals.[48] A study on quercetin supplementation in diabetic patients reported higher impacts on SBP and DBP compared to similarly designed RCT involving non-diabetic participants, indicating higher disease status may be associated with greater impact for the improvement of BP *via* quercetin supplementation.[125] A double-blind crossover RCT includes men and women with prehypertension or stage 1 hypertension in which participants consumed 730 mg per day of quercetin or placebo for 28 days. Results of this study reported no significant effect of quercetin supplementation on prehypertensive, lower-risk individuals, but significant improvements in both SBP, DBP, and mean arterial pressures in stage 1 hypertensive, higher-risk individuals suggesting that hypertension status plays an important role when assessing the effects of flavonol supplementation on vascular health.[44] Egert *et al.* reported significantly greater improvements in hypertensive subjects compared to that experienced by the entire study cohort following 6 weeks 150 mg per day of quercetin supplementation.[45]

In their systematic review related to phenolic acids, Martini *et al.* reported that improvements in blood pressure were observed more frequently when volunteers had a higher baseline SBP significantly. Similarly, improvements in lipid profile were greater when volunteers had greater baseline LDL levels. Interestingly, fasting glucose was only improved in individuals with non-alcoholic fatty liver disease and metabolic syndrome after consumption of hydroxycinnamic acid, while studies including only healthy participants found no such improvements. Participants in studies that reported improvement in HOMA-IR had significantly higher mean baseline HOMA-IR levels compared to those that did not.[56]

All these data suggest that health status is an important parameter to consider when interpreting outcomes of clinical studies. It seems that people at a higher risk of CVD, particularly with elevated blood lipids or baseline hypertension, could benefit more from a berry (poly)phenol intervention.

3.4 Conclusion

Observational studies concur in suggesting that berry (poly)phenols have cardioprotective properties. The evidence from *meta*-analysis tends to indicate that participants with higher cardiovascular risk at baseline have a higher potential to benefit from interventions as it seems that people with overweight or obesity and people with dyslipidemia or higher baseline hypertension benefited more from the dietary interventions. The contribution of age and sex factors seems to exist though the magnitude and direction of the effects have not been clearly established. RCTs support these findings but have also highlighted the importance of the bioavailability of (poly)phenols in relationship with the cardioprotective effects observed. A major determinant involved in the bioavailability lies in the gut microbial composition, as both human trials and mechanistic studies have shown a relationship between metabolite production and specific enterotypes. While interindividual differences in response appear clearly to be affected by factors such as the gut microbial composition, age, sex, BMI, and health status, we are not yet able to predict individual responses. This review has highlighted the complexity of assessing and apprehending the contribution determinants of inter-individual variability in relation to the amplitude of the changes observed in the cardiovascular biomarkers. One explanation is linked to the low number of studies available to perform subgroup analyses due to the fact that most studies are using mixed categories and are not reporting results for each subgroup. In fact, we suggest the necessity to design RCTs specifically aimed at assessing inter-individual variability to get a much clearer idea of the magnitude of the impact of each factor and its direction. It is also important to mention that inter-individual lifestyle habits may also exert an important influence such as smoking status, physical activity, or background diet. Another layer of complexity is the fact that the evidence available for non-anthocyanin intervention is very limited. It has been suggested that the bioavailability of phenolic compounds could be related to their structures and very few studies have specifically used berry flavan-3-ols or berry ellagitannins, which have a very different structure compared to the same class of compounds found in cocoa, tea, or pomegranate, for example. Finally, other external parameters need to be taken into consideration such as the dose given, the duration of the intervention, the health status of the gastrointestinal tract, the food matrix, or the study design itself, in particular, whether or not the diet of the participant is controlled throughout the study.

References

1. A. H. Association, *Cardiovascular Disease: A Costly Burden, For America Projections Through 2035*, 2017.
2. GBD 2017 Causes of Death Collaborators, *Lancet*, 2018, **392**, 1736–1788.
3. S. B. Dunbar, O. A. Khavjou, T. Bakas, G. Hunt, R. A. Kirch, A. R. Leib, R. S. Morrison, D. C. Poehler, V. L. Roger, L. P. Whitsel and American Heart Association, *Circulation*, 2018, **137**, e558–e577.
4. M. C. Corretti, T. J. Anderson, E. J. Benjamin, D. Celermajer, F. Charbonneau, M. A. Creager, J. Deanfield, H. Drexler, M. Gerhard-Herman, D. Herrington, P. Vallance, J. Vita, R. Vogel and

International Brachial Artery Reactivity Task Force, *J. Am. Coll. Cardiol.*, 2002, **39**, 257–265.

5. I. Ikonomidis, G. Makavos and J. Lekakis, *Curr. Opin. Cardiol.*, 2015, **30**, 422–431.
6. W. W. Nichols and B. M. Singh, *Curr. Opin. Cardiol.*, 2002, **17**, 543–551.
7. L. Dauchet, P. Amouyel and J. Dallongeville, *Neurology*, 2005, **65**, 1193–1197.
8. L. Dauchet, P. Amouyel, S. Hercberg and J. Dallongeville, *J. Nutr.*, 2006, **136**, 2588–2593.
9. K. J. Joshipura, F. B. Hu, J. E. Manson, M. J. Stampfer, E. B. Rimm, F. E. Speizer, G. Colditz, A. Ascherio, B. Rosner, D. Spiegelman and W. C. Willett, *Ann. Intern. Med.*, 2001, **134**, 1106–1114.
10. A. Rodriguez-Mateos, C. Heiss, G. Borges and A. Crozier, *J. Agric. Food Chem.*, 2014, **62**, 3842–3851.
11. A. Rodriguez-Mateos, D. Vauzour, C. G. Krueger, D. Shanmuganayagam, J. Reed, L. Calani, P. Mena, D. Del Rio and A. Crozier, *Arch. Toxicol.*, 2014, **88**, 1803–1853.
12. A. Cassidy, K. J. Mukamal, L. Liu, M. Franz, A. H. Eliassen and E. B. Rimm, *Circulation*, 2013, **127**, 188–196.
13. P. J. Mink, C. G. Scrafford, L. M. Barraj, L. Harnack, C. P. Hong, J. A. Nettleton and D. R. Jacobs, Jr, *Am. J. Clin. Nutr.*, 2007, **85**, 895–909.
14. D. Milenkovic, C. Morand, A. Cassidy, A. Konic-Ristic, F. Tomas-Barberan, J. M. Ordovas, P. Kroon, R. De Caterina and A. Rodriguez-Mateos, *Adv. Nutr.*, 2017, **8**, 558–570.
15. Agricultural Research Service. and US Department of Agriculture, in *US Department of Agriculture, Agricultural Research Service, Nutrient Data Laboratory*, 2016.
16. J. M. Koponen, A. M. Happonen, P. H. Mattila and A. R. Torronen, *J. Agric. Food Chem.*, 2007, **55**, 1612–1619.
17. R. Veberic, A. Slatnar, J. Bizjak, F. Stampar and M. Mikulic-Petkovsek, *LWT–Food Sci. Technol.*, 2015, **60**, 509–517.
18. M. P. Kahkonen, A. I. Hopia and M. Heinonen, *J. Agric. Food Chem.*, 2001, **49**, 4076–4082.
19. R. Kimble, K. M. Keane, J. K. Lodge and G. Howatson, *Crit. Rev. Food Sci. Nutr.*, 2019, **59**, 3032–3043.
20. L. Fairlie-Jones, K. Davison, E. Fromentin and A. M. Hill, *Nutrients*, 2017, **9**, 908.
21. L. Yang, W. Ling, Z. Du, Y. Chen, D. Li, S. Deng, Z. Liu and L. Yang, *Adv. Nutr.*, 2017, **8**, 684–693.
22. E. Daneshzad, S. Shab-Bidar, Z. Mohammadpour and K. Djafarian, *Clin. Nutr.*, 2019, **38**, 1153–1165.
23. Y. Zhu, Y. Bo, X. Wang, W. Lu, X. Wang, Z. Han and C. Qiu, *Medicine*, 2016, **95**, e3380.
24. A. J. Stull, K. C. Cash, C. M. Champagne, A. K. Gupta, R. Boston, R. A. Beyl, W. D. Johnson and W. T. Cefalu, *Nutrients*, 2015, **7**, 4107–4123.

25. L. Xie, T. Vance, B. Kim, S. G. Lee, C. Caceres, Y. Wang, P. A. Hubert, J. Y. Lee, O. K. Chun and B. W. Bolling, *Nutr. Res.*, 2017, **37**, 67–77.
26. S. A. Johnson, A. Figueroa, N. Navaei, A. Wong, R. Kalfon, L. T. Ormsbee, R. G. Feresin, M. L. Elam, S. Hooshmand, M. E. Payton and B. H. Arjmandi, *J. Acad. Nutr. Diet.*, 2015, **115**, 369–377.
27. J. A. Novotny, D. J. Baer, C. Khoo, S. K. Gebauer and C. S. Charron, *J. Nutr.*, 2015, **145**, 1185–1193.
28. E. Klewicka, M. Sojka, R. Klewicki, K. Kolodziejczyk, L. Lipinska and A. Nowak, *Molecules*, 2016, **21**, 908.
29. M. V. Selma, A. Gonzalez-Sarrias, J. Salas-Salvado, C. Andres-Lacueva, C. Alasalvar, A. Orem, F. A. Tomas-Barberan and J. C. Espin, *Clin. Nutr.*, 2018, **37**, 897–905.
30. G. Istas, R. P. Feliciano, T. Weber, R. Garcia-Villalba, F. Tomas-Barberan, C. Heiss and A. Rodriguez-Mateos, *Arch. Biochem. Biophys.*, 2018, **651**, 43–51.
31. L. Mele, P. Mena, A. Piemontese, V. Marino, N. Lopez-Gutierrez, F. Bernini, F. Brighenti, I. Zanotti and D. Del Rio, *Arch. Biochem. Biophys.*, 2016, **599**, 42–50.
32. M. T. Garcia-Conesa, K. Chambers, E. Combet, P. Pinto, M. Garcia-Aloy, C. Andres-Lacueva, S. de Pascual-Teresa, P. Mena, A. Konic Ristic, W. J. Hollands, P. A. Kroon, A. Rodriguez-Mateos, G. Istas, C. A. Kontogiorgis, D. K. Rai, E. R. Gibney, C. Morand, J. C. Espin and A. Gonzalez-Sarrias, *Int. J. Mol. Sci.*, 2018, **19**, 694.
33. A. Cassidy and C. Kay, in *Encyclopedia of Human Nutrition*, Elsevier, 2013, pp. 39–46.
34. L. Gu, M. A. Kelm, J. F. Hammerstone, G. Beecher, J. Holden, D. Haytowitz, S. Gebhardt and R. L. Prior, *J. Nutr.*, 2004, **134**, 613–617.
35. E. A. Rue, M. D. Rush and R. B. van Breemen, *Phytochem. Rev.*, 2018, **17**, 1–16.
36. B. L. White, L. R. Howard and R. L. Prior, *J. Agric. Food Chem.*, 2010, **58**, 4037–4042.
37. G. Istas, E. Wood, M. Le Sayec, C. Rawlings, J. Yoon, V. Dandavate, D. Cera, S. Rampelli, A. Costabile, E. Fromentin and A. Rodriguez-Mateos, *Am. J. Clin. Nutr.*, 2019, **110**, 316–329.
38. A. Rodriguez-Mateos, R. P. Feliciano, A. Boeres, T. Weber, C. N. Dos Santos, M. R. Ventura and C. Heiss, *Mol. Nutr. Food Res.*, 2016, **60**, 2130–2140.
39. N. Alshammari, S. A. Palma-Duran, G. Jiang, A. González-Sarrías, P. Pinto, M. Garcia-Aloy, E. Gibney, M. T. García-Conesa, P. Kroon, A. Konić Ristić, A. Katsaltou, C. Morand, A. Rodriguez-Mateos and E. Combet, Proceedings of the Nutrition Society, 2018, **77**, E40.
40. A. González-Sarrías, E. Combet, P. Pinto, P. Mena, M. Dall'Asta, M. Garcia-Aloy, A. Rodríguez-Mateos, E. R. Gibney, J. Dumont, M. Massaro, J. Sánchez-Meca, C. Morand and M.-T. García-Conesa, *Nutrients*, 2017, **9**, 28.

41. K. Ried, P. Fakler and N. P. Stocks, *Cochrane Database Syst Rev*, 2017, **4**, CD008893.
42. P. Knekt, J. Kumpulainen, R. Jarvinen, H. Rissanen, M. Heliovaara, A. Reunanen, T. Hakulinen and A. Aromaa, *Am. J. Clin. Nutr.*, 2002, **76**, 560–568.
43. L. I. Mennen, D. Sapinho, A. de Bree, N. Arnault, S. Bertrais, P. Galan and S. Hercberg, *J. Nutr.*, 2004, **134**, 923–926.
44. R. L. Edwards, T. Lyon, S. E. Litwin, A. Rabovsky, J. D. Symons and T. Jalili, *J. Nutr.*, 2007, **137**, 2405–2411.
45. S. Egert, A. Bosy-Westphal, J. Seiberl, C. Kurbitz, U. Settler, S. Plachta-Danielzik, A. E. Wagner, J. Frank, J. Schrezenmeir, G. Rimbach, S. Wolffram and M. J. Muller, *Br. J. Nutr.*, 2009, **102**, 1065–1074.
46. A. Sahebkar, *Crit. Rev. Food Sci. Nutr.*, 2017, **57**, 666–676.
47. M. C. Serban, A. Sahebkar, A. Zanchetti, D. P. Mikhailidis, G. Howard, D. Antal, F. Andrica, A. Ahmed, W. S. Aronow, P. Muntner, G. Y. Lip, I. Graham, N. Wong, J. Rysz, M. Banach and Lipid and Blood Pressure Meta-analysis Collaboration Group, *J. Am. Heart Assoc.*, 2016, **5**, e002713.
48. R. Menezes, A. Rodriguez-Mateos, A. Kaltsatou, A. González-Sarrías, A. Greyling, C. Giannaki, C. Andres-Lacueva, D. Milenkovic, E. R. Gibney, J. Dumont, M. Schär, M. Garcia-Aloy, S. A. Palma-Duran, T. Ruskovska, V. Maksimova, E. Combet and P. Pinto, *Nutrients*, 2017, **9**, 117.
49. R. M. Alqurashi, L. A. Galante, I. R. Rowland, J. P. Spencer and D. M. Commane, *Am. J. Clin. Nutr.*, 2016, **104**, 1227–1235.
50. R. P. Feliciano, G. Istas, C. Heiss and A. Rodriguez-Mateos, *Molecules*, 2016, **21**, 1120.
51. A. Rodriguez-Mateos, C. Rendeiro, T. Bergillos-Meca, S. Tabatabaee, T. W. George, C. Heiss and J. P. Spencer, *Am. J. Clin. Nutr.*, 2013, **98**, 1179–1191.
52. S. Vendrame, C. Del Bo, S. Ciappellano, P. Riso and D. Klimis-Zacas, *Antioxidants (Basel)*, 2016, **5**, 34.
53. I. Zanotti, M. Dall'Asta, P. Mena, L. Mele, R. Bruni, S. Ray and D. Del Rio, *Food Funct.*, 2015, **6**, 13–31.
54. M. Naruszewicz, I. Laniewska, B. Millo and M. Dluzniewski, *Atherosclerosis*, 2007, **194**, e179–e184.
55. G. Grosso, A. Micek, J. Godos, S. Sciacca, A. Pajak, M. A. Martinez-Gonzalez, E. L. Giovannucci and F. Galvano, *Eur. J. Epidemiol.*, 2016, **31**, 1191–1205.
56. D. Martini, L. Chiavaroli, A. Gonzalez-Sarrias, L. Bresciani, S. A. Palma-Duran, M. Dall'Asta, G. E. Deligiannidou, M. Massaro, E. Scoditti, E. Combet, V. Maksimova, M. Urpi-Sarda, C. A. Kontogiorgis, C. Andres-Lacueva, E. R. Gibney, D. Del Rio, C. Morand, M. Garcia-Aloy, A. Rodriguez-Mateos and P. Mena, *Nutrients*, 2019, **11**, 1805.

57. I. Krga, D. Milenkovic, C. Morand and L. E. Monfoulet, *Food Funct.*, 2016, **7**, 3656–3676.
58. F. A. Tomas-Barberan, M. V. Selma and J. C. Espin, *Curr. Opin. Clin. Nutr. Metab. Care*, 2016, **19**, 471–476.
59. E. R. Gibney, D. Milenkovic, E. Combet, T. Ruskovska, A. Greyling, A. González-Sarrías, B. de Roos, F. Tomás-Barberán, C. Morand and A. Rodriguez-Mateos, *Eur. J. Nutr.*, 2019, **58**, 37–47.
60. C. Manach, D. Milenkovic, T. Van de Wiele, A. Rodriguez-Mateos, B. de Roos, M. T. Garcia-Conesa, R. Landberg, E. R. Gibney, M. Heinonen, F. Tomas-Barberan and C. Morand, *Mol. Nutr. Food Res.*, 2017, **61**, 16.
61. D. Del Rio, A. Rodriguez-Mateos, J. P. Spencer, M. Tognolini, G. Borges and A. Crozier, *Antioxid. Redox Signal.*, 2013, **18**, 1818–1892.
62. G. Williamson and M. N. Clifford, *Biochem. Pharmacol.*, 2017, **139**, 24–39.
63. P. E. Milbury, J. A. Vita and J. B. Blumberg, *J. Nutr.*, 2010, **140**, 1099–1104.
64. R. P. Feliciano, C. E. Mills, G. Istas, C. Heiss and A. Rodriguez-Mateos, *Nutrients*, 2017, **9**, 268.
65. K. K. Filipski, M. A. Pacanowski, A. Ramamoorthy, W. G. Feero and A. N. Freedman, *Pharmacogenet. Genomics*, 2016, **26**, 334–339.
66. M. C. Cornelis, A. El-Sohemy, E. K. Kabagambe and H. Campos, *JAMA*, 2006, **295**, 1135–1141.
67. A. V. Gasper, A. Al-Janobi, J. A. Smith, J. R. Bacon, P. Fortun, C. Atherton, M. A. Taylor, C. J. Hawkey, D. A. Barrett and R. F. Mithen, *Am. J. Clin. Nutr.*, 2005, **82**, 1283–1291.
68. R. J. Miller, K. G. Jackson, T. Dadd, A. E. Mayes, A. L. Brown and A. M. Minihane, *Br. J. Nutr.*, 2011, **105**, 1138–1144.
69. Z. Vinarov, M. Abdallah, J. A. G. Agundez, K. Allegaert, A. W. Basit, M. Braeckmans, J. Ceulemans, M. Corsetti, B. T. Griffin, M. Grimm, D. Keszthelyi, M. Koziolek, C. M. Madla, C. Matthys, L. E. McCoubrey, A. Mitra, C. Reppas, J. Stappaerts, N. Steenackers, N. L. Trevaskis, T. Vanuytsel, M. Vertzoni, W. Weitschies, C. Wilson and P. Augustijns, *Eur. J. Pharm. Sci.*, 2021, **162**, 105812.
70. J. B. Dressman, R. R. Berardi, L. C. Dermentzoglou, T. L. Russell, S. P. Schmaltz, J. L. Barnett and K. M. Jarvenpaa, *Pharm. Res.*, 1990, **7**, 756–761.
71. L. Kalantzi, K. Goumas, V. Kalioras, B. Abrahamsson, J. B. Dressman and C. Reppas, *Pharm. Res.*, 2006, **23**, 165–176.
72. M. Koziolek, M. Grimm, G. Garbacz, J. P. Kuhn and W. Weitschies, *Mol. Pharm.*, 2014, **11**, 1632–1639.
73. O. Petrakis, M. Vertzoni, A. Angelou, F. Kesisoglou, K. Bentz, K. Goumas and C. Reppas, *J. Pharm. Pharmacol.*, 2015, **67**, 56–67.
74. G. S. Smelick, T. P. Heffron, L. Chu, B. Dean, D. A. West, S. L. Duvall, B. L. Lum, N. Budha, S. N. Holden, L. Z. Benet, A. Frymoyer, M. J. Dresser and J. A. Ware, *Mol. Pharm.*, 2013, **10**, 4055–4062.
75. T. Tarko, A. Duda-Chodak and A. Soszka, *Molecules*, 2020, **25**, 5574.

76. A. Cortes-Martin, R. Garcia-Villalba, A. Gonzalez-Sarrias, M. Romo-Vaquero, V. Loria-Kohen, A. Ramirez-de-Molina, F. A. Tomas-Barberan, M. V. Selma and J. C. Espin, *Food Funct.*, 2018, **9**, 4100–4106.
77. S. Hazim, P. J. Curtis, M. Y. Schar, L. M. Ostertag, C. D. Kay, A. M. Minihane and A. Cassidy, *Am. J. Clin. Nutr.*, 2016, **103**, 694–702.
78. M. Romo-Vaquero, A. Cortes-Martin, V. Loria-Kohen, A. Ramirez-de-Molina, I. Garcia-Mantrana, M. C. Collado, J. C. Espin and M. V. Selma, *Mol. Nutr. Food Res.*, 2019, **63**, e1800958.
79. R. Gonzalez-Barrio, P. Truchado, H. Ito, J. C. Espin and F. A. Tomas-Barberan, *J. Agric. Food Chem.*, 2011, **59**, 1152–1162.
80. M. V. Selma, M. Romo-Vaquero, R. Garcia-Villalba, A. Gonzalez-Sarrias, F. A. Tomas-Barberan and J. C. Espin, *Food Funct.*, 2016, **7**, 1769–1774.
81. M. V. Selma, F. A. Tomas-Barberan, D. Beltran, R. Garcia-Villalba and J. C. Espin, *Int. J. Syst. Evol. Microbiol.*, 2014, **64**, 2346–2352.
82. D. Beltran, M. Romo-Vaquero, J. C. Espin, F. A. Tomas-Barberan and M. V. Selma, *Int. J. Syst. Evol. Microbiol.*, 2018, **68**, 1707–1712.
83. K. O. P. Inada, F. A. Tomás-Barberán, D. Perrone and M. Monteiro, *J. Funct. Foods*, 2019, **57**, 299–308.
84. A. Gonzalez-Sarrias, R. Garcia-Villalba, M. Romo-Vaquero, C. Alasalvar, A. Orem, P. Zafrilla, F. A. Tomas-Barberan, M. V. Selma and J. C. Espin, *Mol. Nutr. Food Res.*, 2017, **61**, 1600830.
85. R. K. Singh, H. W. Chang, D. Yan, K. M. Lee, D. Ucmak, K. Wong, M. Abrouk, B. Farahnik, M. Nakamura, T. H. Zhu, T. Bhutani and W. Liao, *J. Transl. Med.*, 2017, **15**, 73.
86. A. Cassidy and A. M. Minihane, *Am. J. Clin. Nutr.*, 2017, **105**, 10–22.
87. S. Roowi, W. Mullen, C. A. Edwards and A. Crozier, *Mol. Nutr. Food Res.*, 2009, **53**, S68–S75.
88. A. L. Simons, M. Renouf, S. Hendrich and P. A. Murphy, *J. Agric. Food Chem.*, 2005, **53**, 4258–4263.
89. M. E. Eker, K. Aaby, I. Budic-Leto, S. R. Brncic, S. N. El, S. Karakaya, S. Simsek, C. Manach, W. Wiczkowski and S. Pascual-Teresa, *Foods*, 2019, **9**, 2.
90. A. Duda-Chodak, *J. Physiol. Pharmacol.*, 2012, **63**, 497–503.
91. F. F. Anhe, D. Roy, G. Pilon, S. Dudonne, S. Matamoros, T. V. Varin, C. Garofalo, Q. Moine, Y. Desjardins, E. Levy and A. Marette, *Gut*, 2015, **64**, 872–883.
92. J. Overall, S. A. Bonney, M. Wilson, A. Beermann, M. H. Grace, D. Esposito, M. A. Lila and S. Komarnytsky, *Int. J. Mol. Sci.*, 2017, **18**, 422.
93. P. Pan, V. Lam, N. Salzman, Y. W. Huang, J. Yu, J. Zhang and L. S. Wang, *Nutr. Cancer*, 2017, **69**, 943–951.
94. G. Paturi, C. A. Butts, J. A. Monro and D. Hedderley, *Plant Foods Hum. Nutr.*, 2018, **73**, 54–60.
95. L. Cui, T. Zhao, H. Hu, W. Zhang and X. Hua, *BioMed Res. Int.*, 2017, **2017**, 3796359.

96. M. Troseid, G. O. Andersen, K. Broch and J. R. Hov, *EBioMedicine*, 2020, **52**, 102649.
97. K. Brown, D. DeCoffe, E. Molcan and D. L. Gibson, *Nutrients*, 2012, **4**, 1095–1119.
98. L. Lavefve, L. R. Howard and F. Carbonero, *Food Funct.*, 2020, **11**, 45–65.
99. S. Wu, R. Hu, H. Nakano, K. Chen, M. Liu, X. He, H. Zhang, J. He and D. X. Hou, *Molecules*, 2018, **23**, 3213.
100. A. Duda-Chodak, T. Tarko, P. Satora and P. Sroka, *Eur. J. Nutr.*, 2015, **54**, 325–341.
101. H. Huang, G. Chen, D. Liao, Y. Zhu and X. Xue, *Sci. Rep.*, 2016, **6**, 23625.
102. C. Heiss, R. Sansone, H. Karimi, M. Krabbe, D. Schuler, A. Rodriguez-Mateos, T. Kraemer, M. M. Cortese-Krott, G. G. Kuhnle, J. P. Spencer, H. Schroeter, M. W. Merx, M. Kelm and Flaviola Consortium, European Union 7th Framework Program, *Age (Dordr)*, 2015, **37**, 9794.
103. N. D. Fisher and N. K. Hollenberg, *J. Hypertens.*, 2006, **24**, 1575–1580.
104. A. Rodriguez-Mateos, T. Cifuentes-Gomez, I. Gonzalez-Salvador, J. I. Ottaviani, H. Schroeter, M. Kelm, C. Heiss and J. P. Spencer, *Mol. Nutr. Food Res.*, 2015, **59**, 1504–1512.
105. L. Mosca, E. Barrett-Connor and N. K. Wenger, *Circulation*, 2011, **124**, 2145–2154.
106. R. Comitato, A. Saba, A. Turrini, C. Arganini and F. Virgili, *Crit. Rev. Food Sci. Nutr.*, 2015, **55**, 227–241.
107. A. Basu, M. Du, M. J. Leyva, K. Sanchez, N. M. Betts, M. Wu, C. E. Aston and T. J. Lyons, *J. Nutr.*, 2010, **140**, 1582–1587.
108. L. Akil and H. A. Ahmad, *J. Health Care Poor Underserved*, 2011, **22**, 61–72.
109. P. W. Wilson, R. B. D'Agostino, L. Sullivan, H. Parise and W. B. Kannel, *Arch. Intern. Med.*, 2002, **162**, 1867–1872.
110. P. M. Solverson, T. R. Henderson, H. Debelo, M. G. Ferruzzi, D. J. Baer and J. A. Novotny, *Nutrients*, 2019, **11**, 2876.
111. Z. Aboo Bakkar, J. Fulford, P. E. Gates, S. R. Jackman, A. M. Jones, B. Bond and J. L. Bowtell, *J. Appl. Physiol. (1985)*, 2019, **126**, 246–254.
112. S. S. Hassellund, A. Flaa, S. E. Kjeldsen, I. Seljeflot, A. Karlsen, I. Erlund and M. Rostrup, *J. Hum. Hypertens.*, 2013, **27**, 100–106.
113. K. S. Stote, M. I. Sweeney, T. Kean, D. J. Baer, J. A. Novotny, N. L. Shakerley, A. Chandrasekaran, P. M. Carrico, J. A. Melendez and K. T. Gottschall-Pass, *BMC Nutr.*, 2017, **3**, 45.
114. A. Basu, K. Sanchez, M. J. Leyva, M. Wu, N. M. Betts, C. E. Aston and T. J. Lyons, *J. Am. Coll. Nutr.*, 2010, **29**, 31–40.
115. P. J. Curtis, V. van der Velpen, L. Berends, A. Jennings, M. Feelisch, A. M. Umpleby, M. Evans, B. O. Fernandez, M. S. Meiss, M. Minnion, J. Potter, A. M. Minihane, C. D. Kay, E. B. Rimm and A. Cassidy, *Am. J. Clin. Nutr.*, 2019, **109**, 1535–1545.

116. C. Del Bo, M. Porrini, D. Fracassetti, J. Campolo, D. Klimis-Zacas and P. Riso, *Food Funct.*, 2014, **5**, 3107–3116.
117. T. C. Wallace, M. Slavin and C. L. Frankenfeld, *Nutrients*, 2016, **8**, 32.
118. S. Kianbakht, B. Abasi and F. Hashem Dabaghian, *Phytother. Res.*, 2014, **28**, 432–436.
119. Y. Qin, M. Xia, J. Ma, Y. Hao, J. Liu, H. Mou, L. Cao and W. Ling, *Am. J. Clin. Nutr.*, 2009, **90**, 485–492.
120. R. Soltani, M. Hakimi, S. Asgary, S. M. Ghanadian, M. Keshvari and N. Sarrafzadegan, *Evidence-Based Complementary Altern. Med.*, 2014, **2014**, 217451.
121. Y. Zhu, W. Ling, H. Guo, F. Song, Q. Ye, T. Zou, D. Li, Y. Zhang, G. Li, Y. Xiao, F. Liu, Z. Li, Z. Shi and Y. Yang, *Nutr., Metab. Cardiovasc. Dis.*, 2013, **23**, 843–849.
122. P. J. Curtis, P. A. Kroon, W. J. Hollands, R. Walls, G. Jenkins, C. D. Kay and A. Cassidy, *J. Nutr.*, 2009, **139**, 2266–2271.
123. A. S. Hansen, P. Marckmann, L. O. Dragsted, I. L. Finne Nielsen, S. E. Nielsen and M. Gronbaek, *Eur. J. Clin. Nutr.*, 2005, **59**, 449–455.
124. O. D. Rangel-Huerta, B. Pastor-Villaescusa, C. M. Aguilera and A. Gil, *Nutrients*, 2015, **7**, 5177–5216.
125. M. Zahedi, R. Ghiasvand, A. Feizi, G. Asgari and L. Darvish, *Int. J. Prev. Med.*, 2013, **4**, 777–785.

CHAPTER 4

Berries: Functional Foods That Promote Beneficial Bacterial Species in the Gut

ALISON LACOMBE AND VIVIAN C. H. WU (0000-0002-1525-1078)*

Produce Safety and Microbiology Research Unit, United States Department Agricultural, Agricultural Research Service, Albany, CA 94710, USA
*Email: vivian.wu@usda.gov

4.1 Introduction

Consumers wish to know how a diet high in bioactive compounds can protect them from various maladies such as metabolic syndrome, urinary tract infections, and digestive disorders.[1] Berries have the potential to protect humans against oxidative stress, which could contribute to inflammation, metabolic disorders, and other chronic ailments.[2–6] Polyphenolic constituents such as anthocyanins, proanthocyanidins, and other monomeric phenolic compounds are abundant in berries and have demonstrated prebiotic capabilities that have putative beneficial downstream effects.[7–9] The gut microbiota is considered a crucial link in the disease prevention mechanism of berries because of their ability to promote beneficial species.[10,11]

Research on functional foods promoting gut health and beneficial microbiota (*i.e.*, prebiotics) has become a topic of interest in preventive medicine and food product development. There is a compilation of evidence that links the composition of gut microbiota, berry phenolics, and metabolic

Food Chemistry, Function and Analysis No. 33
Berries and Berry Bioactive Compounds in Promoting Health
Edited by Dorothy Klimis-Zacas and Ana Rodriguez-Mateos

Published by the Royal Society of Chemistry, www.rsc.org

syndrome.[12] The impacts on intestinal microflora from berry consumption have demonstrated the attenuation of indicators of metabolic syndrome and inflammation.[10,13,14] However, laboratory and clinical demonstrations of berries' influence on microbial ecosystems and the connection to health benefits are difficult to translate into commercial practices. Precise characterization of the putative health benefits gained from the inclusion of berries in the diet requires a detailed description of cultivar viability and administration, such as food matrix or probiotic carriers, which is needed to resolve the effect and mechanism of action of berries on beneficial microorganisms. This obstacle is magnified by the lack of standardization of the target population, study design, and definition of endpoints, making it difficult to validate and compare outcomes with other studies.

Today natural products from berries are being investigated as a new potential arsenal of antimicrobials and prebiotics because of their ability to selectively inhibit enteric pathogens while promoting the growth of beneficial microorganisms.[5,6,15,16] It has been well established that complex mixtures of phytochemicals in various organic plants have complicated interactions and synergistic effects against pathogenic bacteria.[17–20] Berries, in particular, have exhibited various antimicrobial activities *in vitro* including, the prevention of microbial adhesion to the urinary tract,[21] the reduction of biofilm production, and the prevention of foodborne pathogen contaminations.[5,15,22–25] While the exact mechanism of action remains unknown, berry phenolics have a unique chemistry that forms redox reactions damaging the pathogenic bacterial cell membrane.[5,24]

Dietary enrichment of berries demonstrated prebiotic capabilities, impacting gut microbial population dynamics and gastrointestinal tract (GIT) health.[9–11,13,26] There have been many studies on berries and their action in the body that utilized different models to uncover the mechanism of action. The current *meta*-analysis describes clinical trials demonstrating a direct benefit of berry consumption to human health.[7,27,28] This chapter aims to discuss research concerning the bioactivities, quantity, and overall recommended consumption of berries in the diet and the benefits to the gut.

4.2 The Prebiotic Capabilities of Berries

As far back as 1921, several experiments described the enrichment of lactobacilli in humans following consumption of carbohydrates.[29] The first formal definition for prebiotics was established in 1995 as a "non-digestible food ingredient that beneficially affects the host by selectively stimulating the growth and/or activity of one or a limited number bacteria already resident in the colon".[2] The definition of prebiotics has expanded over the years as advances in metagenomics and metabolomics have made it possible to take a snapshot of the gastrointestinal environment.[29] Genomic technologies have developed the definitions of non-digestible food, beneficial bacteria, and the site of interaction.[29] Currently, this definition continues to evolve, and emerging evidence demonstrates that synergistic activities between microorganisms are not limited to the GIT and non-digestible carbohydrates.[29] Here, we will define prebiotics as a food source established to sustain microorganisms. The result is a symbiotic relationship with sustained benefits to the host. This chapter will focus on the specific constituents of berries and microorganisms that support this beneficial relationship.

It has become clear that human beings do not exist as a single entity but rather a composite of organisms living in symbiosis. There are trillions of microorganisms that compose the human microbiome, and the highest density of microbes resides in the GIT.[30] One of the significant advances in GIT microbiology is the realization of the effect diet has on the composition of the community and how it affects the health of the host.[31,32] The gut microbiome's role in digesting foods has been intensively studied. However, GIT microbial process that allowed for berry constituents to contribute to human health remains elusive. Evidence accumulated with *in vitro* research and recent evidence from animal studies and human clinical trials indicate that the phytochemicals, fiber, and other nutrients from berries have prebiotic capabilities.[7,10,13]

The popularity of gut health in the wellness industry has allowed the rapid expansion of products that support beneficial microbes. There are several products on the market of undefined composition that claim prebiotic capabilities.[33] Up to 85% of anthocyanins enter the colon intact and can be used as substrates for microbial metabolism.[34] Therefore, prebiotics from berries must cause changes in gut microbial communities and stimulate the growth of beneficial gut microbes. In addition, prebiotics should be resistant to gastric acidity and mammalian digestive enzymes, allowing them to enter the colon intact. In humans, the intestinal absorption of dietary polyphenols is often slow and incomplete.

4.2.1 Oligosaccharides

Oligosaccharides are classified by their monosaccharide constituents and linkages to lipids (glycolipids) or amino acids (glycoproteins).[35] Oligosaccharides are relatively small chains ranging from 3 to 10 simple saccharides.[35] They elicit many functions in cell biology and host immunity. Numerous food compounds are resistant to digestion and are used for fermentation by gut microflora. Most berries contain low amounts of fructooligosaccharides (FOS) which are composed of linear chains of fructose units linked by beta (2-1) bonds.[35] FOS have many interesting properties, including a low sweetness; they are also calorie free, non-cariogenic, and considered a soluble dietary fiber. Furthermore, FOS have important beneficial physiological effects such as low carcinogenicity, a prebiotic effect, improved mineral absorption, and decreased serum cholesterol levels, triacylglycerols, and phospholipids.[35] A recent analysis found that blueberries contain the most non-digestible oligosaccharides compared to raspberry, strawberry, currant, blackberry, and grapes.[36] Further investigation reveals that blueberries formed statistically significant clusters with garlic, spring garlic, leek, white onion, and scallion for oligosaccharide content.[36]

Fruit pomaces are considered functional food ingredients with promising prebiotic benefits because of the abundance of dietary fiber and pectin.[37,38] Blueberries are often considered the best berries for commercial extractions of prebiotics which can be used as ingredients for functional food products.[38] Blueberry pomace includes the solid remains following the juicing process is rich in prebiotics such as FOS. During blueberry puree commercial processing, most of the anthocyanins adhered to the pomace fiber. Recent research has described blueberry as a complex mixture of insoluble dietary fiber consisting of methyl-esterified and acetylated rhamnogalacturonan I pectin with a 4,5-unsaturated function at the nonreducing end.[38] This flexible random coil-shaped pectin is presumably responsible for the strong ionic anthocyanin pomace fiber interaction.[38] Anthocyanins are also reported to have prebiotic activity. The carbohydrate released from blueberry anthocyanins provides a carbon source for *Bifidobacteria*, and the binding of anthocyanins by blueberry pectin could provide a dual prebiotic effect.[38]

Most gut microbiota members selectively metabolize mono- and oligosaccharides, and bacteria differ in their carbohydrate processing capabilities.[35] Dietary FOS are not digested by small intestinal enzymes and reach the cecum structurally unchanged. Saccharolytic fermentation occurs with non-digestible carbohydrates and favors the growth of some specific bacteria genera such as *Bifidobacterium* and *Lactobacillus*.[39] These saccharolytic bacteria produce short-chain fatty acids (SCFA) in the colon, namely acetate, propionate, or butyrate, that contribute to calories.[39] Then absorption of nutrients by the gut epithelial cells can extend putative beneficial effects on the system. In the large intestine, FOS are metabolized by the intestinal microflora to form short-chain carboxylic acids, L-lactate, CO_2, hydrogen, and other metabolites.[40]

4.2.2 Polyphenols

Polyphenols represent a new category of potential prebiotics. The genome sequence of *Bifidobacterium longum* has many predicted proteins related to the catabolism of non-digestible plant polymers. Until recently, researchers knew little about the metabolic fate of berries in the intestinal tract, and there is still a paucity of evidence that describes the metabolism and bioprotective effect(s) of anthocyanins and proanthocyanins derived from berries.[14,41] Still, recent evidence suggests that bacterial metabolism has a substantial potential for the bioactivation of phenolic compounds. The microbial enzymatic biotransformations may also be relevant for converting flavonoids, isoflavonoids, lignans, phenolic acids, fiber, and tannins into beneficial by-products.[29] Cleavage of the glucosidic bond in the anthocyanin structure is proposed as the first step in bacterial anthocyanin bioconversion, involving β-glucosidase activity.[2] New metabolites, including gallic, syringic, and homogentisic acids, appear due to bacterial enzymatic action on malvidin, some of which have been described as more bioavailable and bioactive than the original molecule.[2] In addition, studies documented a 20% increase in xenobiotic degradation and a two-fold increase in benzoate degradation in the proximal colon of rats fed an 8% blueberry diet.[11] The genome sequence of *Bifidobacterium longum* has many predicted proteins (more than 8%) related to the catabolism of non-digestible plant polymers, including enzymes involved in the degradation of complex polysaccharides and xenobiotics.[11] Therefore, the protective anti-inflammatory effect of blueberries may be credited with positive shifts in the composition of the microbiota and concomitant regulation of microbial metabolism.

Berries are consumed fresh when in season, and the remaining harvest is either frozen or processed into juices, concentrated beverages, and jams. Bioactive compounds in berries tend to be heat labile; therefore, the application of thermal processing is not generally recommended for functional food products. However, the enzymatic degradation of bioactive through present polyphenol oxidase (PPO) can be significantly inhibited using mild blanching prior to frozen storage. In addition, novel non-thermal technologies such as microwave-assisted processing, high-pressure-assisted freezing, high-pressure carbon dioxide processing, and ultrasound could be considered.[37]

4.3 *In Vitro* and *In Vivo* Investigation of Berries' Interaction with Gut Microbiome

Here we will briefly overview the models utilized to study the interactions between berry constituents and gut microbiota (Figure 4.1). The intestinal microbiota has been studied in various *in vitro* and *in vivo* models, each presenting distinct benefits and drawbacks. Bio-protective effects of polyphenols and other nutrients observed *in vitro* cannot be entirely extrapolated to health effects observed *in vivo*. However, all models can provide valuable insights into the mechanism of human digestion of berry prebiotics. Therefore, it is vital to use *in vivo* models to understand mechanisms.

The intestinal microbiota is considered an appropriate target for therapeutic interventions in dietary supplements and food ingredients, with the specific aim of influencing community composition and restoring functional capacity.[36] Health properties attributed to berry constituents include modulation of colonic microbiota by inhibiting a wide range of pathogens, reduction of serum cholesterol, stimulation of the immune system through cytokine stimulus, reinforcement of intestinal epithelial cell tight junctions, and increased mucus secretion.[7,8,13,14,17,41,42] Flavonoids and their derivatives are the largest and most important group of plant phenolics. They have shown various biological effects, including inhibition of low-density lipoprotein (LDL) oxidation, as well as antimicrobial and anticarcinogenic capacities.[43] As the microbiota of the human gut is gaining increasing research attention, the need to develop models for non-human microbiota studies is significant.

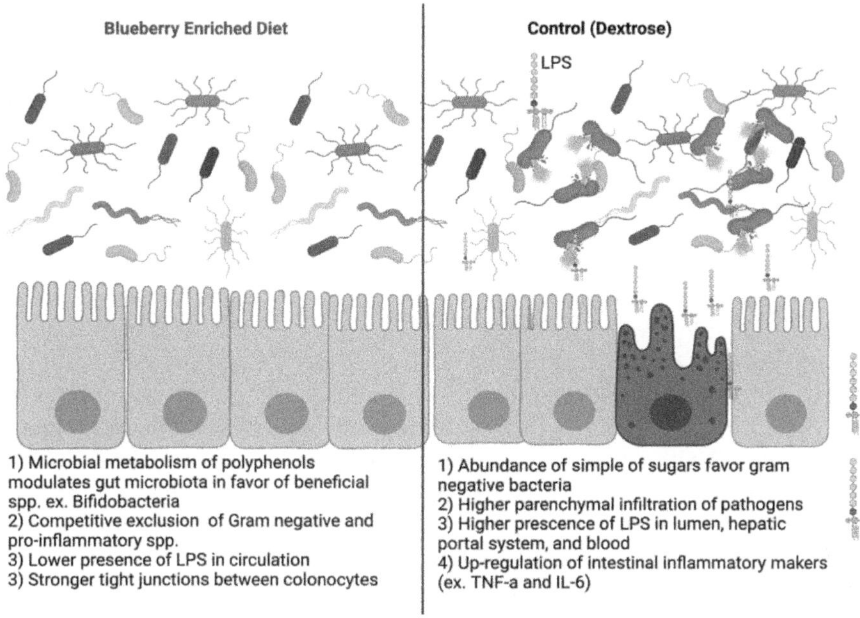

Figure 4.1 The putative mechanism for berries' observed health benefits in the intestinal gut lumen.

4.3.1 *In Vitro* Models

In vitro models constitute an opportunity to study the gut microbiota in cases where an *in vivo* study is considered conceptually not feasible or unethical. In essence, *in vitro* gut models function as chemostats, *i.e.*, bioreactors inoculated with fecal microbiota and operated under physiological temperature, pH, and anaerobic conditions.[44] Experimentation within these models is very flexible regarding studying multiple permutations/combinations of berry constituents and their impact on gut microbial populations.[44] Ultimately, cumulative *in vitro* studies can lead to robust *in vivo* approaches to investigating berry constituents' functionality.[7,8,42,45–47] In theory, each study yields complementary results that strengthen the overall validity of each model and distinguish between the functionality of berry constituents and gut microbiota.

Many microbes that reside within the gut are not cultivable using traditional microbiological techniques such as agar Petri plates.[48] Preliminary human microbiome studies typically start with obtaining donor specimens and culturing the microbes under anaerobic conditions in a bioreactor.[44] The bioreactor setup can be as simple as an Erlenmeyer flask inside a CO_2 incubator on a stir plate or a complex system of pumps and valves meant to closely simulate the physical design of the digestive tract.[44] The berry constituents under investigation affect the complexity of the bioreactor setup, which is usually dictated by the question the investigator is attempting to answer. For example, in determining whether berry constituents are bifidogenic, investigators can quantify the presence of *Bifidobacteria* using agar and simple biochemical tests. *In vitro* human fecal batch cultures have demonstrated the enrichment of *Lactobacillus* and *Bifidobacteria* with 1 $g\,L^{-1}$ gallic acid and 200 $mg\,L^{-1}$ of anthocyanins.[49] These results show the benefits of dietary enrichment of berries and its impact on bacterial communities with unique functional repertoires by promoting their growth in competitive environments. However, such a setup cannot reveal the berry's effect on other critical bacterial species or potential impacts on human physiology.[44,50]

In vitro, human cell models are typically composed of polarized monolayers of single or co-cultured carcinoma cells.[51] These cell lines are distinguished based on their physiological properties and ability to differentiate.[52] The intestinal epithelial consists of several cell types, including enterocytes, goblet cells, stem cells, enteroendocrine cells, Tuft cells, M cells, and Paneth cells, all of which can interact or sense the presence of the gut microbiota and respond accordingly.[52] Caco-2 and mucus-secreting HT29-MTX cell lines predominate the literature regarding the study of probiotics and prebiotics.[52] There are important considerations in selecting a cell culture model, and decisions should be guided by the research question asked.

Caco-2 cell line, derived from colorectal adenocarcinomas, is one of the oldest and most utilized cell cultures in medical research.[52] This cell line can function as an undifferentiated large intestinal cell or spontaneously differentiate into a cell closely resembling a small intestinal enterocyte.[44] This ability to spontaneously differentiate is challenging to control and presents challenges in the reproduction of scientific studies. Despite this, Caco-2 cells are widely used to study the protective effects of probiotics and prebiotics.[52] Caco-2 cells are commonly used in immunological studies since they can express important pattern recognition receptors, Toll-like receptors (TRL), which can provide information about the relative abundance of inflammatory markers, like lipopolysaccharides (LPS), tumor necrosis factors (TNF), and interleukins (IL) known to contribute to the disease process.[53] Phenolic-enriched extracts from cranberry used to treat LPS-stimulated intestinal Caco-2 cells inhibited the release of pro-inflammatory cytokines (TNF-a and IL-6).[54]

Prebiotic oligosaccharides, found in berries, have been shown to positively modulate Caco-2 and HT-29 cells and exhibit reduced inflammatory activity.[35] Co-cultures of Caco-2 and HT-29 render cell culture models more physiologically and functionally relevant by combing mucin secretion and differentiation.[55] For example, the HT-29-MTX cell line, treated with methotrexate to produce more mucin, has been a helpful tool to show that oligosaccharides and a functional mucin layer can reduce bacterial adhesion to the epithelium.[56] Studies with HT-29-MTX demonstrated that blueberry polyphenols could minimize the possibility of an infection by reducing the adhesion, migration, and internalization of pathogens, including *Salmonella* and *E. coli*.[2,57]

Several *in vitro* studies have investigated berry bioactive compounds for anti-inflammatory activity. Bioactive compound-enriched extracts from berries or modulated the levels of inflammatory markers such as endotoxin LPS. In *in vitro* studies, LPS is used to stimulate innate immunity by regulating the production of various inflammatory mediators, including TNF-a, IL-1b, IL-6, and IL-8 found in monocytes/macrophages located in the intestinal lumen.[58] Strawberry extracts demonstrated significant inhibition of IL-8 secretion in TNF-a-treated human gastric epithelial cells by dampening the nuclear factor-κB signaling.[58] Ellagic acid was identified as the active metabolite in strawberries that regulates LPS induce inflammation.[59] The ellagic acid-enriched extracts from strawberries inactivated mitogen-activated protein kinases (MAPKs) signaling in murine macrophage RAW 264.7 cells and the reductions of TNF-a, IL-1b, and nitric oxide synthase iNOS.[59]

Many studies indicate that berries from the *Vaccinium* species, such as cranberry, blackberry, and blueberry extracts, possess an ability to modulate inflammatory markers in various cell types exposed to a variety of stressors *via* an antioxidative pathway. Cranberry phenolic fraction prevented oxidative stress, inflammation, and mitochondrial dysfunction in intestinal cells.[60] Phenolic-enriched extracts from cranberry used to treat LPS-stimulated intestinal Caco-2/15 cells inhibited the release of pro-inflammatory cytokines (TNF-a and IL-6) by activating Nrf2 signaling.[60] *In vitro* studies anthocyanins-enriched fractions from fermented blueberry–blackberry beverages inhibited markers in macrophages and starch-degrading enzyme further protecting the gut lumen from inflammation.[54]

4.3.2 *In Vivo* Models

Berries may be an important component of a healthy diet because of their great content of phenolic acids and flavonoids, especially anthocyanins. In humans, dietary treatment with blueberry increased the population of *Bifidobacteria*, demonstrating the prebiotic activity of berry polyphenols.[7,10,13] These findings are significant for clinicians due to the growing interest in probiotic bacteria and the perceived benefit of increasing their numbers in the GIT to attain their health benefits. Changes in the gut microbiota composition are classically considered one of the many factors involved in the pathogenesis of either inflammatory bowel disease or irritable bowel syndrome (IBS).[17] *In vivo* research in humans and rats demonstrated that blueberry and blackcurrant enriched diets increased *Bifidobacteria* species after dietary treatment and lowered concentrations of pro-inflammatory markers.[7,26,61]

4.3.2.1 Animal Studies

Animal models are important for diet studies to understand the biological roles of specific nutrients before validation in human clinical trials. The murine model, which encompasses mice and rats, is the most utilized in biochemical research because of its genetic similarity.[46] In addition, the murine basal metabolic rate per gram of body weight is seven times greater than in humans, which speeds up the development of diet-induced disease.[46] A common model in diet research is the diet-induced obesity (DIO) model,[2] in which animal is fed high-fat or high-density diets to mimic the most common cause of obesity in humans.[46] Another example is the ob/ob mouse[3] that has a mutation in the hormone leptin, which causes the mice to display severe obesity, insulin resistance, and dyslipidemia.[46] The Sprague-Dawley (SD) rat is used extensively in nutritional research and developed from the laboratory classic Wistar rat.[46] The main advantage of SD rats is their hybrid vigor, longer life spans, higher disease resistance, early fertility, large and frequent litters, low neonatal mortality, rapid growth, and large size than inbred stocks.[46] These approaches can be helpful in providing insights into the potential benefits of berry-derived prebiotics through the observed impact on symptoms of metabolic disorders.

The *in vivo* studies in rats investigate the variety of diseases related to lifestyle and diet and try to associate the downstream beneficial effects of berry-enriched diets.[9–12] Research in rats demonstrated that berry-enriched diets increased the population of *Bifidobacteria* species and lowered the concentration of pro-inflammatory markers. *Enterococcus faecalis* is a commensal bacterium of humans and animals. However, an overabundance of *E. faecalis* is highly correlated to inflammatory bowel disease in IL-10 gene-deficient mice, suggesting that specific enteric microbiota tend to be more opportunistic and induce colon inflammation.[62]

A metagenomic study demonstrated that the genetic signature in the colon of SD rats fed a diet a blueberry-enriched move toward greater species biodiversity. Hierarchical analysis showed a significant reduction in the relative *Enterococcus* after SD rats were fed 6 weeks of an 8% blueberry-enriched diet.[11] A blueberry-enriched diet produced an 8-fold decrease in the number of open reading frames assigned to bacterial invasion of epithelial cells, corroborating evidence of reduced microbial translocation.[11] This might explain the relationship between the consumption of polyphenol-rich foods, the gut microbiota, and the slowdown of the advancement of oxidative stress and aging. In summary, the animal models shed light on the potential health-promoting effects of a berry-enriched diet. These open avenues of human clinical trials are designed to aid in preparing nutraceuticals and functional foods with beneficial effects against diseases and disorders related to oxidative stress.

4.3.2.2 Human Clinical Trials

The *in vivo* evidence supports berries' purported ability and their derived products to confer properties that favorably exert anti-inflammatory benefits. The clinical trials demonstrating these effects used traditional cross-over or parallel study designs, addressing important parameters such as including optimal dose, delivery mode, timing and frequency of intake in different populations, and formulation that maximum benefits in humans. A diet rich in carbohydrates, fats, and proteins of animal origin, combined with a low fiber intake, is common in Western diets. The downstream effect is lower microbial diversity with the decrease in the presence of beneficial microbiota as *Faecalibacterium prausnitzii*, *Akkermansia muciniphila*,

Lactobacillus spp., and *Bifidobacterium* spp.[63] Furthermore, a diet rich in bioactive compounds can promote many health benefits, such as modulating the GIT beneficial microbiota, inhibiting negative bacteria proliferation.[11] Overweight adults who consumed a high-carbohydrate, moderate-fat meal with 305 g of strawberry intake daily for 7 days in a cross-over design significantly attenuated the postprandial inflammatory response measured by high-sensitivity C-reactive protein (CRP) and interleukin-6, two major biomarkers of *in vivo* inflammation.[58] The consumption of 450 mL per day of cranberry beverage for 8 weeks was evaluated in the overweight or obese adults with abdominal adiposity.[64] After 8 weeks of cranberry beverage consumption, fasting CRP and serum insulin were significantly decreased, and HDL cholesterol was increased compared to the placebo group.[64]

LPS associated with the cell wall of gram-negative bacteria are highly inflammatory compounds. LPS are associated with disturbed mucosal integrity and bacterial translocation from the GIT. The permeability of the colonic epithelium is an important aspect of gastrointestinal health, and increases in permeability can allow for gut-derived bacteria and toxins to infiltrate the liver *via* portal circulation.[17] Translocated LPS can cause extensive damage to various organs, including the liver.[57] The action of LPS from gram-negative bacteria is effectuated *via* the toll-like receptor 4 (TLR4), which initiates inflammatory cytokine expression, and expression is up-regulated in colitis-associated cancer lesions from patients with ulcerative colitis (UC).[18] Inactive UC patients, the numbers of fecal lactobacilli decrease, indicating that a reduction in intestinal *Lactobacillus* species may be a sign of mucosal inflammation.[18] *Enterobacteriaceae* and *Clostridium* spp. in feces have been among the most frequently isolated anaerobes from the inflamed mucosa of UC patients.[62] In addition, the biodiversity of the active microbiota was shown to be lower for UC patients than for healthy controls.[62] The dietary enrichment of blueberries increased the viable count of fecal lactobacilli and subsequent lower levels of LPS in the liver and blood plasma and subsequent reduction of inflammatory cytokines.[18]

Developing therapeutic regimens to combat colorectal cancer without significant side effects is of great interest to clinicians and researchers. Recent studies investigated the protective effect of blueberry husks to delay or prevent colon carcinogenesis and pathological abnormalities of the liver.[18,57] The pathogenesis of colorectal carcinogenesis associated with colonic inflammation is believed to involve a progression from inflamed and hyperplastic cryptal cells, through dysplasia, to adenoma and carcinoma.[65] UC-associated dysplasia is a likely precursor lesion or marker of carcinomas.[34] Colorectal carcinomas likely evolve through increasingly severe epithelial dysplasia stages before becoming invasive lesions.[18] Berry constituent intake influences the composition of the gut microbiome through microbial fermentation and unabsorbed components that have antimicrobial effects against less beneficial species. For example, microbial catabolism of anthocyanins, a major berry constituent, occurs through the cleavage of heterocyclic flavylium rings (the C-ring) and subsequent dehydroxylation or decarboxylation to form phenolic acids. These compounds can be subject to further microbial fermentations, serve as antimicrobials against pathogens, or serve as bioavailable antioxidants. While the anticancer pathway has not been fully resolved, the benefit of the berry constituent is apparent.

4.4 Conclusion

The phytochemical constituents that are found in many edible berry fruits have increasingly been linked to the modulation of biomarkers associated with conditions of diabetes, overweight/obesity, and cardiovascular disease (CVD). These conditions have been highly correlated to dysbiosis in the gut. While many berries have long been valued in traditional medicine as health-protective, it is only recently that the ability of berry bioactive compounds to affect human health has been demonstrated. Therefore, the increased consumption of berries represents an important epidemiological target for enrichment in the American diet. The demonstration of berries' influence on microbial ecosystems with respect to health benefits represents a tremendous challenge for science as well as industry. However, clinical observations underline apparent differences at the level of specific genus/species abundances between the murine and human gut microbiota, meaning there may be a direct translation of health benefits to humans. This obstacle is magnified by the lack of standardization of the target population, study design, and definition of endpoints, making it difficult to validate and compare outcomes with other studies. Although absolute comparisons might be challenging and not straightforward, murine models are relevant for studying the processes responsible for microbiota variation and shifts upon perturbations and guided impactful research in humans.

References

1. W. Paul, C. Marta and T. Van de Wiele, *Curr. Opin. Microbiol.*, 2018, **44**, 28–33.
2. S. Mithul Aravind, S. Wichienchot, R. Tsao, S. Ramakrishnan and S. Chakkaravarthi, *Food Res. Int.*, 2021, **142**, 110189.
3. A. Lacombe, C. McGivney, S. Tadepalli, X. Sun and V. C. H. Wu, *Food Microbiol.*, 2013, **34**(2), 352–359.
4. A. Lacombe, A. Breard, C.-A. Hwang, D. Hill, X. Fan, L. Huang, B. K. Yoo, B. A. Niemira, J. B. Gurtler and V. C. H. Wu, *Food Control*, 2017, **73**, 981–985.
5. X. H. Sun, L. R. Hao, Q. C. Xie, W. Q. Lan, Y. Zhao, Y. J. Pan and V. C. Wu, *Food Control*, 2020, **111**, 107020.
6. T. Tong Zhou, C. Hong Wei, W. Qing Lan, Y. Zhao, Y. Jie Pan, X. Hong Sun and V. C. H. Wu, *J. Food Sci.*, 2020, **85**, 1513–1522.
7. S. Vendrame, A. Daugherty, A. S. Kristo, P. Riso and D. Klimis-Zacas, *J. Nutr. Biochem.*, 2013, **24**, 1508–1512.
8. C. Del Bò, S. Ciappellano, D. Klimis-Zacas, D. Martini, C. Gardana, P. Riso and M. Porrini, *J. Agric. Food Chem.*, 2009, **58**, 2491–2497.

9. S. Guglielmctti, D. Fracassetti, V. Taverniti, C. Del Bo', S. Vendrame, D. Klimis-Zacas, S. Arioli, P. Riso and M. Porrini, *J. Agric. Food Chem.*, 2013, **61**, 8134–8140.
10. S. Vendrame, S. Guglielmetti, P. Riso, S. Arioli, D. Klimis-Zacas and M. Porrini, *J. Agric. Food Chem.*, 2011, **59**, 12815–12820.
11. A. Lacombe, R. W. Li, D. Klimis-Zacas, A. S. Kristo, S. Tadepalli, E. Krauss, R. Young and V. C. H. Wu, *PLoS One*, 2013, **8**(6), e67497.
12. S. Guglielmetti, D. Fracassetti, V. Taverniti, C. Del Bo', S. Vendrame, D. Klimis-Zacas, S. Arioli, P. Riso and M. Porrini, *J. Agric. Food Chem.*, 2013, **61**, 8134–8140.
13. C. Del Bo', D. Martini, S. Vendrame, P. Riso, S. Ciappellano, D. Klimis-Zacas and M. Porrini, *Mutat. Res., Genet. Toxicol. Environ. Mutagen.*, 2010, **703**, 158–162.
14. S. Vendrame and D. Klimis-Zacas, *Nutrients*, 2019, **11**(6), 1431.
15. A. Lacombe, S. Tadepalli, C.-A. Hwang and V. C. H. Wu, *Foodborne Pathog. Dis.*, 2013, **10**(11), 944–950.
16. A. Lacombe, V. C. H. Wu, S. Tyler and K. Edwards, *Int. J. Food Microbiol.*, 2010, **139**, 102–107.
17. J. M. Laparra and Y. Sanz, *Pharmacol. Res.*, 2010, **61**, 219–225.
18. Å. HÅkansson, C. Bränning, D. Adawi, G. Molin, M. Nyman, B. Jeppsson and S. Ahrné, *Scand. J. Gastroenterol.*, 2009, **44**, 1213–1225.
19. K. Feghali, M. Feldman, V. D. La, J. Santos and D. Grenier, *J. Agric. Food Chem.*, 2012, **60**, 5728–5735.
20. Y.-I. Kwon, E. Apostolidis, R. G. Labbe and K. Shetty, *Food Biotechnol.*, 2007, **21**, 71–89.
21. Geerlings, *Future Microbiol.*, 2011, **6**, 1385–1386.
22. L. J. Nohynek, H. L. Alakomi, M. P. Kähkönen, M. Heinonen, I. M. Helander, K. M. Oksman-Caldentey and R. H. Puupponen-Pimiä, *Nutr. Cancer*, 2006, **54**, 18–32.
23. M. Heinonen, *Mol. Nutr. Food Res.*, 2007, **51**, 684–691.
24. T. Zhou, C. Wei, W. Lan, Y. Zhao, Y. Pan, X. Sun and V. C. H. Wu, *J. Food Sci.*, 2020, **85**, 1513–1522.
25. X. H. Sun, T. T. Zhou, C. H. Wei, W. Q. Lan, Y. Zhao, Y. J. Pan and V. C. Wu, *Food Control*, 2018, **94**, 155–161.
26. A. L. Molan, M. A. Lila, J. Mawson and S. De, *World J. Microbiol. Biotechnol.*, 2009, **25**, 1243–1249.
27. S. W. Josewin, M. J. Kim and H. G. Yuk, *Food Microbiol.*, 2018, **76**, 219–225.
28. G. Ma and Y. Chen, *J. Funct. Foods*, 2020, **66**, 103829.
29. N. B. Danneskiold-Samsøe, H. Dias de Freitas Queiroz Barros, R. Santos, J. L. Bicas, C. B. B. Cazarin, L. Madsen, K. Kristiansen, G. M. Pastore, S. Brix and M. R. Maróstica Júnior, *Food Res. Int.*, 2019, **115**, 23–31.
30. P. J. Turnbaugh, V. K. Ridaura, J. J. Faith, F. E. Rey, R. Knight and J. I. Gordon, *Sci. Transl. Med.*, 2009.
31. S. P. Claus, S. L. Ellero, B. Berger, L. Krause, A. Bruttin, J. Molina, A. Paris, E. J. Want, I. de Waziers, O. Cloarec, S. E. Richards, Y. Wang, M. E. Dumas,

A. Ross, S. Rezzi, S. Kochhar, P. van Bladeren, J. C. Lindon, E. Holmes and J. K. Nicholson, *mBio*, 2011, 2, e00271-10.
32. K. P. Scott, J. C. Martin, C. Chassard, M. Clerget, J. Potrykus, G. Campbell, C. D. Mayer, P. Young, G. Rucklidge, A. G. Ramsay and H. J. Flint, *Proc. Natl. Acad. Sci. U. S. A.*, 2011, **108**, 4672–4679.
33. M. Shoaib, A. Shehzad, M. Omar, A. Rakha, H. Raza, H. R. Sharif, A. Shakeel, A. Ansari and S. Niazi, *Carbohydr. Polym.*, 2016, **3**, 55–57.
34. K. Kahle, M. Kraus, W. Scheppach, M. Ackermann, F. Ridder and E. Richling, *Mol. Nutr. Food Res.*, 2006, **50**, 418–423.
35. G. Kelly, *Altern. Med. Rev.*, 2008, **13**, 315–329.
36. J. Juśkiewicz, J. Milala, A. Jurgoński, B. Król and Z. Zduńczyk, *Nutrition*, 2011, **27**, 351–357.
37. S. Salaheen, C. Nguyen, D. Hewes and D. Biswas, *Food Control*, 2014, 174–181. Complete.
38. A. T. Hotchkiss, H. K. Chau, G. D. Strahan, A. Nuñez, S. Simon, A. K. White, S. Dieng, E. R. Heuberger, M. P. Yadav and J. Hirsch, *Food Hydrocolloids*, 2021, **116**, 106572.
39. S. Wang, Y. Xiao, F. Tian, J. Zhao, H. Zhang, Q. Zhai and W. Chen, *J. Funct. Foods*, 2020, **66**, 103838.
40. R. Jovanovic-Malinovska, S. Kuzmanova and E. Winkelhausen, *Int. J. Food Prop.*, 2014, **17**, 949–965.
41. T. K. Bloedon, R. E. Braithwaite, I. A. Carson, D. Klimis-Zacas and R. A. Lehnhard, *Nutr. Rev.*, 2019, **77**, 630–645.
42. P. Tsakiroglou, J. Weber, S. Ashworth, C. Del Bo' and D. Klimis-Zacas, *J. Med. Food*, 2021, **24**, 226–235.
43. A. Basu, M. Du, M. J. Leyva, K. Sanchez, N. M. Betts, M. Wu, C. E. Aston and T. J. Lyons, *J. Nutr.*, 2010, **140**, 1582–1587.
44. S. Ahmadi, S. Wang, R. Nagpal, R. Mainali, S. Soleimanian-Zad, D. Kitzman and H. Yadav, *J. Vis. Exp.*, 2019, 59524.
45. J. C. Litten-Brown, A. M. Corson and L. Clarke, *Animal*, 2010, **4**, 899–920.
46. T. L. A. Nguyen, S. Vieira-Silva, A. Liston and J. Raes, *Dis. Models Mech.*, 2015.
47. A. Ray and B. N. Dittel, *Immunology*, 2015, **146**, 359–368.
48. N. A. Nagalingam, J. Y. Kao and V. B. Young, *Inflammatory Bowel Dis.*, 2011, **4**, 917–926.
49. M. Hidalgo, M. J. Oruna-Concha, S. Kolida, G. E. Walton, S. Kallithraka, J. P. E. Spencer, G. R. Gibson and S. De Pascual-Teresa, *J. Agric. Food Chem.*, 2012, **60**, 3882–3890.
50. A. Vrieze, P. F. De Groot, R. S. Kootte, M. Knaapen, E. Van Nood and M. Nieuwdorp, *Best Pract. Res. Clin. Gastroenterol.*, 2013, **27**, 127–137.
51. J. Z. H. von Martels, M. Sadaghian Sadabad, A. R. Bourgonje, T. Blokzijl, G. Dijkstra, K. N. Faber, H. J. M. Harmsen, W. Paul, C. Marta, V. de, W. Tom, A. N. Payne, A. Zihler, C. Chassard, C. Lacroix, L. Cassani, A. Gomez-Zavaglia, J. Simal-Gandara, M. Patel, H. J. Lee, S. Park, Y. Kim, B. Jeong, N. Ashammakhi, R. Nasiri, N. R. de Barros, P. Tebon, J. Thakor, M. Goudie, A. Shamloo, M. G. Martin, A. Khademhosseni, A. Vrieze,

P. F. De Groot, R. S. Kootte, M. Knaapen, E. Van Nood, M. Nieuwdorp, S. C. Pearce, H. G. Coia, J. P. Karl, I. G. Pantoja-Feliciano, N. C. Zachos, K. Racicot, G. Roeselers, M. Ponomarenko, S. Lukovac, H. M. Wortelboer, S. Wang, Y. Xiao, F. Tian, J. Zhao, H. Zhang, Q. Zhai, W. Chen, E. Saygili, A. A. Dogan-Gurbuz, O. Yesil-Celiktas and M. S. Draz, *Best Pract. Res. Clin. Gastroenterol.*, 2020, **27**, 127–137.

52. S. C. Pearce, H. G. Coia, J. P. Karl, I. G. Pantoja-Feliciano, N. C. Zachos and K. Racicot, *Front. Physiol.*, 2018.
53. A. N. Payne, A. Zihler, C. Chassard and C. Lacroix, *Trends Biotechnol.*, 2012, **30**, 17–25.
54. N. Pap, M. Fidelis, L. Azevedo, M. A. V. do Carmo, D. Wang, A. Mocan, E. P. R. Pereira, D. Xavier-Santos, A. S. Sant'Ana, B. Yang and D. Granato, *Curr. Opin. Food Sci.*, 2021, **42**, 167–186.
55. J. Colom, M. Cano-Sarabia, J. Otero, J. Aríñez-Soriano, P. Cortés, D. Maspoch and M. Llagostera, *Sci. Rep.*, 2017, **7**, 41441.
56. S. O'Connor, S. Chouinard-Castonguay, C. Gagnon and I. Rudkowska, *Maturitas*, 2017, **104**, 11–18.
57. C. Bränning, Å. Håkansson, S. Ahrné, B. Jeppsson, G. Molin and M. Nyman, *Br. J. Nutr.*, 2009, **101**, 859–870.
58. M. Fumagalli, E. Sangiovanni, U. Vrhovsek, S. Piazza, E. Colombo, M. Gasperotti, F. Mattivi, E. De Fabiani and M. Dell'Agli, *Pharmacol. Res.*, 2016, **111**, 703–712.
59. J. Lee, S. Kim, H. Namgung, Y.-H. Jo, C. Bao, H.-K. Choi, J.-H. Auh and H. J. Lee, *J. Agric. Food Chem.*, 2013, **62**, 3954–3962.
60. M. C. Denis, Y. Desjardins, A. Furtos, V. Marcil, S. Dudonné, A. Montoudis, C. Garofalo, E. Delvin, A. Marette and E. Levy, *Clin. Sci.*, 2015, **128**, 197–212.
61. A. L. Molan, Z. Liu and M. Kruger, *World J. Microbiol. Biotechnol.*, 2010, **26**, 1735–1743.
62. I. A. Eydelnant and N. Tufenkji, *Langmuir*, 2008, **24**, 10273–10281.
63. P. J. Turnbaugh, R. E. Ley, M. A. Mahowald, V. Magrini, E. R. Mardis and J. I. Gordon, *Nature*, 2006, **444**, 1027–1031.
64. B. Chew, B. Mathison, L. Kimble, D. McKay, K. Kaspar, C. Khoo, C.-Y. O. Chen and J. Blumberg, *Eur. J. Nutr. 2018 583*, 2018, **58**, 1223–1235.
65. B. Bisha, J. A. Adkins, J. C. Jokerst, J. C. Chandler, A. Pérez-Méndez, S. M. Coleman, A. O. Sbodio, T. V. Suslow, M. D. Danyluk, C. S. Henry and L. D. Goodridge, *J. Vis. Exp.*, 2014, **88**, 51414.

CHAPTER 5

The Effects of Berry Bioactive Compounds on Vascular Function

ANA RODRIGUEZ-MATEOS,[a] MELANIE LE SAYEC,[a] GEOFFREY ISTAS[a] AND SARAH A. JOHNSON*[b]

[a] Department of Nutritional Sciences, School of Life Course Sciences, Faculty of Life Sciences and Medicine, King's College London, London, SE1 9NH, UK; [b] Department of Food Science and Human Nutrition, Colorado State University, 502 West Lake Street, Gifford Room 206, Fort Collins, CO 80524-1571, USA
*Email: sarah.johnson@colostate.edu

5.1 Introduction

The health benefits of fruits, vegetables, and plant-based diets are increasingly being recognized, particularly for the prevention of cardiometabolic diseases.[1] A growing number of human studies and *meta*-analyses are demonstrating health benefits associated specifically with the consumption of berries, which are believed to be beneficial largely due to their high flavonoid content, in particular anthocyanins.[2–4] Epidemiological studies generally indicate that higher consumption of anthocyanins and other berry (poly)phenols is associated with a reduced risk of cardiovascular disease (CVD). A number of well-conducted randomized controlled trials have shown improvements in biomarkers of cardiometabolic disease risk related to vascular function such as blood pressure, endothelial function, and arterial

Food Chemistry, Function and Analysis No. 33
Berries and Berry Bioactive Compounds in Promoting Health
Edited by Dorothy Klimis-Zacas and Ana Rodriguez-Mateos

Published by the Royal Society of Chemistry, www.rsc.org

stiffness after short-term consumption of berries in different study populations. However, mixed results exist, and the mechanisms of action have not been completely elucidated. The aim of this chapter is to give an overview of the effects of berry (poly)phenol consumption on human vascular function. The potential mechanisms driving these effects will be briefly mentioned, as they will be discussed in Chapter 6.

5.2 Vascular Function and Dysfunction in Cardiovascular Disease Development

5.2.1 Cardiovascular Disease Prevalence and Global Impact

According to the Global Burden of Disease study, in 2015, an estimated 31% of all deaths were caused by CVD, making it the leading cause of death worldwide.[5] The study identified about 422 million prevalent cases (~1 in 17 of the global population) for CVD globally and found that when adjusted for age, more cases occur in countries with low sociodemographic index than in high-income countries. The study concluded that, globally, CVD is more prevalent than maternal-child and infectious diseases, even in low-income countries. The World Health Organization proposes population-wide strategies to reduce the burden of CVD by focusing from high to low priority on tobacco control policies, healthy diet, increased physical activity, reduced alcohol consumption, and eduction of children in schools.[6]

5.2.2 Overview of Cardiovascular Disease

CVD is a group of disorders that affect the blood vessels and the heart, many of which are related to atherosclerosis. Aging is the primary risk factor for CVD, largely due to the development of vascular dysfunction, which includes vascular endothelial dysfunction and arterial stiffness.[7,8] Vascular dysfunction is implicated in the development of atherosclerosis and hypertension, which may not only result in CVD but also in adverse events such as myocardial infarction, stroke, and death. The underlying mechanisms contributing to the development of vascular dysfunction are complex, but known central mechanisms include oxidative stress and chronic inflammation, while the gut microbiome is emerging as an important mediator as well. Indeed, oxidative stress and chronic inflammation within the vasculature, and gut microbial dysbiosis, are associated with aging.[7,9] While vascular dysfunction occurs as a by-product of vascular aging, certain diseases and conditions such as obesity and metabolic syndrome, among other factors, can accelerate the development of vascular dysfunction and thus CVD. Certain lifestyle behaviors, such as healthy dietary patterns and consumption of foods rich in plant bioactive compounds, can attenuate the development of, and may reverse, vascular dysfunction, therefore, reducing the risk for CVD.[7,10]

5.2.3 Vascular Endothelial Function and Dysfunction

The endothelial monolayer is the first tissue of contact with blood and responds to a plethora of physical and chemical stimuli to maintain a healthy vascular homeostasis. One of the crucial properties of the endothelium is the ability to release vasoactive molecules to modulate the vascular tone and diameter, which consequently regulate oxygen supply and metabolic demand.[11] Numerous endothelium-derived molecules influence endothelial homeostasis and function, including nitric oxide (NO), endothelin-1, angiotensin II, prostaglandins, and endothelium-derived hyperpolarizing factors. The identification of NO as a vasorelaxant factor released by the endothelium marked a turning point in deciphering the mechanisms behind endothelial homeostasis.[12] NO is arguably the most critical endothelium-derived molecule for cardiovascular health as it promotes vasodilation and blood flow and inhibits coagulation, adhesion, proliferation, and inflammation. NO is produced from L-arginine by the action of endothelial NO synthase (eNOS) and diffuses from the endothelial cells to surrounding vascular smooth muscle cells (VSMC) to induce relaxation[13] (see Figure 5.1).

In healthy arteries, the constant laminar shear stress keeps levels of NO high enough to inhibit coagulation, adhesion, proliferation, and inflammation.[14,15] With increasing age and/or the presence of CVD risk factors

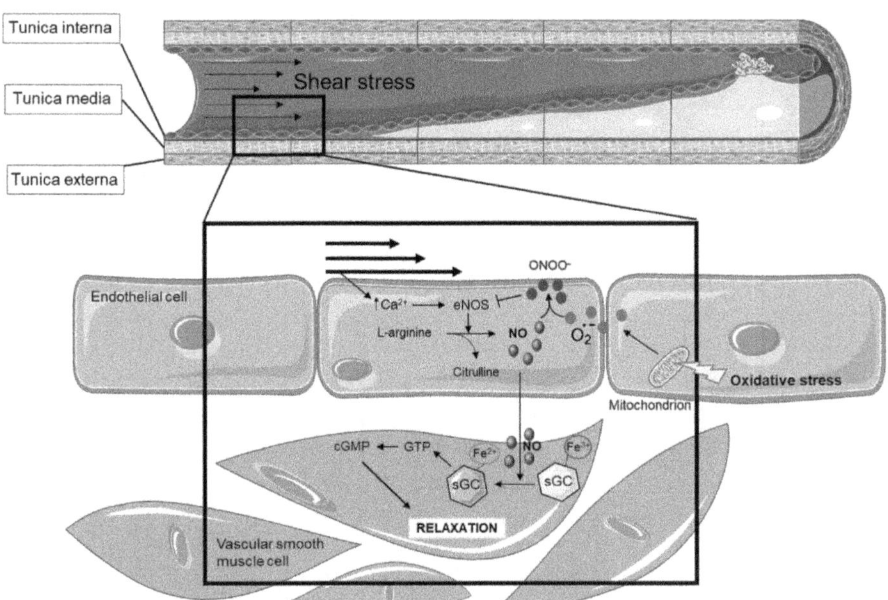

Figure 5.1 Representation of the accumulation of fatty deposits during artery thrombus formation and the NO/cGMP pathway. Image produced by Servier Medical Art, www.servier.com, under the terms of the CC BY 3.0 license, https://creativecommons.org/licenses/by/3.0/.

such as obesity, high blood pressure, and dyslipidemia, there is an increase in vasoconstriction, coagulation, thrombosis, and proliferation relative to vasodilation.[16] This imbalance is referred to as endothelial dysfunction and is characterized by impaired NO-mediated endothelium-dependent dilation (EDD).[16–18] A reduction in NO bioavailability is the major driving factor causing endothelial dysfunction and is largely caused by excessive superoxide-driven oxidative stress within the vasculature. Superoxide scavenges NO, thereby reducing its bioavailability and forming peroxynitrite, a reactive nitrogen species that damages cellular components and inactivates the eNOS cofactor tetrahydrobiopterin (BH_4), causing eNOS uncoupling. This results in superoxide production instead of NO by eNOS.[19] Chronic inflammation promotes oxidative stress through a bidirectional relationship.[19] Endothelial dysfunction is central to the initiation and progression of atherosclerosis, which contributes to the development of CVD.[20]

5.2.4 Arterial Stiffness

Healthy large elastic arteries, such as the aorta and carotid artery, expand and recoil to buffer the energy of the pulse wave and move blood distally to peripheral vascular beds and bodily tissues.[10] When the heart contracts, a forward propagated wave is generated that interacts with the impedance of the aorta and other regions of the vasculature, resulting in a low-velocity pressure wave. With aging and in the presence of CVD risk factors, changes occurring within the arterial wall include increased collagen deposition, degradation and fragmentation of elastin, and the formation of advanced glycation end-products. These structural changes lead to decreased distensibility and increased stiffness of the arterial wall leading to higher impedance in the aorta and other vascular tissues that increase pressure wave velocity.[21–23] As with endothelial dysfunction, excessive vascular oxidative stress and inflammation are the major culprits.[24,25] Arterial stiffness occurs with aging, and its development and progression can be accelerated by CVD risk factors.[22] Arterial stiffness requires the heart to work harder, which increases systolic blood pressure, often leading to isolated systolic hypertension in middle-aged/older adults[26] and left ventricular hypertrophy,[22,27] and to increases pulsatile blood flow that causes damage to cerebral and renal blood vessels, thereby increasing the risk of stroke and impairments in cognitive and renal function.[28] In addition, arterial stiffness is independently associated with the risk for CVD, including atherosclerotic CVD, CVD events in various populations, and all-cause mortality.[29–36]

5.3 Berry (Poly)phenols and Vascular Function

5.3.1 Berry (Poly)phenols

(Poly)phenols are the largest group of plant-derived chemicals that comprise one or more phenolic rings and have a typical mass range between 100 and

3000 Da. More than 8000 structures are known, of which more than 4000 have been identified as flavonoids.[37] Plants synthetize (poly)phenols as secondary metabolites to fight off microbial infections and herbivores, protect against UV-mediated oxidative stress, attract pollinators, and strengthen the cell wall.[38,39] These phenolic compounds can be divided into two major groups: flavonoids (two aromatic rings bound by a three-carbon bridge) and nonflavonoids (one or more phenolic rings arranged differently). The flavonoids can be further divided into seven subgroups: flavonols, flavones, isoflavones, flavanones, anthocyanidins, flavan-3-ols, and dihydrochalcones.[40] They occur naturally in plants as glycosides and to a lesser extent as aglycones. With respect to berries, examples are provided here but the reader is referred to Chapter 2 for more detail. Flavanols or flavan-3-ols are found in berries as polymers (proanthocyanidins) and monomers such as (+)-catechin, (−)-epicatechin, (−)-epigallocatechin, or (−)-epigallocatechin-3-*O*-gallate.[41] The most abundant flavonols in berries are quercetin, isorhamnetin, kaempferol, and myricetin.[42] Most colored berries are rich in anthocyanidins, of which the major ones are identified as cyanidin, malvidin, peonidin, pelargonidin, delphinidin, and petunidin.[43] These compounds are responsible for the red, blue, and purple pigmentations of berries.

Unlike flavonoids, nonflavonoids do not necessarily contain two aromatic structures linked with a three-carbon bridge but can have one (or more) phenolic rings arranged differently. Nonflavonoids include phenolic acid derivatives of ellagitannins, benzoic acids, cinnamic acids, lignans, and stilbenes.[40] Ellagic acid polymerized into condensed ellagitannins is found in raspberries.[44] Hydroxycinnamic acids and hydroxybenzoic acids are phenolic acids commonly found in berries.[45] In particular, chlorogenic acid, a cinnamic acid derivative, is very abundant in berries is one of the major phenolic acids consumed by humans. Stilbenes contain two aromatic rings linked by a double bond and are found in berries as monomers, dimers, and oligomers. Resveratrol, pterostilbene, and piceatannol are common stilbenes found in berries and predominately accumulate in berry skins.[46] Blueberries, in particular, are known for their high stilbene concentrations, especially pterostilbene.[47–49]

5.3.1.1 *Overview of Berry (Poly)Phenol Bioavailability and Metabolism*

A short overview of berry polyphenol bioavailability and metabolism is provided here; however, the reader is referred to Chapter 2 for a detailed overview of berry (poly)phenol bioavailability. Most berry (poly)phenols are attached to sugars and are ingested as glycosides. After passing through the stomach, these glycosides undergo hydrolysis reactions where sugars are removed by brush border enzymes in the small intestine (*i.e.*, glucosidases). Lactase phlorizin hydrolase is a key enzyme with broad substrate specificity for flavonoid-3-β-D-glucosides.[50] Upon hydrolysis, aglycones passively diffuse in the epithelial cells due to increased lipophilicity and proximity to the intestinal membrane.[50] Flavonoid glycosides might also be transported into epithelial cells with the aid of an active sodium-dependent glucose transporter, sodium/glucose cotransporter 1.[51] Once inside the cytosol, flavonoid glycosides are cleaved by cytosolic β-glucosidase into aglycones.[52] Most (poly)phenol glycosides are cleaved in the intestinal lumen or in the cytosol of intestinal epithelial cells, with a few exceptions such as anthocyanins and flavonols, which have been previously found in plasma and urine as glycosides.[53,54] The majority of (poly)phenols are not absorbed in the small intestine but rather travel down the intestine where they encounter the gut microbiota. The gut microbiota metabolizes (poly)phenols into smaller phenolic acids that can be absorbed in the colon.[55] Thus, the gut microbial composition is an important factor influencing the bioavailability of many berry (poly)phenols.

5.3.2 Effects of Berry (Poly)phenols on Vascular Function

5.3.2.1 Berry Poly(Phenols) and CVD Risk: Evidence from Epidemiological Studies

Epidemiological studies are useful to assess the long-term effects of (poly)-phenols on CVD risk at a population level. Recent evidence indicates that intake of dietary (poly)phenols, particularly those from blueberries and strawberries, may be beneficial for cardiovascular health.[56–60] In a prospective study with ∼35 000 postmenopausal women followed over 16 years, a higher intake of anthocyanins in the diet was associated with a 10% reduction in CVD mortality.[59] Similarly, in an older population of 100 000 men and women, total flavonoid intake was associated with a 18% reduced risk of mortality from CVD. In addition, anthocyanin and proanthocyanidin intake was associated with ∼15% lower risk.[60] A recent *meta*-analysis of 15 prospective studies on flavonoid intake and CVD mortality published in 2017 showed a relative CVD risk reduction of 14% for people with the highest flavonoid intake.[61] A total of 386 610 individuals were included and subgroup analysis revealed that flavan-3-ols, flavones, flavanones, anthocyanidins, and proanthocyanidins were inversely associated with CVD mortality by 11–14%.[61] In an epidemiological study carried out in a younger population, a higher intake of anthocyanins was associated with a 32% decreased risk for myocardial infarction in 100 000 women after an 18 year follow-up.[56] Despite the promising findings described here, several epidemiological studies failed to report significant associations of flavonoid intake on CVD risk.[62–65] These discrepancies may be due to the nature of the methodology used to assess dietary (poly)phenol intake. For example, the lack of food composition databases makes it difficult to assess food intake accurately due to the high variability of (poly)phenol content in plant-based foods. For this reason, two large databases of (poly)phenol content in foods have been developed. The first one is Phenol-Explorer,[66] the second one is the USDA database for the flavonoid content of foods.[67] These databases accumulate published data on (poly)phenol content in foods based on HPLC and HPLC-MS methods. However, the variations in (poly)phenol content due to processing, different varieties, seasonal changes, and storage conditions are reflected as limitations in the existing databases. Another major limitation of food intake assessment in prospective studies is the food frequency questionnaires (FFQ), as they do not accurately measure food intake. In the study by Cassidy *et al.*, for example, it is worth considering that the anthocyanin intake was very low, with the highest quintile being 25 mg and the lowest quintile being 2.5 mg per day.[56] This partly reflects the inaccuracy of FFQs used, where only blueberries and strawberries were estimated, but not other berries. In addition, FFQs have an important self-reporting bias which induces a high degree of measurement errors.[68] To avoid these discrepancies in the future, prospective studies using biomarkers of flavonoid intake in plasma, urine, or stool samples are required to confirm the current findings. Although the available evidence indicates that a diet rich in (poly)phenols might reduce the risk for CVD development, it most certainly does not provide a causal relationship between (poly)phenol intake and reduced CVD risk.

5.3.2.2 Randomized Controlled Trials

5.3.2.2.1 Aronia Berries. Aronia berry (poly)phenols primarily include anthocyanins (*e.g.*, cyanidin glycosides), proanthocyanidins, flavonols, and phenolic acids (*e.g.*, hydroxycinnamic acids).[69,70] Few clinical studies have been performed to evaluate the effects of aronia berries or aronia berry (poly)phenols on endothelial function, blood pressure, and arterial stiffness (see Table 5.1).

5.3.2.2.1.1 Effect on Endothelial Function. Two studies demonstrated significant improvements in flow-mediated dilation (FMD) following chronic aronia berry consumption (6 and 12 weeks in duration), one of which in middle-aged hypercholesterolemic males,[71] and the other in young healthy males.[72] Istas *et al.*[72] evaluated the parameters of vascular function, including FMD, arterial stiffness, blood pressure, and heart rate before and after consuming a (poly)phenol-rich aronia extract (containing 116 mg (poly)phenols; equivalent to 75 g aronia berry but with fibers and organic acids removed), whole fruit powder (containing 12 mg (poly)phenols; equivalent to 10 g aronia berries), or placebo (containing 0 mg (poly)-phenols) daily for 12 weeks in 66 healthy young men. They also evaluated FMD 2 hours after consumption at zero weeks. Improvements in FMD following chronic consumption of a (poly)phenol-rich aronia extract and the whole fruit powder were found to be correlated with increases in plasma phenolic metabolites and specific gut microbiota genera. Significant improvements in FMD were noted 2 hours following consumption of the

Table 5.1 Characteristics of human studies evaluating the effects of aronia berry (poly)phenols on vascular function.[a]

Ref.	Study design	Population	Vascular function and related outcome measures	Treatment	Results
Chronic studies					
Poreba *et al.* (2009)[71]	One-arm uncontrolled quasi-designed trial.	35 men with mild hypercholesterolemia (mean age: 53.9 ± 5.8 years, mean BMI: 27.4 ± 2.9 $kg\,m^{-2}$).	Brachial artery FMD at 0 and 6 weeks during each 6 week treatment period.	250 mL aronia berry juice per day for 6 weeks.	FMD ↑ after each 6 week intervention compared to week 0, and the final 6 week measurement ↑ compared to the initial 6 week measurement. NO metabolites ↑ 6 weeks after repeated intervention compared to week 0 and the first 6 week time point.
Naruszewicz *et al.* (2007)[78]	Randomized, double-blind, placebo-controlled, parallel trial.	44 Patients (11 women and 33 men, mean age 66 years) who survived myocardial infraction and have received statin therapy for at least 6 months.	Ox-LDL, hsCRP, SBP, and DBP, measured at baseline and after 6 weeks of supplementation.	85 mg 3× per day of aronia berry extract or matched placebo.	Ox-LDL levels ↓ by 29%, hsCRP ↓ by 23%, SBP and DBP ↓ by 11 and 7.2 mmHg, respectively, after the 6 week intervention.
Skoczynska *et al.* (2007)[80]	One-arm uncontrolled quasi-designed trial.	58 Healthy men diagnosed with mild hypercholesterolemia without pharmacological treatment.	HDL, LDL, hsCRP, SBP, and DBP, measured at baseline, after 6 weeks of regular aronia berry juice drinking, after 6 weeks without the juice drinking, then after 6 weeks of aronia berry juice drinking.	250 mL 100% aronia berry juice per day.	LDL cholesterol ↓ (−27.8 $mg\,dL^{-1}$), and SBP and DBP ↓ (−13.5 and −7 mmHg, respectively) following the intervention.

Table 5.1 (*Continued*)

Ref.	Study design	Population	Vascular function and related outcome measures	Treatment	Results
Broncel *et al.* (2010)[79]	One-arm uncontrolled quasi-designed trial	47 Middle-aged adults: 22 healthy volunteers and 25 patients with metabolic syndrome.	BP, hsCRP, CAT, SOD, GPX, and lipids, measured at baseline and after 2 months.	Aronia berry extract (100 mg 3× per day) for 2 months.	SBP ↓ (−11.6 mmHg), DBP ↓ (5 mmHg), total cholesterol (−14.9 mg dL^{-1}), LDL-C ↓ (−2 mg dL^{-1}), TG ↓ (−28.3 mg dL^{-1}), CAT ↓ (−48 U per mg-Hb). SOD ↑ (+686 U per g-Hb), GPX ↑ (+6.6 U per g-Hb).
Kardum *et al.* (2014)[81]	One-arm uncontrolled quasi-designed trial.	20 Middle-aged women with central obesity.	BP and antioxidant enzymes status, at baseline and after 4 weeks.	100 mL glucomannan-enriched, aronia berry juice-based supplement daily for 4 weeks.	GPX ↑ (+4.65 U per g-Hb). SBP ↓ (−11.2 mmHg). No change in DBP.
Sikora *et al.* (2014)[82]	Three-arm, uncontrolled quasi-designed trial	70 Subjects divided into three groups: I—patients with MS who received aronia berry extract supplements, II—healthy controls, and III—patients with MS treated with angiotensin I-converting enzyme inhibitors.	BP, hsCRP, and blood lipids, measured at baseline, after 1 and 2 months.	Subjects with MS from group I were treated with capsules containing 100 mg of aronia berry extract 3× daily during the 2 month study period.	SBP and DBP ↓ (−10.5 and −6.3 mmHg, respectively), and ↓ in LDL-C (−12.5 mg dL^{-1}) after 2 months of daily intake.

Kardum *et al.* (2015)[84]	One-arm uncontrolled quasi-designed trial.	23 Middle-aged adults with elevated BP (130–159/85–99 mmHg).	24 hour SBP and DBP, measured at baseline and after 4 weeks.	200 mL of aronia berry juice daily for 4 weeks.	↓ in 24 h SBP of 8.5 mmHg and a ↓ in awake SBP and DBP of 10 and 5.9 mmHg, respectively.
Loo *et al.* (2016)[83]	Randomized single-blinded crossover trial.	38 Middle-aged adults with elevated BP (130–159/85–99 mmHg).	24 hour BP, blood lipids, measured at baseline and after 8 weeks.	Cold-pressed 100% aronia berry juice (300 mL per day) and oven-dried aronia berry powder (3 g per day), or matched placebo.	↓ in awake DBP (−1.64 mmHg).
Xie *et al.* (2017)[73]	Randomized, placebo-controlled trial.	49 Healthy adult former smokers.	BP, blood lipids, biomarkers of inflammation and oxidative stress, measured at baseline and after 12 weeks.	500 mg aronia berry extract per day.	No significant changes following the intervention in any of the parameters tested.
Milutinović *et al.* (2019)[76]	Open-label trial.	35 Middle-aged subjects with type 2 diabetes mellitus and taking antidiabetic drugs therapy for at least 6 months.	BP and blood lipids, measured at baseline, after 3 months of supplementation and after a further 3 month follow-up period.	Aronia berry juice, 150 mL per day, 3× per day for 3 months.	No significant changes following the intervention in any of the parameters tested.
Pokimica *et al.* (2019)[75]	Randomized double-blinded three-parallel groups trial.	84 Adults "at cardiovascular risk."	Total cholesterol levels and SBP.	100 mL per day for 4 weeks of one of the following three drinks: aronia berry juice ($n=27$); matched placebo drink ($n=29$), and mix of the 2 drinks ($n=28$).	No significant changes following the intervention in any of the parameters tested.

Table 5.1 (*Continued*)

Ref.	Study design	Population	Vascular function and related outcome measures	Treatment	Results
Ahles *et al.* (2020)[77]	Randomized, double-blind, placebo-controlled, parallel trial.	101 Healthy, middle-aged, overweight adults (mean ± SEM age: 53 ± 1 years, BMI between 25 and 35 $kg\,m^{-2}$.	Brachial BP, arterial elasticity, ankle-brachial index, and carotid intima-media thickness measured at 0, 6, 12, 18, and 24 weeks.	90 mg aronia berry extract (16 mg anthocyanins), 150 mg aronia berry extract (27 mg anthocyanins), or placebo daily for 24 weeks.	150 mg aronia berry extract ↓ brachial diastolic BP compared to 90 mg aronia berry extract with no effect on the other parameters.
Acute and chronic studies					
Istas *et al.* (2019)[72]	Randomized, double-blind, placebo-controlled, parallel trial.	66 Healthy men (mean age: 24 ± 5.3 years and mean BMI: 23 ± 2.1 $kg\,m^{-2}$).	BP, PWV, AIx, and FMD at 0 weeks/0 hours, 0 weeks/2 hours, and 12 weeks/0 hours.	(Poly)phenol-rich aronia berry extract [12 mg (poly)phenols]; whole fruit powder [12 mg (poly)phenols], or placebo [0 mg (poly)phenols] per day for 12 weeks.	Aronia berry extract and whole fruit powder ↑ FMD at 12 weeks compared to control, and aronia berry extract ↑ FMD 2 hours after consumption at week 0. Plasma phenolic metabolites ↑ and correlated with FMD. Aronia extract ↑ growth of *Anaerostipes* and aronia whole fruit ↑ *Bacteroides*, and specific gut microbial genera correlated with FMD.

[a]**Abbreviations:** AIx, augmentation index; BMI, body mass index; BP, blood pressure; CAT, catalase; CRP, C-reactive protein; hsCRP, high sensitivity C-reactive protein; FMD, flow-mediated dilation; GPX, glutathione peroxidase; HDL, high-density lipoprotein; LDL, low-density lipoprotein; PWA, pulse wave analysis; PWV, pulse wave velocity; SOD, superoxide dismutase.

(poly)phenol-rich aronia extract, but not the whole fruit powder. These data suggest that aronia berry (poly)phenols can modulate endothelial function acutely, but that the gut microbiota and its impact on (poly)-phenol bioavailability is important for observing chronic effects of the whole fruit powder. Poręba *et al.*[71] assessed FMD and blood biomarkers before and after 6 weeks of 250 mL per day of aronia berry juice consumption, followed by a 6 week washout period, and then a repeated 6 week exposure to 250 mL per day of aronia berry juice consumption in middle-aged hypercholesterolemic men. They found that FMD increased (+4.5%) after the initial and the repeated 6 week intervention period compared to week zero and that the final 6 week measurement was higher than the initial 6 week measurement. NO metabolites (nitrate/nitrite) were increased 6 weeks after the repeated 6 week intervention compared to week zero and the first 6 week time point, suggesting that NO bioavailability was modulated. They also found reductions in total cholesterol, low-density lipoprotein cholesterol, and triglycerides, which may have contributed to the observed improvements in endothelial function. However, their study did not include a placebo/control group, so findings must be interpreted with caution.

5.3.2.2.1.2 Effect on Blood Pressure. To date, a total of twelve studies have investigated the effects of aronia berry (poly)phenol consumption on blood pressure, including eight randomized controlled trials and five uncontrolled quasi-designed studies. Among these studies, five trials did not find any significant differences in blood pressure following daily intake of aronia-rich products for four to 24 weeks,[73–77] while three observed a significant decrease in both office systolic blood pressure and diastolic blood pressure after the aronia berry product supplementation.[78–80] Precisely, Naruszewicz and colleagues[78] observed a decrease of 11 and 7.2 mmHg in office systolic blood pressure and diastolic blood pressure, respectively, in 44 middle-aged survivors of myocardial infarction following 6 weeks of aronia berry extract supplementation. A similar magnitude of effect was reported in the uncontrolled quasi-designed studies led by Broncel and Skoczyńska.[79] In fact, while the first study observed a decrease in systolic blood pressure and diastolic blood pressure of 11.6 and 5.0 mmHg, respectively, in a population of 58 men with CVD risk, the second study involving subjects with mild hypercholesterolemia showed a reduction in systolic blood pressure and diastolic blood pressure of 13.5 and 7 mmHg, respectively. Additionally, three studies have found an improvement in office systolic blood pressure but not in diastolic blood pressure following the intervention.[81,82] This is the case for a trial by Sikora and colleagues,[81] which included middle-aged females with obesity and noted a significant reduction in office systolic blood pressure of 11.2 mmHg following a crossover intervention with a 4 week supplementation of aronia berry juice.

Most studies used office blood pressure, while only two trials have investigated the impact of aronia berries on 24 hour ambulatory blood pressure, showing a significant decrease in blood pressure during daytime.[83,84] Indeed, in prehypertensive individuals with elevated baseline blood pressure (140.8 ± 9.3 and 86.6 ± 7.0 mmHg for 24 hour systolic blood pressure and diastolic blood pressure, respectively), Kardum *et al.*[81] observed a reduction in 24 hour systolic blood pressure of 8.5 mmHg and a decrease in awake systolic blood pressure and diastolic blood pressure of 10 and 5.9 mmHg, respectively, following a 4 week daily intake of 200 mL of aronia berry juice containing 386 ± 9.7 mg of (poly)phenols per 100 g. The other trial[85] showed a significant decrease of 1.64 mmHg in awake diastolic blood pressure following daily intake of 300 mL of aronia berry juice and 3 g of aronia berry powder containing a total of 2194 mg of (poly)phenols.

A large variation can be noted in the study designs of these studies, as per the nature of the interventional product (juice or extract), the duration of the intervention (from four to 24 weeks) or the type of population included (young individuals *versus* middle-aged participants, healthy *versus* at risk of CVD). A recent and small *meta*-analysis including four of the studies described above concluded that aronia berry extract supplementation for 2 to 3 months led to a significant mean size effect of −0.331 for systolic blood pressure, while the analysis did not find any significant effect on diastolic blood pressure.[86] The authors highlighted that the effect on systolic blood pressure was even stronger when considering subjects aged over 50 years only. More studies will be needed to confirm the findings of this *meta*-analysis which, due to the small number of studies included, should be interpreted with caution.

Lastly, Ahles *et al.*[77] evaluated the parameters of vascular function (brachial blood pressure, but also arterial elasticity, ankle-brachial index, and carotid intima-media thickness) at zero, six, twelve, 18, and 24 weeks following consumption of 90 mg aronia berry extract (containing 16 mg anthocyanins), 150 mg aronia berry extract (containing 27 mg anthocyanins), or placebo (containing 0 mg anthocyanins) daily for 24 weeks in healthy, middle-aged adults. They found that 150 mg aronia berry extract decreased brachial diastolic blood pressure compared to 90 mg aronia berry extract, but not placebo, with no effect on the other parameters assessed.

5.3.2.2.1.3 Effect on Pulse Wave Velocity and Augmentation Index. While almost 30 trials investigating the biological effects of aronia berry can be reported in the literature, only one randomized controlled trial documented the impact of the berry on pulse wave velocity (PWV) and augmentation index (Aix) so far. Istas *et al.*[72] measured changes in arterial stiffness after 12 week daily aronia berry extract or whole fruit powder consumption. No significant differences in PWV or AIx were reported after acute or chronic consumption of the extract or the whole fruit powder when compared with placebo. This could be due to the young age of the population investigated (average age 24 ± 5.3 years old).

Overall, these data suggest the potential for aronia berries and their (poly)phenol-rich derivatives to improve vascular function, particularly endothelial function, in healthy populations and those at risk for CVD. However, further research is needed to confirm these findings and explore their potential in other populations.

5.3.2.2.2 Blackberries. Blackberries are rich in anthocyanins (*e.g.*, cyanidin and pelargonidin glycosides), proanthocyanins, and ellagitannins.[87,88] No clinical studies evaluating the impact of blackberry consumption on vascular function, or cardiovascular health in general, were identified. In fact, no work has been done in this area in animal models, and only two studies using (poly)phenols or (poly)phenol-rich extracts in cardiovascular cell models were identified.[89,90] This is a major gap in the knowledge that would benefit from further investigation, as both *in vitro* studies suggest a potential role for blackberry (poly)phenols, and potentially blackberry consumption, in improving vascular function.

5.3.2.2.3 Blackcurrant. Blackcurrants are predominately rich in flavonoid (poly)phenols, specifically anthocyanins (*e.g.*, delphinidin and cyanidin glycosides) and proanthocyanidins (*e.g.*, procyanidins and prodelphinidins).[91] Limited clinical research studies have been performed to evaluate the impact of blackcurrant (poly)phenols on vascular function, and those that have are heterogeneous (see Table 5.2). In terms of the acute effects of blackcurrant (poly)phenols, three studies were identified, only one of which demonstrated vascular modulatory effects. In that study, Tomisawa *et al.*[92] assessed FMD in young healthy male nonsmokers and smokers before and 1 and 2 hours following smoking-induced acute endothelial dysfunction with 50 mg blackcurrant anthocyanins or 50 mg blackcurrant anthocyanins with vitamin E, or no supplement. In the group of smokers, they found that FMD was decreased at 1 hour after smoking and returned to baseline at 2 hours after smoking and blackcurrant anthocyanin intake, but not following blackcurrant anthocyanin with vitamin E intake or no supplement. Additionally, endothelial dysfunction was not induced in the nonsmoker group, nor were there any acute effects of the treatments. Another study by Castro-Acosta *et al.*[93] evaluated the effects of low-sugar fruit drinks containing blackcurrant extract providing 150, 300, or 600 mg total anthocyanins, or no extract, on measures of arterial stiffness and wave reflection measured by digital volume pulse (DVP) and blood pressure when provided immediately before a high-fat meal in healthy men and postmenopausal women. No effects on vascular function parameters were observed, despite seeing improvements in postprandial glycemia, insulinemia, and incretin secretion following consumption of blackcurrant extract treatment containing 600 mg anthocyanins. In a

Table 5.2 Characteristics of human studies evaluating the effects of blackcurrant (poly)phenols on vascular function.[a]

Ref.	Study design	Population	Vascular function and related outcome measures	Treatment	Results
Acute studies					
Castro-Acosta *et al.* (2016)[93]	Randomized, controlled, double-blind, crossover trial with an acute meal.	Healthy men aged 20–60 years and postmenopausal women aged 45–60 years with a BMI 18–35 kg m^{-2}.	DVP, systolic BP, and diastolic BP measured at 0 and 120 minutes.	Low-sugar fruit drinks containing blackcurrant extract (150 mg, 300 mg, and 600 mg total anthocyanins) or no blackcurrant extract (control) provided immediately before a high-fat meal.	No effect on DVP or BP.
Jin *et al.* (2011)[94]	Randomized, crossover, double-blind, placebo-controlled trial with an acute treatment	Healthy men and women (mean ± SD age: 44.55 ± 13.34 years, mean BMI: 23.81 ± 2.46 kg m^{-2}).	Vascular reactivity to acetylcholine (endothelium dependent) and sodium nitroprusside (endothelium independent) using LDI with iontophoresis measured at 0 and 120 minutes.	250 mL of 20% blackcurrant juice drink or a control drink.	No effect on endothelial function.

Tomisawa *et al.* (2019)[92]	Randomized controlled, crossover trial with acute treatment and smoking-induced acute endothelial dysfunction	Young, healthy, male nonsmokers (mean age: 22 ± 2 years) and smokers (mean age: 21 ± 1 years).	FMD, measured at baseline, 1 and 2 hours post-consumption.	50 mg of blackcurrant anthocyanin, 50 mg of blackcurrant anthocyanin and vitamin E, or no treatment.	During the no treatment consumption test day, FMD was unchanged during the 2 hour period after smoking in the nonsmokers' group but was ↓ during the 2 hour period after smoking in the smoking group. Following intake of the blackcurrant (but not blackcurrant with vitamin E) treatment, FMD was ↓ 1 hour after smoking and returned to the baseline level 2 hours after smoking in the smoking group.
Chronic Studies					
Cook *et al.* (2017)[95]	Randomized, controlled, crossover trial.	Endurance-trained male cyclists (aged 26–50 years).	Systolic BP, diastolic BP, mean arterial BP, HR, ejection time, cardiac output, stroke volume, and total peripheral resistance.	300, 600, and 900 mg per day of New Zealand blackcurrant extract containing 105 mg of anthocyanins per each 300 mg extract or no blackcurrant extract for 7 days.	Dose-dependent effect observed for cardiac output, stroke volume, and total peripheral resistance. A trend for a dose-dependent effect was observed for mean arterial BP. Cardiac output ↑ and total peripheral resistance ↓ following intake of 600 and 900 mg.

Table 5.2 (*Continued*)

Ref.	Study design	Population	Vascular function and related outcome measures	Treatment	Results
Khan *et al.* (2014)[96]	Randomized, double-blind, placebo-controlled, parallel-group trial.	Healthy adults (mean age of 53 years).	FMD.	250 ml of placebo (flavored water) or low (6.4% juice) or high (20% juice) blackcurrant juice drink four times a day for 6 weeks.	FMD ↑ in the high blackcurrant juice drink group (5.873.1 at baseline to 6.973.1% at 6 weeks, $P = 0.022$) compared to the placebo group (6.072.2 at baseline to 5.172.4% at 6 weeks). No change in EID observed.
Willems *et al.* (2018)[97]	Randomized, double-blind, placebo-controlled, crossover trial.	Healthy Thai men (mean age: 22 ± 3 years; mean BMI: 21.8 ± 1.1 kg m^{-2}).	Systemic vascular resistance, HR, stroke volume, cardiac output.	600 mg New Zealand blackcurrant extract containing 105 mg anthocyanins per each 300 mg extract dose or placebo for 7 days.	No effect on systemic vascular resistance, heart rate, stroke volume, cardiac output at rest.

[a]**Abbreviations:** AIx, augmentation index; AIx normalized to heart rate of 75 bpm; BMI, body mass index; BP, blood pressure; DPV, digital pulse volume; EID, endothelium-independent dilation; FMD, flow-mediated dilation; HR, heart rate; LDI, laser Doppler imaging; PWA, pulse wave analysis; PWV, pulse wave velocity; RHI, reactive hyperemia index.

group of healthy men and women, Jin *et al.*[94] evaluated 250 mL of a drink containing 20% blackcurrant juice or a control rink on vascular reactivity to acetylcholine (EDD) and sodium nitroprusside (endothelium-independent dilation [EID]) using laser Doppler imaging (LDI) with iontophoresis. They did not observe an effect on endothelial function at 120 minutes post-treatment consumption, though they observed an increase in microbial metabolites of anthocyanins at 180 minutes. The authors attributed the lack of an effect on the endothelial function to the low dose treatment they provided; however, it is also possible that the timing of endothelial function assessment was too early to observe the effects of microbial metabolites from the large intestine.

With respect to the chronic effects of blackcurrant (poly)phenols on vascular function, three studies were identified, two of which demonstrated vascular modulatory effects. In a study by Cook *et al.*[95] endurance-trained male cyclists were given 300, 600, and 900 mg of New Zealand blackcurrant extract containing 105 mg anthocyanins per each 300 mg dose or no blackcurrant extract for 7 days and evaluated for the impact on brachial and central blood pressure parameters, ejection time, cardiac output, stroke volume, and total peripheral resistance. They observed dose-dependent effects on cardiac output, stroke volume, and total peripheral resistance, and a trend for a dose-dependent effect on mean arterial pressure, while total peripheral resistance was reduced following intake of the 600 and 900 mg blackcurrant extract doses. Khan *et al.*[96] evaluated the effects of consuming 250 mL of low (6.4% juice) or high (20% juice) blackcurrant juice drink or placebo four times per day for 6 days on FMD. They reported that FMD increased following high blackcurrant juice consumption compared to placebo, which was paralleled by increased plasma vitamin C concentrations and reductions in F_2-isoprostane concentrations, and thus, the authors suggested reductions in oxidative stress were likely responsible for improvements in endothelial function. Willems *et al.*[97] examined the impact of daily intake for 7 days of 600 mg New Zealand Blackcurrant extract containing a total of 210 mg anthocyanins or a placebo on systemic vascular resistance, heart rate, stroke volume, and cardiac output in healthy Thai men at rest and during moderate-intensity exercise. No effects were observed at rest, while improvements in cardiac output and total peripheral resistance, and a trend for an improvement in stroke volume, were observed during exercise, suggesting vasodilatory effects of the extract during moderate-intensity exercise.

Altogether, it is difficult to evaluate the impact of blackcurrant (poly)-phenol intake on vascular function due to the heterogeneity of the published studies in this area. Nonetheless, though several studies observed no effects of blackcurrant (poly)phenols on vascular function, this is outweighed by the number of studies that observed an effect. The reasons for the discrepancy in findings are likely to include the population studied, treatment dose and duration of treatment, as well as the outcomes assessed. More research is needed to better understand the vascular modulatory effects of blackcurrant (poly)phenols, including considerations to optimize efficacy.

5.3.2.2.4 Blueberries. Blueberries are rich in (poly)phenols, including anthocyanins (*i.e.*, glucosides, galactosides, and arabinosides of delphinidin, malvidin, petunidin, cyanidin, and peonidin), procyanidins, flavonols (*e.g.*, chlorogenic acid), and phenolic acids.[98] With respect to vascular function, blueberries have been studied more than any other berry type, and numerous studies have been performed to evaluate the acute and chronic impacts of blueberry (poly)phenols on vascular function (see Table 5.3). Seven studies designed to evaluate the acute effects of blueberries and blueberry (poly)phenols on vascular function were identified. Using smokers as a model developed to study the vasoactive properties of food,[99] two studies by Del Bo *et al.*[100,101] performed in male smokers and nonsmokers found that blueberries exerted beneficial effects on vascular function, namely endothelial function. In the first study,[101] 16 healthy male smokers consumed 300 g of blueberries or a control prior to smoking a cigarette, or only smoked a cigarette, and vascular function measures were performed 20 minutes following. They found that blueberry consumption counteracted the decrease in endothelial function as determined by reactive hyperemia index (RHI) and the increase in systolic blood pressure induced by smoking, while no other effects on vascular function were noted. In the second study,[100] 24 men with peripheral arterial dysfunction (12 smokers and 12 nonsmokers) were evaluated in two experimental designs. The first design evaluated the impact of 300 g blueberries or a control on vascular function in nonsmokers 2 hours following consumption, while the second study design evaluated 300 g blueberries plus smoking, a control plus smoking, or smoking alone on vascular function in smokers 2 hours following consumption. In the first study design, they found that blueberry consumption improved RHI in nonsmokers with peripheral arterial dysfunction but observed no impact on other vascular function parameters. In the second study design, acute cigarette smoking increased blood pressure and heart rate but did not affect other vascular function parameters, while consumption of both the blueberry and control treatments increased RHI compared with the smoking-only treatment. Conversely, a randomized, controlled crossover trial by Del Bo *et al.*[102] investigating the impact of 300 g blueberries or a control jelly in 10 healthy males found no acute effect of blueberries on vascular function. Four studies by Rodriguez-Mateos *et al.*[103–105] demonstrated the acute vascular-protective effects of blueberries, specifically on endothelial function, in healthy men. The studies investigated different doses of blueberries and their (poly)phenols, ranging from 150 mg to 1791 mg total (poly)phenols, at time points up to 6 hours following blueberry consumption and found that FMD increased at 1–2 and 6 hours. Additionally, two of the studies evaluating the dose-dependent effects of blueberry (poly)phenols found that maximal effects occurred with doses containing 766 mg blueberry (poly)phenols (equivalent to 240 g fresh blueberries/34 g freeze-dried blueberries) and plateaued with higher doses.[105] Consumption of as low as 11 g freeze-dried blueberries providing 150 mg total (poly)phenols exerted

Table 5.3 Characteristics of human studies evaluating the effects of blueberry (poly)phenols on vascular function.[a]

Ref.	Study design	Population	Vascular function and related outcome measures	Treatment	Results
Acute studies					
Del Bo *et al.* (2013)[102]	Randomized, controlled crossover trial.	10 healthy males (mean age: 20.8 ± 1.6, mean BMI: 22.5 ± 2.1 $kg\,m^{-2}$).	BP, RHI before and at 1 hour after consumption.	300 g blueberries or control jelly.	No impact on vascular function parameters.
Del Bo *et al.* (2014)[101]	Randomized, single-blind, controlled crossover trial.	16 healthy male smokers (~15 cigarettes per day, mean age: 24 ± 3, mean BMI: 23.0 ± 1.9 $kg\,m^{-2}$).	BP, HR (at $t = 100$ min and $t = 105$ min), RHI, digital AIx and digital AIx@75 (at $t = 120$ min).	300 g blueberry (~348 mg anthocyanins) + smoking, 300 mL water with sugar + smoking, or smoking alone.	The ↓ in RHI and ↑ systolic blood pressure induced by smoking counteracted by blueberry treatment. No impact of smoking or treatments observed for remaining parameters.
Del Bo *et al.* (2017)[100]	Randomized, single-blind, controlled crossover trial.	24 men with peripheral arterial dysfunction (12 smokers, 12 nonsmokers, mean age: 24 ± 1, RHI < 1.67, mean BMI: 22.5 ± 1.2 $kg\,m^{-2}$).	RHI, BP, HR, digital AIx and digital AIx@75.	300 g blueberries with or without smoking, or 300 mL water + sugar without smoking for nonsmokers and 300 mL water + sugar + smoking for smokers.	Blueberry treatment ↑ RHI in nonsmokers with peripheral arterial dysfunction with no effects on other parameters. Acute cigarette smoking ↑ BP and HR but did not affect vascular function parameters. In smokers, blueberry and control treatments ↑ RHI compared with the smoking only treatment.
Rodriguez-Mateos *et al.* (2013)[105]	Randomized, double-blind, crossover, controlled trial	10 healthy males (mean age: 27 ± 1, mean BMI: 22 ± 0.9 $kg\,m^{-2}$).	FMD, PWV, PWA, and DVP at 0 hours and 1 (FMD only) 2, 4, and 6 hours after consumption.	Blueberry drinks with 766, 1278, and 1791 mg total (poly)phenols (equivalent to 240, 400, and 560 g fresh blueberries, respectively) or a macro-/micronutrient-matched control drink matched to the 1278 mg blueberry dose.	FMD ↑ after consuming the blueberry drink containing 766 mg total (poly)phenols and peaked at 1–2 and 6 hours post-consumption, but changes in the 1278 and 1791 mg total (poly)phenol blueberry drinks were not significantly different from the 766 mg dose. No impact on remaining vascular function parameters.

Table 5.3 (*Continued*)

Ref.	Study design	Population	Vascular function and related outcome measures	Treatment	Results
Rodriguez-Mateos *et al.* (2013)[105]	Randomized, double-blind, crossover, controlled trial.	11 healthy males (mean age: 27 ± 1, mean BMI: 22 ± 0.9 kg m^{-2}).	FMD at 0 hours and 1 hour after consumption.	Blueberry drinks with 319, 639, 766, 1278, and 1791 mg total (poly)phenols (equivalent to 240, 400, and 560 g fresh blueberries, respectively) or a macro-/micronutrient-matched control drink.	FMD ↑ at 1 hour dose-dependently up to the 766 mg total (poly)phenols dose and plateaued.
Rodriguez-Mateos *et al.* (2014)[103]	Randomized, double-blind, crossover, controlled trial.	10 healthy males (mean age: 27 ± 1, mean BMI: 25 ± 0.8 kg m^{-2}).	FMD, brachial BP, and HR at 0 hours and 1, 2, 3, 4, 5, and 6 hours after consumption.	Baked product containing 34 g freeze-dried blueberry powder (196 mg anthocyanins) or 34 mg unprocessed freeze-dried blueberry powder (containing 339 mg anthocyanins).	↑ FMD at 1, 2 and 6 hours (maximal ↑ at 2 hours) following baked product consumption. Similar results in the blueberry powder group but maximal ↑ in FMD was at 1 hour. No effect on BP or HR.
Rodriguez-Mateos *et al.* (2019)[104]	Randomized, double-blind, crossover, controlled trial.	5 healthy males (mean age: 23 ± 3, mean BMI: 24 ± 3 kg m^{-2})	FMD at 0 hours and 1, 2, and 6 hours after consumption.	11 g wild blueberry powder drink (containing 150 mg anthocyanins), control drink, control drink plus fiber, control drink plus a mix of minerals and vitamins, 160 mg pure anthocyanins.	FMD ↑ 2 and 6 hours following blueberry and anthocyanin consumption with no difference between the two treatments.
Chronic Studies					
Basu *et al.* (2010)[107]	Randomized, controlled trial.	48 men and women with metabolic syndrome (mean age: 50 ± 3).	Brachial BP.	50 g freeze-dried blueberry powder or 960 mL water (control) to match water intake in the treatment group.	Systolic and diastolic BP ↓ 8 weeks following daily blueberry consumption.

Curtis *et al.* (2019)[112]	Randomized, double-blind, parallel, controlled trial.	115 men and women with metabolic syndrome (68% male, mean age: 63 ± 7, mean BMI: 31.2 ± 3.0 kg m^{-2}).	FMD, AIx@75, cfPWV, brachial BP.	13 g freeze-dried blueberry powder plus 13 g placebo powder (containing 879 mg total (poly)phenols and 364 mg anthocyanins), 26 g freeze-dried blueberry powder (containing 439 mg total (poly)phenols and 182 mg anthocyanins), or 26 g placebo control powder (containing 0 mg total (poly)phenols or anthocyanins) daily for 6 months.	FMD ↑ and AIx ↓ 6 months following 26 g daily blueberry consumption. No impact on remaining vascular function parameters.
Johnson *et al.* (2015)[106]	Randomized, double-blind, parallel, controlled trial.	40 postmenopausal women with pre- and stage 1-hypertension (mean age in blueberry group: 59.7 ± 4.58, mean BMI in blueberry group: 30.1 ± 5.94 kg m^{-2}, mean age in control group: 57.3 ± 4.76, mean BMI in control group: 32.7 ± 6.79 kg m^{-2}).	Brachial BP, baPWV, cfPWV, mean arterial pressure, and heart rate at baseline/0 weeks and 4 and 8 weeks after daily consumption.	22 g freeze-dried blueberry powder or 22 g calorie-matched placebo control powder.	Systolic and diastolic BP, and baPWV ↓ 8 weeks following blueberry consumption. No impact on remaining vascular function parameters.
McAnulty *et al.* (2014)[108]	Randomized, double-blind, parallel controlled trial	25 sedentary men and postmenopausal women (mean age in blueberry group: 46.15 ± 11.92, mean BMI in blueberry group: 27.8 ± 5.46 kg m^{-2}, mean age in control group: 39.92 ± 13.38, mean BMI in control group: 24.23 ± 3.44 kg m^{-2}).	AIx, aortic systolic pressure, cfPWV.	38 g freeze-dried blueberry powder or 38 g placebo powder per day for 6 weeks.	AIx and aortic systolic pressure ↓ 6 weeks following blueberry consumption. Reduction in diastolic BP in subset of participants with prehypertension following 6 weeks blueberry consumption. No impact on remaining vascular function parameters.
Rodriguez-Mateos *et al.* (2019)[104]	Randomized, double-blind,	40 healthy males (mean age: 33 ± 6, mean BMI: 24 ± 3 kg m^{-2}).	FMD, PWV, aortic AIx, 24 hour ambulatory BP at baseline/0	22 g freeze-dried blueberry powder or 22 g macro-/micronutrient-matched	FMD ↑ 4 weeks after blueberry consumption, and acutely at 2 hours following

Table 5.3 (*Continued*)

Ref.	Study design	Population	Vascular function and related outcome measures	Treatment	Results
	parallel controlled trial.		weeks and after 4 weeks of daily consumption at 0 hours and 2 hours post-consumption.	control powder per day for 4 weeks (11 g twice per day) and acute consumption at baseline/0 weeks and 4 week visit (for acute and acute on chronic component).	consumption at the baseline/0 week time point but not at the 4 week time point, and 24 hour systolic BP was decreased after 4 weeks blueberry consumption. No impact on remaining vascular function parameters.
Rodriguez-Mateos *et al.* (2019)[104]	Uncontrolled, single-arm, pilot trial.	5 healthy males (mean age: 23 ± 3, mean BMI: 24 ± 3 kg m^{-2}).	FMD at baseline, and 7, 14, 21, and 28 days following daily consumption.	11 g wild blueberry powder drink (containing 150 mg anthocyanins).	FMD ↑ after 1 week and further increased after 2 weeks, and plateaued thereafter.
Riso *et al.* (2013)[110]	Randomized, crossover, controlled trial.	18 healthy men with at least one CVD risk factor (mean age: 47.8 ± 9.7, mean BMI: 24.8 ± 2.6 kg m^{-2}).	RHI, Framingham RHI, digital AIx, digital AIx@75.	25 g freeze-dried blueberry powder (containing 400 mg anthocyanins) or a placebo control drink (containing 0 mg anthocyanins) daily for 6 weeks.	No impact on vascular function parameters.
Stote *et al.* (2017)[109]	Randomized, single-blind, controlled, crossover trial.	19 women at risk for type 2 diabetes (mean age: 53 ± 6.3, mean BMI: 31.4 ± 2.9 kg m^{-2}).	RHI, brachial BP.	240 mL blueberry juice [containing 2138 mg total (poly)phenols and 314 mg anthocyanins] or 240 mL calorie-matched placebo control beverage per day for 7 days.	Trend for ↓ systolic BP after 7 days blueberry consumption. No impact on other vascular function parameters.
Stull *et al.* (2015)[111]	Randomized, double-blind, parallel, controlled trial.	44 men and women with metabolic syndrome	RHI, brachial BP, 24 hour ambulatory BP	22.5 g freeze-dried blueberry powder in a smoothie twice daily (45 g per day total) or a macronutrient-matched placebo control smoothie twice daily.	RHI ↑ after 6 weeks of daily blueberry consumption. No impact on other vascular function parameters.

[a]**Abbreviations:** AIx, augmentation index; AIx normalized to heart rate of 75 bpm; baPWV, brachial-ankle pulse wave velocity; BP, blood pressure; cfPWV, carotid-femoral pulse wave velocity; DPV, digital pulse volume; FMD, flow-mediated dilation; HR, heart rate; PWA, pulse wave analysis; PWV, pulse wave velocity; RHI, reactive hyperemia index.

acute vasodilatory effects,[104] and consumption of a baked product containing 34 g freeze-dried blueberry powder providing 196 mg anthocyanins or 34 mg unprocessed freeze-dried blueberry powder containing 339 mg anthocyanins did not produce differential effects on FMD.[103] Overall, these data strongly support the acute vasodilatory properties of blueberry (poly)phenols in men and suggest that blueberry consumption may protect against the detrimental effects of cigarette smoking on endothelial function and blood pressure. One of the studies by Rodriguez-Mateos *et al.* found that increases in FMD were highly correlated with increases in circulating (poly)phenol metabolites and decreases in neutrophil NADPH oxidase activity. These data suggest that blueberry (poly)phenols exert their acute vasodilatory effects through inhibition of oxidative stress and likely improved NO bioavailability. Importantly, no studies have evaluated the acute effects of blueberries and their (poly)phenols on vascular function in women, and thus research is needed to understand if these findings translate to women and if there are differential effects in menstruating women *versus* postmenopausal women.

Nine studies evaluating the chronic vascular-protective effects of blueberries and their (poly)phenols in various populations include men and women, postmenopausal women with pre- and stage 1-hypertension, and individuals with metabolic syndrome. The studies performed to date have been heterogeneous in design, with a duration of treatment ranging from 7 days to 6 months and blueberry treatment doses ranging from 13 g to 50 g per day of freeze-dried blueberry powder, while one study evaluated blueberry juice consumption. Two studies found that blueberry consumption reduced blood pressure following 8 weeks of daily blueberry consumption, with similar effects observed with a 22 g dose and a 50 g dose of freeze-dried blueberry powder. The study by Johnson *et al.*[106] evaluating blueberry consumption (22 g per day) for 8 weeks on postmenopausal women with pre- and stage 1-hypertension documented the decrease in brachial blood pressure was accompanied by a decrease in brachial-ankle PWV (measure of systemic arterial stiffness), and an increase in plasma NO metabolites (measure of NO). The study by Basu *et al.*[107] evaluating blueberry consumption (50 g per day) for 8 weeks on obese men and women with metabolic syndrome showed the decrease in brachial blood pressure was accompanied by reductions in products of oxidative stress in the blood, including oxidized low-density lipoprotein, malondialdehyde, and hydroxynonenal. Data from these studies suggest vascular-protective effects of blueberries and potential mediation through reductions in oxidative stress and improved NO bioavailability. Another study by McAnulty *et al.*[108] evaluated blueberry consumption (38 g per day) for 6 weeks on sedentary men and women and observed reductions in aortic systolic pressure and augmentation index (measure of arterial stiffness and wave reflection), as well as reductions in brachial diastolic blood pressure in a subset of prehypertensive participants. Their findings were not accompanied by increased plasma antioxidant status, but the authors suggested the overall study findings were likely due to increased NO bioavailability and greater arterial distensibility.

Six studies evaluated the impact of blueberries and blueberry (poly)-phenols on endothelial function, as well as other measures of vascular function. The study shortest in duration was by Stote *et al.*[109] and evaluated daily consumption of 240 mL blueberry juice containing 2138 mg total (poly)phenols and 314 mg anthocyanins compared with a 240 mL calorie-matched placebo control beverage for 7 days in women at risk for type 2 diabetes. They found a trend for a reduction in systolic blood pressure but no significant impacts on other parameters, including endothelial function as measured by RHI. They did observe an increase in NO metabolites. The authors suggest this may have contributed to the trend for a lowering effect on systolic blood pressure. It is likely that the duration of treatment was not long enough to observe major effects on blood pressure or endothelial function in that study. In support of this, Rodriguez-Mateos[104] assessed FMD at baseline and 7, 14, 21, and 28 days following daily blueberry consumption (22 g per day) containing 150 mg anthocyanins in an uncontrolled pilot study. They found that FMD increased after 1 week and further increased after 2 weeks, but plateaued thereafter, suggesting 2 weeks of daily blueberry consumption is needed to achieve a sustained improvement in endothelial function. Rodriguez-Mateos *et al.*[104] also evaluated the impact of acute and chronic blueberry consumption (22 g per day) for 4 weeks on healthy men and observed an acute increase in FMD 2 hours following consumption, and a chronic increase 4 weeks following daily consumption, but not acutely 2 hours following consumption after chronic daily consumption for 4 weeks. They also found that 24 hour ambulatory blood pressure decreased after 4 weeks of daily blueberry consumption but noted no changes in other measures of vascular function. Two studies evaluated the impact of 6 weeks of daily blueberry consumption per day of freeze-dried blueberry powder) on RHI and other measures of vascular function and had discrepant findings. Riso *et al.*[110] found no impact of daily consumption of 25 g freeze-dried blueberry powder on vascular function in healthy men with at least one CVD risk factor, though they did observe reductions in DNA damage. On the other hand, Stull *et al.*[111] found that RHI increased 6 weeks after daily consumption of 45 g freeze-dried blueberry powder incorporated into smoothies in men and women with metabolic syndrome, but no effects on other parameters of vascular function. Lastly, Curtis *et al.*[112] performed the longest intervention study with blueberries to date, where they evaluated their dose-dependent effects (13 and 26 g of freeze-dried blueberry powder) on vascular function in men and women with metabolic syndrome. Daily consumption of 26 g, but not 13 g, led to improvements in FMD and AIx and increases in serum and urine anthocyanin metabolites at 6 months, with no effect on other vascular function parameters. Additionally, plasma cyclic GMP concentrations were increased relative to the other groups, supporting enhanced NO bioavailability.

Overall, the results of the studies evaluating the chronic effects of blueberries and their (poly)phenols on vascular function suggest that the strongest evidence is in the area of improved endothelial function. Improvements in endothelial function appear to be linked to enhanced NO bioavailability, likely due to circulating (poly)phenol metabolites and their impact on oxidative stress. The data suggest that a minimum of a 2 week intervention of daily blueberry consumption is needed to observe sustained improvements in endothelial function and that the effects can be sustained for at least 6 months. Not all studies noted improvements in endothelial function, which may be due to the study population or the length of the treatment duration. The effects of blueberry consumption on measures of arterial stiffness are less conclusive, but the data suggest that a 6 week treatment intervention is needed to observe a significant effect. This is supported by other data regarding arterial stiffness and is particularly true with measures of arterial stiffness not found to be improved by daily blueberry consumption, namely cfPWV.[110,112–115] The effects of blueberries on blood pressure are also less conclusive, but the data suggest an anti-hypertensive effect of blueberries. Further research is needed to better understand the impact of blueberries on blood pressure and the reasons for the discrepancies in findings.

5.3.2.2.5 Cranberries. Cranberries are rich in anthocyanins (*e.g.*, glucosides, galactosides, and arabinosides of cyanidin and peonidin), flavan-3-ols (*e.g.*, proanthocyanidins, epicatechin), flavonols (*e.g.*, quercetin, myricetin), and phenolic acids (*e.g.*, benzoic acid, hydroxycinnamic acid).[116] Limited clinical research studies have been performed to evaluate the effects of cranberry (poly)phenols on vascular function, but all used cranberry juice or cranberry juice derivatives as the intervention (see Table 5.4). With respect to endothelial function, two identified clinical studies suggest an acute effect of cranberry (poly)phenols,[117,118] while a chronic effect on endothelial function has not been observed.[117,119,120] Conversely, limited research also suggests a chronic effect on arterial stiffness but a minimal or no acute effect. No impact on blood pressure has been observed. In an open-label, uncontrolled, single-arm pilot trial, Dohadwala *et al.*[117] observed improvements in FMD and the natural logarithm of RHI at 4 hours and 2 and 4 hours, respectively, following consumption of 480 mL cranberry juice in 15 men and women with coronary artery disease with no effects on blood pressure. Rodriguez-Mateos *et al.*[118] assessed FMD, AIx, and PWV at baseline/0 hours and 1, 2, 4, 6, and 8 hours post-consumption of 450 mL cranberry juice containing 409, 787, 1238, 1534, or 1910 mg total (poly)phenols or an isocaloric macro-/micronutrient matched control in healthy men. They observed a dose-dependent increase in FMD with peak effects occurring 4 hours after consumption and maximal effects occurring with the 1238 mg (poly)phenol dose. They also noted a decreased blood pressure 6 hours following consumption of the 1910 mg dose compared to baseline and decreased AIx at 1.5 and 6

Table 5.4 Characteristics of human studies evaluating the effects of cranberry (poly)phenols on vascular function.[a]

Ref.	Study design	Population	Vascular function and related outcome measures	Treatment	Results
Acute studies					
Dohadwala *et al.* (2011)[117]	Open-label, uncontrolled, single-arm, pilot trial.	15 men and women with coronary artery disease (mean age: 62 ± 8).	FMD, lnPat ratio (lnRHI), brachial BP.	480 mL cranberry juice.	↑ FMD 4 hours after cranberry juice consumption and ↑ lnPAT ratio 2 and 4 hours after consumption. No effects on other vascular function parameters.
Rodriguez-Mateos *et al.* (2016)[118]	Randomized, double-blind, controlled, dose-response, crossover study.	10 healthy men (mean age: 24 ± 2, mean BMI: 24 ± 2 kg m^{-2}).	FMD, AIx, PWV measured at 1, 2, 4, 6, and 8 hours following treatment consumption.	450 mL cranberry juice containing 409, 787, 1238, 1534, and 1910 mg total (poly)phenols or an isocaloric macro-/micronutrient matched control.	Dose-dependent ↑ in FMD with peak effects at 4 hours and maximal effects at the 1238 mg dose. ↓ blood pressure at 6 hours compared to baseline at the 1910 mg dose, and ↓ AIx at 1.5 and 6 hours compared to baseline at the 1534 mg dose and 4 and 6 hours compared to baseline at the 409 mg dose.

Chronic Studies					
Dohadwala *et al.* (2011)[117]	Randomized, double-blind, controlled, crossover trial.	44 men and women with coronary artery disease (mean age in cranberry group: 61 ± 11, mean BMI in cranberry group: 30 ± 5 kg m^{-2}, mean age in control group: 63 ± 9, mean BMI in control group: 29 ± 4 kg m^{-2}).	FMD, lnPat ratio (lnRHI), brachial BP, cfPWV, crPWV, EID.	480 mL double-strength cranberry juice [containing 835 mg total (poly)phenols and 94 mg anthocyanins] or a calorie-matched placebo control beverage containing no (poly)phenols per day for 4 weeks.	cfPWV ↓ 4 weeks after cranberry juice consumption. No change in remaining vascular function parameters.
Hsia *et al.* (2020)[119]	Randomized, double-blind, parallel, controlled trial.	35 men and women with obesity (mean age in cranberry group: 47 ± 16, mean BMI in cranberry group: 37.5 ± 4.5 kg m^{-2}, mean age in control group: 48 ± 12, mean BMI in control group: 36.4 ± 4.8 kg m^{-2}).	RHI, brachial BP.	450 mL low-calorie cranberry juice beverage containing 158 mg total (poly)phenols, 143.6 mg proanthocyanidins, and 6.5 mg anthocyanins, or 480 mL placebo control drink for 8 weeks.	No impact on vascular function parameters.
Ruel *et al.* (2013)[120]	Randomized, double-blind, controlled, crossover trial.	35 sedentary overweight men with a high waist circumference (mean age: 45 ± 10, mean BMI: 28.3 ± 2.4 kg m^{-2}).	Brachial BP, AIx, AIx following oral administration of salbutamol, AIx following oral administration of glyceryl trinitrite, global endothelial function.	500 mL low-calorie cranberry juice cocktail or 500 mL placebo control juice per day for 4 weeks.	AIx ↓ 4 weeks after cranberry juice consumption compared to baseline in a subgroup with metabolic syndrome. No impact on other vascular function parameters.

[a]**Abbreviations:** AIx, augmentation index; BP, blood pressure; cfPWV, carotid-femoral pulse wave velocity; crPWV, carotid-radial pulse wave velocity; EID, endothelium-independent dilation; FMD, flow-mediated dilation; lnPAT ratio, natural logarithm of reactive hyperemia index; PWA, pulse wave analysis; PWV, pulse wave velocity.

hours post-consumption of the 1534 mg dose compared to baseline, and 4 and 6 hours following consumption of the 409 mg dose compared to baseline. Additionally, 12 (poly)phenol metabolites including cinnamic acid and benzoic acid derivatives, a conjugated flavonoid (*i.e.*, quercetin 3-*O*-β-D-glucuronide), and a proanthocyanidin/flavan-3-ol metabolite (*i.e.*, γ-valerolactone sulfate) correlated with improvements in FMD, supporting acute vasodilatory effects of cranberry (poly)phenols and their metabolites.

Three clinical studies have evaluated the chronic effects of cranberry (poly)phenols on vascular function, with effects limited to measures of arterial stiffness. Dohadwal *et al.*[117] observed reductions in carotid-femoral PWV 4 weeks following daily consumption of 480 mL double-strength cranberry juice containing 835 mg total (poly)phenols and 94 mg anthocyanins in men and women with coronary artery disease. They did not observe improvements in measures of endothelial function, EID, or brachial blood pressure. Hsia *et al.*[119] observed no effects of daily consumption of 450 mL low-calorie cranberry juice containing 158 mg total (poly)phenols, 143.6 mg proanthocyanidins, and 6.5 mg anthocyanins for 8 weeks on measures of vascular function in men and women with obesity but did note some improvements in certain blood biomarkers of oxidative stress (*e.g.*, 8-isoprostane) and triglycerides. It is likely the dose of (poly)phenols in the cranberry juice beverage was not high enough to modulate vascular function. Lastly, Ruel *et al.*[120] reported that consuming 500 mL of a low-calorie cranberry juice cocktail containing 400 mg total (poly)phenols and 20.8 mg anthocyanins per day for 4 weeks reduced AIx compared to baseline, but not control, in a subgroup of sedentary overweight men with metabolic syndrome but not in the full study cohort. No effects on other vascular function parameters, including endothelial function and blood pressure or blood biomarkers, were noted. Overall, these data suggest the potential for cranberry juice to reduce arterial stiffness, but the impact appears to be greater with interventions using higher doses of (poly)phenols and in those with more advanced disease states (*e.g.*, coronary artery disease, metabolic syndrome).

5.3.2.2.6 Raspberries. Black raspberries are rich in anthocyanins, including cyanidin-3-rutinoside, cyanidin-3-glucoside, and cyanidin-3-xylosylrutinoside.[121] Limited clinical research has been performed to examine the vascular-protective effects of black raspberries (Table 5.5). Three clinical trials were identified, all performed by the same research group, evaluating the chronic effects of black raspberry consumption. In the first study, Jeong *et al.*[122] observed increased FMD 12 weeks following daily consumption of 750 mg black raspberry powder in men and women with metabolic syndrome. They noted no effects on measures of arterial stiffness or carotid intima-media thickness did observe improvements in cholesterol and inflammatory cytokines. In a later study by Jeong *et al.*[123] evaluating the impact of daily consumption of 750 mg black raspberry powder in men and women with metabolic syndrome, radial AIx decreased after 12 weeks, improved blood-inflammatory markers, and increased

Table 5.5 Characteristics of human studies evaluating the effects of black raspberry (poly)phenols on vascular function.[a]

Ref.	Study design	Population	Vascular function and related outcome measures	Treatment	Results
Chronic studies					
Jeong *et al.* (2014)[122]	Randomized, controlled, parallel trial.	77 men and women with metabolic syndrome (mean age in black raspberry group: 58.0 ± 9.2, mean BMI in black raspberry group: 26.3 ± 4.3 kg m^{-2}, mean age in control group: 60.1 ± 9.5, mean BMI in control group: 25.1 ± 4.0 kg m^{-2}).	Brachial artery diameter (a measure of FMD), CIMT, baPWV, ABI.	750 mg black raspberry powder or placebo powder in capsules per day for 12 weeks.	↑ FMD following 12 weeks black raspberry consumption. No impact on other parameters of vascular function.
Jeong *et al.* (2016)[123]	Randomized, controlled, parallel trial.	51 men and women with metabolic syndrome (mean age in black raspberry group: 56.4 ± 9.2, mean BMI in black raspberry group: 25.9 ± 4.6 kg m^{-2}, mean age in control group: 60.7 ± 10.4, mean BMI in control group: 24.7 ± 3.9 kg m^{-2}).	Brachial BP, central BP, HR, radial AIx.	750 mg black raspberry powder or placebo powder in capsules per day for 12 weeks.	Radial AIx ↓ 12 weeks after daily black raspberry consumption compared to baseline. No significant impact on other vascular function parameters.

Table 5.5 (*Continued*)

Ref.	Study design	Population	Vascular function and related outcome measures	Treatment	Results
Jeong *et al.* (2016)[124]	Randomized, double-blind, controlled, parallel trial.	45 men and women with prehypertension (mean age in moderate-dose black raspberry group: 60.2 ± 11.2, mean BMI in moderate-dose black raspberry group: 24.5 ± 2.9 kg m^{-2}, mean age in high-dose black raspberry group: 55.5 ± 12.3, mean BMI in high-dose black raspberry group: 23.5 ± 2.4 kg m^{-2} mean age in placebo control group: 55.9 ± 12.8, mean BMI in placebo control group: 25.8 ± 3.0 kg m^{-2}).	24 hour BP, baPWV, central BP, Aix.	1500 mg black raspberry (moderate-dose), 2500 mg black raspberry (high-dose), or placebo control in capsules daily for 8 weeks.	24 hour systolic BP and nighttime systolic BP ↓ 8 weeks after daily high-dose black raspberry consumption. No significant impact on other vascular function parameters.

[a]**Abbreviations:** ABI, ankle-brachial index; AIx, augmentation index; baPWV, brachial-ankle pulse wave velocity; BP, blood pressure; CIMT, carotid intima-media thickness; FMD, flow-mediated dilation; HR, heart rate.

endothelial progenitor cells. No effects were noted on measures of blood pressure. Finally, Jeong *et al.*[124] showed that daily consumption of 2500 mg black raspberry powder, but not 1500 mg black raspberry powder, for 8 weeks reduced 24 hour systolic blood pressure and nighttime systolic blood pressure in men and women with prehypertension. There was no significant impact on measures of arterial stiffness, central blood pressure, or blood biomarkers. Overall, these data suggest the potential for black raspberries to improve vascular function in adults with metabolic syndrome and blood pressure in adults with prehypertension. However, more research is needed to confirm these findings in these populations and others.

Only one clinical study has been conducted to evaluate the vascular-protective effects of red raspberries. In a randomized, double-blind, controlled, crossover trial by Istas *et al.*,[125] consumption of 200 g and 400 g red raspberries [containing 201 or 403 mg of total (poly)phenols, respectively] by 10 healthy men led to increased FMD two and 24 hours post-consumption in healthy men. No other impacts on vascular function were noted. They reported that plasma ellagic acid, urolithin A-3-glucuronide, and urolithin A-sulfate correlated with improvements in FMD, suggesting a direct effect of (poly)phenol metabolites on endothelial function. These data are promising but further research is needed to confirm these findings and to better understand the impact of red raspberry consumption on vascular function.

5.3.2.2.7 Strawberries. Strawberries are rich in (poly)phenols, particularly in anthocyanins and especially pelargonidin-glucosides, and in phenolic acids such as hydroxycinnamic and hydroxybenzoic acids.[126] Limited evidence of the effect of strawberry consumption on cardiovascular outcomes is presented in the literature (Table 5.6). A total of four clinical studies were identified, including one acute study[127] and three chronic studies.[128–130] While the acute study did not report any changes from control following the consumption of 40 g of freeze-dried strawberry powder in overweight and obese adults, the chronic trials observed some significant improvements in vascular outcomes. First, in their randomized controlled trial involving 25 overweight or obese males and the consumption of 50 g of freeze-dried strawberry powder, Djurica *et al.*[128] showed an improvement of the RHI in a subgroup of the population with improved plasma nitrate levels. Furthermore, another trial including 60 prehypertensive women and studying the impact of 4 and 8 weeks consumption of 25 g or 50 g of freeze-dried strawberry powder on PWV and blood pressure observed an improvement of these parameters following 4 weeks of treatment, but not after 8 weeks of daily intake of the freeze-dried powder. Finally, a recent trial by Huang *et al.*[130] documented that 50 g of freeze-dried strawberry powder did not improve FMD or blood pressure in a population of 34 adults with mild hypercholesterolemia but did highlight a significant improvement of these factors when considering an intervention-by-hour interaction.

Table 5.6 Characteristics of human studies evaluating the effects of strawberry (poly)phenols on vascular function.[a]

Ref.	Study design	Population	Vascular function and related outcome measures	Treatment	Results
Acute studies					
Richter *et al.* (2017)[127]	Randomized, double-blind, controlled, crossover trial.	30 overweight or obese adults.	Blood samples before intervention and at 0.5, 1, 2, and 4 hours post-consumption. Central BP and arterial stiffness at baseline, 2 hours and 4 hours post-consumption.	40 g freeze-dried strawberry powder (~1 lb. in 0.45 kg fresh strawberries) in a high-fat (50 g total fat) meal.	There were no significant differences between the strawberry and control meals for any outcomes.
Chronic Studies					
Djurica *et al.* (2016)[128]	Randomized, double-blind, controlled, crossover trial.	25 overweight or obese males.	RHI, Framingham RHI, and BP measured at baseline and 2 hours after acute treatment consumption before and after the 1 week intervention.	50 g freeze-dried strawberry powder or 50 g placebo control powder daily for 1 week.	↑ in RHI noted in those with improved plasma nitrate levels over the 1 week period but not in the full cohort. No acute effects observed, nor any effects on BP.
Feresin *et al.* (2017)[129]	Randomized, double-blind, controlled, parallel trial.	60 postmenopausal women with pre- and stage 1-hypertension.	baPWV, faPWV, and brachial BP measured at baseline and 4 and 8 weeks following treatment consumption.	25 g or 50 g freeze-dried strawberry powder or 50 g placebo control powder (the 25 g freeze-dried strawberry powder treatment and the 50 g placebo control treatment were calorie-matched to the 50 g freeze-dried strawberry powder intervention).	BaPWV, faPWV, and systolic BP ↓ after 8 weeks of 25 g, but not 50 g, freeze-dried strawberry consumption.

Huang *et al.* (2021)[130]	Randomized, controlled, double-blind, 2-arm, 2-period (4 weeks per period) crossover trial.	34 adults with moderate hypercholesterolemia.	FMD, BP, at baseline and after 4 week consumption, but also acutely 1 and 2 hours post-consumption of a bolus.	Freeze-dried strawberry powder (2×25 g) or energy-, volume-matched control powder.	No change in FMD and BP after 4 weeks, but significant intervention-by-hour interaction for FMD and BP revealed increased FMD at 1 hour after strawberry compared with control by $1.5 \pm 0.38\%$ and attenuated systolic BP at 2 hours by 3.1 ± 0.99 mmHg.

[a]**Abbreviations:** AIx, augmentation index; baPWV, brachial-ankle pulse wave velocity; BP, blood pressure; faPWV, femoral-ankle pulse wave velocity; FMD, flow-mediated dilation; RHI, reactive hyperemia index.

5.4 Overview of the Effects of Berry (Poly)phenols on Vascular Function and Potential Mechanisms

The totality of the underlying mechanisms by which berry (poly)phenols improve vascular function is complex, but some mechanisms of action are being elucidated. An overview of potential mechanisms by which berry (poly)phenols improve vascular function is provided here; however, the reader is referred to Chapter 6 for more in-depth information.

5.4.1 (Poly)phenols as Direct Antioxidants: Early Misconceptions

Researchers initially hypothesized that (poly)phenols improved cardiovascular health through direct antioxidant activity *in vivo*, due to their *in vitro* free radical scavenging activities.[131] The field has strongly evolved over the past 30 years and the direct antioxidant theory has been disproven for several reasons. First, plasma (poly)phenol concentrations are 1000 to 100 000 times lower than in food extracts traditionally used in *in vitro* experiments.[40] Second, ingested (poly)phenols are extensively metabolized and only a small fraction of the parent compounds as they occur in the food reach circulation.[40] In addition, (poly)phenol metabolites have a lower antioxidant capacity, due to the blocking of the radical scavenging of hydroxy groups by methylation, sulfation, and glucuronidation reactions.[132] The previous direct antioxidant mechanism attributed to (poly)phenols *in vivo* has changed to one that is indirect and more complex, where (poly)phenol metabolites are suggested to function as signaling molecules.

5.4.2 Indirect Antioxidant Activity of Berry (Poly)phenols and the Vascular Endothelium

Dietary (poly)phenols and their metabolites may regulate NADPH oxidase (NOX) activity, thus lowering O_2^- generation and in turn increasing NO bioavailability.[133] For instance, Petersen *et al.*[134] showed that improvements in vascular function and inflammation in a diabetic db/db mice model that consumed a diet supplemented with 2.35% freeze-dried strawberry powder for 10 weeks were likely mediated through serum (poly)phenol metabolites, as co-incubation of the mice aortic endothelial cells with serum from strawberry fed mice reduced NOX. Rodriguez-Mateos *et al.*[105] showed acute increases in FMD correlated with increases in circulating (poly)phenol metabolites and decreases in neutrophil NOX activity. Johnson *et al.*[106] noted increases in NO metabolites 8 weeks following daily blueberry consumption in postmenopausal women with pre- and stage 1-hypertension, suggesting improved NO bioavailability. Another possible mechanism responsible for increased NO bioavailability is the activation of eNOS enzyme. (Poly)phenols were shown to improve vessel function by inducing eNOS *via* phosphorylation of adenosine monophosphate-activated protein kinase (AMPK).[135] In addition, activation of an antioxidant response through the nuclear E2-related factor 2 (Nrf2) pathway has emerged as a potential target for (poly)phenols.[136] For instance, ellagic acid (30 $mg\,kg^{-1}$ of body weight) had a protective effect on vascular dysfunction and atherosclerosis in wild-type ApoE deficient mice fed with a high-fat diet through increased Nrf2 activity and heme oxygenase 1 (HO-1) protein expression.[137] In a feeding experiment where ApoE-deficient mice were given bilberry anthocyanin-rich extracts, atherosclerotic lesion development was attenuated.[138] They also showed the phenolic acids gallic acid, *p*-coumaric acid, and ferulic acid (100 $mg\,kg^{-1}$ of body weight) enhanced antioxidant enzyme (superoxide dismutase [SOD], glutathione peroxidase) and Nrf2 expression in Sprague-Dawley rats.[139] Lastly, ellagic acid prevented hypochlorous acid-induced cellular damage by activating Nrf2 signaling and by increasing the antioxidant HO-1 protein expression in human umbilical vein endothelial cells (HUVECs).[137]

5.4.3 Anti-inflammatory Properties of Berry (Poly)phenols and the Vascular Endothelium

(Poly)phenols have been shown to have anti-inflammatory properties such as decreasing inflammatory cytokine production and expression of adhesion molecules on the cell surface.[140] Karlsen *et al.*[141] showed that supplementation of 300 mg purified anthocyanidin in healthy volunteers for 3 weeks decreased plasma concentrations of pro-inflammatory cytokines. In a secondary analysis[142] of their initial study,[111] Nair *et al.* found that 6 weeks of blueberry consumption decreased levels of pro-inflammatory cytokines, superoxide, and total reactive oxygen species in adults with metabolic syndrome. Basu *et al.*[143] demonstrated that 8 weeks of strawberry consumption in adults with metabolic syndrome reduced vascular cell adhesion molecule-1 (VCAM-1).

5.4.4 The Gut Microbiota in Determining the Vascular-protective Effects of Berry (Poly)phenols

The composition of the gut microbiome may be an important factor in determining the vascular-protective effects of berry (poly)phenols, as they are largely metabolized by the gut microbiota to smaller compounds, such as phenolic acids, prior to their absorption and further metabolism by the liver.[45] In addition, the gut microbiome has emerged as a regulator of cardiovascular health, and modulation of the gut microbiota may improve vascular function.[9] Berries and their (poly)phenols have been demonstrated to favorably modulate the gut microbiota in animal models and humans, likely due in large part to the prebiotic effects of (poly)phenols.[72,144–148] In the first human study linking improvements in vascular function following berry consumption to the gut microbiota, Istas *et al.*[72] demonstrated that acute and chronic (12 weeks) improvements in FMD following consumption of an aronia berry dietary supplement were linked to increases in plasma phenolic metabolites and specific gut microbiota genera.

5.5 Research Gaps and Future Directions

While studies continue to demonstrate the beneficial effect of dietary (poly)phenols for the prevention of CVD,[149] the literature is mixed regarding the magnitude of effects attributed to the consumption of phenolic compounds.[150] Following recent studies,[125,151,152] the importance of compounds' bioavailability appears to be key in the understanding of metabolites' bioactivity. In the case of vascular function assessment, it was reported that plasma (poly)phenol concentration and magnitude of the biological effect are directly related.[105] There is a need for the development of time courses and dose–response analysis to assess the optimal daily intake of berry (poly)phenols to exert beneficial effects. As a result, human studies involving state-of-the-art techniques to assess both bioactivity and bioavailability are needed.

Mixed results reported from the literature could be related to the inter-individual variability in response to a (poly)phenol intervention. The work of the European Scientific Cooperative COST Action POSITIVe highlighted that the gut microbiota, background genetics, age, sex, ethnicity, physical activity, smoking status, and pathophysiological status could be linked to the variability in the biological response to (poly)phenol supplementation.[153] The importance of subgroup examination based on these factors and analysis of the responders *versus* nonresponders' responsiveness is thus key and could explain the inconclusive or divergent results of some randomized controlled trials. This topic is addressed in detail in Chapter 3.

As described previously, very few randomized controlled trials have investigated the effect of berry (poly)phenols on cardiovascular health, in particular endothelial function and arterial stiffness. Most trials focusing on the effect of berry (poly)phenol supplementation present a low number of participants included, which is prone to bias.[86] Future research on the health effects of berries will need to use a stronger methodology, including larger sample sizes justified by power calculation, detailed criteria for the inclusion of volunteers as well as the consideration of confounding factors such as sex, BMI, age, or baseline (poly)phenol intake. A 6–8 week intervention has been shown to be sufficient to induce significant improvements in systolic blood pressure, based on four randomized controlled trials.[86] Nevertheless, accounting for the possibility of bias described above, future randomized controlled trials should investigate the potential of longer durations of supplementation to induce greater vascular enhancements.

Recently, gut microbiome composition and diversity have been gaining attention in the fields of nutritional and food sciences as the number of studies demonstrating the importance of the gut microbiota in the health benefits of foods, and in particular plant foods, is growing.[154–157] However, the impact of berry consumption on the gut microbiota diversity and composition, and in particular the impact of the gut microbiota on berry (poly)phenol bioavailability and bioactivity, is not fully elucidated yet and will need further investigation. The use of well-designed randomized controlled trials involving "omics" technologies such as targeted and untargeted nutrigenomics, metabolomics, and microbiomics will be useful tools to uncover these aspects.[158]

5.6 Conclusion

Based on the current literature, berry (poly)phenols are promising candidates to improve human vascular function. The evidence available shows that berry (poly)phenols are bioavailable and can improve vascular function measured as endothelial function and arterial stiffness (and indirectly as blood pressure) in various segments of the population. The strongest evidence exists for the effects of blueberries on endothelial function measured as FMD and on blood pressure and arterial stiffness, although these findings are not consistent across all studies and study populations. Aronia berries appear to be a strong candidate for the prevention of hypertension as several randomized controlled trials showed the potential of these berries to decrease blood pressure. Cranberries appear to significantly improve arterial stiffness, with some evidence on blood pressure-lowering effects, although not consistently in all studies. The evidence of the vascular health benefits of strawberries, blackcurrant, blackberries, black raspberries, and red raspberries is scarce and divergent, and there is a need for more studies investigating these berry sources. Although more evidence from longer and larger clinical studies and adequate epidemiological studies using more accurate biomarkers of berry intake are needed, overall, the data provided in this chapter provides a solid basis for the further investigation of (poly)-phenol-rich berries as potential candidates in the primary and secondary prevention of CVD progression.

References

1. D. Aune, E. Giovannucci, P. Boffetta, L. T. Fadnes, N. Keum, T. Norat, D. C. Greenwood, E. Riboli, L. J. Vatten and S. Tonstad, *Int. J. Epidemiol.*, 2017, **46**, 1029–1056.
2. A. Rodriguez-Mateos, C. Heiss, G. Borges and A. Crozier, *J. Agric. Food Chem.*, 2014, **62**, 3842–3851.
3. E. Wood, S. Hein, C. Heiss, C. Williams and A. Rodriguez-Mateos, *Food Funct.*, 2019, **10**, 7621–7633.
4. K. Miller, W. Feucht and M. Schmid, *Nutrients*, 2019, **11**(7), 1510.
5. G. A. Roth, C. Johnson, A. Abajobir, F. Abd-Allah, S. F. Abera, G. Abyu, M. Ahmed, B. Aksut, T. Alam, K. Alam, F. Alla, N. Alvis-Guzman, S. Amrock, H. Ansari, J. Ärnlöv, H. Asayesh, T. M. Atey, L. Avila-Burgos,

A. Awasthi, A. Banerjee, A. Barac, T. Bärnighausen, L. Barregard, N. Bedi, E. Belay Ketema, D. Bennett, G. Berhe, Z. Bhutta, S. Bitew, J. Carapetis, J. J. Carrero, D. C. Malta, C. A. Castañeda-Orjuela, J. Castillo-Rivas, F. Catalá-López, J.-Y. Choi, H. Christensen, M. Cirillo, L. Cooper, M. Criqui, D. Cundiff, A. Damasceno, L. Dandona, R. Dandona, K. Davletov, S. Dharmaratne, P. Dorairaj, M. Dubey, R. Ehrenkranz, M. El Sayed Zaki, E. J. A. Faraon, A. Esteghamati, T. Farid, M. Farvid, V. Feigin, E. L. Ding, G. Fowkes, T. Gebrehiwot, R. Gillum, A. Gold, P. Gona, R. Gupta, T. D. Habtewold, N. Hafezi-Nejad, T. Hailu, G. B. Hailu, G. Hankey, H. Y. Hassen, K. H. Abate, R. Havmoeller, S. I. Hay, M. Horino, P. J. Hotez, K. Jacobsen, S. James, M. Javanbakht, P. Jeemon, D. John, J. Jonas, Y. Kalkonde, C. Karimkhani, A. Kasaeian, Y. Khader, A. Khan, Y.-H. Khang, S. Khera, A. T. Khoja, J. Khubchandani, D. Kim, D. Kolte, S. Kosen, K. J. Krohn, G. A. Kumar, G. F. Kwan, D. K. Lal, A. Larsson, S. Linn, A. Lopez, P. A. Lotufo, H. M. A. El Razek, R. Malekzadeh, M. Mazidi, T. Meier, K. G. Meles, G. Mensah, A. Meretoja, H. Mezgebe, T. Miller, E. Mirrakhimov, S. Mohammed, A. E. Moran, K. I. Musa, J. Narula, B. Neal, F. Ngalesoni, G. Nguyen, C. M. Obermeyer, M. Owolabi, G. Patton, J. Pedro, D. Qato, M. Qorbani, K. Rahimi, R. K. Rai, S. Rawaf, A. Ribeiro, S. Safiri, J. A. Salomon, I. Santos, M. Santric Milicevic, B. Sartorius, A. Schutte, S. Sepanlou, M. A. Shaikh, M.-J. Shin, M. Shishehbor, H. Shore, D. A. S. Silva, E. Sobngwi, S. Stranges, S. Swaminathan, R. Tabarés-Seisdedos, N. Tadele Atnafu, F. Tesfay, J. S. Thakur, A. Thrift, R. Topor-Madry, T. Truelsen, S. Tyrovolas, K. N. Ukwaja, O. Uthman, T. Vasankari, V. Vlassov, S. E. Vollset, T. Wakayo, D. Watkins, R. Weintraub, A. Werdecker, R. Westerman, C. S. Wiysonge, C. Wolfe, A. Workicho, G. Xu, Y. Yano, P. Yip, N. Yonemoto, M. Younis, C. Yu, T. Vos, M. Naghavi and C. Murray, *J. Am. Coll. Cardiol.*, 2017, **70**, 1–25.
6. World Health Organization, 2018.
7. S. A. Johnson, N. S. Litwin and D. R. Seals, *J. Acad. Nutr. Diet.*, 2019, **119**, 1790–1796.
8. M. J. Rossman, T. J. LaRocca, C. R. Martens and D. R. Seals, *J. Appl. Physiol.*, 2018, **125**, 1888–1900.
9. M. L. Battson, D. M. Lee, T. L. Weir and C. L. Gentile, *J. Nutr. Biochem.*, 2018, **56**, 1–15.
10. T. J. LaRocca, C. R. Martens and D. R. Seals, *Ageing Res. Rev.*, 2017, **39**, 106–119.
11. A. N. Schechter and M. T. Gladwin, *N. Engl. J. Med.*, 2003, **348**, 1483–1485.
12. R. F. Furchgott and J. V. Zawadzki, *Nature*, 1980, **288**, 373–376.
13. U. Forstermann and T. Munzel, *Circulation*, 2006, **113**, 1708–1714.
14. S. Moncada and J. D. Erusalimsky, *Nat. Rev. Mol. Cell Biol.*, 2002, **3**, 214.
15. M. A. Gimbrone Jr, *Am. J. Pathol.*, 1999, **155**, 1–5.
16. D. R. Seals, K. L. Jablonski and A. J. Donato, *Clin. Sci.*, 2011, **120**, 357–375.

17. A. J. Flammer and T. F. Lüscher, *Pflügers Arch. Eur. J. Physiol.*, 2010, **459**, 1005–1013.
18. A. Virdis, L. Ghiadoni and S. Taddei, *Pflügers Archiv -. Eur. J. Physiol.*, 2010, **459**, 1015–1023.
19. D. R. Seals, K. L. Jablonski and A. J. Donato, *Clin. Sci.*, 2011, **120**, 357–375.
20. M. A. Gimbrone, Jr and G. García-Cardeña, *Circ. Res.*, 2016, **118**, 620–636.
21. B. S. Fleenor, *Aging Dis.*, 2013, **4**, 76–83.
22. S. J. Zieman, V. Melenovsky and D. A. Kass, *Arterioscler., Thromb., Vasc. Biol.*, 2005, **25**, 932–943.
23. E. G. Lakatta and D. Levy, Part I: Aging Arteries: A "Set Up" for Vascular Disease, *Circulation*, 2003, **107**, 139–146.
24. M. Wang, L. Jiang, R. E. Monticone and E. G. Lakatta, *Trends Endocrinol. Metab.*, 2014, **25**, 72–79.
25. S. J. Zieman, V. Melenovsky and D. A. Kass, *Arterioscler., Thromb., Vasc. Biol.*, 2005, **25**, 932–943.
26. S. M. L. Wallace, C. M. Yasmin, K. M. McEniery, A. D. Mäki-Petäjä, J. R. Booth, Cockcroft and I. B. Wilkinson, *Hypertension*, 2007, **50**, 228–233.
27. U. Quinn, L. A. Tomlinson and J. R. Cockcroft, *JRSM Cardiovasc. Dis.*, 2012, **1**, 18.
28. M. F. O'Rourke and M. E. Safar, *Cause and Logic of Therapy*, 2005, **46**, 200–204.
29. J. J. Oliver and D. J. Webb, *Arterioscler., Thromb., Vasc. Biol.*, 2003, **23**, 554–566.
30. F. U. S. Mattace-Raso, T. J. M. van der Cammen, A. Hofman, N. M. van Popele, M. L. Bos, M. A. D. H. Schalekamp, R. Asmar, R. S. Reneman, A. P. G. Hoeks, M. M. B. Breteler and J. C. M. Witteman, *Rotterdam Study*, 2006, **113**, 657–663.
31. J. Chiha, P. Mitchell, B. Gopinath, G. Burlutsky, A. Plant, P. Kovoor and A. Thiagalingam, *PLoS One*, 2016, **11**, e0168598.
32. H. E. Lim, C. G. Park, S. H. Shin, J. C. Ahn, H. S. Seo and D. J. Oh, *Blood Pressure*, 2004, **13**, 369–375.
33. A. P. Guerin, J. Blacher, B. Pannier, S. J. Marchais, M. E. Safar and G. M. London, *Circulation*, 2001, **103**, 987–992.
34. P. Boutouyrie, A. I. Tropeano, R. Asmar, I. Gautier, A. Benetos, P. Lacolley and S. Laurent, *A Longitudinal Study*, 2002, **39**, 10–15.
35. S. Laurent, J. Cockcroft, L. Van Bortel, P. Boutouyrie, C. Giannattasio, D. Hayoz, B. Pannier, C. Vlachopoulos, I. Wilkinson and H. Struijker-Boudier, *Eur. Heart J.*, 2006, **27**, 2588–2605.
36. C. Vlachopoulos, K. Aznaouridis and C. Stefanadis, *J. Am. Coll. Cardiol.*, 2010, **55**, 1318–1327.
37. V. Cheynier, *Am. J. Clin. Nutr.*, 2005, **81**, 223s–229s.
38. R. Gunnaiah, A. C. Kushalappa, R. Duggavathi, S. Fox and D. J. Somers, *PLoS One*, 2012, **7**, e40695.
39. J. B. Harborne and C. A. Williams, *Phytochemistry*, 2000, **55**, 481–504.

40. C. Manach, A. Scalbert, C. Morand, C. Remesy and L. Jimenez, *Am. J. Clin. Nutr.*, 2004, **79**, 727–747.
41. C. Lakenbrink, S. Lapczynski, B. Maiwald and U. H. Engelhardt, *J. Agric. Food Chem.*, 2000, **48**(7), 2848–2852.
42. A. Crozier, M. E. J. Lean, M. S. McDonald and C. Black, *J. Agric. Food Chem.*, 1997, **45**, 590–595.
43. M. N. Clifford, *J. Sci. Food Agric.*, 2000, **80**, 1063–1072.
44. J. M. Landete, *Food Res. Int.*, 2011, **44**, 1150–1160.
45. L. Lavefve, L. R. Howard and F. Carbonero, *Food Funct.*, 2020, **11**, 45–65.
46. A. Błaszczyk, S. Sady and M. Sielicka, *Phytochem. Rev.*, 2019, **18**, 37–67.
47. O. Paredes-López, M. L. Cervantes-Ceja, M. Vigna-Pérez and T. Hernández-Pérez, *Plant Foods Hum. Nutr.*, 2010, **65**, 299–308.
48. H. Kim, K. H. Seo and W. Yokoyama, *J. Agric. Food Chem.*, 2020, **68**, 12836–12841.
49. Y. Liu, Y. You, J. Lu, X. Chen and Z. Yang, *Molecules*, 2020, **25**(21), 5166.
50. A. J. Day, F. J. Canada, J. C. Diaz, P. A. Kroon, R. McLauchlan, C. B. Faulds, G. W. Plumb, M. R. Morgan and G. Williamson, *FEBS Lett.*, 2000, **468**, 166–170.
51. S. Wolffram, M. Block and P. Ader, *J. Nutr.*, 2002, **132**, 630–635.
52. A. J. Day, M. S. DuPont, S. Ridley, M. Rhodes, M. J. Rhodes, M. R. Morgan and G. Williamson, *FEBS Lett.*, 1998, **436**, 71–75.
53. C. D. Kay, G. Mazza and B. J. Holub, *J. Nutr.*, 2005, **135**, 2582–2588.
54. P. C. Hollman, J. M. van Trijp, M. N. Buysman, M. S. van der Gaag, M. J. Mengelers, J. H. de Vries and M. B. Katan, *FEBS Lett.*, 1997, **418**, 152–156.
55. M. V. Selma, J. C. Espín and F. A. Tomás-Barberán, *J. Agric. Food Chem.*, 2009, **57**, 6485–6501.
56. A. Cassidy, K. J. Mukamal, L. Liu, M. Franz, A. H. Eliassen and E. B. Rimm, *Circulation*, 2013, **127**, 188–196.
57. A. Cassidy, É. J. O'Reilly, C. Kay, L. Sampson, M. Franz, J. P. Forman, G. Curhan and E. B. Rimm, *Am. J. Clin. Nutr.*, 2011, **93**, 338–347.
58. A. Jennings, A. A. Welch, S. J. Fairweather-Tait, C. Kay, A.-M. Minihane, P. Chowienczyk, B. Jiang, M. Cecelja, T. Spector, A. Macgregor and A. Cassidy, *Am. J. Clin. Nutr.*, 2012, **96**, 781–788.
59. P. J. Mink, C. G. Scrafford, L. M. Barraj, L. Harnack, C.-P. Hong, J. A. Nettleton and D. R. Jacobs, *Am. J. Clin. Nutr.*, 2007, **85**, 895–909.
60. M. L. McCullough, J. J. Peterson, R. Patel, P. F. Jacques, R. Shah and J. T. Dwyer, *Am. J. Clin. Nutr.*, 2012, **95**, 454–464.
61. Y. Kim and Y. Je, *Clin. Nutr. ESPEN*, 2017, **20**, 68–77.
62. M. G. Hertog, E. J. Feskens and D. Kromhout, *Lancet*, 1997, **349**, 699.
63. J. Lin, K. M. Rexrode, F. Hu, C. M. Albert, C. U. Chae, E. B. Rimm, M. J. Stampfer and J. E. Manson, *Am. J. Epidemiol.*, 2007, **165**, 1305–1313.
64. E. B. Rimm, M. B. Katan, A. Ascherio, M. J. Stampfer and W. C. Willett, *Ann. Intern. Med.*, 1996, **125**, 384–389.
65. H. D. Sesso, J. M. Gaziano, S. Liu and J. E. Buring, *Am. J. Clin. Nutr.*, 2003, 77, 1400–1408.

66. V. Neveu, J. Perez-Jiménez, F. Vos, V. Crespy, L. du Chaffaut, L. Mennen, C. Knox, R. Eisner, J. Cruz, D. Wishart and A. Scalbert, *Database*, 2010, **2010**, bap024.
67. USDA, USDA, Beltsville, 2014, Release 3.1.
68. R. Zamora-Ros, M. Touillaud, J. A. Rothwell, I. Romieu and A. Scalbert, *Am. J. Clin. Nutr.*, 2014, **100**, 11–26.
69. S. E. King and W. B. Bolling, *J. Food Bioactives*, 2020, **11**, DOI: 10.31665/JFB.2020.11235.
70. A. Kokotkiewicz, Z. Jaremicz and M. Luczkiewicz, *J. Med. Food*, 2010, **13**, 255–269.
71. R. Poreba, A. Skoczynska, P. Gac, M. Poreba, I. Jedrychowska, A. Affelska-Jercha, B. Turczyn, A. Wojakowska, J. Oszmianski and R. Andrzejak, *Ann. Agric. Environ. Med.*, 2009, **16**, 305–308.
72. G. Istas, E. Wood, M. Le Sayec, C. Rawlings, J. Yoon, V. Dandavate, D. Cera, S. Rampelli, A. Costabile, E. Fromentin and A. Rodriguez-Mateos, *Am. J. Clin. Nutr.*, 2019, **110**, 316–329.
73. L. Xie, T. Vance, B. Kim, S. G. Lee, C. Caceres, Y. Wang, P. A. Hubert, J. Y. Lee, O. K. Chun and B. W. Bolling, *Nutr. Res.*, 2017, **37**, 67–77.
74. G. Istas, E. Wood, M. Le Sayec, C. Rawlings, J. Yoon, V. Dandavate, D. Cera, S. Rampelli, A. Costabile, E. Fromentin and A. Rodriguez-Mateos, *Am. J. Clin. Nutr.*, 2019, **110**(2), 316–329, DOI: 10.1093/ajcn/nqz075.
75. B. Pokimica, M. T. García-Conesa, M. Zec, J. Debeljak-Martačić, S. Ranković, N. Vidović, G. Petrović-Oggiano, A. Konić-Ristić and M. Glibetić, *Nutrients*, 2019, **11**(4), 850.
76. M. Milutinovic, R. Velickovic Radovanovic, K. Savikin, S. Radenkovic, M. Arvandi, M. Pesic, M. Kostic, B. Miladinovic, S. Brankovic and D. Kitic, *J. Appl. Biomed.*, 2019, **17**, 218–224.
77. S. Ahles, Y. R. Stevens, P. J. Joris, D. Vauzour, J. Adam, E. de Groot and J. Plat, *Nutrients*, 2020, **12**(8), 2475.
78. M. Naruszewicz, I. Laniewska, B. Millo and M. Dluzniewski, *Atherosclerosis*, 2007, **194**, e179–e184.
79. M. Broncel, M. Kozirog, P. Duchnowicz, M. Koter-Michalak, J. Sikora and J. Chojnowska-Jezierska, *Med. Sci. Monitor: Int. Med. J. Exp. Clin. Res.*, 2010, **16**, Cr28-34.
80. A. Skoczyńska, I. Jedrychowska, R. Poręba, A. Affelska-Jercha, B. Turczyn, A. Wojakowska and R. Andrzejak, *Pharmacol Rep.*, 2007, **59**, 177–182.
81. N. Kardum, A. Konić-Ristić, K. Savikin, S. Spasić, A. Stefanović, J. Ivanišević and M. Miljković, *J. Med. Food*, 2014, **17**, 869–874.
82. J. Sikora, M. Broncel and E. Mikiciuk-Olasik, *Oxid. Med. Cell. Longevity*, 2014, **2014**, 739721.
83. B. M. Loo, I. Erlund, R. Koli, P. Puukka, J. Hellstrom, K. Wahala, P. Mattila and A. Jula, *Nutr. Res.*, 2016, **36**, 1222–1230.
84. N. Kardum, B. Milovanovic, K. Savikin, G. Zdunic, S. Mutavdzin, T. Gligorijevic and S. Spasic, *J. Med. Food*, 2015, **18**, 1231–1238.

85. B.-M. Loo, I. Erlund, R. Koli, P. Puukka, J. Hellström, K. Wähälä, P. Mattila and A. Jula, *Nutr. Res.*, 2016, **36**, 1222–1230.
86. J. Hawkins, C. Hires, C. Baker, L. Keenan and M. Bush, *J. Diet. Suppl.*, 2020, 517–530.
87. D. Del Rio, A. Rodriguez-Mateos, J. P. Spencer, M. Tognolini, G. Borges and A. Crozier, *Antioxid. Redox Signaling*, 2013, **18**, 1818–1892.
88. H. Land Lail, R. G. Feresin, D. Hicks, B. Stone, E. Price and D. Wanders, *Nutrients*, 2021, **13**(2), 334.
89. I. Serraino, L. Dugo, P. Dugo, L. Mondello, E. Mazzon, G. Dugo, A. P. Caputi and S. Cuzzocrea, *Life Sci.*, 2003, **73**, 1097–1114.
90. R. G. Feresin, J. Huang, D. S. Klarich, Y. Zhao, S. Pourafshar, B. H. Arjmandi and G. Salazar, *Food Funct.*, 2016, **7**, 4175–4187.
91. R. E. Cortez and E. Gonzalez de Mejia, *J. Food Sci.*, 2019, **84**, 2387–2401.
92. T. Tomisawa, N. Nanashima, M. Kitajima, K. Mikami, S. Takamagi, H. Maeda, K. Horie, F. C. Lai and T. Osanai, *Molecules*, 2019, **24**, 4295.
93. M. L. Castro-Acosta, L. Smith, R. J. Miller, D. I. McCarthy, J. A. Farrimond and W. L. Hall, *J. Nutr. Biochem.*, 2016, **38**, 154–161.
94. Y. Jin, D. Alimbetov, T. George, M. H. Gordon and J. A. Lovegrove, *Eur. J. Clin. Nutr.*, 2011, **65**, 849–856.
95. M. D. Cook, S. D. Myers, M. L. Gault, V. C. Edwards and M. E. Willems, *Eur J. Appl. Physiol.*, 2017, **117**, 247–254.
96. F. Khan, S. Ray, A. M. Craigie, G. Kennedy, A. Hill, K. L. Barton, J. Broughton and J. J. Belch, *Free Radical Biol. Med.*, 2014, **72**, 232–237.
97. M. E. T. Willems, N. Parktin, W. Widjaja and A. Ajjimaporn, *Nutrients*, 2018, **10**(11), 1732.
98. E. Wood, S. Hein, C. Heiss, C. Williams and A. Rodriguez-Mateos, *Food Funct.*, 2019, **10**, 7621–7633.
99. C. Del Bo, J. Campolo, M. Porrini, D. Fracassetti, M. Parolini, D. Klimis-Zacas and P. Riso, *PharmaNutrition*, 2014, **2**, 1–7.
100. C. Del Bo, V. Deon, J. Campolo, C. Lanti, M. Parolini, M. Porrini, D. Klimis-Zacas and P. Riso, *Food Funct.*, 2017, **8**, 4108–4117.
101. C. Del Bo, M. Porrini, D. Fracassetti, J. Campolo, D. Klimis-Zacas and P. Riso, *Food Funct.*, 2014, **5**, 3107–3116.
102. C. Del Bo, P. Riso, J. Campolo, P. Moller, S. Loft, D. Klimis-Zacas, A. Brambilla, A. Rizzolo and M. Porrini, *Nutr. Res.*, 2013, **33**, 220–227.
103. A. Rodriguez-Mateos, R. Del Pino-Garcia, T. W. George, A. Vidal-Diez, C. Heiss and J. P. Spencer, *Mol. Nutr. Food Res.*, 2014, **58**, 1952–1961.
104. A. Rodriguez-Mateos, G. Istas, L. Boschek, R. P. Feliciano, C. E. Mills, C. Boby, S. Gomez-Alonso, D. Milenkovic and C. Heiss, *J. Gerontol., Ser. A*, 2019, **74**(7), 967–976.
105. A. Rodriguez-Mateos, C. Rendeiro, T. Bergillos-Meca, S. Tabatabaee, T. W. George, C. Heiss and J. P. Spencer, *Am. J. Clin. Nutr.*, 2013, **98**, 1179–1191.
106. S. A. Johnson, A. Figueroa, N. Navaei, A. Wong, R. Kalfon, L. T. Ormsbee, R. G. Feresin, M. L. Elam, S. Hooshmand, M. E. Payton and B. H. Arjmandi, *J. Acad. Nutr. Diet.*, 2015, **115**, 369–377.

107. A. Basu, M. Du, M. J. Leyva, K. Sanchez, N. M. Betts, M. Wu, C. E. Aston and T. J. Lyons, *J. Nutr.*, 2010, **140**, 1582–1587.
108. L. S. McAnulty, S. R. Collier, M. J. Landram, D. S. Whittaker, S. E. Isaacs, J. M. Klemka, S. L. Cheek, J. C. Arms and S. R. McAnulty, *Nutr. Res.*, 2014, **34**, 577–584.
109. K. S. Stote, M. I. Sweeney, T. Kean, D. J. Baer, J. A. Novotny, N. L. Shakerley, A. Chandrasekaran, P. M. Carrico, J. A. Melendez and K. T. Gottschall-Pass, *BMC Nutr.*, 2017, **3**, 45.
110. P. Riso, D. Klimis-Zacas, C. Del Bo, D. Martini, J. Campolo, S. Vendrame, P. Moller, S. Loft, R. De Maria and M. Porrini, *Eur. J. Nutr.*, 2013, **52**, 949–961.
111. A. J. Stull, K. C. Cash, C. M. Champagne, A. K. Gupta, R. Boston, R. A. Beyl, W. D. Johnson and W. T. Cefalu, *Nutrients*, 2015, **7**, 4107–4123.
112. P. J. Curtis, V. van der Velpen, L. Berends, A. Jennings, M. Feelisch, A. M. Umpleby, M. Evans, B. O. Fernandez, M. S. Meiss, M. Minnion, J. Potter, A. M. Minihane, C. D. Kay, E. B. Rimm and A. Cassidy, *Am. J. Clin. Nutr.*, 2019, **109**, 1535–1545.
113. A. Rodriguez-Mateos, G. Istas, L. Boschek, R. P. Feliciano, C. E. Mills, C. Boby, S. Gomez-Alonso, D. Milenkovic and C. Heiss, *J. Gerontol. Ser.A*, 2019.
114. S. A. Johnson, A. Figueroa, N. Navaei, A. Wong, R. Kalfon, L. T. Ormsbee, R. G. Feresin, M. L. Elam, S. Hooshmand, M. E. Payton and B. H. Arjmandi, *J. Acad. Nutr. Diet.*, 2015, **115**, 369–377.
115. L. S. McAnulty, S. R. Collier, M. J. Landram, D. S. Whittaker, S. E. Isaacs, J. M. Klemka, S. L. Cheek, J. C. Arms and S. R. McAnulty, *Nutr. Res.*, 2014, **34**, 577–584.
116. D. L. McKay and J. B. Blumberg, *Nutr. Rev.*, 2007, **65**, 490–502.
117. M. M. Dohadwala, M. Holbrook, N. M. Hamburg, S. M. Shenouda, W. B. Chung, M. Titas, M. A. Kluge, N. Wang, J. Palmisano, P. E. Milbury, J. B. Blumberg and J. A. Vita, *Am. J. Clin. Nutr.*, 2011, **93**, 934–940.
118. A. Rodriguez-Mateos, R. P. Feliciano, A. Boeres, T. Weber, C. N. Dos Santos, M. R. Ventura and C. Heiss, *Mol. Nutr. Food Res.*, 2016, **60**, 2130–2140.
119. D. S. Hsia, D. J. Zhang, R. S. Beyl, F. L. Greenway and C. Khoo, *Br. J. Nutr.*, 2020, **124**, 577–585.
120. G. Ruel, A. Lapointe, S. Pomerleau, P. Couture, S. Lemieux, B. Lamarche and C. Couillard, *Nutr. Res.*, 2013, **33**, 41–49.
121. T. Lim, J. Ryu, K. Lee, S. Y. Park and K. T. Hwang, *Nutrients*, 2020, **12**, 2448.
122. H. S. Jeong, S. J. Hong, T. B. Lee, J. W. Kwon, J. T. Jeong, H. J. Joo, J. H. Park, C. M. Ahn, C. W. Yu and D. S. Lim, *Phytother. Res.*, 2014, **28**, 1492–1498.
123. H. S. Jeong, S. Kim, S. J. Hong, S. C. Choi, J. H. Choi, J. H. Kim, C. Y. Park, J. Y. Cho, T. B. Lee, J. W. Kwon, H. J. Joo, J. H. Park, C. W. Yu and D. S. Lim, *J. Med. Food*, 2016, **19**, 346–352.

124. H. S. Jeong, S. J. Hong, J. Y. Cho, T. B. Lee, J. W. Kwon, H. J. Joo, J. H. Park, C. W. Yu and D. S. Lim, *Nutrition*, 2016, **32**, 461–467.
125. G. Istas, R. P. Feliciano, T. Weber, R. Garcia-Villalba, F. Tomas-Barberan, C. Heiss and A. Rodriguez-Mateos, *Arch. Biochem. Biophys.*, 2018, **651**, 43–51.
126. J. A. Rothwell, J. Perez-Jimenez, V. Neveu, A. Medina-Remon, N. M'Hiri, P. Garcia-Lobato, C. Manach, C. Knox, R. Eisner, D. S. Wishart and A. Scalbert, *Database*, 2013, **2013**, bat070.
127. C. K. Richter, A. C. Skulas-Ray, T. L. Gaugler, J. D. Lambert, D. N. Proctor and P. M. Kris-Etherton, *Am. J. Clin. Nutr.*, 2017, **105**, 313–322.
128. D. Djurica, R. R. Holt, J. Ren, A. W. Shindel, R. M. Hackman and C. L. Keen, *Br. J. Nutr.*, 2016, **116**, 639–647.
129. R. G. Feresin, S. A. Johnson, S. Pourafshar, J. C. Campbell, S. J. Jaime, N. Navaei, M. L. Elam, N. S. Akhavan, S. Alvarez-Alvarado, G. Tenenbaum, K. Brummel-Smith, G. Salazar, A. Figueroa and B. H. Arjmandi, *Food Funct.*, 2017, **8**, 4139–4149.
130. L. Huang, D. Xiao, X. Zhang, A. K. Sandhu, P. Chandra, C. Kay, I. Edirisinghe and B. Burton-Freeman, *J. Nutr.*, 2021, **151**, 1517–1526.
131. E. Niki, *J. Berry Res.*, 2011, **1**(4), 169–176.
132. S. E. Pollard, G. G. Kuhnle, D. Vauzour, K. Vafeiadou, X. Tzounis, M. Whiteman, C. Rice-Evans and J. P. Spencer, *Biochem. Biophys. Res. Commun.*, 2006, **350**, 960–968.
133. Y. Steffen, T. Schewe and H. Sies, *Biochem. Biophys. Res. Commun.*, 2007, **359**, 828–833.
134. C. Petersen, D. Bharat, B. R. Cutler, S. Gholami, C. Denetso, J. E. Mueller, J. M. Cho, J. S. Kim, J. D. Symons and P. V. Anandh Babu, *Int. J. Cardiol.*, 2018, **263**, 111–117.
135. Y. Shen, K. D. Croft, J. M. Hodgson, R. Kyle, I.-L. E. Lee, Y. Wang, R. Stocker and N. C. Ward, *Biochem. Pharmacol.*, 2012, **84**(8), 1036–1044.
136. S. Satta, A. M. Mahmoud, F. L. Wilkinson, M. Yvonne Alexander and S. J. White, *Oxid. Med. Cell. Longevity*, 2017, **2017**, 9237263.
137. Y. Ding, B. Zhang, K. Zhou, M. Chen, M. Wang, Y. Jia, Y. Song, Y. Li and A. Wen, *Int. J. Cardiol.*, 2014, **175**, 508–514.
138. A. Mauray, C. Felgines, C. Morand, A. Mazur, A. Scalbert and D. Milenkovic, *Nutr. Metab. Cardiovasc. Dis.*, 2012, **22**, 72–80.
139. C. T. Yeh, L. C. Ching and G. C. Yen, *J. Nutr. Biochem.*, 2009, **20**, 163–171.
140. K. Goszcz, G. G. Duthie, D. Stewart, S. J. Leslie and I. L. Megson, *Br. J. Pharmacol.*, 2017, **174**(11), 1209–1225.
141. A. Karlsen, L. Retterstol, P. Laake, I. Paur, S. K. Bohn, L. Sandvik and R. Blomhoff, *J. Nutr.*, 2007, **137**, 1951–1954.
142. A. R. Nair, N. Mariappan, A. J. Stull and J. Francis, *Food Funct.*, 2017, **8**, 4118–4128.
143. A. Basu, D. X. Fu, M. Wilkinson, B. Simmons, M. Wu, N. M. Betts, M. Du and T. J. Lyons, *Nutr. Res.*, 2010, **30**, 462–469.

144. M. C. Rodriguez-Daza, M. Roquim, S. Dudonne, G. Pilon, E. Levy, A. Marette, D. Roy and Y. Desjardins, *Front. Microbiol.*, 2020, **11**, 2032.
145. A. Ntemiri, T. S. Ghosh, M. E. Gheller, T. T. T. Tran, J. E. Blum, P. Pellanda, K. Vlckova, M. C. Neto, A. Howell, A. Thalacker-Mercer and P. W. O'Toole, *Nutrients*, 2020, **12**(9), 2800.
146. Z. Ezzat-Zadeh, S. M. Henning, J. Yang, S. L. Woo, R. P. Lee, J. Huang, G. Thames, I. Gilbuena, C. H. Tseng, D. Heber and Z. Li, *Nutr. Res.*, 2021, **85**, 60–70.
147. A. Gomes, C. Oudot, A. Macia, A. Foito, D. Carregosa, D. Stewart, T. Van de Wiele, D. Berry, M. J. Motilva, C. Brenner and C. N. Dos Santos, *Nutrients*, 2019, **11**(11), 2634.
148. J. Gu, J. M. Thomas-Ahner, K. M. Riedl, M. T. Bailey, Y. Vodovotz, S. J. Schwartz and S. K. Clinton, *Mol. Nutr. Food Res.*, 2019, **63**, e1800636.
149. K. L. Ivey, J. M. Hodgson, K. D. Croft, J. R. Lewis and R. L. Prince, *Am. J. Clin. Nutr.*, 2015, **101**, 1012–1020.
150. J. I. Dower, J. M. Geleijnse, L. Gijsbers, P. L. Zock, D. Kromhout and P. C. H. Hollman, *Am. J. Clin. Nutr.*, 2015, **101**, 914–921.
151. R. Feliciano, G. Istas, C. Heiss and A. Rodriguez-Mateos, *Molecules*, 2016, **21**, 1120, 2016, 21, 1120-1120.
152. R. P. Feliciano, C. E. Mills, G. Istas, C. Heiss and A. Rodriguez-Mateos, *Nutrients*, 2017, **9**(3), 268.
153. E. R. Gibney, D. Milenkovic, E. Combet, T. Ruskovska, A. Greyling, A. González-Sarrías, B. de Roos, F. Tomás-Barberán, C. Morand and A. Rodriguez-Mateos, *Eur. J. Nutr.*, 2019, **58**, 37–47.
154. K. Kawabata, Y. Yoshioka and J. Terao, *Molecules*, 2019, **24**(2), 370.
155. F. A. Tomas-Barberan, M. V. Selma and J. C. Espin, *Curr. Opin. Clin. Nutr. Metab. Care*, 2016, **19**, 471–476.
156. A. Cassidy and A. M. Minihane, *Am. J. Clin. Nutr.*, 2017, **105**, 10–22.
157. E. S. Chambers, T. Preston, G. Frost and D. J. Morrison, *Curr. Nutr. Rep.*, 2018, **7**, 198–206.
158. B. Bayram, A. González-Sarrías, G. Istas, M. Garcia-Aloy, C. Morand, K. Tuohy, R. García-Villalba and P. Mena, *J. Agric. Food Chem.*, 2018, **66**, 10686–10692.

CHAPTER 6

Mechanisms of Action of Berry Bioactives on the Vascular System

PON VELAYUTHAM ANANDH BABU*[a] AND
DOROTHY KLIMIS-ZACAS*[b]

[a] Department of Nutrition and Integrative Physiology, College of Health, University of Utah, Salt Lake City, UT 84112, USA; [b] School of Food and Agriculture, University of Maine, Orono, ME 04469, USA
*Emails: anandh.velayutham@utah.edu; dorothea@maine.edu

6.1 Introduction

Cardiovascular disease (CVD) is the leading cause of death for men and women in the US.[1] CVD accounted for one in every four deaths in the United States and 655 000 Americans die from CVD each year.[2] Lifestyle factors (such as smoking, lack of physical activity, nutrient-poor diet) and health factors (such as hyperglycemia, dyslipidemia, and hypertension) contribute to the development of CVD.[1] Treating cardiovascular complications contributes to increasing medical costs that burden the patient and society. CVD is mainly initiated through atherosclerosis that has a major impact on organs leading to heart attack, stroke, and peripheral artery disease. Atherosclerosis is the build-up of plaques within the arterial walls, and over time, this accumulation can form an elevated plaque that protrudes and narrows the vascular lumen leading to restricted blood flow. Plaques are composed of an accumulation of lipids, cells, fibrous tissue, and debris.

Food Chemistry, Function and Analysis No. 33
Berries and Berry Bioactive Compounds in Promoting Health
Edited by Dorothy Klimis-Zacas and Ana Rodriguez-Mateos

Published by the Royal Society of Chemistry, www.rsc.org

In plaques that become weak and rupture in severe cases, clotting factors such as tissue factor-FVIIa, FXa, and thrombin adhere to the rupture site.[3,4] Subsequently, blood flow is restricted, leading to complications like heart attack and stroke.[3] Studies indicate that approximately 1.5 million American adults have a heart attack or stroke each year.[5]

Diet has a major impact on a variety of cardiovascular risk factors, including dyslipidemia, oxidative stress, inflammation, and blood pressure.[6] Epidemiological and clinical studies have documented that a healthy diet including a variety of fruits and vegetables can prevent the development of CVD and reduce premature morbidity and mortality.[1,7–11] Indeed, the Mediterranean-style diet and dietary approaches to stop hypertension (DASH) diet, which emphasize fruits and vegetables, were shown to improve cardiovascular health by reducing blood pressure, triglycerides, and body fat.[12,13] Berries are an important part of the diet as they are a rich source of a wide variety of non-nutritive bioactive compounds (such as flavonoids, polyphenols, and derivatives of stilbenes) in addition to nutritive compounds (such as carbohydrates, vitamins, and minerals). One of the major flavonoids and bioactive components of berries is anthocyanins. Berry anthocyanins are galactoside, glucoside, and arabinoside derivatives of delphinidin, cyanidin, malvidin, petunidin, and peonidin that imbue berries with dark colors.[14] The anthocyanin content of berries depends on the size of the fruit, ripening stage, environment, climate, and storage.[14] The mean dietary intake of anthocyanins in the United States is approximately 12.5 mg per day, and it is mostly from strawberries, blueberries, blackberries, and raspberries that are the most commonly consumed berries in the United States.[15,16] Berries contribute significantly to the human diet and berry consumption has shown positive effects on cardiovascular health.[14,17–20] Epidemiological and clinical studies indicate the beneficial effects of berries on cardiovascular surrogate markers including blood lipids, endothelial function, arterial stiffness, and blood pressure.[14,17–21] The reported vascular effects of berry anthocyanins are attributed to their antioxidant, anti-inflammatory, anti-atherosclerotic, and anti-hypertensive properties.[19,20]

This chapter highlights recent advancements toward our understanding of the vascular effects of dietary berries such as blueberries, strawberries, and raspberries, with special emphasis on the molecular mechanisms involved.

6.2 Metabolism of Berry Bioactives

The biological activity of berry bioactives is linked to their absorption and bioavailability.[22,23] Many factors such as pH, presence of other nutrients, intestinal microbiota, and chemical structure influence the bioavailability of berry bioactives such as anthocyanins.[23] Following the consumption of berries, less than 2% of parent anthocyanins reach the circulation and target organs. Evidence indicates that most of the anthocyanins are not metabolized in the upper gastrointestinal tract but reach the cecum and colon.[24,25] Anthocyanins are metabolized by digestive enzymes in the small intestine and by the action of intestinal microbiota in the large intestine. The metabolism of berry bioactives includes removal of the sugar moiety, conjugation (glucuronidation, sulfation, and methylation), catabolic breakdown, and structural rearrangements.[26] In humans, anthocyanins are extensively metabolized by digestive enzymes and gut microbiota, suggesting that the reported vascular benefits of berries are mediated by their circulating metabolites.[27] Anthocyanins are biotransformed into diverse phenolic metabolites and intermediates that retain the flavonoid backbone structure.[22,28] Further, the resident time of these metabolites in the circulation is sufficient to mediate their biological activity in humans.[22,28] These studies indicate that the anthocyanins are far more bioavailable than previously proposed.[22] Evidence from several laboratories indicates that the reported vascular beneficial effects of berry bioactives could be mediated by the metabolites of berry bioactives as opposed to their parent structure.[29–32] Indeed, in a human study, blueberry intake (freeze-dried powder at a dose that is equivalent to 240 g of whole blueberries) increased flow-mediated dilation, which was accompanied by an increase in the plasma concentrations of phenolic metabolites such as homovanillic acid, vanillic acid, benzoic acid, and hippuric acid.[28] Recently, in a blueberry intervention study, 13 metabolite signatures were shown to be independent predictors of genes involved in cell migration, adhesion, and differentiation as well as immune response and concluded that blueberry anthocyanin metabolites are critical for vascular bioactivity.[33] However, only limited studies assessed the endothelial and vascular effects of the circulating metabolites as many of these metabolites are not commercially available.[31,34,35]

6.3 Vascular Effects of Berries

Evidence from epidemiological, clinical, and animal studies supports the beneficial effects of dietary berries on cardiovascular health.[28,36,37] Epidemiological studies indicate that the consumption of two to three servings of blueberries or strawberries reduces the risk of myocardial infarction.[38] Clinical studies have documented that dietary blueberry improves vascular function in healthy men, reduces blood pressure in individuals with Metabolic Syndrome (MetS) or hypertension, and ameliorates peripheral arterial dysfunction in individuals with endothelial dysfunction.[28,33,37,39–42] Blueberry supplementation also increased reactive hyperemia index (a measure of endothelial function) in healthy male smokers, non-smokers with peripheral artery dysfunction, and individuals with MetS.[41,43,44] A study conducted in postmenopausal women with pre- and stage-1 hypertension indicated a reduction in systolic and diastolic blood pressures and branchial-ankle pulse wave velocity (a measure of systemic arterial stiffness) following 8 weeks of blueberry consumption.[40] Consistently, blueberry supplementation for 6 weeks reduced arterial stiffness in sedentary males and females, as shown by a reduced augmentation index.[45] Strawberry intake was shown to improve vascular function and reduce vascular adhesion molecules in humans with cardiovascular risk factors.[46,47] Further, dietary strawberry increased plasma nitric oxide (NO) metabolites level in pre- and stage-1 hypertensive postmenopausal women.[48] Consumption of red raspberries at nutritional dosage was reported to improve endothelial function as shown by an increase in flow-mediated dilation in healthy men, which is associated with an increase in the phenolic metabolites of raspberry bioactives.[49] Animal studies indicate that dietary blueberries suppress α-adrenergic-induced vasoconstriction and reduce blood pressure in spontaneously hypertensive rats (SHRs).[50,51] Strawberry supplementation at nutritional doses suppresses vascular inflammation, reduces arterial pressure, and improves endothelial-dependent vessel relaxation in diabetic mice.[52] Interestingly, evidence indicates that the metabolites of berry bioactives play a pivotal role in mediating the vascular beneficial effects of dietary berries.[22,28,31,49] A detailed overview of these studies is discussed in Chapter 5.

6.4 Molecular Mechanisms Underlying the Vascular Effects of Berries

Emerging evidence indicates that berry bioactives exert vascular beneficial effects by acting on multiple targets such as (i) activating endothelial nitric oxide synthase (eNOS)/NO/cyclic guanosine monophosphate (cGMP) signaling pathway, (ii) suppressing endothelial inflammation, (iii) modulating redox signaling, (iv) altering lipid metabolism, and (v) modifying vascular glycocalyx (Figure 6.1).

6.4.1 Activation of eNOS Signaling Pathway

The endothelium is made up of a monolayer of endothelial cells (ECs) and plays a critical role in controlling vascular homeostasis.[53] The proper functioning of endothelium is required for a healthy vasculature and an impairment in endothelium could lead to irreversible loss of endothelial function, which contributes to vascular dysfunction. Indeed, endothelial dysfunction is a characteristic phenotype of vascular complications associated with metabolic diseases that can lead to atherosclerosis and hypertension.[53,54] eNOS is a homodimeric enzyme that catalyzes the production of NO and L-citrulline from L-arginine in the endothelium.[55] NO maintains vascular homeostasis by acting as a vasodilatory agent and a signaling molecule. Endothelial NO activates soluble guanylate cyclase in the vascular smooth muscle cells that results in the release of second messenger cGMP, which then causes the smooth muscle cells of the vessels to relax.[56] Further, endothelial-derived NO exhibits anti-inflammatory, anti-platelet, and antiproliferative properties.[53] An impaired eNOS and lack of endogenous NO production leads to the development of atherosclerosis by accelerating leukocyte rolling and binding to the aortic cells, arterial stiffness, and an increase in blood pressure.[57] The activation of eNOS/NO/cGMP signaling in

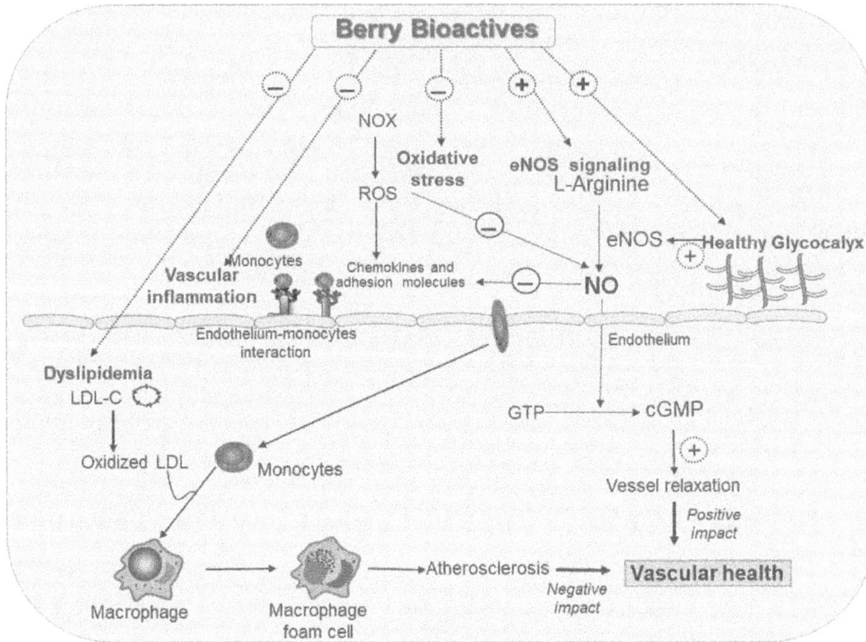

Figure 6.1 Molecular mechanisms involved in the vascular effects of dietary berries.

the vasculature is critical for proper vessel relaxation and to maintain vascular tone.

Emerging evidence indicates that berry bioactives improve endothelial dysfunction by acting on eNOS/NO/cGMP signaling. Blueberry consumption (22 g freeze-dried blueberry for 8 weeks) decreased blood pressure and reduced arterial stiffness in postmenopausal women with pre- and stage-1 hypertension in a randomized, double-blind, placebo-controlled clinical trial.[40] In this study, an increase in NO was reported in participants who consumed blueberry, indicating that the vascular beneficial effect may be mediated through NO signaling. Further, blueberry consumption (240 mL wild blueberry juice for 7 days) increased the concentration of nitrates and nitrites (an index of NO production) in adults at risk for type 2 diabetes.[58] Animal studies indicated that dietary supplementation of blueberry (8% w/w wild blueberry diet for 7 or 8 weeks) improves vascular reactivity in young and old SHRs.[59,60] These studies used NO and cyclooxygenase (COX) inhibitors and showed two pathways may be involved in exerting the vascular beneficial effects of blueberries – activation of the NO-mediated pathway and suppression of COX-dependent vasoconstriction pathway.[59,60] Another study from this group further showed that dietary blueberries (8% w/w wild blueberry diet for 8 weeks) suppress α-adrenergic-induced vasoconstriction in SHRs, which is associated with the increased concentration of cGMP that mediates vessel relaxation.[50] To determine the effect of blueberry bioactives on the NO-cGMP pathway, the investigators used inhibitors that target guanylate cyclase (an enzyme that synthesizes cGMP from GTP) and cGMP-specific phosphodiesterase-5 (an enzyme that breakdown cGMP). This study suggested that blueberry bioactives preserve vascular cGMP by directly acting on guanylate cyclase and/or preventing the degradation of cGMP by reducing phosphodiesterase. Berry bioactives are extensively metabolized in humans, and hence an *in vitro* study assessed the effect of metabolites on palmitate-induced impaired NO production. In this study, a blueberry metabolite cocktail at physiologically relevant concentrations restored insulin-stimulated NO production in human aortic ECs exposed to lipotoxicity.[31] Further, blueberry metabolites improved palmitate-induced endothelial dysfunction in aortic vessels by increasing endothelial-dependent vasorelaxation, indicating the possible involvement of the eNOS/NO pathway.[31] The effect of strawberry on eNOS/NO signaling is reported in a few studies. Strawberry intake (50 g freeze-dried powder equivalent to 450 g of whole strawberries) improved vascular function in overweight adolescents, which is associated with increased plasma nitrate/nitrate.[46] Consistent with this study, dietary strawberry (50 g freeze-dried powder for 4 and 8 weeks) increased plasma NO metabolites (nitrate and nitrite) level in pre- and stage-1 hypertensive postmenopausal women.[48] Strawberry supplementation (2.35% freeze-dried strawberry in diet for 10 weeks) was also shown to increase endothelial-dependent vasorelaxation without altering endothelial-independent vasorelaxation in diabetic db/db mice. This study further indicates strawberry may improve endothelial dysfunction in diabetic mice by activating eNOS/NO pathway.[52]

Collectively, *in vitro* and *in vivo* studies indicate that berry bioactives may exert vascular effects by activating eNOS/NO/cGMP signaling. However, studies are focused on the role of berry bioactives on eNOS phosphorylation are limited. Furthermore, the specific metabolites responsible for the activation of eNOS/NO signaling need to be further determined.

6.4.2 Suppression of Endothelial Inflammation

Endothelial inflammation plays a pivotal role in the development of atherosclerosis by inducing endothelial dysfunction. The healthy endothelium resists prolonged contact and binding of circulating blood monocytes. Conversely, CVD risk factors such as diabetes, dyslipidemia, hypertension, and smoking activate the endothelium. An activated endothelium secrets pro-inflammatory mediators such as tumor necrosis factor-α (TNF-α), monocyte chemotactic protein-1 (MCP-1), interleukin-8 (IL-8), vascular cell adhesion molecule-1 (VCAM-1), intracellular adhesion molecule-1 (ICAM-1), and E-selectin.[61–64] These inflammatory factors accelerate leukocyte rolling, increase their recruitment from the circulation to the vascular ECs, and their subsequent migration to the subendothelial layer, a key event in the development of atherosclerosis.[65–67] In the innermost layer of the vasculature, the migrated leukocytes differentiate into macrophage foam cells by their uptake of several lipid and lipoprotein molecules.[68] Further, foam cells secret a number of pro-inflammatory molecules such as TNF-α, IL-8, and E-selectin to amplify the inflammatory response and consequently lead to vascular complications.[68,69] The sites of the arteries near the branches and bends are prone to atherosclerosis because of the low shear stress on vascular ECs in these regions. Indeed, low shear stress is a potential activator of endothelial inflammation.[70]

Nuclear factor κB (NFκB), a pro-inflammatory transcription factor, promotes vascular inflammation by upregulating inflammatory chemokines and adhesion molecules and it is activated by NOX-derived reactive oxygen species (ROS).[71] The most abundant form of NFκB is p50/p65. Under homeostatic conditions, it is retained in the cytoplasm by binding to IκBα (an inhibitor of NFκB).[72] However, inflammatory conditions activate IκB kinase (IκKβ), which phosphorylates IκBα that leads to the proteasomal degradation of IκBα and subsequent nuclear translocation of p50/p65.[72] P50/p65 dimer in the nucleus binds to the promoter regions of NFκB-dependent inflammatory genes and upregulates the genes involved in vascular inflammation and atherosclerosis.[73,74] Studies indicate that p65 nuclear translocation in ECs and aortic vessels are involved in diabetes-related vascular complications.[75,76]

Evidence indicates that dietary berries and metabolites of berry bioactives improve cardiovascular complications by suppressing endothelial and vascular inflammation *via* the NFκB signaling cascade.[77] Dietary supplementation of wild blueberry (8% blueberry in diet for 8 weeks) was shown to improve pro-inflammatory status associated with metabolic syndrome in obese Zucker rats, which is an experimental model for human metabolic syndrome.[78] In this study, blueberry supplementation reduced plasma inflammatory markers (TNF-α, IL-6, and C-reactive protein [CRP]) and downregulated NFκB expression in obese rats. Even the lean control rats exhibited reduced NFκB, CRP, IL-6, and TNF-α in the liver with blueberry supplementation. In another short-term study, blueberry supplementation (1% freeze-dried blueberries in diet for 5 weeks) reduced the production of TNF-α and IL-6 in macrophages of Apo E mice by inhibiting the activation of NFκB.[79] In addition, dietary blueberry reduced serums TNF-α and IL-6, which are associated with downregulated aortic TNF-α expression. Furthermore, wild blueberry intervention resulted in attenuation of local inflammation in perivascular adipose tissue indicated by downregulation of pro-inflammatory markers and upregulation of anti-inflammatory markers.[80] In another study, strawberry supplementation (2.35% freeze-dried strawberry, which is equivalent to two human servings of fresh strawberries for 8 weeks) reduced vascular inflammation in diabetic db/db mice as shown by reduced monocyte binding to the aortic vessel, reduced expression of aortic inflammatory markers (VCAM-1, MCP-1/JE, and KC/IL-8), and reduced serum inflammatory molecules (MCP-1/JE and KC/IL-8).[52] In addition, downregulation of MCP-1/JE, KC/IL-8, VCAM-1, IκKβ, and NOX2 was observed in ECs isolated from strawberry supplemented diabetic mice as compared to untreated diabetic mice. The findings indicate an endothelial-specific anti-inflammatory effect of strawberry bioactives. This study further investigated whether the endothelial effects of strawberry are mediated through their metabolites.[52] Mouse aortic ECs exposed to high glucose and palmitate were incubated with serum obtained from strawberry-fed mice (metabolite serum) or standard diet-fed mice (control serum). Metabolites in serum significantly reduced high glucose and palmitate-induced inflammation in ECs, as shown by reduced monocyte binding to them and reduced secretion of KC/IL-8. This study suggests that the endothelial benefits of dietary strawberry could be due to the metabolites of strawberry bioactives present in the serum.[52] An *in vitro* study assessed the effect of blueberry metabolites on lipotoxicity-induced endothelial inflammation.[31] In this study, human aortic ECs were treated with selected parent anthocyanins and metabolites of blueberries (at physiologically relevant concentrations) with or without palmitate. Blueberry metabolites but not the parent anthocyanins suppressed the palmitate-induced endothelial inflammation as shown by reduced monocyte binding to ECs, reduced expression of inflammatory markers (IL-8, ICAM-1, and VCAM-1), and NOX4 expression.[31] Consistently, these blueberry metabolites suppressed monocyte binding and reduced IL-8 and VCAM-1 in ECs isolated from type 2 diabetic individuals.[32]

In a human study, dietary supplementation of blueberry (45 g freeze-dried blueberry equivalent to two cups of whole blueberries) for 6 weeks improved insulin sensitivity without changing the serum inflammatory markers (IL-6, inducible nitric oxide synthase (iNOS), and MCP-1).[81] The difference in inflammatory marker findings between human and animal studies could be due to the method of evaluation. Animal studies measured the gene expression, whereas human studies measured the serum inflammatory markers.[81] Strawberry supplementation was shown to reduce the circulating levels of ICAM-1 and VCAM-1 in humans with cardiovascular risk factors.[46,47] Supplementation of raspberries (250 g red raspberries for 4 weeks) was shown to reduce IL-6 and high sensitivity TNF-α in adults with type 2 diabetes.[82]

From the above studies, we observe that the limited evidence points to the proteomic and genomic role of berry bioactives in modulating the NFκB pathway. Collectively, the anti-inflammatory effects of berry bioactives may be one of the reasons for their vascular protective effects. However, studies are needed to determine the specific signaling pathways that are activated or inhibited by berry bioactives to exert these anti-inflammatory effects.

6.4.3 Modification of Redox Signaling

Cardiovascular risk factors such as diabetes, dyslipidemia, hypertension, aging, and smoking upregulate ROS and oxidative stress.[83] Oxidative stress plays a pivotal role in mediating endothelial inflammation and subsequent atherosclerosis.[83] One of the important sources of ROS in the vasculature is nicotinamide adenine dinucleotide phosphate (NADPH) oxidases (NOXs).

NOXs play an important role in the maintenance of normal physiological function of the vasculature as well as in vascular pathology.[84] NOXs-mediated vascular redox signaling is involved in vascular homeostasis, endothelial dysfunction, hypertension, and atherosclerosis.[85] NOX is one of the sources of ROS that creates superoxide molecules and affects the bioavailability of NO. In healthy endothelium, eNOS produces NO but under pathological conditions, an uncoupled eNOS acts as a potent ROS generator.[83] In addition, ROS interacts with NO to produce a highly oxidizing peroxynitrite ($ONOO^-$). Hence, ROS mediates endothelial dysfunction by uncoupling eNOS, reducing the bioavailability of NO, producing peroxynitrite ($ONOO^-$), and stimulating endothelin expression.[27,83,86] In addition, endothelium activated by ROS attracts monocytes to the vasculature. Further, ROS mediates the production of oxidized low-density lipoprotein (LDL) particles that enter the subendothelial space and mediate the formation of macrophage-derived foam cells and the subsequent formation of atherogenic plaques. Hence reducing ROS and inhibiting ROS generating sources is one of the potential targets to prevent atherosclerosis.

Blueberry supplemented diet (2% w/w for 12 weeks) was shown to improve redox status and reduce blood pressure in SHRs.[51] In a human study, blueberry intake (single portion of 300 g blueberries) improved cell antioxidant defense against DNA damage but did not alter NO levels and peripheral artery function.[87] In another study, blueberry consumption (25 g freeze-dried powder for 6 weeks) reduced endogenously oxidized DNA bases and increased the resistance to oxidative DNA damage in subjects with risk factors for CVD.[88] Blueberry consumption (freeze-dried powder equivalent to 75 g of fresh blueberry) with a high-carbohydrate, low-fat diet suppressed the postprandial serum markers and increased serum oxygen radical absorbance capacity in the participants.[89] Further, blueberry intake (freeze-dried powder equivalent to 350 g of fresh blueberries) reduced the levels of biomarkers of lipid and lipoprotein oxidation such as plasma oxidized LDL and serum malondialdehyde in obese men and women, suggesting that the vascular effects may be due to their role on redox signaling.[37] Dietary blueberry (300 g single serving) improved smoking-induced arterial dysfunction in young participants but did not improve the markers of oxidative stress and antioxidant status in smokers.[43,90]

Wild blueberry supplementation (8% freeze-dried in diet for 8 weeks) partially restored phenylephrine-induced constriction and attenuated exaggerated response to acetylcholine-induced vasorelaxation, which is associated with downregulation of inducible iNOS in obese Zucker rats.[91] This study suggests that blueberry can suppress the overproduction of NO, which mediates peroxynitrite radical production. Another study assessed the effect of blueberries on the antioxidant defense system.[92] In this study, dietary blueberry (8% freeze-dried blueberry in diet) increased the activity of hepatic glutathione *S*-transferase in high-cholesterol fed guinea pigs without altering the levels of glutathione, superoxide dismutase, or glutathione peroxidase.[92] Blueberry intake was shown to decrease neutrophil NOX activity, which is associated with an increase in the concentration of anthocyanin metabolites and increased flow-mediated dilation in healthy humans.[85] In a recent study, strawberry supplementation (2.35% in diet) reduced the endothelial expression of NOX2 and IκKβ in the carotid artery, which is associated with an improvement in vascular inflammation and dysfunction in diabetic db/db mice. Studies were also conducted with metabolites of berry bioactives. A blueberry metabolite cocktail containing circulating metabolites at physiologically relevant concentrations reduced lipotoxicity-induced NOX4 expression and ROS production in human aortic ECs, which was associated with an improved endothelial inflammation.[31]

Evidence indicates that berry bioactives and their metabolites improve vascular complications by suppressing NOX-mediated signaling. However, further studies are needed to identify the specific targets of berry bioactives on NOX signaling and the molecular mechanisms involved.

6.4.4 Alteration of Lipid Metabolism

Dyslipidemia is one of the major risk factors for CVD. Increased levels of plasma total cholesterol, triglycerides, and non-low-density lipoprotein (non-HDL) contribute to the pathogenesis of vascular disease by inducing endothelial inflammation, impairing NO synthesis, reducing NO bioavailability, and enhancing the formation of foam cells.[93] Both human and animal studies suggest that the vascular beneficial effects of berry bioactives may be mediated by improving dyslipidemia.

Evidence shows that pigs develop atherosclerotic plaque in their aorta and carotid arteries similar to humans and few studies used pigs to study the effect of blueberries on vascular complications, specifically focusing on lipid metabolism.[92,94] Blueberry supplementation (8% blueberry in diet) for 75 days improved aortic atherosclerotic lesions in high-cholesterol-fed guinea pigs, which are associated with a reduced serum cholesterol, LDL-cholesterol, and aortic cholesterol levels.[92] Another study reported that blueberry supplementation (1%, 2%, and 4% in diet) for 4 and 8 weeks reduced total, LDL-, and HDL-cholesterol in the plasma of pigs.[94] Blueberry juice (*ad libitum* consumption of blueberry juice instead of water for 12 weeks) was shown to decrease serum cholesterol, reduce lipid accumulation, and increase β-oxidation of fatty acid in high-fat diet-fed C57BL/6 mice.[95]

The effect of dietary blueberry (8% wild blueberry diet for 8 weeks) on lipid metabolism in obese Zucker rats, which is a model of the MetS with severe dyslipidemia, was reported.[96] In this study, blueberry supplementation decreased total cholesterol and triglycerides, which was associated with decreased levels of sterol regulatory element-binding protein 1 (transcription factor that promotes the synthesis and accumulation of triglycerides) and reduced expression of fatty acid synthase in obese Zucker rats. In addition, blueberry upregulated the expression of PPARα, which promotes lipid catabolism through activation of enzymes involved in fatty acid uptake and β-oxidation of fatty acid. This study indicated that the bioactives in blueberries may improve dyslipidemia by modulating the expression of key enzymes and transcription factors of lipid metabolism.[96] Blueberry consumption (150 g blueberries for 4 weeks) improved HDL- and LDL-cholesterol in healthy subjects.[97] However, few human studies failed to show improvement in lipid profile following blueberry consumption.[37,88] Strawberry intake (25 g freeze-dried powder for 4 weeks) reduced total cholesterol and LDL-cholesterol levels in women with MetS.[98] Another study documented that dietary strawberry (freeze-dried powder equivalent to four servings of strawberries for 3 weeks) reduced plasma cholesterol and small HDL-cholesterol particles with an increase in LDL particle size in obese subjects.[99]

Together, though studies indicate the beneficial effects of berry bioactives on dyslipidemia, further studies with similar protocols are needed to specifically focus on signaling pathways and genes involved in lipid metabolism.

6.4.5 Modification of Vascular Glycocalyx

The glycocalyx is a carbohydrate-rich layer and biologically active barrier that covers the luminal side of the vascular endothelium. Glycocalyx is comprised of proteoglycans [PGs; core proteins with glycosaminoglycans (GAG) side chains], glycoproteins, and plasma proteins.[100,101] GAGs are polysaccharides composed of alternating disaccharide units of hexosamine (glycosamine or galactosamine) and uronic acid.[102] An intact glycocalyx of healthy vasculature forms the first line of defense of blood vessels against atherogenic stimuli because it (i) prevents endothelial inflammation by blocking the binding of monocyte to endothelium, (ii) increases NO bioavailability and vasodilation by translating shear stress to eNOS activation (mechanotransduction), (iii) maintains EC permeability and plasma membrane transport, and (iv) is critical in lipoprotein binding and retention.[100,103] Heparan sulfate proteoglycans (HSPG), which represent 50–90% of PGs in endothelial glycocalyx, mediate various vascular functions.[101] HSPGs are composed of several negatively charged side chains of heparan sulfate (HS) (a type of GAG side chain) and a core protein.[100,103,104] The sulfation pattern, chain length, and domain organization determine the functions of HS.[105] Hyaluronic acid (HA) and HS of aortic glycocalyx translate shear stress to activate flow-induced NO production in ECs.[106] The degree of GAG sulfation is biosynthetically regulated and provides great structural complexity. Indeed, an increased GAG sulfation enhances the LDL-binding affinity to proteoglycans.[107] Removal of the glycocalyx enhances the binding of leukocyte to endothelium.[108] Hence, a compromised endothelial glycocalyx makes blood vessels more vulnerable to vascular inflammation and atherosclerosis.[100,101,103,109–111] The structure and function of the glycocalyx are severely compromised in diabetes, hypertension, and metabolic disorders leading to vascular complications.[101,112–114] Specifically, the GAGs of glycocalyx undergo remodeling during diabetes and hypertension that negatively impact vascular function.[106,115] The amount of HS decreases in the arteries of diabetic individuals and the anti-atherogenic effects of HSPG are well documented.[110,116–118] Emerging evidence suggests that the preservation and restoration of glycocalyx is one of the novel strategies for the treatment of chronic vascular complications.[119,120] Studies on the effect of berry bioactives on endothelial glycocalyx are limited. However, a few recent studies suggest that the reported vascular benefits of berry bioactives could be due to their effects on vascular glycocalyx.

Animal studies have documented that dietary supplementation of blueberries modulates the structure and function of GAGs in the aortic vessels of Sprague-Dawley rats and SHRs.[102,106] Dietary blueberry was shown to increase total GAGs and decrease the sulfation of GAGs in Sprague-Dawley rats indicating the reported vascular beneficial effects of blueberry may be mediated through GAGs.[102] SHRs exhibit an abnormal GAG profile that is associated with a thicker subendothelial matrix and an increased blood pressure.[106] Blueberry supplementation for 8 weeks normalized the abnormal GAG profile associated with hypertension by redistributing GAG populations with an increased amount of HA and HS and reduced the GAG sulfation in the aorta of SHR.[106] This study suggests that the reported antihypertensive effect of blueberries is mediated through an improved GAG profile. Blueberry anthocyanins are extensively metabolized in humans, suggesting that the bioactivities of blueberries are mediated through their circulating metabolites.[28] In a recent *in vitro* study, diabetic ECs were exposed to a blueberry metabolite cocktail at physiologically relevant concentrations. This blueberry cocktail restored normal levels of cell surface GAGs in the diabetic ECs, which is associated with reduced endothelial inflammation.[32] These studies suggest that the cardiovascular benefits of blueberry consumption may be mediated through the effects of their metabolites on GAGs. However, future studies should be conducted to uncover the role of dietary berries on the vascular glycocalyx, how it relates to vascular function alterations observed in *in vivo, ex vivo*, and clinical studies and the molecular mechanisms involved.[121]

6.5 Conclusion

Epidemiological, clinical, and animal studies support the vascular beneficial effects of dietary berries. Limited *in vitro* studies indicate the role of circulating metabolites in exerting the vascular effects of berry bioactives. Berry bioactives act on multiple targets in the vasculature that include: activating eNOS/NO/cGMP signaling, suppressing endothelial inflammation, improving redox signaling, positively regulating lipid metabolism, and modifying vascular glycocalyx. However, studies are limited on the specific targets of berry bioactives and the role of circulating metabolites in mediating the vascular effects of dietary berries. Further, studies on the structure–activity relationships of berry metabolites are required to understand the interaction between metabolites and biomolecules in the vasculature. In addition, the role of gut microbiota in producing the metabolites and mediating the vascular effects of berry bioactives are needed. Finally, well-designed human clinical trials may be required to evaluate the potential vascular effects of dietary berries and to recommend them to improve vascular health.

References

1. S. S. Virani, A. Alonso, E. J. Benjamin, M. S. Bittencourt, C. W. Callaway, A. P. Carson, A. M. Chamberlain, A. R. Chang, S. Cheng, F. N. Delling, L. Djousse, M. S. V. Elkind, J. F. Ferguson, M. Fornage, S. S. Khan, B. M. Kissela, K. L. Knutson, T. W. Kwan, D. T. Lackland, T. T. Lewis, J. H. Lichtman, C. T. Longenecker, M. S. Loop, P. L. Lutsey, S. S. Martin, K. Matsushita, A. E. Moran, M. E. Mussolino, A. M. Perak, W. D. Rosamond, G. A. Roth, U. K. A. Sampson, G. M. Satou, E. B. Schroeder, S. H. Shah, C. M. Shay, N. L. Spartano, A. Stokes, D. L. Tirschwell, L. B. VanWagner, C. W. Tsao and American Heart Association Council on Epidemiology and Prevention Statistics Committee and Stroke Statistics Subcommittee, *Circulation*, 2020, **141**, e139–e596.
2. A. S. Go, D. Mozaffarian, V. L. Roger, E. J. Benjamin, J. D. Berry, M. J. Blaha, S. Dai, E. S. Ford, C. S. Fox, S. Franco, H. J. Fullerton, C. Gillespie, S. M. Hailpern, J. A. Heit, V. J. Howard, M. D. Huffman, S. E. Judd, B. M. Kissela, S. J. Kittner, D. T. Lackland, J. H. Lichtman, L. D. Lisabeth, R. H. Mackey, D. J. Magid, G. M. Marcus, A. Marelli, D. B. Matchar, D. K. McGuire, E. R. Mohler, 3rd, C. S. Moy, M. E. Mussolino, R. W. Neumar, G. Nichol, D. K. Pandey, N. P. Paynter, M. J. Reeves, P. D. Sorlie, J. Stein, A. Towfighi, T. N. Turan, S. S. Virani, N. D. Wong, D. Woo, M. B. Turner, American Heart Association Statistics Committee and Stroke Statistics Subcommittee, *Circulation*, 2014, **129**, e28–e292.
3. E. Galkina and K. Ley, *Annu. Rev. Immunol.*, 2009, **27**, 165–197.
4. S. P. Grover and N. Mackman, *Arterioscler., Thromb., Vasc. Biol.*, 2018, **38**, 709–725.
5. M. D. Ritchey, H. K. Wall, C. Gillespie, M. G. George, A. Jamal, Division for Heart Disease and Stroke Prevention CDC, *Morb. Mortal. Wkly. Rep.*, 2014, **63**, 462–467.
6. D. Mozaffarian, *Circulation*, 2016, **133**, 187–225.
7. L. Hooper, P. A. Kroon, E. B. Rimm, J. S. Cohn, I. Harvey, K. A. Le Cornu, J. J. Ryder, W. L. Hall and A. Cassidy, *Am. J. Clin. Nutr.*, 2008, **88**, 38–50.
8. E. E. Mulvihill and M. W. Huff, *Can. J. Cardiol.*, 2010, **26**(Suppl A), 17A–21A.
9. J. W. Erdman Jr, D. Balentine, L. Arab, G. Beecher, J. T. Dwyer, J. Folts, J. Harnly, P. Hollman, C. L. Keen, G. Mazza, M. Messina, A. Scalbert, J. Vita, G. Williamson and J. Burrowes, *J. Nutr.*, 2007, **137**, 718S–737S.
10. G. R. Beecher, *J. Nutr.*, 2003, **133**, 3248S–3254S.

11. D. Del Rio, G. Borges and A. Crozier, *Br. J. Nutr.*, 2010, **104**(Suppl 3), S67–90.
12. F. Sofi, M. Dinu, G. Pagliai, F. Cesari, A. M. Gori, A. Sereni, M. Becatti, C. Fiorillo, R. Marcucci and A. Casini, *Circulation*, 2018, **137**, 1103–1113.
13. R. Estruch, M. A. Martinez-Gonzalez, D. Corella, J. Salas-Salvado, M. Fito, G. Chiva-Blanch, M. Fiol, E. Gomez-Gracia, F. Aros, J. Lapetra, L. Serra-Majem, X. Pinto, P. Buil-Cosiales, J. V. Sorli, M. A. Munoz, J. Basora-Gallisa, R. M. Lamuela-Raventos, M. Serra-Mir, E. Ros and PREDIMED Study Investigators, *Lancet Diabetes Endocrinol.*, 2019, **7**, e6–e17.
14. A. Michalska and G. Lysiak, *Int. J. Mol. Sci.*, 2015, **16**, 18642–18663.
15. X. Wu, G. R. Beecher, J. M. Holden, D. B. Haytowitz, S. E. Gebhardt and R. L. Prior, *J. Agric. Food Chem.*, 2006, **54**, 4069–4075.
16. A. Basu, M. Rhone and T. J. Lyons, *Nutr. Rev.*, 2010, **68**, 168–177.
17. F. Giampieri, T. Y. Forbes-Hernandez, M. Gasparrini, J. M. Alvarez-Suarez, S. Afrin, S. Bompadre, J. L. Quiles, B. Mezzetti and M. Battino, *Food Funct.*, 2015, **6**, 1386–1398.
18. G. A. Manganaris, V. Goulas, A. R. Vicente and L. A. Terry, *J. Sci. Food Agric.*, 2014, **94**, 825–833.
19. D. Li, P. Wang, Y. Luo, M. Zhao and F. Chen, *Crit. Rev. Food Sci. Nutr.*, 2017, **57**, 1729–1741.
20. A. Basu and T. J. Lyons, *J. Agric. Food Chem.*, 2012, **60**, 5687–5692.
21. A. Rodriguez-Mateos, C. Heiss, G. Borges and A. Crozier, *J. Agric. Food Chem.*, 2014, **62**, 3842–3851.
22. M. A. Lila, B. Burton-Freeman, M. Grace and W. Kalt, *Annu. Rev. Food Sci. Technol.*, 2016, **7**, 375–393.
23. G. B. Celli, A. Ghanem and M. S. Brooks, *Crit. Rev. Food Sci. Nutr.*, 2017, **57**, 3197–3207.
24. C. A. Morais, V. V. de Rosso, D. Estadella and L. P. Pisani, *J. Nutr. Biochem.*, 2016, **33**, 1–7.
25. A. Faria, I. Fernandes, S. Norberto, N. Mateus and C. Calhau, *J. Agric. Food Chem.*, 2014, **62**, 6898–6902.
26. L. Marin, E. M. Miguelez, C. J. Villar and F. Lombo, *BioMed Res. Int.*, 2015, **2015**, 905215.
27. B. R. Cutler, C. Petersen and P. V. Anandh Babu, *Mol. Nutr. Food Res.*, 2017, **61**, 1600271.
28. A. Rodriguez-Mateos, C. Rendeiro, T. Bergillos-Meca, S. Tabatabaee, T. W. George, C. Heiss and J. P. Spencer, *Am. J. Clin. Nutr.*, 2013, **98**, 1179–1191.
29. R. M. de Ferrars, A. Cassidy, P. Curtis and C. D. Kay, *Mol. Nutr. Food Res.*, 2014, **58**, 490–502.
30. R. M. de Ferrars, C. Czank, Q. Zhang, N. P. Botting, P. A. Kroon, A. Cassidy and C. D. Kay, *Br. J. Pharmacol.*, 2014, **171**, 3268–3282.
31. D. Bharat, R. R. M. Cavalcanti, C. Petersen, N. Begaye, B. R. Cutler, M. M. A. Costa, R. Ramos, M. R. Ferreira, Y. Li, L. P. Bharath,

E. Toolson, P. Sebahar, R. E. Looper, T. Jalili, N. S. Rajasekaran, Z. Jia, J. D. Symons and P. V. Anandh Babu, *Mol. Nutr. Food Res.*, 2018, **62**, 1700601.
32. B. Cutler, S. Gholami, J. S. Chua, B. Kuberan and P. V. Anandh Babu, *Int. J. Cardiol.*, 2018, **261**, 155–158.
33. A. Rodriguez-Mateos, G. Istas, L. Boschek, R. P. Feliciano, C. E. Mills, C. Boby, S. Gomez-Alonso, D. Milenkovic and C. Heiss, *J. Gerontol., Ser. A*, 2019, **74**, 967–976.
34. H. P. Amin, C. Czank, S. Raheem, Q. Zhang, N. P. Botting, A. Cassidy and C. D. Kay, *Mol. Nutr. Food Res.*, 2015, **59**, 1095–1106.
35. M. Edwards, C. Czank, G. M. Woodward, A. Cassidy and C. D. Kay, *J. Agric. Food Chem.*, 2015, **63**, 2423–2431.
36. A. Cassidy, K. J. Mukamal, L. Liu, M. Franz, A. H. Eliassen and E. B. Rimm, *Circulation*, 2013, **127**, 188–196.
37. A. Basu, M. Du, M. J. Leyva, K. Sanchez, N. M. Betts, M. Wu, C. E. Aston and T. J. Lyons, *J. Nutr.*, 2010, **140**, 1582–1587.
38. A. Cassidy, I. De Vivo, Y. Liu, J. Han, J. Prescott, D. J. Hunter and E. B. Rimm, *Am. J. Clin. Nutr.*, 2010, **91**, 1273–1280.
39. A. Rodriguez-Mateos, R. Del Pino-Garcia, T. W. George, A. Vidal-Diez, C. Heiss and J. P. Spencer, *Mol. Nutr. Food Res.*, 2014, **58**, 1952–1961.
40. S. A. Johnson, A. Figueroa, N. Navaei, A. Wong, R. Kalfon, L. T. Ormsbee, R. G. Feresin, M. L. Elam, S. Hooshmand, M. E. Payton and B. H. Arjmandi, *J. Acad. Nutr. Diet.*, 2015, **115**, 369–377.
41. C. Del Bo, V. Deon, J. Campolo, C. Lanti, M. Parolini, M. Porrini, D. Klimis-Zacas and P. Riso, *Food Funct.*, 2017, **8**, 4108–4117.
42. E. Wood, S. Hein, C. Heiss, C. Williams and A. Rodriguez-Mateos, *Food Funct.*, 2019, **10**, 7621–7633.
43. C. Del Bo, M. Porrini, D. Fracassetti, J. Campolo, D. Klimis-Zacas and P. Riso, *Food Funct.*, 2014, **5**, 3107–3116.
44. A. J. Stull, K. C. Cash, C. M. Champagne, A. K. Gupta, R. Boston, R. A. Beyl, W. D. Johnson and W. T. Cefalu, *Nutrients*, 2015, **7**, 4107–4123.
45. L. S. McAnulty, S. R. Collier, M. J. Landram, D. S. Whittaker, S. E. Isaacs, J. M. Klemka, S. L. Cheek, J. C. Arms and S. R. McAnulty, *Nutr. Res.*, 2014, **34**, 577–584.
46. D. Djurica, R. R. Holt, J. Ren, A. W. Shindel, R. M. Hackman and C. L. Keen, *Br. J. Nutr.*, 2016, **116**, 639–647.
47. A. Basu, D. X. Fu, M. Wilkinson, B. Simmons, M. Wu, N. M. Betts, M. Du and T. J. Lyons, *Nutr. Res.*, 2010, **30**, 462–469.
48. R. G. Feresin, S. A. Johnson, S. Pourafshar, J. C. Campbell, S. J. Jaime, N. Navaei, M. L. Elam, N. S. Akhavan, S. Alvarez-Alvarado, G. Tenenbaum, K. Brummel-Smith, G. Salazar, A. Figueroa and B. H. Arjmandi, *Food Funct.*, 2017, **8**, 4139–4149.
49. G. Istas, R. P. Feliciano, T. Weber, R. Garcia-Villalba, F. Tomas-Barberan, C. Heiss and A. Rodriguez-Mateos, *Arch. Biochem. Biophys.*, 2018, **651**, 43–51.

50. A. S. Kristo, A. Z. Kalea, D. A. Schuschke and D. Klimis-Zacas, *Int. J. Food Sci. Nutr.*, 2013, **64**, 979–987.
51. C. M. Elks, S. D. Reed, N. Mariappan, B. Shukitt-Hale, J. A. Joseph, D. K. Ingram and J. Francis, *PLoS One*, 2011, **6**, e24028.
52. C. Petersen, D. Bharat, B. R. Cutler, S. Gholami, C. Denetso, J. E. Mueller, J. M. Cho, J. S. Kim, J. D. Symons and P. V. Anandh Babu, *Int. J. Cardiol.*, 2018, **263**, 111–117.
53. J. B. Su, *World J. Cardiol.*, 2015, **7**, 719–741.
54. D. A. Chistiakov, A. N. Orekhov and Y. V. Bobryshev, *Front. Physiol.*, 2015, **6**, 365.
55. J. F. Gielis, J. Y. Lin, K. Wingler, P. E. Van Schil, H. H. Schmidt and A. L. Moens, *Free Radical Biol. Med.*, 2011, **50**, 765–776.
56. A. Quillon, B. Fromy and R. Debret, *Nitric oxide*, 2015, **45**, 20–26.
57. P. Ponnuswamy, A. Schrottle, E. Ostermeier, S. Gruner, P. L. Huang, G. Ertl, U. Hoffmann, B. Nieswandt and P. J. Kuhlencordt, *PLoS One*, 2012, **7**, e30193.
58. K. S. Stote, M. I. Sweeney, T. Kean, D. J. Baer, J. A. Novotny, N. L. Shakerley, A. Chandrasekaran, P. M. Carrico, J. A. Melendez and K. T. Gottschall-Pass, *BMC Nutr*, 2017, **3**, 45.
59. A. Z. Kalea, K. Clark, D. A. Schuschke, A. S. Kristo and D. J. Klimis-Zacas, *J. Nutr. Biochem.*, 2010, **21**, 14–22.
60. A. S. Kristo, A. Z. Kalea, D. A. Schuschke and D. J. Klimis-Zacas, *J. Agric. Food Chem.*, 2010, **58**, 11600–11605.
61. T. A. Pearson, G. A. Mensah, R. W. Alexander, J. L. Anderson, R. O. Cannon, 3rd, M. Criqui, Y. Y. Fadl, S. P. Fortmann, Y. Hong, G. L. Myers, N. Rifai, S. C. Smith, Jr, K. Taubert, R. P. Tracy and F. Vinicor, *Circulation*, 2003, **107**, 499–511.
62. S. G. Thompson, J. Kienast, S. D. Pyke, F. Haverkate and J. C. van de Loo, *N. Engl. J. Med.*, 1995, **332**, 635–641.
63. J. Danesh, J. G. Wheeler, G. M. Hirschfield, S. Eda, G. Eiriksdottir, A. Rumley, G. D. Lowe, M. B. Pepys and V. Gudnason, *N. Engl. J. Med.*, 2004, **350**, 1387–1397.
64. P. M. Ridker, C. H. Hennekens, J. E. Buring and N. Rifai, *N. Engl. J. Med.*, 2000, **342**, 836–843.
65. C. Menzaghi, S. Bacci, L. Salvemini, C. Mendonca, G. Palladino, A. Fontana, C. De Bonis, A. Marucci, E. Goheen, S. Prudente, E. Morini, S. Rizza, A. Kanagaki, G. Fini, D. Mangiacotti, M. Federici, S. De Cosmo, F. Pellegrini, A. Doria and V. Trischitta, *PLoS One*, 2013, **8**, e64729.
66. A. Whelton, *Am. J. Card.*, 2006, **97**, 3–9.
67. T. A. Pearson, G. A. Mensah, Y. Hong and S. C. Smith, Jr, *Circulation*, 2004, **110**, e543–544.
68. P. Libby, *Nature*, 2002, **420**, 868–874.
69. P. Nallasamy, H. Si, P. V. Babu, D. Pan, Y. Fu, E. A. Brooke, H. Shah, W. Zhen, H. Zhu, D. Liu, Y. Li and Z. Jia, *J. Nutr. Biochem.*, 2014, **25**, 824–833.

70. M. Zakkar, K. Van der Heiden, A. Luong le, H. Chaudhury, S. Cuhlmann, S. S. Hamdulay, R. Krams, I. Edirisinghe, I. Rahman, H. Carlsen, D. O. Haskard, J. C. Mason and P. C. Evans, *Arterioscler., Thromb., Vasc. Biol.*, 2009, **29**, 1851–1857.
71. C. Savoia and E. L. Schiffrin, *Clin. Sci.*, 2007, **112**, 375–384.
72. T. Bouwmeester, A. Bauch, H. Ruffner, P. O. Angrand, G. Bergamini, K. Croughton, C. Cruciat, D. Eberhard, J. Gagneur, S. Ghidelli, C. Hopf, B. Huhse, R. Mangano, A. M. Michon, M. Schirle, J. Schlegl, M. Schwab, M. A. Stein, A. Bauer, G. Casari, G. Drewes, A. C. Gavin, D. B. Jackson, G. Joberty, G. Neubauer, J. Rick, B. Kuster and G. Superti-Furga, *Nat. Cell Biol.*, 2004, **6**, 97–105.
73. K. K. Yerneni, W. Bai, B. V. Khan, R. M. Medford and R. Natarajan, *Diabetes*, 1999, **48**, 855–864.
74. W. Bartchewsky, Jr, M. R. Martini, M. Masiero, A. C. Squassoni, M. C. Alvarez, M. S. Ladeira, D. Salvatore, M. Trevisan, J. Pedrazzoli, Jr and M. L. Ribeiro, *Scand. J. Gastroenterol.*, 2009, **44**, 153–161.
75. P. V. Babu, H. Si and D. Liu, *Mol. Nutr. Food Res.*, 2012, **56**, 1424–1432.
76. E. Maloney, I. R. Sweet, D. M. Hockenbery, M. Pham, N. O. Rizzo, S. Tateya, P. Handa, M. W. Schwartz and F. Kim, *Arterioscler., Thromb., Vasc. Biol.*, 2009, **29**, 1370–1375.
77. S. Vendrame and D. Klimis-Zacas, *Nutr. Rev.*, 2015, **73**, 348–358.
78. S. Vendrame, A. Daugherty, A. S. Kristo, P. Riso and D. Klimis-Zacas, *J. Nutr. Biochem.*, 2013, **24**, 1508–1512.
79. C. Xie, J. Kang, M. E. Ferguson, S. Nagarajan, T. M. Badger and X. Wu, *Mol. Nutr. Food Res.*, 2011, **55**, 1587–1591.
80. S. Vendrame, P. Tsakiroglou, A. S. Kristo, D. A. Schuschke and D. Klimis-Zacas, *Appl. Physiol., Nutr., Metab.*, 2016, **41**, 1045–1051.
81. A. J. Stull, K. C. Cash, W. D. Johnson, C. M. Champagne and W. T. Cefalu, *J. Nutr.*, 2010, **140**, 1764–1768.
82. J. Schell, N. M. Betts, T. J. Lyons and A. Basu, *Ann. Nutr. Metab.*, 2019, **74**, 165–174.
83. K. H. Park and W. J. Park, *J. Korean Med. Sci.*, 2015, **30**, 1213–1225.
84. A. Konior, A. Schramm, M. Czesnikiewicz-Guzik and T. J. Guzik, *Antioxid. Redox Signaling*, 2014, **20**, 2794–2814.
85. E. De Marchi, F. Baldassari, A. Bononi, M. R. Wieckowski and P. Pinton, *Oxid. Med. Cell. Longevity*, 2013, **2013**, 564961.
86. J. D. Symons, *Diabetes*, 2013, **62**, 1818–1820.
87. C. Del Bo, P. Riso, J. Campolo, P. Moller, S. Loft, D. Klimis-Zacas, A. Brambilla, A. Rizzolo and M. Porrini, *Nutr. Res.*, 2013, **33**, 220–227.
88. P. Riso, D. Klimis-Zacas, C. Del Bo, D. Martini, J. Campolo, S. Vendrame, P. Moller, S. Loft, R. De Maria and M. Porrini, *Eur. J. Nutr.*, 2013, **52**, 949–961.
89. B. C. Blacker, S. M. Snyder, D. L. Eggett and T. L. Parker, *Br. J. Nutr.*, 2013, **109**, 1670–1677.
90. C. Del Bo, M. Porrini, J. Campolo, M. Parolini, C. Lanti, D. Klimis-Zacas and P. Riso, *Mutagenesis*, 2016, **31**, 215–224.

91. S. Vendramc, A. S. Kristo, D. A. Schuschke and D. Klimis-Zacas, *Appl. Physiol., Nutr., Metab.*, 2014, **39**, 255–261.
92. J. Coban, B. Evran, F. Ozkan, A. Cevik, S. Dogru-Abbasoglu and M. Uysal, *Biosci. Biotechnol. Biochem.*, 2013, **77**, 389–391.
93. P. N. Hopkins, *Physiol. Rev.*, 2013, **93**, 1317–1542.
94. W. Kalt, K. Foote, S. A. Fillmore, M. Lyon, T. A. Van Lunen and K. B. McRae, *Br. J. Nutr.*, 2008, **100**, 70–78.
95. T. Wu, Q. Tang, Z. Gao, Z. Yu, H. Song, X. Zheng and W. Chen, *PLoS One*, 2013, **8**, e77585.
96. S. Vendrame, A. Daugherty, A. S. Kristo and D. Klimis-Zacas, *Br. J. Nutr.*, 2014, **111**, 194–200.
97. S. Nyberg, E. Gerring, S. Gjellan, M. Vergara, T. Lindstrom and F. H. Nystrom, *Upsala J. Med. Sci.*, 2013, **118**, 247–255.
98. A. Basu, M. Wilkinson, K. Penugonda, B. Simmons, N. M. Betts and T. J. Lyons, *Nutr J*, 2009, **8**, 43.
99. S. J. Zunino, M. A. Parelman, T. L. Freytag, C. B. Stephensen, D. S. Kelley, B. E. Mackey, L. R. Woodhouse and E. L. Bonnel, *Br. J. Nutr.*, 2012, **108**, 900–909.
100. J. A. Kolsen-Petersen, *Acta Anaesthesiol. Scand.*, 2015, **59**, 137–139.
101. H. Kolarova, B. Ambruzova, L. Svihalkova Sindlerova, A. Klinke and L. Kubala, *Mediat. Inflamm.*, 2014, **2014**, 694312.
102. A. Z. Kalea, F. N. Lamari, A. D. Theocharis, P. Cordopatis, D. A. Schuschke, N. K. Karamanos and D. J. Klimis-Zacas, *J. Nutr. Biochem.*, 2006, **17**, 109–116.
103. B. F. Becker, M. Jacob, S. Leipert, A. H. Salmon and D. Chappell, *Br. J. Clin. Pharmacol.*, 2015, **80**, 389–402.
104. H. L. Mooij, P. Cabrales, S. J. Bernelot Moens, D. Xu, S. D. Udayappan, A. G. Tsai, M. A. van der Sande, E. de Groot, M. Intaglietta, J. J. Kastelein, G. M. Dallinga-Thie, J. D. Esko, E. S. Stroes and M. Nieuwdorp, *J. Am. Heart Assoc.*, 2014, **3**, e001274.
105. P. Babu, X. V. Victor, E. Nelsen, T. K. Nguyen, K. Raman and B. Kuberan, *Anal. Bioanal. Chem.*, 2011, **401**, 237–244.
106. A. S. Kristo, C. J. Malavaki, F. N. Lamari, N. K. Karamanos and D. Klimis-Zacas, *J. Nutr. Biochem.*, 2012, **23**, 961–965.
107. L. R. Tannock and V. L. King, *Rev. Endocr. Metab. Disord.*, 2008, **9**, 289–300.
108. A. H. Salmon and S. C. Satchell, *J. Pathol.*, 2012, **226**, 562–574.
109. B. J. Eskens, C. J. Zuurbier, J. van Haare, H. Vink and J. W. van Teeffelen, *Cardiovasc. Diabetol.*, 2013, **12**, 175.
110. F. Wasty, M. Z. Alavi and S. Moore, *Diabetologia*, 1993, **36**, 316–322.
111. L. N. Broekhuizen, B. A. Lemkes, H. L. Mooij, M. C. Meuwese, H. Verberne, F. Holleman, R. O. Schlingemann, M. Nieuwdorp, E. S. Stroes and H. Vink, *Diabetologia*, 2010, **53**, 2646–2655.
112. L. M. Hiebert, J. Han and A. K. Mandal, *Antioxid. Redox Signaling*, 2014, **21**, 1032–1043.
113. W. Yen, B. Cai, J. Yang, L. Zhang, M. Zeng, J. M. Tarbell and B. M. Fu, *PLoS One*, 2015, **10**, e0117133.

114. M. I. Noble, A. J. Drake-Holland and H. Vink, *Qjm*, 2008, **101**, 513–518.
115. V. Gowd, A. Gurukar and N. D. Chilkunda, *World J. Diabetes*, 2016, **7**, 67–73.
116. R. K. Vikramadithyan, Y. Kako, G. Chen, Y. Hu, E. Arikawa-Hirasawa, Y. Yamada and I. J. Goldberg, *J. Lipid Res.*, 2004, **45**, 1806–1812.
117. K. Tran-Lundmark, P. K. Tran, G. Paulsson-Berne, V. Friden, R. Soininen, K. Tryggvason, T. N. Wight, M. G. Kinsella, J. Boren and U. Hedin, *Circ. Res.*, 2008, **103**, 43–52.
118. F. Wang, Y. Wang, D. Zhang, P. Puthanveetil, J. D. Johnson and B. Rodrigues, *Arterioscler., Thromb., Vasc. Biol.*, 2012, **32**, 406–414.
119. O. Yilmaz, B. Afsar, A. Ortiz and M. Kanbay, *Clin. Kidney J.*, 2019, **12**, 611–619.
120. U. Schott, C. Solomon, D. Fries and P. Bentzer, *Scand. J. Trauma Resusc. Emerg. Med.*, 2016, **24**, 48.
121. D. Klimis-Zacas and A. S. Kristo, Emerging Trends in Dietary Components for Preventing and Combating Disease, in *ACS Publication Series*, 2012, pp. 151–166.

CHAPTER 7

Berry Bioactives and Cognition

L. BELL,[a] B. SHUKITT-HALE[b] AND C. M. WILLIAMS*[a]

[a] University of Reading, Reading, UK; [b] USDA-ARS, Human Nutrition Research Center on Aging at Tufts University, Boston, MA, USA
*Email: claire.williams@reading.ac.uk

7.1 Introduction

Although nutritional psychology is a relatively new field of study, there has already been much investigation into the impact of what we eat on our brain health. The specific impact of berries on cognitive function has already been extensively reviewed, particularly the findings of randomised-controlled trials (RCTs),[1–3] and research into the underlying physiological mechanisms of action.[4,5] Indeed, berry bioactive polyphenols such as anthocyanins, flavanols, flavonols, and proanthocyanidins have been linked with improvements to both physiological health and cognitive function. Here we aim to consolidate the findings of this field of scientific endeavour. We consider a range of supporting evidence, starting with epidemiological observations across large populations, before giving an overview of evidence from smaller-scale experimental studies in both animals and humans. In addition to studies directly investigating the impact of berries on cognition, we also present some of the evidence for physiological benefits of berries that might facilitate the improvements in brain function seen after these types of intervention. Finally, we propose directions for future research.

Food Chemistry, Function and Analysis No. 33
Berries and Berry Bioactive Compounds in Promoting Health
Edited by Dorothy Klimis-Zacas and Ana Rodriguez-Mateos

Published by the Royal Society of Chemistry, www.rsc.org

7.2 Epidemiological Evidence

Investigation of the relationships between dietary patterns and health outcomes in large populations (cohort studies) is useful to identify the specific food types and/or nutrients that convey the greatest benefits for public health. These studies typically track a large group of subjects over time while monitoring diet and health at fixed periods. A number of such studies have included cognitive outcomes or physiological outcomes that may indirectly relate to brain function in their health measures. Behavioural endpoints relating to cognition may include measures of memory, executive function, and subjective mood, and also rates of pathology from neurodegenerative disease. Typical physiological outcome measures in cohort studies include incidence of physical disease but may also include endpoints directly or indirectly related to brain function, such as glucose and insulin regulation, vascular health, or even, more recently, gut health. Glucose regulation and vascular health are particularly important to cognitive function as the brain is reliant upon the continuous supply of glucose *via* the vascular system and the tightly controlled uptake of glucose into neuronal cells *via* the endocrine system. Vascular disease and metabolic disorders such as type 2 diabetes mellitus (T2DM) are therefore strongly associated with cognitive dysfunction.[6,7]

An example of the association between berry consumption and metabolic health is demonstrated by a cross-sectional study of nearly 2000 female twins who supplied dietary information using a food frequency questionnaire and were also tested for biomarkers of insulin resistance.[8] The study found that those with higher intake of anthocyanins (and flavones) had lower Homeostatic Model Assessment for Insulin Resistance (HOMA-IR) ratings, lower levels of fasting insulin, and higher levels of fasting adiponectin (a protein hormone responsible for glucose regulation). These markers all indicate healthy glucose regulation and those subjects consuming a high anthocyanins diet presented a lower risk for developing T2DM and associated cognitive disorders. Indeed, a similar cohort study of more than 70 000 nurses (Nurses' Health Study) also identified a link between anthocyanin-rich fruit consumption (particularly blueberries and grapes) and a lower risk of T2DM.[9,10] The Nurses' Health Study also included measures of cognition during the final stages of the study as subjects reached the age of 70 or more years.[11] Long-term berry intake (particularly of strawberries and blueberries) was associated with slower cognitive decline. Indeed, age-related cognitive decline was estimated to be delayed by up to 2.5 years in those consuming the greatest amounts of berries. In the same cohort, high berry intake was also significantly associated with a greater likelihood of healthy cognitive ageing, avoiding both neurodegenerative and mental health disorders such as depression and anxiety.[12] An additional analysis of the Nurses' Health Study, including data from the Health Professionals' Follow-up Study, also observed a significantly lower risk for Parkinson's disease (PD) in those individuals consuming higher quantities of berries.[13] While PD is predominantly associated with deficits in motor function, it is also commonly associated with mild cognitive impairment (MCI) due to associated reductions in the neurotransmitter dopamine. Further data from the later Framingham Offspring Cohort also revealed that a high intake of anthocyanins (and some other flavonoid groups) was significantly associated with a lower risk for Alzheimer's disease (AD) and related dementias.[14] Clear associations have also been made between anthocyanins intake and reduced risk of cardiovascular disease (CVD).[15] Blueberry and strawberry consumption, in particular, have been associated with a lower risk of hypertension[16] and myocardial infarction (MI)[17] following separate analyses of the Nurses' Health Study data and data from the Health Professionals' Follow-up Study. Cognitive dysfunction is commonly associated with these vascular conditions due to a reduction in blood supply to the brain.

A number of other cohort studies note a significant association between general fruit and vegetable consumption and cognition, including measures of memory, executive function, and subjective mood.[18–20] While these studies do not investigate the role of specific fruits or bioactives such as berry anthocyanins, it is likely that these dietary components may also contribute to the observed positive associations with cognitive health. However, while compelling, the epidemiological evidence is purely correlational and so direct inferences of causation cannot be drawn. Therefore, many experimental intervention studies have been conducted to provide evidence for a direct link between consumption of berries and improvements to cognitive function.

7.3 Experimental Evidence

Experimental studies may be *in vitro* investigations of the impact of bioactives on cells or may administer a quantity of the fruit or bioactive extract *in vivo* to the human or animal participants rather than measuring habitual intake. Here, we have largely considered animal and human studies that have administered whole fruits (including juiced or freeze-dried) rather than isolated extracts of bioactives. Interestingly, cell studies suggest that the whole berry may be more effective than any single polyphenolic compound, possibly because different polyphenols may exert their effects through different and/or independent mechanisms.[21,22] RCTs are the gold standard in intervention studies; animal or human participants are randomly allocated to receive either the active intervention or a control intervention, and the outcome differences between the two groups are compared. Experimental studies, including RCTs, are particularly important when trying to establish a causal relationship. Experimental studies may also be used to determine likely mechanisms of action, through measurement of physiological and neurological changes that may accompany any cognitive change.

One metric that is particularly pertinent here is the dose used in experimental studies, as it is helpful to know how much of a whole food or bioactive is needed to observe a health effect. Careful consideration of dose is needed when investigating dose ranges in animal studies, as it is important to be able to translate rodent intakes into equivalent human dose sizes. For example, in rodent studies, the dose is often expressed as a percentage of the standard weight of the feed. The use of 2% supplementation in a rodent diet equates to approximately 4 g per day of berry consumption for a rat, which is roughly equivalent to $\frac{1}{2}$–1 cup per day for humans.[23] This dose has been used for a large number of animal studies as it was shown to have beneficial effects in early supplementation studies.[23,24] Quantities of whole foods are helpful to understand whether effects are achievable through normal dietary intake amounts; however, the levels of bioactives can vary considerably (*e.g.*, due to growing conditions, length of storage, and processing). Therefore, quantification of the bioactives present is particularly helpful for drawing comparisons between studies (although some studies only supply a weight or volume). Anthocyanins are the main bioactive flavonoids found in berries and aggregate fruits, and typically, studies provide an anthocyanin content to allow for dose comparisons. However, monomers and oligomers of flavanols (proanthocyanidins) are also present in significant quantities in these fruits, along with several other non-flavonoid, polyphenol compounds, so it is helpful to quantify all these bioactives. Indeed, these additional compounds are quantified in some studies, either as individual polyphenols or as a total polyphenol content. In addition to flavonoids, this quantification generally encompasses other polyphenols such as caffeic acid, ferulic acid, quinic acid, associated esters such as chlorogenic acids, and ellagitannins. Another important feature of RCTs is the length of duration of supplementation. Some studies investigate the impact of a single acute dose, while others provide a chronic daily dose for a fixed period of time that may span days, weeks, or months. The cognitive tests used and any accompanying physiological measures may also differ between acute and chronic intervention study designs, reflecting differences in the likely mechanisms of action. The target population of the intervention also influences the type of measures selected. For example, memory tests are often investigated in ageing-related studies, whereas executive functions may be of greater interest when investigating younger demographics.

7.3.1 Rodent Studies

In vitro data from cellular and molecular studies have shown that oxidative stress and inflammation, as well as declines in cellular protective mechanisms, may be significant contributors to behavioural changes, particularly age-related declines in cognitive function. When considered alongside human epidemiological data, this suggests that consumption of foods with antioxidant and anti-inflammatory properties, such as berries, could forestall or reverse age-related changes in cognitive performance. A number of rodent trials have sought to investigate this possibility.[25] In such rodent studies, it is common for cognitive and physiological outcomes to be measured concurrently. Physiological investigations are also permitted to be invasive and so detailed observations of underlying changes to brain structure or neurochemistry are possible. This research allows direct investigation of the key behavioural changes and potential mechanisms previously identified through correlational observations in human epidemiological studies.

7.3.1.1 Physiological and Cognitive Evidence in Rodents

A majority of rodent berry intervention (feeding) studies have investigated the benefits of blueberries and strawberries. In one of the first berry feeding trials whose endpoint was cognition, male Fischer 344 (F344) rats (a general rodent model of ageing) ate a control diet (CD) (a standard rodent "chow"), or one of three antioxidant-rich diets, one of which contained 0.95% strawberry.[24] The rats consumed the diets from adulthood (6 months) to middle age (15 months), and at the end of the supplementation period, learning and memory were measured through performance in a Morris water maze (MWM). Although strawberry did not improve cognition, this longitudinal study showed that strawberry improved a number of different neuronal parameters such as receptor sensitivity, calcium buffering capacity, and signal transduction, all known to be sensitive to ageing and oxidative stress.[24] A subsequent experiment from this lab found that strawberry (1.48%) and blueberry (1.86%) diets were effective in reversing age-related deficits in brain and cognitive function in old-aged (19-month-old) F344 rats when fed for 8 weeks.[26] These berry-fed rats showed improved working memory performance in a modified version of the MWM, when compared with rats fed the CD. A similar study found that the protective effects of blueberries (supplemented at 2% of diet) were still observed if the standard rodent chow was substituted with a natural cereal-based diet more representative of a balanced human diet.[27] In this case, the blueberries elicited improvement in MWM reference memory performance (a type of memory formed over multiple repeat trials).[27]

Extending the results found in male rats to female rats, one study fed 12- and 16-month-old female Long-Evans rats (a model less prone to cancer, so commonly used in ageing studies) either a standard diet or a standard diet supplemented with either freeze-dried organic strawberries or spinach for 8 weeks before assessing MWM performance.[28] Both strawberry and spinach supplemented diets were observed to delay the mild cognitive declines in spatial learning that occurred in the standard diet female rats as they entered late middle age. This suggests that the period of transition from early to late middle age is a promising target for dietary intervention in healthy adults.[28]

Recognition memory, as measured by the novel object recognition (NOR) test, is another cognitive domain where performance decreases are observed with age, but these age-related memory deficits have been shown to be slowed by blueberry supplementation.[29] In this study, young and old F344 rats were presented with two like objects during a training session, and then one of the two objects was replaced by a new object during the test phase. When the delay between training and test phases was set to 60 minutes, old-aged (19-month-old) rats fed a 2% blueberry diet (for 4 months) showed improved memory as demonstrated by spending significantly more time near the novel object compared to age-matched control rats fed a normal diet.[29] Indeed, the performance of the blueberry-fed rats was similar to that of much younger (4-month-old) rats. Reduced levels of the transcription factor nuclear factor-kappa B (Nf-κB, a marker of inflammation) were also observed in multiple brain regions of the aged, blueberry-fed rats, compared to aged, control-fed rats. Those aged rats who spent the most time investigating the novel object also had the lowest levels of Nf-κB, suggesting that reduction of ageing-induced oxidative stress and inflammation is the likely mechanism of action for the observed memory performance benefits.

Other possible mechanisms for the beneficial effects of blueberries were examined in a study that showed that supplementation with a 2% blueberry diet for 12 weeks improved the spatial working memory performance of aged, Lister hooded outbred rats (a rodent model commonly used for investigating behaviour). The treated rats made fewer errors and displayed faster performance times in a modified food-motivated plus-shaped maze.[30] These improvements in memory were found to correlate with the activation of cyclic adenosine monophosphate (cAMP)-response element-binding protein (CREB) and with increases in levels of brain-derived neurotrophic factor (BDNF) in the hippocampus. Increases in the phosphorylation state of the extracellular signal-related kinase (ERK) were also observed.[30] The ERK–CREB–BDNF pathway is known to regulate neural processes, including synaptic plasticity, neuronal survival and differentiation, neurotransmitter release, and learning and memory, and therefore suggests that berry supplementation may improve cognition through exerting direct effects on the brain.[31] Several studies have also shown that blueberries can increase the neuroprotective trophic factor, insulin-like growth factor-1 (IGF-1) and these increases are similarly associated with improvements in cognition.[32–35]

Blueberry[32] and strawberry[34] have both been documented to increase the growth of new neurons (neurogenesis) in aged rats following 8 weeks of supplementation with a 2% berry diet. However, the ways in which each type of berry impacts neurogenesis may be different. For blueberry-fed rats, the proliferation of precursor cells was facilitated in the dentate gyrus (DG) and the resulting increase in cells was associated with improved spatial memory performance.[32,34] Furthermore, the strawberry diet increased survival rates in the precursor cells of the DG, relative to controls, but was not associated with improvements in cognition.[34] Similarly, another study suggested that the different bioactives in blueberries and strawberries might be acting in different brain regions and *via* multiple mechanisms to produce their beneficial effects. This study evaluated the efficacy of an 8 week feeding with diets supplemented with 2% strawberry or blueberry on irradiation-induced deficits, which are similar to those seen in ageing.[36] Berry-fed rats exposed to whole-body irradiation were better protected than control-diet-fed irradiated rats in MWM performance. However, the strawberry diet offered better protection against spatial memory deficits in the maze because strawberry-fed animals were better able to retain place information (mediated by the hippocampus) compared to controls. Furthermore, the blueberry diet seemed to improve reversal learning (where the target location learned over a number of trials is changed and rats must forget this location and learn a new one) and mental flexibility, behaviours more dependent on the striatum rather than the hippocampus.[36]

The time course of dietary blueberry's effects, *i.e.*, how long one has to eat the blueberry diet, were investigated in another study.[37] Findings showed that although age-related declines in object recognition memory were reversed and prevented by a 2% blueberry-enriched diet for 1 month, feeding the blueberry-enriched diet for 2 months resulted in a more prolonged benefit after the end of the diet. Specifically, rats that had consumed the blueberry diet for 2 months exhibited no decline in performance after four weeks when switched to the CD, whereas rats supplemented with the blueberry diet for 1 month showed a decline in performance after 2 weeks and no benefit at all after 4 weeks on the CD.[37] Another experiment in the same paper reported that a blueberry diet lasting one month was enough to reverse pre-existing declines in object memory performance in 19-month-old F344 rats.[37] These findings show that deficits in cognition may be reversed in aged rats after as little as one month of dietary supplementation with blueberry; however, at least 2 months of supplementation are required to maintain the benefits long term.

In addition to blueberry and strawberry research, red raspberries have been shown to mitigate cognitive deficits, this time associated with the consumption of a high-fat diet (HFD) in young mice.[38] In this study, 8-week-old mice ate an HFD (containing 60% calories from fat) or a CD with and without 4% freeze-dried red raspberry. Following 12 weeks on the diets, only mice fed the 4% raspberry supplemented diet were able to maintain their baseline level of performance for NOR memory, with significant declines in performance observed in the control mice. After approximately 20 weeks on the diets, raspberry supplementation reduced HFD-induced spatial learning and memory impairments, as measured by the time taken to complete a Barnes (circular) maze. Raspberry feeding was also found to reduce interleukin-6 (IL-6, a pro-inflammatory cytokine) and increase BDNF in the brain. These chemical changes may, therefore, have contributed to the preservation of learning and memory in the raspberry-fed mice compared to those fed a CD.[38] Similarly, açai-supplemented diets fed to 19-month-old rats have also been associated with improved cognition and neuromodulatory effects in the brain.[39,40] Interestingly, the *Euterpe precatoria* (EP) species of the açai fruit has a higher total polyphenol content compared to the *Euterpe oleracea* (EO) species.[41] In the MWM, both the EO and EP diets improved working memory, relative to controls; however, only the EO diet improved reference memory.[39] Both açai species reduced markers of reactive oxygen species (ROS) and inflammation, activated endogenous antioxidant enzymes, and upregulated markers of autophagy (evidence for the removal of damaged cells) in the hippocampus and cortex, but the improvements in these markers were more pronounced for EP than EO.[40] These findings provide support for the neuromodulatory effects of açai diet supplementation in critical brain regions involved in memory, cognition, and overall brain function.

Other berry fruits have also been investigated and researchers have found that different berries may impact different neurochemical pathways. For example, old-aged (19-month-old) F344 rats were fed 2% blueberry, cranberry, blackcurrant, or boysenberry for 2 months and were subsequently tested on an inclined screen task (a measure of psychomotor function) before analysis of their brain chemistry.[35] The blackcurrant and cranberry diets enhanced the release of the neurotransmitter dopamine in the striatum, an indicator of neuronal signal transduction. However, only the blueberry and cranberry diets were effective in ameliorating deficits in inclined screen performance, suggesting increased muscle tone and control following feeding with these berries. Performance on the task was positively correlated with observed levels of hippocampal heat shock protein 70 (HSP70, responsible for neuroprotection). No effects were observed after feeding with boysenberry. Therefore, cognitive and neurochemical effects may differ depending on the exact profile of bioactives present in different fruit types.

In addition to cognitive benefits, berries have also been observed to impact the motor function of rats, as measured using balance tests such as the accelerating rotarod, wire suspension, and the small plank walk. In particular, age-related declines in motor performance have been slowed or even reversed following feeding with blackberries[42] or red raspberries.[43,44] Given that falls are the number one health hazard for otherwise healthy older adults, these results may have important implications for benefitting healthy ageing. It is also important to note that there are individual differences in the rate and severity of cognitive and motor decline with age. To answer whether the brain health benefits of berries are a global phenomenon or more beneficial to a specific subset of the population, two studies examined the differential, beneficial effects of berries depending on baseline ability. Both blueberries[45] and raspberries[44] improved cognitive and psychomotor performance, respectively, in animals that performed poorly at baseline, while blueberries also preserved cognitive function in animals that performed well at baseline. Additionally, animals with higher levels of inflammation were observed to show poorer performance.[44,45]

7.3.1.2 Mechanisms of Action in Rodents

The possible mechanisms of action behind the neuronal and behavioural improvements observed in rodents following feeding with berries have been reviewed in several publications[4,46–49] and are described in more detail in Chapter 8. The beneficial effects of berry fruit seem to be due to multiple mechanisms of action, and it appears to be the polyphenolic compounds in the berries that account for the positive effects observed;[26] however, different polyphenolic profiles may elicit different neurochemical effects. Berries seem to improve learning and memory through mechanisms linked to changes in important structural and synaptic plasticity markers in the brain.[46] The possible mechanisms include direct effects on signalling to enhance neuronal communication, the ability to buffer against excess calcium, increased neurogenesis, alterations in neuronal morphology (the branching of neurons), enhancement of neuroprotective stress shock proteins, alteration of oxidative and inflammatory gene expression and protection against neurodegeneration following excitotoxic stress, and inhibition of signalling pathways related with the activation of inflammation and other stress signals. Therefore, berry bioactives facilitate a myriad of changes to the structure and chemistry of the brain. In rodents, these changes accompany behavioural changes, including benefits to cognitive and motor function, but it is also important to understand whether these behavioural effects observed in rodents translate to humans.

7.3.2 Human Studies

Human intervention studies are more problematic than rodent studies as humans are more diverse in physiology than their lab-bred counterparts. It is also more difficult to tightly control any potentially confounding behaviour that may impact cognitive function. Some studies opt to match participants between the active and control intervention groups according to factors such as age, level of education, or BMI. Other studies use a crossover design, where each participant receives both the active and control intervention, separated by a suitable washout period. Studies often also try to control for confounding effects of diet by restricting intake of alcohol, caffeine, nicotine, other dietary supplements, or foods high in the bioactive of interest, both before and during the course of the intervention study. However, it remains impossible to eliminate some confounding variables such as sleep attainment or exposure to external stresses that could impact physiological or cognitive outcome measures. Therefore, it is important that studies include sufficient numbers of participants to ensure adequate statistical power to observe the effects of the intervention.

Human cognition can be measured using many different metrics. A range of cognitive tests can be used to measure performance across different cognitive domains, including memory (verbal memory, spatial memory, working memory) and executive functions (attention, decision making, planning, switching ability, and inhibition of responses). Subjective measures of mood are also often included in cognitive test batteries, as mood and cognitive function are closely interlinked. Cognitive tests may be domain-specific but more commonly engage multiple aspects of cognition within a single test. Cognitive tests may also vary in their level of cognitive load or difficulty. Therefore, it can be challenging to directly compare the findings of different studies if they use different cognitive measures. For that reason, here we have summarised cognitive findings with reference to the broader domain classifications rather than reporting specific cognitive tests. This contrasts with the description of rodent research used in the previous section, where a small number of standardised tests have been used and are reported in more detail.

In addition to the use of cognitive test batteries, neuroimaging techniques including electroencephalography (EEG), near-infrared reflectance spectroscopy (NIRS), and functional magnetic resonance imaging (fMRI) are often used to monitor neural activity changes during cognitive testing, or in a resting state, following nutritional interventions. In much of the human literature, there is often a distinction between the investigation of physiological changes and investigation of cognition, and so here we consider them separately.

7.3.2.1 Physiological Evidence in Humans

A number of studies have investigated changes to neurological or physiological function following berry supplementation that may impact cognitive function. In the first instance, there is considerable evidence for the absorption and metabolism of bioactives from berries. Indeed, anthocyanins and associated metabolites have been observed in blood plasma or serum as quickly as 1–3 hours following consumption of strawberry,[50] cranberry,[51] chokeberry,[52] and blueberry.[53] Plasma or serum levels of these bioactives have also been observed to persist well beyond 24 hours.[54,55] There is some debate about whether, once in circulation, these bioactives are able to cross the blood–brain barrier in order to directly interact with neural processes or whether they simply exert their effects peripherally. It is not possible to investigate the presence of berry bioactives in a live human brain; however, the question has been investigated within the rodent brain. Indeed, a study was conducted that showed that the anthocyanins in blueberries did enter parts of the brain, including the cortex and hippocampus, and their concentrations were correlated with cognitive performance.[56] Specifically, the total level of anthocyanins found in a sample of rats correlated significantly with MWM behaviour. The rats with more hippocampal anthocyanins exhibited enhanced spatial learning, and the rats with more cortical anthocyanins performed faster on reversal learning trials.[56] Therefore, dietary anthocyanins from berries are neuroavailable and localise in various brain regions important for learning and memory. However, more work needs to be done to identify which other classes of flavonoid or polyphenolic compounds cross the blood–brain barrier, and in what amounts, as well as their contributions to improvements in cognition. This topic is covered in more detail in Chapter 8.

Flavonoids such as anthocyanins are known vasodilators due to their facilitation effect on the synthesis of endothelial nitric oxide, which augments arterial elasticity. Therefore, vascular effects have been commonly observed following berry consumption. Vasodilation can be measured directly using flow-mediated dilation (FMD), an ultrasound technique that visually determines changes to the diameter of an artery (usually the brachial artery). Following a range of acute doses of freeze-dried blueberry powder and a blueberry-free control, FMD response was observed to significantly increase in the blueberry conditions after 1–2 hours and again after 6 hours in a group of healthy adults.[57] The most effective dose contained 766 mg total polyphenols and 310 mg anthocyanins. A similar dose of açai berry (containing 694 mg polyphenols) was also observed to elicit significant FMD benefits at 2 hour and 6 hour postprandial time points,[58] this time in overweight adults.

In addition to FMD, there are a number of other related indicators of vascular health (covered in more detail in Chapter 5). Arterial tone can be determined using peripheral arterial tonometry (PAT) to measure elasticity within the endothelium in a resting state (through measurement of pulse wave amplitude [PWA]) or following restriction of blood flow (through measurement of amplitude reactive hyperaemia index [RHI]). Pulse wave velocity (PWV) provides a further measure of arterial stiffness and can be determined by comparing the timing and amplitude of a pulse wave at different arterial locations in the body.[59] Along with FMD, these are fairly sensitive measures of vascular function. RHI improvements have been shown following 1 week of daily supplementation with freeze-dried strawberry powder (containing approximately 200 mg anthocyanins) in overweight adolescents, but only in those participants where fasting nitrate/nitrite levels (an index of nitric oxide production) were concurrently observed to rise.[60] No benefits to RHI were observed in a sample of adults with vascular dysfunction and risk factors for CVD following 4 months of supplementation with cranberry juice (containing 69 mg anthocyanins).[61] However, following a 4 week period of daily supplementation with cranberry juice (containing 94 mg anthocyanins) in older adults with CVD, significant decreases in carotid-femoral PWV were observed compared to a control condition.[62] Therefore, vascular effects may be strongly dependent on factors such as population and dose.

Blood pressure is a less sensitive measure of vascular function but is probably the most well known. Significant changes in vascular tone may not necessarily translate into significant blood pressure effects, as observed in a study investigating blueberry smoothie effects in adults with Metabolic Syndrome over a 6 week period.[63] However, a number of studies have observed blood pressure benefits following berry consumption, both long term and in the immediate postprandial period. For example, following daily supplementation with blueberries (1 cup daily) for 8 weeks, significant reductions in systolic and diastolic blood pressure were observed in a group of obese adults.[64] Similar benefits to diastolic blood pressure have also been observed in middle-aged adults following daily consumption of cranberry juice (containing 173 mg total polyphenols) for 8 weeks.[65] In this case, blood pressure changes were also accompanied by improvements in other physiological measures, including fasting levels of glucose and measures of insulin resistance. Vascular improvements have also been observed in conjunction with benefits to cognitive function. Following daily consumption of freeze-dried blueberry powder (containing 260 mg anthocyanins), older adults saw benefits to FMD and systolic blood pressure following a 12 week period of supplementation.[66] These vascular changes were also accompanied by improvements to memory and executive function (word recall and task switching, respectively). Therefore, it is likely that the observed vascular changes may, at least in part, be responsible for changes to cognitive function. In addition to vascular changes, neurochemical changes have also been identified peripherally in plasma samples following berry consumption. For example, monoamine oxidase (MAO) is an enzyme involved in the degradation of monoamines, including some neurotransmitters. Levels of this circulating enzyme have been observed to decrease between 15 minutes and 4 hours following an acute dose of blackcurrant juice (containing 500 mg total polyphenols) in healthy young adults.[67] This inhibition of MAO is likely to result in an increase in monoamine neurotransmitter levels within the brain.

In addition to peripheral physiological measures, studies have also investigated changes in blood flow or levels of blood oxygenation directly within the brain. A study using magnetic resonance imaging (MRI) during a working memory task observed enhanced neural responses in participants who received a daily blueberry supplementation for 16 weeks.[68] In this instance, task performance did not appear significantly improved but, nevertheless, indicates that neurological changes can be elicited by blueberry consumption. Indeed, a similar study with a slightly higher daily dose (387 mg anthocyanins *versus* 269 mg) observed significant increases in brain activity that were accompanied by improvements to working memory task performance,[69] potentially highlighting the importance of dose in the effectiveness of berry consumption.

The studies documented above mainly consider vasodilatory benefits following berry supplementation that are likely to also result in increased cerebral blood flow. However, it is also important to consider the consistent delivery and uptake of glucose to the brain. Berry anthocyanins and other polyphenols are known to influence the rate of absorption, metabolism, and uptake of glucose to cells by a number of different mechanisms[70] and are described in more detail in Chapter 8. In healthy young adults, freeze-dried blueberry powder has been observed to slow the postprandial release of glucose into the bloodstream, in a dose-dependent manner, compared with a sugar-matched control.[71] Higher doses of up to 724 mg anthocyanins resulted in a significantly longer period of glucose availability. Similar observations have been made in a group of obese adults in the immediate postprandial period following ingestion of freeze-dried strawberry powder, with doses up to three cups of fresh strawberry equivalent.[72] In this case, glucose effects were also accompanied by a tempering of the peak insulin response, desirable for maintaining metabolic health. Similar patterns of effects have also been observed following 480 mL sweetened cranberry juice[73] and following a 150 g portion of mixed berry purée.[74] These immediate postprandial effects appear to also have a cumulative effect on glucose regulation, particularly in those with metabolic disorders. For example, following an 8 week blueberry-enriched diet, fasting levels of anthocyanin metabolites were found to be significantly elevated, and the presence of these metabolites correlated with fasting levels of glucose and a measure of insulin secretion in participants with Metabolic Syndrome.[75] However, in this case, the findings did not extend to a concurrently investigated diet condition rich in strawberries, raspberries, and cloudberries with a much lower anthocyanins content (71 mg *vs.* 1300 mg). Indeed, the majority of studies that observe a significant reduction in markers of the metabolic disease involve moderate levels of anthocyanins (equating to a daily dose of around one to two cups of fresh berries) and these are often obtained from blueberries or bilberries.[76–78] A detailed overview of berry benefits in metabolic disease can be found in Chapter 10.

While the outcomes relating to clinical conditions such as metabolic and CVDs demonstrate an impact of berries on mechanisms that relate to cognition, the data directly linking these conditions to cognitive function is often derived from epidemiological data and so provides correlational rather than causal evidence. A clearer indication of the inter-relationship between glucose regulation, vascular function, and cognitive function was shown by a study that administered an insulin nasal spray to participants with T2DM and observed significant increases in vasodilation within the brain, with concurrent improvements to cognitive function.[79] It is similarly possible to measure the impact of berries on cognitive function directly. Indeed, a majority of intervention studies investigate effects on cognitive behaviour independently of any additional physiological measure and these cognitive benefits are considered in more detail in the next section.

7.3.2.2 Behavioural Evidence in Humans

Human studies have investigated a range of berry intervention types, including strawberry, raspberry, blackberry, cranberry, and blackcurrant, in whole, freeze-dried, juiced, or extract forms. However, the most researched berry appears to be the blueberry. Blueberry studies have investigated cognitive and mood benefits in children, adolescents, young adults, and older adults. As described in the previous section, the physiological benefits of berry consumption appear to peak around 2 hours postprandially, and a large number of cognitive benefits have been observed at the same time point. For example, in primary school-aged children, faster reaction times and enhanced verbal memory performance were observed at 2 hours following consumption of a blueberry drink prepared from freeze-dried blueberry powder, containing approximately 250 mg anthocyanins.[80] Improvements were observed in comparison to performance following a sugar-matched control drink consumed by a separate group of children but did not extend to measures of reading ability. Although reported separately, significant improvements to positive mood were also observed in the children who consumed the blueberry drink.[81] Improvements to verbal memory were also observed in similar-aged children at 2 hours following a slightly lower dose (approximately 140 mg anthocyanins), obtained from a smoothie combining fresh blueberries and milk.[82] In this case, no additional benefits to reaction times or other executive functions such as response inhibition, or visuospatial memory, were observed. However, a further study observed benefits to verbal memory and executive function performance between 1 and 6 hours following two separate doses (approximately 130 mg and 250 mg anthocyanins, respectively) in a dose-dependent manner.[83] Therefore, the dose may play a crucial role in the effectiveness of an intervention. An additional study in 8- to 10-year-old children investigating only executive function performance observed significantly faster reaction times at 3 hours following a blueberry intervention containing approximately 250 mg anthocyanins.[84] In this case, benefits were most evident for cognitively demanding test trials, suggesting that the benefits of berry anthocyanins may only be detectable during sufficiently difficult tasks with a high cognitive load. A likely explanation is that when a task or test is not particularly challenging, a ceiling for cognitive performance is achieved irrespective of the intervention received, with no room for additional improvement due to the presence of an additional dietary bioactive.

Evidence for the cognitive benefits of blueberries in older children, adolescents, and young adults is currently lacking. However, in a study of mood effects in young adults aged 18–21 years, improvements to positive mood were observed 2 hours following a blueberry beverage containing approximately 250 mg anthocyanins.[81] In a group of adolescents, decreases in depressive symptoms were observed following 4 weeks of daily supplementation with a similar strength blueberry beverage.[85] It may be that cognitive benefits are less likely to be observed in such age groups that are already operating at a cognitive peak, but berries may still be effective in raising general levels of mood.

The majority of blueberry research has focused on older adults but has tended to investigate the impact of longer-term supplementation following daily treatment for a number of days or weeks. Significant findings have been observed for both memory and executive function. For example, fewer repetitive memory recall errors and improved task-switching ability were observed in older adults after a daily blueberry dose containing around 460 mg anthocyanins for a period of 90 days.[86] Similar benefits to task switching ability were observed to correlate with changes to serum levels of the anthocyanin malvidin-3-galactoside following 90 day supplementation with blueberries.[87] In the same study, changes to repetition errors during a verbal memory task were also correlated with changes to other blueberry polyphenol metabolites such as hippuric acid,[87] suggesting a clear link between the digestion and absorption of blueberry bioactives and improvements to cognitive function. In a small study of older adults with MCI, significant benefits to a measure of learning were observed following 12 weeks of daily supplementation with blueberry juice containing 425–600 mg anthocyanins.[88] In a similar study, older adults with MCI reported reduced cognitive symptoms during everyday activities following a daily blueberry dose, containing approximately 270 mg anthocyanins, for 24 weeks.[89]

Indeed, the reduction in symptoms was maintained for a further 24 weeks demonstrating long-lasting, everyday relevance of blueberry effects in cognitively impaired groups. Practical benefits to functional mobility have also been observed following comparable daily doses of freeze-dried blueberries for a period of 6 weeks,[90] suggesting that cognitive benefits of blueberry may also extend to psychomotor function (the cognitive control of movement). Similar effects have been previously noted in rodents.[42–44] In older adults, then, blueberry-induced cognitive benefits to memory and learning appear most prevalent, with some benefits to executive functions such as task switching. The size and cognitive range of the effects appear relevant to the performance of daily activities. Blueberries, therefore, appear to exert some level of cognitive benefit across different age groups. Particularly at specific developmental periods such as in childhood and during ageing.[1]

Other berries have not been investigated in such a wide range of populations, but nonetheless, similar benefits are evident. In a blackcurrant juice study, with an anthocyanin dose slightly higher than those used in blueberry studies (approximately 550 mg), cognitive benefits have been observed in young adults following a single acute dose.[91] Specifically, during the cognitively fatiguing battery of repeated cognitive testing, blackcurrant juice elicited faster reaction times on an attention task, compared with a sugar-matched control. The improvements to attention were accompanied by sustained availability of blood glucose, inhibition of monoamine oxidase (MAO-B), a reduction in plasma concentrations of normetadrenaline (a neurotransmitter metabolite), and an increase in dihydroxyphenylglycine concentration (DHPG, an amino acid implicated in the enhancement of cognitive function through activation of glutamate neurotransmitter receptors). These additional outcome measures provide further evidence for possible mechanisms of action for observed cognitive changes.

The effects of cranberry on cognition appear under-investigated despite the high levels of anthocyanins and flavanols they contain, potentially due to taste and acidity issues. An investigation of the long-term effects of cranberry juice supplementation in older adults found no benefits to cognition;[92] however, a very dilute concentration of the juice with added sweetener was used. Formulation of the juice to maximise palatability may have significantly reduced the content of cranberry bioactives (these were not reported). Other berries such as strawberries, raspberries, or blackberries have not been investigated individually but have been included in mixed berry interventions. In the immediate postprandial period, a mixed berry smoothie containing blueberry, strawberry, raspberry, and blackberry, with a total anthocyanins content of approximately 250 mg, was given to a group of fasted young adults along with a light breakfast.[93] Cognitive function was tracked for a period of 6 hours (a light lunch was also provided). Interestingly, the results showed that executive function performance on measures of task switching, and inhibition was maintained throughout the 6 hour period following the berry smoothie, whereas performance was observed to decline throughout the day following an energy-matched control drink.

This demonstrates that even in a cognitively healthy population, where possible cognitive gains are likely to be small, berries may be helpful in reducing fatigue-induced deficits. Over the longer term, mixed berry benefits to working memory have also been observed in middle-aged adults,[94] following daily supplementation with an active beverage containing blueberry, elderberry, lingonberry, strawberry, blackcurrant, and tomato containing around 250 mg anthocyanins.

While there is plenty of experimental evidence in support of the impact of berries on cognitive function, there are some caveats. Findings from human RCTs are often inconsistent between studies. Differences between different berry interventions may be attributable to the different profiles of anthocyanins and other bioactive compounds present. Indeed, as noted in the previous section on rodent research, different physiological and behavioural effects have similarly been observed in rats following feeding with different berry types. Differences in the polyphenolic content of berry varieties are notable. For example, when considering anthocyanins profiles, strawberries have been shown to be rich in pelargonidin,[95] and red raspberries and blackberries are rich in cyanidin,[95] but cultivated blueberries are often richer in peonidin,[96] while wild blueberries typically contain delphinidin and either petunidin[96] or malvidin.[97] Cranberries have also been shown to be rich in peonidin and cyanidin[95] but generally contain lower levels of anthocyanins compared to other berries.[98] Cranberries do, however, contain high levels of proanthocyanidins.[99] Therefore, it is important to consider other polyphenols in addition to anthocyanins when profiling berry bioactives. However, irrespective of berry type, choice of methodology may also play a role, such as differences in type or sensitivity of the cognitive tests used. Studies generally incorporate large test batteries that include a range of different cognitive tests measuring different cognitive domains, often with multiple outcome measures for each. Moderately sized effects are observed sporadically across a range of cognitive outcomes in both healthy and cognitively impaired populations, but comparison across studies is often difficult due to methodological differences, and even where studies are closely matched in design or population, findings may differ. Typically, significant results are only observed for a few outcome measures across a larger selection of tests, raising the additional possibility of type I error, where false positives may occur randomly due to chance. Meta-analysis (the statistical combination of the outcomes of many research studies) can be used to assess the impact of berries on cognitive function in humans by consolidating the findings of many studies. Currently, there are no published meta-analyses investigating the effects of berries on cognition; however, evidence from a systematic review suggests tentative support for the cognitive benefits of polyphenol bioactives.[100] Indeed, results of a recent meta-analysis investigating polyphenol effects (from multiple sources including berry) on cognition[101] concluded that there were moderate to large effects on the Trail Making task (a measure of psychomotor processing speed). However, this was a very small meta-analysis that included only six studies that had used the same cognitive task. Such meta-analysis findings may be subject to publication bias, where non-significant findings are less likely to be published. Therefore, the true effect size is likely to be small rather than moderate or large, as suggested by much-published research. The size of the studies discussed in this chapter range from around 10 to 55 participants. Therefore, much current research is likely to be statistically underpowered, and much larger intervention studies may be required in the future. To confirm this, future meta-analyses should seek to include data from a wide range of studies, perhaps classifying outcomes by cognitive domain rather than trying to target specific cognitive tests. Other factors such as dose and population should also be investigated. The findings of such meta-analyses could then be used to better inform future experimental design.

7.3.2.3 Mechanisms of Action in Humans

Mechanistic pathways are covered in additional detail in Chapter 8. From the evidence, it appears that the most likely mechanisms underlying berry induced behavioural changes following a single acute dose are related to transient changes to the supply of oxygen and glucose to the brain, whereas longer-term supplementation is likely to more permanently influence vascular function and glucose regulation, while also altering the chemistry and structure of areas of the brain such as the hippocampus. A general limitation of the human literature is that physiological changes are frequently not accompanied by observable behavioural changes. Often this is because the physiology and cognitive function have been measured separately. In cases where they are measured together, it is sometimes possible to see a physiological or neurological response to a berry intervention, but cognitive behaviour does not change.[68] This may, perhaps, be due to different dose sensitivities for physiological and cognitive outcomes, or cognitive effects may be less robust and more likely to be confounded by additional factors, for example, stress levels or sleepiness at the time of cognitive testing, or a lack of sensitivity in the cognitive measures used. Irrespective of the cause, this disconnect makes it difficult to provide direct evidence for mechanisms of action. Further studies are therefore needed that target specific mechanisms of action and use cognitive tests that are most likely to be sensitive to behavioural changes *via* the mechanism of interest. Such studies will require careful design considerations. A recent example of such a targeted design sought to investigate the effect of cocoa flavanols on a bespoke pattern recognition task that was specifically designed to investigate DG function within the brain.[102] In this study, neuroimaging (fMRI) was first used to aid in the development of a cognitive task that elicited activation in the DG (an area known to be susceptible to age-related dysfunction). Neuroimaging was then used to observe any changes in neural activity in the DG after consuming a cocoa-rich diet for three months. Both DG activity and task performance were observed to improve following the cocoa intervention. This type of study design could be extended to berry research.

7.4 Conclusion

While there are clear similarities between the findings of rodent and human research, such as the impact of berries on spatial working memory, and neurochemical mechanisms of action, including BDNF pathways, there are also clear differences between the methodologies used. The majority of rodent research focuses on ageing, and studies have often investigated long-term supplementation with berries with a view to mitigating the impact of age-related decline in cognitive functions such as memory. While some of this research is echoed in the human trials, much human research has targeted the immediate postprandial response (cognitive and physiological) following a single acute dose of berries. In these instances, it is often a much younger population that has been investigated. Therefore, in humans, we see immediate benefits to memory and executive function, accompanied by physiological changes to glucose regulation and vascular function that are more likely to provide viable mechanisms of action over short periods of time, where changes to neurochemistry and neural structure are unlikely to have had time to take effect.

Overall, the epidemiological evidence for the benefits of berries in humans is strong, and the experimental evidence for physiological benefits related to brain function is similarly compelling. Behavioural benefits are frequently observed in rodent studies. However, the findings of human RCT studies are often mixed. This result may be due to a number of confounding factors, such as unreliable cognitive measures, statistical power, and competing factors (*e.g.*, sleep, caffeine, mood) that may have a greater impact than any dietary interventions on cognitive performance, potentially masking dietary benefits if not adequately controlled. A better approach may be needed to more rigorously investigate the cognitive benefits of berries. Future research should seek to design larger, or perhaps collaborative, multi-centre studies and should aim to use a standardised range of cognitive tests that are suitably cognitively demanding for the population tested. Indeed, a key difference between current rodent and human research is that rodent studies use a small range of standardised cognitive tests to investigate specific cognitive domains such as spatial working memory and psychomotor function. Conversely, human studies have used a wide range of different non-standardised tasks, varying in difficulty, to assess a wider range of cognitive functions, including episodic memory, working memory, mood, and different executive functions such as planning, inhibition, and switching abilities. This difference has likely contributed to the mixed findings of human experimental studies, along with the many other confounding factors described above that arise when trying to assess human behaviour. Indeed, careful consideration should be given to all these factors in order to minimise their impact on experimental findings when designing future studies. In addition to the type of cognitive measures used, dose appears as a particularly relevant variable to the effectiveness of berry interventions, and so future dose–response studies will also be important to ascertain whether cognitive benefits are achievable within normal dietary ranges. The field of nutritional psychology is currently in its infancy, but there is clear potential for dietary manipulations, such as increased berry intake, to make a significant difference to cognitive outcomes throughout the lifespan. Additionally, it is important to develop an appropriate methodology to assess the effects of berry bioactives on cognition.

References

1. L. Bell and C. M. Williams, *Int. J. Food Sci. Nutr.*, 2020, 1–3.
2. S. Hein, A. R. Whyte, E. Wood, A. Rodriguez-Mateos and C. M. Williams, *J. Gerontol., Ser. A*, 2019, **74**, 984–995.
3. N. Travica, N. M. D'Cunha, N. Naumovski, K. Kent, D. D. Mellor, J. Firth, E. N. Georgousopoulou, O. M. Dean, A. Loughman and F. Jacka, *Brain, Behav., Immun.*, 2020, **85**, 96–105.
4. M. G. Miller and B. Shukitt-Hale, *J. Agric. Food Chem.*, 2012, **60**, 5709–5715.
5. B. Yang and M. Kortesniemi, *Curr. Opin. Food Sci.*, 2015, **2**, 36–42.
6. R. J. McCrimmon, C. M. Ryan and B. M. Frier, *Lancet*, 2012, **379**, 2291–2299.
7. J. T. O'Brien, T. Erkinjuntti, B. Reisberg, G. Roman, T. Sawada, L. Pantoni, J. V. Bowler, C. Ballard, C. DeCarli and P. B. Gorelick, *Lancet Neurol.*, 2003, **2**, 89–98.
8. A. Jennings, A. A. Welch, T. Spector, A. Macgregor and A. Cassidy, *J. Nutr.*, 2014, **144**, 202–208.
9. N. M. Wedick, A. Pan, A. Cassidy, E. B. Rimm, L. Sampson, B. Rosner, W. Willett, F. B. Hu, Q. Sun and R. M. van Dam, *Am. J. Clin. Nutr.*, 2012, **95**, 925–933.
10. I. Muraki, F. Imamura, J. E. Manson, F. B. Hu, W. C. Willett, R. M. van Dam and Q. Sun, *BMJ*, 2013, **347**, f5001.
11. E. E. Devore, J. H. Kang, M. M. Breteler and F. Grodstein, *Ann. Neurol.*, 2012, **72**, 135–143.
12. C. Samieri, Q. Sun, M. K. Townsend, E. B. Rimm and F. Grodstein, *Am. J. Clin. Nutr.*, 2014, **100**, 1489–1497.
13. X. Gao, A. Cassidy, M. Schwarzschild, E. Rimm and A. Ascherio, *Neurology*, 2012, **78**, 1138–1145.
14. E. Shishtar, G. T. Rogers, J. B. Blumberg, R. Au and P. F. Jacques, *Am. J. Clin. Nutr.*, 2020, **112**, 343–353.
15. X. Wang, Y. Y. Ouyang, J. Liu and G. Zhao, *Br. J. Nutr.*, 2014, **111**, 1–11.
16. A. Cassidy, É. J. O'Reilly, C. Kay, L. Sampson, M. Franz, J. Forman, G. Curhan and E. B. Rimm, *Am. J. Clin. Nutr.*, 2011, **93**, 338–347.
17. A. Cassidy, K. J. Mukamal, L. Liu, M. Franz, A. H. Eliassen and E. B. Rimm, *Circulation*, 2013, **127**, 188–196.
18. L. Letenneur, C. Proust-Lima, A. Le Gouge, J.-F. Dartigues and P. Barberger-Gateau, *Am. J. Epidemiol.*, 2007, **165**, 1364–1371.
19. X. Ye, S. N. Bhupathiraju and K. L. Tucker, *Br. J. Nutr.*, 2013, **109**, 503–510.

20. R. Mujcic and A. J. Oswald, *Am. J. Public Health*, 2016, **106**, 1504–1510.
21. A. N. Carey, D. R. Fisher, A. M. Rimando, S. M. Gomes, D. F. Bielinski and B. Shukitt-Hale, *J. Agric. Food Chem.*, 2013, **61**, 5979–5986.
22. J. A. Joseph, B. Shukitt-Hale, G. J. Brewer, K. A. Weikel, W. Kalt and D. R. Fisher, *J. Agric. Food Chem.*, 2010, **58**, 8196–8204.
23. L. M. Willis, B. Shukitt-Hale and J. A. Joseph, *Am. J. Clin. Nutr.*, 2009, **89**, 1602S–1606S.
24. J. A. Joseph, B. Shukitt-Hale, N. A. Denisova, R. L. Prior, G. Cao, A. Martin, G. Taglialatela and P. C. Bickford, *J. Neurosci.*, 1998, **18**, 8047–8055.
25. B. Shukitt-Hale, F. C. Lau and J. A. Joseph, *J. Agric. Food Chem.*, 2008, **56**, 636–641.
26. J. A. Joseph, B. Shukitt-Hale, N. A. Denisova, D. Bielinski, A. Martin, J. J. McEwen and P. C. Bickford, *J. Neurosci.*, 1999, **19**, 8114–8121.
27. K. Youdim, B. Shukitt-Hale, A. Martin, H. Wang, N. Denisova, P. Bickford and J. Joseph, *Nutr. Neurosci.*, 2000, **3**, 383–397.
28. P. M. Millin and G. M. Rickert, *Antioxidants*, 2019, **8**, 1.
29. P. Goyarzu, D. H. Malin, F. C. Lau, G. Taglialatela, W. D. Moon, R. Jennings, E. Moy, D. Moy, S. Lippold and B. Shukitt-Hale, *Nutr. Neurosci.*, 2004, **7**, 75–83.
30. C. M. Williams, M. A. El Mohsen, D. Vauzour, C. Rendeiro, L. T. Butler, J. A. Ellis, M. Whiteman and J. P. Spencer, *Free Radical Biol. Med.*, 2008, **45**, 295–305.
31. P. H. L. Tran and T. T. D. Tran, *Biomolecules*, 2021, **11**, 102.
32. G. Casadesus, B. Shukitt-Hale, H. M. Stellwagen, X. Zhu, H. G. Lee, M. A. Smith and J. A. Joseph, *Nutr. Neurosci.*, 2004, **7**, 309–316.
33. B. Shukitt-Hale, F. C. Lau, A. N. Carey, R. L. Galli, E. L. Spangler, D. K. Ingram and J. A. Joseph, *Nutr. Neurosci.*, 2008, **11**, 172–182.
34. B. Shukitt-Hale, D. F. Bielinski, F. C. Lau, L. M. Willis, A. N. Carey and J. A. Joseph, *Br. J. Nutr.*, 2015, **114**, 1542–1549.
35. B. Shukitt-Hale, R. L. Galli, V. Meterko, A. Carey, D. F. Bielinski, T. McGhie and J. A. Joseph, *Age (Dordr)*, 2005, **27**, 49–57.
36. B. Shukitt-Hale, A. N. Carey, D. Jenkins, B. M. Rabin and J. A. Joseph, *Neurobiol. Aging*, 2007, **28**, 1187–1194.
37. D. H. Malin, D. R. Lee, P. Goyarzu, Y.-H. Chang, L. J. Ennis, E. Beckett, B. Shukitt-Hale and J. A. Joseph, *Nutrition*, 2011, **27**, 338–342.
38. A. N. Carey, G. I. Pintea, S. Van Leuven, K. R. Gildawie, L. Squiccimara, E. Fine, A. Rovnak and M. Harrington, *Nutr. Neurosci.*, 2019, 1–11.
39. A. N. Carey, M. G. Miller, D. R. Fisher, D. F. Bielinski, C. K. Gilman, S. M. Poulose and B. Shukitt-Hale, *Nutr. Neurosci.*, 2017, **20**, 238–245.
40. S. M. Poulose, D. F. Bielinski, A. Carey, A. G. Schauss and B. Shukitt-Hale, *Nutr. Neurosci.*, 2017, **20**, 305–315.
41. J. Kang, K. M. Thakali, C. Xie, M. Kondo, Y. Tong, B. Ou, G. Jensen, M. B. Medina, A. G. Schauss and X. Wu, *Food Chem.*, 2012, **133**, 671–677.
42. B. Shukitt-Hale, V. Cheng and J. A. Joseph, *Nutr. Neurosci.*, 2009, **12**, 135–140.

43. R. L. Galli, A. N. Carey, K. A. Luskin, D. F. Bielinski and B. Shukitt-Hale, *J. Berry Res.*, 2016, **6**, 97–103.
44. B. Shukitt-Hale, N. Thangthaeng, M. E. Kelly, D. E. Smith and M. G. Miller, *Food Funct.*, 2017, **8**, 4752–4759.
45. B. Shukitt-Hale, N. Thangthaeng, M. G. Miller, S. M. Poulose, A. N. Carey and D. R. Fisher, *J. Gerontol., Ser. A*, 2019, **74**, 977–983.
46. C. Rendeiro, J. D. Guerreiro, C. M. Williams and J. P. Spencer, *Proc. Nutr. Soc.*, 2012, **71**, 246–262.
47. K. R. Gildawie, R. L. Galli, B. Shukitt-Hale and A. N. Carey, *Curr. Nutr. Rep.*, 2018, **7**, 39–48.
48. F. Sarubbo, S. Esteban, A. Miralles and D. Moranta, *Curr. Neuropharmacol.*, 2018, **16**, 126–136.
49. F. Sarubbo, D. Moranta and G. Pani, *Neurosci. Biobehav. Rev.*, 2018, **90**, 456–470.
50. K. Banaszewski, E. Park, I. Edirisinghe, J. C. Cappozzo and B. M. Burton-Freeman, *J. Berry Res.*, 2013, **3**, 113–126.
51. P. E. Milbury, J. A. Vita and J. B. Blumberg, *J. Nutr.*, 2010, **140**, 1099–1104.
52. C. D. Kay, G. Mazza, B. J. Holub and J. Wang, *Br. J. Nutr.*, 2004, **91**, 933–942.
53. A. Rodriguez-Mateos, R. P. Feliciano, T. Cifuentes-Gomez and J. P. Spencer, *J. Berry Res.*, 2016, **6**, 137–148.
54. A. K. Sandhu, M. G. Miller, N. Thangthaeng, T. M. Scott, B. Shukitt-Hale, I. Edirisinghe and B. Burton-Freeman, *Food Funct.*, 2018, **9**, 96–106.
55. W. Kalt, Y. Liu, J. E. McDonald, M. R. Vinqvist-Tymchuk and S. A. Fillmore, *J. Agric. Food Chem.*, 2014, **62**, 3926–3934.
56. C. Andres-Lacueva, B. Shukitt-Hale, R. L. Galli, O. Jauregui, R. M. Lamuela-Raventos and J. A. Joseph, *Nutr. Neurosci.*, 2005, **8**, 111–120.
57. A. Rodriguez-Mateos, C. Rendeiro, T. Bergillos-Meca, S. Tabatabaee, T. W. George, C. Heiss and J. P. Spencer, *Am. J. Clin. Nutr.*, 2013, **98**, 1179–1191.
58. R. M. Alqurashi, L. A. Galante, I. R. Rowland, J. P. Spencer and D. M. Commane, *Am. J. Clin. Nutr.*, 2016, **104**, 1227–1235.
59. P. Boutouyrie, M. Briet, C. Collin, S. Vermeersch and B. Pannier, *Artery Res.*, 2009, **3**, 3–8.
60. D. Djurica, R. R. Holt, J. Ren, A. W. Shindel, R. M. Hackman and C. L. Keen, *Br. J. Nutr.*, 2016, **116**, 639–647.
61. A. J. Flammer, E. A. Martin, M. Gössl, R. J. Widmer, R. J. Lennon, J. A. Sexton, D. Loeffler, S. Khosla, L. O. Lerman and A. Lerman, *Eur. J. Nutr.*, 2013, **52**, 289–296.
62. M. M. Dohadwala, M. Holbrook, N. M. Hamburg, S. M. Shenouda, W. B. Chung, M. Titas, M. A. Kluge, N. Wang, J. Palmisano and P. E. Milbury, *Am. J. Clin. Nutr.*, 2011, **93**, 934–940.
63. A. J. Stull, K. C. Cash, C. M. Champagne, A. K. Gupta, R. Boston, R. A. Beyl, W. D. Johnson and W. T. Cefalu, *Nutrients*, 2015, **7**, 4107–4123.

64. A. Basu, M. Du, M. J. Leyva, K. Sanchez, N. M. Betts, M. Wu, C. E. Aston and T. J. Lyons, *J. Nutr.*, 2010, **140**, 1582–1587.
65. J. A. Novotny, D. J. Baer, C. Khoo, S. K. Gebauer and C. S. Charron, *J. Nutr.*, 2015, **145**, 1185–1193.
66. S. Hein, E. Wood, N. Abhayaratne, R. Mesnage, F. Fernandes, C. Williams and A. Rodriguez-Mateos, *Curr. Dev. Nutr.*, 2021, **5**, 318.
67. A. W. Watson, A. Scheepens, D. O. Kennedy, J. M. Cooney, T. M. Trower and C. F. Haskell-Ramsay, *Nutr. Neurosci.*, 2020, **23**, 516–525.
68. E. L. Boespflug, J. C. Eliassen, J. A. Dudley, M. D. Shidler, W. Kalt, S. S. Summer, A. L. Stein, A. N. Stover and R. Krikorian, *Nutr. Neurosci.*, 2018, **21**, 297–305.
69. J. L. Bowtell, Z. Aboo-Bakkar, M. E. Conway, A.-L. R. Adlam and J. Fulford, *Appl. Physiol., Nutr., Metab.*, 2017, **42**, 773–779.
70. B. Burton-Freeman, *Br. J. Nutr.*, 2010, **104**, S1–S14.
71. L. Bell, D. J. Lamport, L. T. Butler and C. M. Williams, *Food Funct.*, 2017, **8**, 3104–3110.
72. E. Park, I. Edirisinghe, H. Wei, L. P. Vijayakumar, K. Banaszewski, J. C. Cappozzo and B. Burton-Freeman, *Mol. Nutr. Food Res.*, 2016, **60**, 1099–1109.
73. T. Wilson, A. P. Singh, N. Vorsa, C. D. Goettl, K. M. Kittleson, C. M. Roe, G. M. Kastello and F. R. Ragsdale, *J. Med. Food*, 2008, **11**, 46–54.
74. R. Törrönen, E. Sarkkinen, N. Tapola, E. Hautaniemi, K. Kilpi and L. Niskanen, *Br. J. Nutr.*, 2010, **103**, 1094–1097.
75. V. D. de Mello, M. A. Lankinen, J. Lindström, R. Puupponen-Pimiä, D. E. Laaksonen, J. Pihlajamäki, M. Lehtonen, M. Uusitupa, J. Tuomilehto and M. Kolehmainen, *Mol. Nutr. Food Res.*, 2017, **61**, 1700019.
76. A. J. Stull, K. C. Cash, W. D. Johnson, C. M. Champagne and W. T. Cefalu, *J. Nutr.*, 2010, **140**, 1764–1768.
77. H.-M. Lehtonen, J. Suomela, R. Tahvonen, B. Yang, M. Venojärvi, J. Viikari and H. Kallio, *Eur. J. Clin. Nutr.*, 2011, **65**, 394–401.
78. K. S. Stote, M. M. Wilson, D. Hallenbeck, K. Thomas, J. M. Rourke, M. I. Sweeney, K. T. Gottschall-Pass and A. R. Gosmanov, *Curr. Dev. Nutr.*, 2020, **4**, nzaa030.
79. V. Novak, W. Milberg, Y. Hao, M. Munshi, P. Novak, A. Galica, B. Manor, P. Roberson, S. Craft and A. Abduljalil, *Diabetes Care*, 2014, **37**, 751–759.
80. K. L. Barfoot, G. May, D. J. Lamport, J. Ricketts, P. M. Riddell and C. M. Williams, *Eur. J. Nutr.*, 2019, **58**, 2911–2920.
81. S. Khalid, K. L. Barfoot, G. May, D. J. Lamport, S. A. Reynolds and C. M. Williams, *Nutrients*, 2017, **9**, 158.
82. A. R. Whyte and C. M. Williams, *Nutrition*, 2015, **31**, 531–534.
83. A. R. Whyte, G. Schafer and C. M. Williams, *Eur. J. Nutr.*, 2016, **55**, 2151–2162.
84. A. R. Whyte, G. Schafer and C. M. Williams, *Food Funct.*, 2017, **8**, 4129–4138.

85. J. Fisk, S. Khalid, S. A. Reynolds and C. M. Williams, *Br. J. Nutr.*, 2020, **124**, 181–188.
86. M. G. Miller, D. A. Hamilton, J. A. Joseph and B. Shukitt-Hale, *Eur. J. Nutr.*, 2018, **57**, 1169–1180.
87. G. A. Rutledge, A. K. Sandhu, M. G. Miller, I. Edirisinghe, B. B. Burton-Freeman and B. Shukitt-Hale, *Food Funct.*, 2020, 107–118.
88. R. Krikorian, M. D. Shidler, T. A. Nash, W. Kalt, M. R. Vinqvist-Tymchuk, B. Shukitt-Hale and J. A. Joseph, *J. Agric. Food Chem.*, 2010, **58**, 3996–4000.
89. R. K. McNamara, W. Kalt, M. D. Shidler, J. McDonald, S. S. Summer, A. L. Stein, A. N. Stover and R. Krikorian, *Neurobiol. Aging*, 2018, **64**, 147–156.
90. M. A. Schrager, J. Hilton, R. Gould and V. E. Kelly, *Appl. Physiol., Nutr., Metab.*, 2015, **40**, 543–549.
91. A. W. Watson, C. F. Haskell-Ramsay, D. O. Kennedy, J. M. Cooney, T. Trower and A. Scheepens, *J. Funct. Foods*, 2015, **17**, 524–539.
92. W. D. Crews Jr, D. W. Harrison, M. L. Griffin, K. Addison, A. M. Yount, M. A. Giovenco and J. Hazell, *J. Altern. Complementary Med.*, 2005, **11**, 305–309.
93. A. R. Whyte, N. Cheng, L. T. Butler, D. J. Lamport and C. M. Williams, *Nutrients*, 2019, **11**, 2685.
94. A. Nilsson, I. Salo, M. Plaza and I. Björck, *PLoS One*, 2017, **12**, e0188173.
95. J. M. Harnly, R. F. Doherty, G. R. Beecher, J. M. Holden, D. B. Haytowitz, S. Bhagwat and S. Gebhardt, *J. Agric. Food Chem.*, 2006, **54**, 9966–9977.
96. A. Bunea, O. D. Rugina, A. M. Pintea, Z. SCONŢA, C. I. Bunea and C. Socaciu, *Not. Bot. Horti Agrobot. Cluj-Napoca*, 2011, **39**, 70–76.
97. A. Rodriguez-Mateos, T. Cifuentes-Gómez, S. Tabatabaee, C. Lecras and J. P. Spencer, *J. Agric. Food Chem.*, 2012, **60**, 5772–5778.
98. G. Borges, A. Degeneve, W. Mullen and A. Crozier, *J. Agric. Food Chem.*, 2010, **58**, 3901–3909.
99. J. K. Hellstrom, A. R. Torronen and P. H. Mattila, *J. Agric. Food Chem.*, 2009, **57**, 7899–7906.
100. D. J. Lamport and C. M. Williams, *Brain Plast.*, 2020, **6**, 137.
101. A. Ammar, K. Trabelsi, P. Müller, B. Bouaziz, O. Boukhris, J. M. Glenn, N. Bott, T. Driss, H. Chtourou and N. Müller, *J. Clin. Med.*, 2020, **9**, 835.
102. A. M. Brickman, U. A. Khan, F. A. Provenzano, L.-K. Yeung, W. Suzuki, H. Schroeter, M. Wall, R. P. Sloan and S. A. Small, *Nat. Neurosci.*, 2014, **17**, 1798.

CHAPTER 8

Neuroprotective Mechanisms of Berry Bioavailable Polyphenol Metabolites

NATASA LONCAREVIC-VASILJKOVIC, DIOGO CARREGOSA AND CLÁUDIA NUNES DOS SANTOS*

CEDOC, Chronic Diseases Research Centre, NOVA Medical School, Universidade NOVA de Lisboa, Campo dos Mártires da Pátria, 130, 1169-056 Lisboa, Portugal
*Email: claudia.nunes.santos@nms.unl.pt

8.1 Main Molecular and Cellular Hallmarks of Neurodegenerative Diseases

Neurodegenerative diseases (NDDs) represent a family of heterogeneous disorders that affect millions of people worldwide. The prevalence of these disorders is increasing due, in part, to the extension in human lifespan.[1]

NDDs are characterized by a progressive and irreversible loss of function and/or death of neurons from specific brain regions, leading to severe cognitive and functional decline. The most known NDDs are Alzheimer's disease (AD), Parkinson's disease (PD), Huntington's disease (HD), and amyotrophic lateral sclerosis (ALS), which share common cellular and molecular mechanisms of disease, as can be seen in Figure 8.1, including protein misfolding and aggregation, neuroinflammation, oxidative stress, and neuronal degeneration.[2]

Food Chemistry, Function and Analysis No. 33
Berries and Berry Bioactive Compounds in Promoting Health
Edited by Dorothy Klimis-Zacas and Ana Rodriguez-Mateos

Published by the Royal Society of Chemistry, www.rsc.org

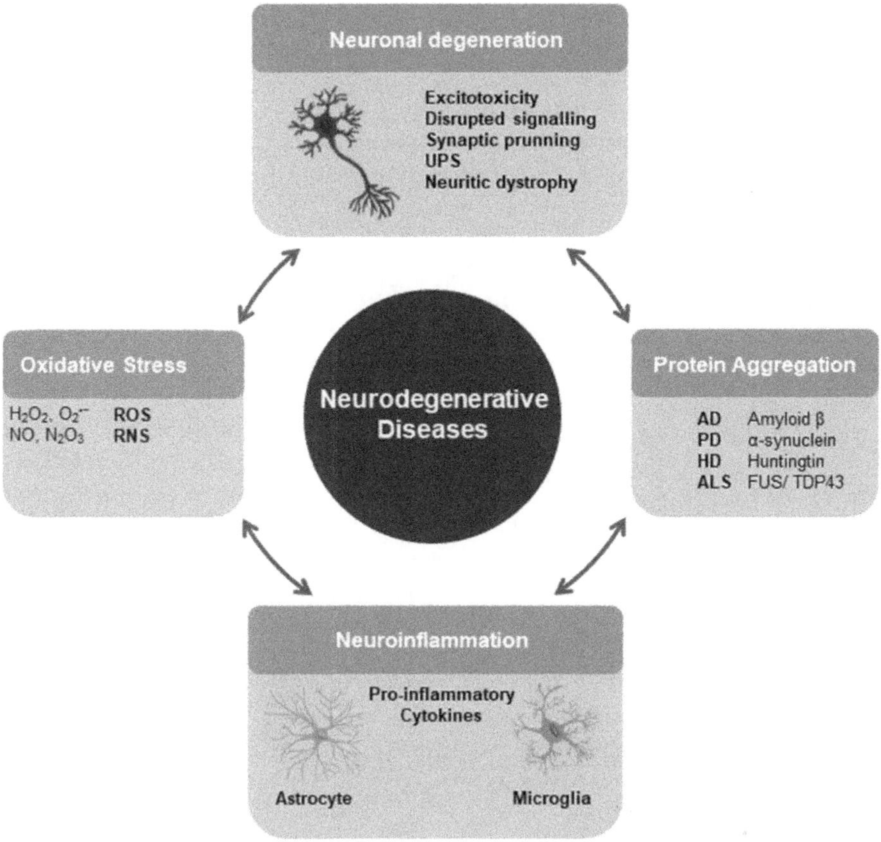

Figure 8.1 Neurodegenerative diseases such as Alzheimer's disease (AD), Parkinson's disease (PD), Huntington's disease (HD), and amyotrophic lateral sclerosis (ALS), all have common cellular and molecular mechanisms including protein misfolding, aggregation, and inclusion body formation, chronic neuroinflammation and oxidative stress that culminates in neuronal degeneration and cell death. ROS, reactive oxygen species; RNS, reactive nitrogen species; UPS, ubiquitin–proteasome system; CX3CR1, C-X3-C motif chemokine receptor 1; FUS, fused in sarcoma; TDP43, transactive response DNA binding protein 43.

8.1.1 Protein Misfolding and Aggregation

Intrinsically misfolded proteins exist in cells in several forms: unfolded, intermediately folded, and correctly folded species.[3] Healthy cells degrade or try to refold misfolded proteins by chaperone proteins that are involved in protein folding and trafficking as well as intermediate stabilization.[3] The exact mechanism by which protein misfolding and aggregation becomes pathological is still unclear but it is clear that cells lose the ability to refold or degrade misfolded proteins.

Despite the involvement of distinct proteins in different NDDs, the process of protein misfolding and aggregation is remarkably similar in all mentioned diseases (Figure 8.1).[4] AD is the most common NDD (or the most common of NDDs). It is an irreversible, progressive brain disorder that represents the most common cause of dementia. Pathologically, it is characterized by atrophy and/or loss of cortical neurons, extracellular deposition of amyloid-beta (Aβ) plaques, and neurofibrillary tangles composed of hyperphosphorylated and aggregated tau protein that promotes synaptic dysfunction, neuroinflammation, neurodegeneration, and neuronal cell death early and consistently.[5] The most affected regions of the brain in terms of neurodegeneration are the hippocampus, frontal, temporal, and parietal lobes. Neurodegeneration within these regions develops over the course of the disease progression.

PD, the second most common NDD after AD, is a progressive nervous system disorder that mainly affects the motor system. The most obvious early symptoms are tremor, rigidity, slowness of movement, postural instability, and difficulty with walking, but cognitive problems may also occur. Neuropathologically, PD is characterized by the presence of intracellular inclusions known as Lewy bodies (LBs), the main constituent of which are α-synuclein (α-syn) fibrils.[6] Although α-syn belongs to the family of natively unfolded proteins, which demonstrate little or no ordered structure, the protein has the intrinsic propensity to fibrillate, giving rise to insoluble fibrils similar to the ones detected in LBs within dopaminergic neuronal cells.[7–11] Overall, the most affected areas of the brain are the substantia nigra, (SN) areas of the midbrain, basal forebrain, and the neocortex. Notwithstanding, as is the case with other disease, the presence of aggregates, LBs in the case of PD might not be enough to diagnose PD, since dementia with LBs shows similar pathological features.[12]

HD is a rare, inherited disease that causes the progressive degeneration of neuronal cells in the brain. This results in uncontrolled movements, loss of intellectual abilities, and emotional disturbances. HD is caused by a mutation in the gene that encodes a protein called huntingtin (Htt), which is reflected in the abnormal increase in the number of CAG nucleotide repeats. As a result, the translated protein Htt contains disease-causing expansions of glutamines (polyQ) that make it prone to misfold and aggregate. A hallmark pathological feature of HD is intracellular aggregates of mutant Htt called inclusion bodies that lead to neurodegeneration and neuronal cell death in different brain regions.[13,14] The most obvious neuropathology in HD occurs within the neostriatum, comprising gross atrophy of the caudate nucleus and putamen, accompanied by selective neuronal loss and astrogliosis. Marked neuronal loss is also seen in deep layers of the cerebral cortex.

ALS is a devastating NDD characterized by the selective and progressive loss of upper and lower motor neurons of the cerebral cortex, brainstem, and spinal cord. The result of this loss is progressive and irreversible paralysis leading to a complete incapacity of movements and finally to respiratory failure, but the cognitive functions are not affected.[15] Three proteins are known to be responsible for the development of ALS: fused in sarcoma/translocated in liposarcoma (FUS/TLS), TAR DNA-binding protein 43 (TDP43), and UBQLN2 that encodes a protein called ubiquilin 2, which is very important as it plays a major role in recycling damaged proteins from neurons in the spinal cord and neurons in the cortex and hippocampus of the brain. Variations in FUS/TLS, TDP43, and UBQLN2 genes cause an accumulation of proteins in the cytoplasm of neurons, which is suspected of contributing to neuronal dysfunction.[16]

8.1.2 Neuroinflammation

Although different NDDs have distinct triggering events, a common feature is chronic immune activation, in particular of microglia cells, the resident macrophages of the central nervous system (CNS) (Figure 8.1). Apart from playing critical roles in the normal development and function of neural circuits and the pathogenic role of immune responses, microglial cells are activated by neuronal injury or infection to produce pro-inflammatory or anti-inflammatory factors, depending on the activation state of these cells. Maintaining the right balance of anti-inflammatory mediators, which allow tissue repair and healing, and pro-inflammatory mediators, to clear cellular debris and aggregated misfolded proteins, is essential to maintaining a healthy brain. While acute inflammatory response is beneficial for ridding the brain of dying cells or foreign pathogens, prolonged or chronic inflammation, which has been implicated as a factor underlying neurodegeneration in several diseases, can lead to the death of surrounding neurons.[17] Furthermore, chronic neuroinflammatory response in the brain is believed to play a crucial role in the development of NDDs and an acceleration of disease progression.[18,19] The activation of microglia in AD, PD, and ALS seems to reflect this pro-inflammatory phenotype. Moreover, microglial cells are reported to accumulate around affected brain regions in AD, PD, and HD, reinforcing their involvement NDDs.[20,21] Large numbers of reactive microglial cells have been observed in the substantia nigra pars compacta in post-mortem tissue from patients with PD[22] as well as patients suffering from 1-methyl-4-phenyl-1,2,3,6-tetrahydropyridine (MPTP)-induced PD,[23] indicating that this inflammatory process might aggravate neurodegeneration. By releasing various kinds of damaging factors such as pro-inflammatory cytokines or pro-inflammatory molecules (*i.e.*, proteolytic enzymes, reactive oxygen intermediates, or nitric oxide [NO]), microglia propagate and intensify the damage to CNS.[24–27] Another glial type within the CNS is astrocytes (Figure 8.1). They represent the most abundant type of glial cells in the brain, regulating the maturation and maintenance of neuronal function. They are highly sensitive to their environment and respond quickly to injury. Like microglia, they exist in various states of activation. In some cases, they can be neuroprotective by reducing inflammation and stimulating repair, but in others, they are neurotoxic, promoting inflammation and contributing to neuronal cell death. Activated astrocytes that release pro-inflammatory factors have been reported in AD, PD, and ALS.[28] Since the astrocytes are the important part of the blood–brain barrier (BBB), with their processes forming the involucre around the blood vessels in the brain, their dysfunction could be detrimental for maintenance of the permeability control of the BBB.

Although great strides have been made toward understanding how the interactions between neurons and glia are disrupted in NDD, many aspects of inflammation-driven neurodegeneration remain unclear. It is still a theme of debate whether glial activation is the cause or consequence of NDDs. But even if inflammation is not a primary causative process, its presence may greatly contribute to the continued loss of neurons. Therefore, strategies to detect and modulate the activity of microglia and astrocytes could help tilt the inflammatory balance in the right direction to treat or mitigate NDDs more effectively.

8.1.3 Oxidative Stress

Oxidative stress was described as one of the major mechanisms linked to all NDDs, which has made it an appealing therapeutic target (Figure 8.1). In neurons, the primary sources of oxidative stress are dysfunctional mitochondria. Under normal conditions, small endogenous antioxidants, such as glutathione (GSH) and Coenzyme Q10 (CoQ10), as well as a number of antioxidant enzymes, such as catalase (CAT) and superoxide dismutases (SOD), are responsible for detoxifying reactive oxygen species (ROS) and reactive nitrogen species (RNS) generated by cellular processes like mitochondrial respiration. However, evidence suggests that in the context of neurodegeneration, the activity of antioxidant enzymes is decreased, and pools of GSH and CoQ10 are depleted.[29–31] Loss of intrinsic antioxidant defenses allows ROS and RNS to accumulate and cause oxidative damage to vital cellular macromolecules such as proteins, lipid membranes, DNA, and other cellular components, as well as disrupt normal mechanisms of cellular signaling.[32,33] In particular, ROS produced at the mitochondria are capable of causing oxidative damage and subsequent mutations in mitochondrial DNA, which encodes major components of mitochondrial respiratory complexes and proteins involved in electron transport, leading to impaired function of the mitochondrial respiratory chain. This, in turn, results in perturbations in cellular energy, further production of ROS, and perpetuation of oxidative stress on a cellular level, ultimately culminating in cell death.[31] Mitochondria also play an important role in apoptosis by sequestering cytochrome c, which is tethered to the inner mitochondrial membrane by cardiolipin, and normally acts as an electron shuttle between complexes III and IV in the electron transport chain.[34] It is thought that oxidative damage to cardiolipin leads to a loss of its interaction with cytochrome c, which can then be released into the cytosol, where it is able to activate caspase-dependent apoptosis.

Because of the low activity of antioxidant defense systems compared with the high dependence of mitochondria for energy, the brain is susceptible to oxidative stress more than other organs.[35,36] Moreover, the high levels of neurotransmitters that are autoxidized to generate ROS further increase this problem.[37] In agreement with these observations, there is evidence that increased oxidative stress plays an important role in the pathogenesis of NDDs such as AD, PD, brain ischemic disease, and aging.[38,39] Compelling evidence supports a tight link between oxidative stress and protein aggregation processes, which are noticeably involved in the development of proteinopathies, such as AD, PD, HD, ALS, and prion diseases. The literature is tremendously rich in studies that establish a functional link between both processes, revealing that oxidative stress can be either causative or consecutive, to protein aggregation.[40] Additionally, the relationship between oxidative stress and inflammation has been well documented,[41] with some evidence indicating that oxidative stress may play a pathogenic role in chronic inflammatory diseases.[42] ROS generated in brain tissues can modulate synaptic and non-synaptic communication between neurons that result in neuroinflammation and consequently in neurodegeneration and memory loss.[43] Therefore, the inhibition of oxidative stress-induced neuronal injury could function as the underlying strategy in the treatment or prevention of aging and NDDs.[44]

8.1.4 Neuronal Degeneration

NDDs are characterized by the loss or dysfunction of groups of neurons. Neuronal dysfunction may not necessarily be the result of neuronal cell death. It can be caused by neuritic dystrophy, synaptic loss, impairment of long-term potentiation, or disruption of neuronal signaling as a result of disease pathology (Figure 8.1). These events often precede neuronal cell loss in chronic NDDs.[45] The pathology of AD includes the loss of medium and large pyramidal neurons from hippocampal regions.[46–48] It is uncertain what leads to neuronal cell death in AD, but it has been suggested that accumulation of β-amyloid leads to increased sensitivity to excitotoxicity in AD[49] and excessive activation of *N*-methyl-D-aspartate (NMDA).[50,51]

In PD, the degeneration of dopaminergic neurons in the substantia nigra pars compacta occurs, leading to dopamine depletion in the striatum, which is the direct cause of the clinical features of PD: bradykinesia, resting tremor, and rigidity. The clinical symptoms of PD become very evident when there is a loss of about 50–70% of the dopaminergic neurons within the substantia nigra, with initial motor signs in PD occurring as early as around 30% of total substantia nigra (SN) neurons are lost.[52] Thus, PD is a good example of an NDD where neuronal cell loss is directly responsible for the clinical features observed. The exact molecular mechanisms that initiate dopaminergic neuron loss in PD, however, are still elusive, but these include some of the previously mentioned mechanisms like oxidative stress, defects in mitochondrial complex I, neuroinflammation, protein aggregation, and impairment of the ubiquitin–proteasome system (UPS) that degrades damaged or misfolded proteins.[53–56] Interestingly in the case of PD, several toxins, capable of interfering with the mitochondria respiratory chain, are known to induce PD or PD-like symptoms. These include, for example, MPTP, paraquat, and rotenone.[57]

Neuronal cell death also occurs in HD, mainly in the striatum.[58] The exact cause of neurodegeneration is still not known yet has been linked to the accumulation of toxic mutated Htt fragments and caspase activation in HD brains.[59,60] Excitotoxicity has also been suggested to play a role in the pathogenesis of HD. The medium spiny neurons, found in the striatum, that are lost in HD are innervated by glutamate-releasing neurons from the cortex, and they selectively degenerate upon the administration of glutamate receptor agonists into the rat striatum.[61]

In ALS, neuronal death happens in motor neurons located in the cortex and lower motor neurons located in the brainstem and spinal cord.[62] In ALS, symptoms may start in a single spinal cord region and progress to additional regions over time, giving it a prion-like disease status.[63]

It is likely that the accumulation of protein aggregates is linked in some way across all NDDs; however, our understanding of aggregate formation or size and neuronal death has not been fully established so far. In addition to protein aggregation, two other main processes observed in all NDDs, chronic inflammation and oxidative stress, greatly contribute to the progression of NDDs, ultimately leading to neurodegeneration and neuronal cell death. The role of the other cells in the CNS is crucial for neuronal homeostasis and survival. The already mentioned microglia cells and astrocytes together with microvascular endothelial cells can play a number of vital roles besides inflammatory roles like maintaining neurotransmitter, ion, energy homeostasis and lead to tissue repair and neurogenesis, and as such, dysfunction on any of these well-controlled systems could be harmful to neuronal viability. Also, BBB integrity represents a crucial step to maintain the balance of substances in the brain and their flow to the cerebrospinal fluid, and it is also proposed that this integrity might diminish during aging, leading to a pathological brain micro-environment. All these factors and the intense number of investigations in the last decades about the mechanisms by which neurons die in NDDs have led to the conclusion that the process of neurodegeneration is multifactorial, which could explain why so many drugs have so far failed to show significant results.[64] The failure to develop effective therapeutic agents for NDDs may be in large part because most current treatments target only single aspects of disease pathology. As such, more recently, efforts to develop new therapeutic strategies have focused on targeting multiple factors contributing to disease progression. In this sense, dietary (poly)phenols arose as promising candidates due to their described pleiotropic effects, since they are able to modulate many of the features underlying neurodegenerative processes.[65]

8.2 Berry (Poly)phenols as Potential Modulators of Neurodegenerative Diseases

Berries are among one of the most well-documented fruits for their potential health benefits, including NDDs.[66–68] Dozens of berry species are present in the normal human diet, including strawberries, blueberries, blackberries, or cranberries. Berries possess a vast array of small molecules like phenolic acids, stilbenes, ascorbic acid, and flavonoids, and in particular, large amounts of anthocyanins.[68,69] Anthocyanins are appealing as therapeutic agents for many reasons, most notable being generally recognized as safe molecules and having several interesting pleiotropic effects.[70] Such effects include anti-inflammatory action and protection against oxidative insults.[70]

Anthocyanins have been reported as an anti-inflammatory agent able to suppress microgliosis and astrogliosis and to reduce overall neuroinflammation *via* suppression of nuclear factor kappa-light-chain-enhancer of activated B cells (NF-κB) interleukin-1β (IL-1β), cyclooxygenase-2 (COX-2), inducible nitric oxide synthase (iNOS), and tumor necrosis factor-alpha (TNF-α) in the hippocampus and cortex regions of rat's brain.[44,71] It has also been observed that anthocyanins ameliorate anti-ocular inflammation in male Lewis rats. This effect was evident by the direct blocking of the expression of the iNOS.[72] Anthocyanins have also been shown to abrogate neuroapoptosis *via* c-Jun N-terminal kinase (JNK) suppression and improved deregulated synaptic proteins, including synaptosomal synaptophysin.[71] Moreover, anthocyanins lowered p-JNK.[44]

This has led to the proposal that anthocyanins may be promising agents for the prevention and treatment of NDDs. Interestingly these are largely present in berries and could easily and safely being incorporated into a day-to-day diet. Moreover, they possess a therapeutic potential as both a dietary intervention and as the basis for future drug discovery efforts, making them an interesting class of compounds to consider in the context of pharmacological development.

Yet as is the case for all molecules, their bioavailability and pharmacokinetic properties are critical aspects for the success of a molecule *in vivo* (see Chapter 7). To exert the potential neuroprotective effect, anthocyanins or their metabolites must reach circulation at physiologically relevant concentrations and eventually the brain. Interestingly, anthocyanins, like other flavonoids, undergo extensive metabolism, in particular by the microbiota, generating a vast array of low molecular weight (LMW) (poly)phenol metabolites, including benzoic acids, benzaldehydes, hydroxybenzenes, phenylpropionic acids, phenylacetic acids, cinnamic acids, and hippuric acids that can appear in circulation in concentrations higher to those found for anthocyanins themselves.[73] In a study using 500 mg $^{13}C_5$-cyanidin-3-glucoside, Czank *et al.* showed that only 9.99 ± 3.90 ng of $^{13}C_5$-cyanidin-3-glucoside was recovered in serum.[74] Meanwhile, metabolites such as protocatechuic acid (PCA) (3,4-dihydroxybenzoic acid, a benzoic acid) and phloroglucinaldehyde (2,4,6-trihydroxybenzaldehyde, a benzaldehyde metabolite) reach a total recovery of 149.88 ± 93.20 ng, with other benzoic acids reaching 321.73 ± 130.89 ng, phenylpropionic and phenylacetic acids reach a recovery of 153.32 ± 60.95 ng and hippuric acid reaching a total recovery of 348.04 ± 246.32 ng. Moreover, in circulation, anthocyanins and anthocyanin-glucosides stay for brief periods of time,[75] ranging from 0 to 2 hours up to 4 hours, depending upon the anthocyanin and concentration in the food matrix.[76] In contrast, LMW (poly)phenol metabolites can stay in circulation for periods for 2 up to 48 hours.[74] As such, it is critical to consider that such LMW (poly)phenol metabolites may be responsible for many of the systemic effects of berries.

Driven by the beneficial effects of berries on brain disorders, several studies have focused on the brain permeability of berry (poly)phenols, mainly flavonoids, using *in vitro* and *in vivo* studies.[77] The most common practice in *in vitro* studies relies on the transwell system using brain endothelial cell cultures as a barrier to create two chambers mimicking the luminal (blood) and abluminal (brain) sides.[78] Using this system with hCMEC/D3 cells (immortalized human cerebral micro vessel endothelial cell line), the transport of flavan-3-ols and anthocyanins together with some of their methoxy and glucuronide conjugates was observed. In this case, the authors concluded that the conjugates presented higher permeation.[79] Furthermore, catechin and epicatechin were also shown to cross the BBB in RBE-4 (rat brain endothelial cells) and hCMEC/D3.[79] Regarding the LMW (poly)phenol metabolites, several hydroxybenzenes, benzoic acids have been tested in HBMEC cells (human brain microvascular endothelial cells) and RBT-24 cell system while cinnamic acid phenylpropionic acids tested in BBB-RTU transwell unit, altogether showed that LMW (poly)-phenol metabolites can also cross the blood–brain barrier.[77]

Although *in vitro* studies can be used to suggest the potential of (poly)-phenols and their metabolites to reach the brain, the use of *in vivo* experiments is crucial to explore the full complexity of the BBB. As such, several studies have been able to confirm the brain bioavailability of flavonoids in different brain regions in rats[80,81] and pigs.[82,83] Notwithstanding, the presence of several (poly)phenols seems to depend on the route of administration. Epigallocatechin gallate,[84,85] epicatechin,[86–89] anthocyanins,[81,90] quercetin,[91,92] and naringenin,[93] were detected in the brain after oral administration, while naringenin and naringenin-glucuronide were detected after intravenous injection.[94] Regarding anthocyanins, they could not be detected in rat brains after supplementation with raspberry juice in a dose equivalent to 700 mL to a 70 kg human.[95] However, they can be observed in other studies using higher doses and/or continuous administration.[81,96,97] For example, cyanidin-glucoside and malvidin-glucoside have been found in the rat brain after a 2% blueberry diet for 10 weeks.[96] Nevertheless, in most cases, concentrations of anthocyanins and other flavonoids are low. For instance, cyanidin-3-glucoside reached concentrations up to 21 nmol per gram of rat brain.[81] As such, studies on the brain permeability of anthocyanins indicate that low amounts of anthocyanins and other flavonoids may reach the brain, suggesting they are not the ones solely responsible for all the proven effects of berries. Nevertheless, studies in rats have shown that the LMW (poly)phenol metabolites can be recovered in the brain, even minutes after entering circulation.[77,98] Thus, it is possible that the proven effects of berries may be attributed to more than just anthocyanins alone.

While anthocyanins have proven to be promising therapeutic candidates in an *in vitro* context, the overall bioavailability of parent anthocyanins is very low.[99] Indeed, the concentrations that are needed to achieve neuroprotection in cultured neurons far exceed physiological concentrations of anthocyanins observed in brain tissue from supplemented animals.[70] Since the bioavailability studies led to the hypothesis that anthocyanin metabolites, but not the parent compounds, are likely to mediate many of the biological effects observed following anthocyanin ingestion *in vivo*, it is of great importance to investigate the nutritional and therapeutic properties of these metabolites in order to better understand their potential role in attenuating NDDs.

Overall, there are already some clues of the mechanistic actions of some metabolites obtained in cellular models of NDDs. Although most of these studies still rely on the effects of the compounds in levels so high that are not brain relevant, they are of great importance since they represent a tool for understanding mechanisms underlying the pleiotropic effects of polyphenol metabolites in the NDDs scenario. In the past years, studies using lower concentrations of metabolites (1–10 μM range) started to emerge, referring to a more realistic (nutritional) scenario.[100] These studies are of a great importance in terms of translating the data from the basic research to potential clinical trials.

We will summarize the current evidence highlighting the unique activity of not only anthocyanins but also their metabolites against oxidative stress, neuroinflammation, and protein aggregation as the processes that contribute to the development and progression of NDDs. Moreover, we will also explore the current evidence highlighting the use of these compounds in a therapeutic context in various pre-clinical models of NDD and aging.

8.2.1 Protein Aggregation

As a main feature of all NDDs, protein aggregation has drawn a lot of attention in terms of different therapeutic approaches, all having a goal to alleviate, restore, or even prevent this process. Many (poly)phenolic compounds, especially anthocyanin metabolites, have been shown as potent inhibitors of α-syn and Aβ aggregation as they interfere with the oligomerization of these proteins (Table 8.1). Although not always physiologically relevant concentrations are considered in the studies, there is a clear indication of the effects of the metabolites in protein aggregation. A considerable number of studies within this field were done *in vitro* using Aβ or α-syn fibril formation and destabilization assay (ThT assay). Within these studies, the potential of chosen polyphenol metabolite to inhibit fibril formation or to destabilize pre-formed fibrils of Aβ or α-syn is estimated using different metabolite concentrations and different incubation paradigms and times (Table 8.1). Both PCA and gallic acid (GA), 3,4,5-trihydroxybenzoic acid, were found to inhibit the formation of toxic α-syn fibrils, destabilize pre-formed fibrils, and preserve cellular viability in cells treated with α-syn oligomers and protofibrils.[101–104] Though it is not clear how PCA carries out this function, GA was found to bind to soluble, non-toxic oligomers with no β-sheet content and to stabilize their structure.[105] The binding of GA to the oligomers may represent a potential mechanism of action. Moreover, GA was shown not only to inhibit α-syn fibrillation and toxicity but also to disaggregate pre-formed α-syn fibrils.[106] GA has also been shown to potently inhibit aggregation of proteins induced by

Table 8.1 Insights on the potential mechanisms of action of (poly)phenol metabolites on protein aggregation.[a]

(Poly)phenol metabolite	Model	Protein misfolding model	Metabolite dose and time	Mechanistic evidence	Ref.
Benzoic acids					
Protocatechuic acid/ 3,4-dihydroxybenzoic acid (PCA)	PC12 cells	Treatment with MPP^+	330, 650, and 1300 μM for 6 h	↓ abnormal oligomerization of α-synuclein	97
	In vitro	Aβ and α-syn fibril formation and destabilization assay	Aβ condition: Treatment using 10, 20, 50, and 100 μM for 2 h up to 48 h α-syn condition: Treatment using 10, 20, 50, and 100 μM for 24 h up to 48 h	↓ Aβ and α-syn aggregation and destabilization of their pre-formed fibrils	95
	PC12 cells	Aβ and α-syn induced toxicity	1, 5, 25, 50, and 100 μM for 24 h	↑ viability of cells exposed to aggregates	95
	AβPP/PS1 mice	Animal transgenic model	100 mg kg^{-1} orally for 4 weeks	↓ Aβ deposition in hippocampal tissue ↓ APP expression	104
Gallic acid/3,4,5-trihydroxybenzoic acid (GA)	*In vitro*	α-syn fibril formation	40 μM (1 : 2 molar ratio to A53TαS) from 4 h up to 100 h	Interaction with A53TαS through weak, transient interactions inhibiting early stages of α-syn aggregation	98
	BE (2)-M17 human neuroblastoma cells	α-syn aggregates	25,50, and 100 μM (1 : 1, 2 : 1, 1 : 1 ratio to α-syn) up to 6 days prior to cells treatment	↓ α-syn fibrillation and toxicity disaggregation of pre-formed α-syn fibrils	99

	In vitro	Aβ induced aggregation	500 and 1000 μM (1 : 5 and 1 : 10 ratio to amyloids) 5 days of incubation	↓ metal-induced Aβ aggregation by chelating the metal ions	100
	In vitro	Aβ fibril formation	10 μM for 24 h	↓ Aβ aggregation	102
	Mixed glial/neuronal hippocampal cells	Aβ-induced toxicity	1, 5, 10, 20 μM for 24 h	↓ Aβ-induced apoptosis	102
	APP/PS1 mouse model of AD	Animal transgenic model	30 $mg\,kg^{-1}$ per day administration through gavage for 30 days	↓ $A\beta_{1-42}$ plaque size in hippocampus and cortex	103
	Primary cortical neuron	Aβ-induced calcium change	1 and 5 μM for 15 min	↓ (Ca^{2+}) induction caused by oligomeric Aβ1–42	103
	In vitro	Aβ fibril formation	Aβ : GA ratio of 1 : 2 from 2 h up to 7 days	↓ of $A\beta_{1-42}$ aggregate particle sizes at the concentration ratio (1 : 2)	103
Phenylacetic acids					
3′,4′-Dihydroxyphenyl acetic acid (DOPAC)	*In vitro*	α-syn aggregation assay (α-syn and its mutants Y39W, Y125W, and Y136W)	7, 14, 28, 35, 42, 49, 100 μM and 1 mM for 0, 19, 48, and 72 h	Non-covalent binding to α-syn originating specific α-syn oligomers that do not form fibrils. At high concentrations, there was covalent binding to α-syn	101

[a] **Abbreviations:** PC12 – pheochromocytoma-derived rat adrenal medulla cell line; MPP+ – 1-methyl-4-phenylpyridinium (neurotoxin); Aβ – amyloid-beta; α-syn – alpha-synuclein; APP – amyloid precursor protein; AβPP/PS1 mice – double transgenic mice expressing a chimeric mouse/human APP and a mutant human presenilin 1 (PS1); Ca^{2+} – calcium ion; BE (2)-M17 – human neuroblastoma cells; Y39W, Y125W, Y136W – different mutant forms of α-syn protein.

perturbations in metal ion levels, which are a hallmark of several NDDs, by acting as a metal chelator to preserve protein homeostasis.[107] A study by Zhou and colleagues showed that another polyphenol metabolite, 3′,4′-dihydroxyphenylacetic acid (DOPAC), can inhibit the fibrillation of α-syn, due to non-covalent binding of DOPAC to α-syn monomer (Table 8.1).[108] In addition to its effects on α-syn fibrillation, GA was shown to interfere with Aβ fibril formation in primary rat hippocampal cells, which was thought to contribute to its protective effects against Aβ toxicity.[109] These results were further corroborated in primary cortical neurons, and further study has revealed that GA may destabilize Aβ and prevent its aggregation by interacting directly with the peptide to block sites that allow Aβ to form stable oligomeric structures.[110] A similar observation was made for PCA, which both inhibited the oligomerization of Aβ and destabilized pre-formed fibrils, which resulted in enhanced viability in a neuronal cell line.[102]

In the APP/PS1 mutant mouse model of AD, orally administered PCA was found to reduce Aβ deposition in hippocampal tissue, in addition to reducing expression of APP, from which Aβ is generated.[111] Moreover, PCA has been shown to significantly increase learning and memory ability.[111] In addition to PCA effects on amyloid deposition, GA treatment was also proven to be effective, as it was capable of decreasing the size of amyloid plaques within the hippocampus of APP/PS1 mice.[110] Moreover, oral administration of GA not only improved the spatial reference memory and spatial working memory of 4 month-old APP/PS1 mice but also significantly reduced the more severe deficits developed in the 9 month-old APP/PS1 mice in terms of spatial learning, reference memory, short-term recognition, and spatial working memory.[110] While further exploration is needed to determine if these effects can be reproduced with other anthocyanin-derived metabolites, it is a promising finding for their use as food supplements or therapeutic agents in neurodegeneration for which protein aggregation is a major contributing factor.

8.2.2 Neuroinflammation

Although there is some evidence mainly highlighting the ability of several circulating metabolites to reduce oxidative stress, lipid peroxidation, and cytotoxicity in neuronal models, few studies have explored their anti-neuroinflammatory properties.[73,100,112] Cellular models to access the anti-inflammatory potential of (poly)phenols metabolites for neurodegeneration are in general based on using microglia cells and exploring the therapeutic potential of metabolites by using metabolite concentrations much higher than what could be brain relevant. Since microglia are the resident innate immune cells of the CNS, microglial cell lines such as N9, MG6, and BV-2 have been used to evaluate the potential of a compound to mitigate inflammation induced by lipopolysaccharide (LPS) or TNF-α (Table 8.2). Although the evidence for modulatory effects of LMW (poly)phenol metabolites in microglial cells is still very scarce, the ability of a few of them to reduce neuroinflammatory markers has been reported (Table 8.2) and the underlying pathways involved have been proposed. More recently, some studies have been highlighting this potential to attenuate neuroinflammation by the metabolites at lower concentrations (<10 μM) that are brain relevant.[100,113,114] One of those is a study by Figueira and colleagues that shown that pre-conditioning with phenolic sulfates, pyrogallol (benzene-1,2,3-triol)-sulfate, 4-methylcatechol (4-methylbenzene-1,2-diol)-sulfate, 1-methylpyrogalol (methoxybenzene-2,3-diol)-sulfate, in physiological concentrations (5 μM) improved cellular responses to oxidative, excitotoxicity and inflammatory injuries and this attenuation of neuroinflammation was achieved *via* modulation of NF-κB pathway.[100] Pretreatment with an anthocyanin parent compound, cyanidin, also strongly inhibited the NF-κB signaling pathway of Aβ-induced neuroinflammation in cells evidenced by suppressing the degradation of IκBα, translocation of p65 subunit of NF-κB from the cytoplasm to the nucleus, and thereby reducing the expression of iNOS protein and the production of NO.[115] Pretreatment with cyanidin also decreased the expression of Toll-like receptor 4 (TLR4) (an LPS receptor), directly improved intracellular ROS levels, and regulated the activity of inflammation-related downstream pathways including SOD activity through TLR4/NADPH oxidase 4 (NOX4) signaling in human neuroblastoma cell line (SK-N-SH).[115] These results suggest that inhibition of TLR4 by cyanidin could be beneficial in preventing neuroinflammation-induced neurodegeneration in AD; however, it remains to be confirmed that its circulating metabolites keep the same activity assuming they reach the brain. Interestingly, in the evaluation of the potential to reduce LPS-induced TNF-α secretion in THP-1 monocytes, some metabolites of flavonoids, individually and in combination, appear to present more anti-inflammatory effects than their dietary parent compounds.[116] Moreover, raspberry (poly)phenol metabolites resulting from the *in vitro* digestion of raspberry extract significantly inhibited microglial pro-inflammatory activation by LPS through the inhibition of ionized calcium-binding adapter molecule 1 (Iba 1) expression, TNF-α release, and NO production.[113] Benzoic acids conjugates were also studied as attenuators of neuroinflammation (Table 8.2). This is a very important class not only because the majority of flavonoids' metabolism converge to these metabolites, the benzoic acid derivatives, but also because they are very well represented in diverse foods, including berries, and therefore could be ingested directly (*e.g.*, GA and PCA in beer, wine, berries, cereals, grape, apple, kiwi, olive oil, vinegar, chicory).[68,73,114] Kim and colleagues have demonstrated the neuroprotective effects of GA using the BV-2 cell line and primary microglia cells prior to Aβ stimulation.[117] Conditioned media was transferred to Neuro-2A cells, and in both situations, GA prevented Aβ neurotoxicity by inhibiting the expression levels of iNOS, COX-2, IL-1β, and TNF-α by blocking NF-κB acetylation and that way inhibiting NF-κB-mediated neuroinflammatory signaling.[117] It has been shown that VA treatment also decreased $A\beta_{1-42}$-induced neuronal apoptosis and neuroinflammation and improved synaptic and cognitive deficits.[118] Intriguingly, both GA and VA were shown to have additional neuroprotective and

Table 8.2 Insights on the potential mechanisms of (poly)phenol metabolites on models of induced inflammation.[a]

(Poly)phenol metabolite	Model	Inflammation model	Metabolite dose and time	Mechanistic evidence	Ref.
Benzenes					
Pyrogallol-sulfate	N9 cells	LPS	5 μM for 6 h	↓ TNF-α cell release ↓ NF-κB translocation ↓ IκB degradation	105
1-Methyl pyrogallol-sulfate 4-Methyl catechol-sulfate	N9 cells	LPS	5 μM for 6 h	↓ TNF-α cell release ↓ CD40 membrane presence	105
Cinnamic acids					
Cinnamic acid	BV-2 cells	LPS	100, 200, and 300 μM for 24 h	↑ SOCS3 level ↓↓ iNOS, TNFα, IL-1β, and IL-6	116
Ferulic acid/4-hydroxy-3-methoxycinnamic acid (FA)	BV-2 cells C57BL/6N mice	LPS	10 and 100 μM for 2 h 20 mg kg^{-1} orally for 11 days	↓ glial cell activation ↓ p-JNK, p-NF-κB ↓ iNOS, COX-2, TNF-α, and IL-1β in the mouse hippocampus and BV-2 microglial cells	117
	MG6 cells	LPS	10, 50, 250 μM for 24 h	↓ Indoleamine 2,3-dioxygenase (IDO) mRNA expression ↓ nuclear translocation of NF-κB ↓ phosphorylation of p38 MAPK No effect on inflammatory mediators and phosphorylation of JNK.	115
	Wistar rats	Rotenone-induced Parkinson's disease	50 mg kg^{-1} i.p. injection for 4 weeks	↓ Iba-1 and GFAP activation ↓ iNOS and COX-2 expression ↓ IL-1b, IL-6, and TNF-α levels	118

Benzoic acids					
Vanillic acid/4-hydroxy-3-methoxybenzoic acid (VA)	Primary glial cells co-culture	Lysolecithin	1 μM for 72 h	Promoted remyelination ↓ deposition of chondroitin sulfate proteoglycans and tenascins ↑ neuronal regeneration Restored neuronal function and firing rate	113
	Sprague-Dawley rat	Cerebral ischemia-reperfusion (I/R)	50 and 100 mg kg^{-1} gavage for 14 days	↓ NF-κB p65 protein levels ↓ IL-6, IL-1β, and TNF-α	119
	Wistar rat	Carotid artery occlusion and reperfusion in rat	100 mg kg^{-1}, gavage for 14 days	↓ IL-6, TNF-α, and TUNEL positive cells ↑ IL-10 levels in the hippocampus	112
	C57BL/6N mice	ICV. injection of $A\beta_{1-42}$	30 mg kg^{-1} i.p. for 3 weeks	↓ number of activated micro- and astroglia cells ↓ iNOS expression ↓ NF-κB	122
Gallic acid/3,4,5-trihydroxybenzoic acid (GA)	BV-2, neuro-2A and primary cells	Aβ-stimulated	5, 25, and 50 μM for 12 h	↓ pro-inflammatory cytokine levels through NF-κB p65 subunit deacetylation and inhibition	111
	Primary glial co-culture	Lysolecithin	1 μM for 72 h	Promoted remyelination ↓ deposition of chondroitin sulfate proteoglycans and tenascins ↑ neuronal regeneration Restored neuronal function and firing rate	113

Table 8.2 (*Continued*)

(Poly)phenol metabolite	Model	Inflammation model	Metabolite dose and time	Mechanistic evidence	Ref.
Protocatechuic acid/3,4-dihydroxybenzoic acid (PCA)	BV-2 cells	LPS	5, 10, 20 μM for 1 h	↓ TNF-α, IL6, IL-1b, and PGE2 production ↓ TLR4 expression ↓ NF-κB and MAPKs activation	108
	BV-2 cells	LPS	10, 25, 50 and 100 μM for 24 h	↓ NO production	114
	AβPP/PS1	Animal transgenic model	100 mg kg^{-1} orally for 4 weeks	↓ in Aβ deposition ↓ TNF-α, IL-1β, IL-6, and IL-8 expression ↑ learning and memory performance ↑ BDNF	104
	Balb/cA mice	D-galactose treatment	0.5%, 1%, or 2% enriched food content for 8 weeks	↓ IL-1β, TNF-α, IL-6, and prostaglandin E2 ↓↓ COX-2 activity and expression ↓ mRNA expression and protein production of NF-κB p65	120
	Sprague-Dawley rat	Ischemia-induced hippocampal neuronal death	30 mg kg^{-1} per day orally for 7 days	↓ microglial activation ↓ astroglia activation	121

[a]**Abbreviations:** N9 – murine microglial cell line; BV-2 – murine microglial cell line; LPS – lipopolysaccharide; TNF-α – tumor necrosis factor-alpha; NF-κB – nuclear factor kappa B; p-NF-κB – phosphorylated nuclear factor kappa-B; NF-κB p65 – p65 subunit of nuclear factor kappa-B; IκB – (IKK) I kappa B kinase; CD40 – cluster of differentiation 40; SOCS – suppressor of cytokine signaling 3 protein; iNOS – inducible nitric oxide synthase; IL1β – interleukin-1 beta; IL-6 – interleukin-6; IL-8 – interleukin-8; IL-10 – interleukin-10; p-JNK – phosphor-c-Jun *N*-terminal kinase; COX-2 – cyclooxygenase-2; p38 MAPK – p38 mitogen-activated protein kinase; Iba-1 – ionized calcium-binding adapter molecule 1; GFAP – glial fibrillary acidic protein; BDNF – brain-derived neurotrophic factor; TLR4 – Toll-like receptor 4; NO – nitric oxide; TUNEL – terminal deoxynucleotidyl transferase dUTP nick end labeling; Aβ – amyloid-beta.

anti-inflammatory abilities beyond simply preserving cellular viability and reducing inflammatory markers. A recent work by Siddiqui and colleagues suggests that both phenolic acids are capable of stimulating neurite outgrowth in primary hippocampal neurons.[119] Moreover, in a mixed primary glial and neuronal cultures, lysolecithin-induced demyelination of neurons and oligodendrocytes and induction of inflammatory markers, such as NF-κB, COX-2, and glial fibrillar acidic protein (GFAP) in astrocytes, was abrogated by treatment with both GA and VA. Indeed, both GA and VA promoted remyelination, significantly reduced deposition of chondroitin sulfate proteoglycans and tenascins by glial cells, which act as an inhibitor to neuronal regeneration, and restored neuronal function and neuronal firing rate.[119] In addition, PCA, another benzoic acid (Table 8.2), has also presented potent anti-inflammatory effects in LPS-stimulated BV-2 microglia cells. Mechanistic assays have revealed its ability to inhibit the production of TNF-α, IL-6, IL-1β, and prostaglandin E2 (PGE2). PCA also suppressed the activation of NF-κB and mitogen-activated protein kinases (MAPKs), and it was also able to inhibit the TLR4 expression.[114] Likewise, PCA has also been reported to reduce microglial inflammation in BV-2 cells, significantly decreasing the amount of NO produced by these cells following treatment with LPS.[120] The data gathered reveals ferulic acid (FA) potential for attenuating pro-inflammatory mediators in microglia (Table 8.2). The general trend observed for FA was to reduce more than one pro-inflammatory mediator in microglial cells. In the MG6 microglia cell line, however, FA was ineffective in reducing LPS stimulated production of NO, TNF-α, and IL-6 and was only effective in other mediators such as 2,3-dioxygenase, a tryptophan catabolic enzyme closely involved in NDD pathogenesis, but at concentrations far beyond physiological ones.[121] Cinnamic acid is another example of an effective metabolite tested for reducing several pro-inflammatory markers, although at high levels only suitable for pharmacological approaches.[122]

The data from the animal studies further corroborate the anti-inflammatory action of different metabolites and confirm the mechanisms observed in cells. Anti-inflammatory effects of FA in the LPS-induced neuroinflammation and rotenone-induced rat model of PD have been suggested. In the mentioned models, FA exerts its effects by inhibiting p-JNK, p-NF-κB, and downstream signaling molecules, such as iNOS, COX-2, TNF-α, and IL-1β, which were accompanied by a remarkable reduction of Iba-1 and GFAP levels and attenuation of microglial and astrocytic activation.[123,124] Findings by Rehman and colleagues also demonstrated that FA treatment interfered with the TLR4/myeloid differentiation factor 2 (MD-2) complex binding site, which is crucial for evoking neuroinflammation *via* microglial activation and inhibited NF-κB likely *via* a JNK-dependent mechanism, implicating a therapeutic application for neuroinflammation-induced neurodegeneration.[123] Two weeks of orally administered VA as a pretreatment significantly restored the spatial memory, while the levels of IL-6, IL-1β, TNF-α, NF-κB p65 subunit and TUNEL positive cells were remarkably decreased in the hippocampi of rats upon VA supplementation in the model of transient bilateral common carotid artery occlusion and reperfusion.[118,125] Protocatechuic acid also reduced the presence of inflammatory markers in tissue homogenates from the hippocampus and cerebral cortex of aged AβPP/PS1 double transgenic mice, including several inflammatory cytokines such as TNF-α, IL-1β, IL-6, and IL-8, while increasing the production of pro-survival signaling factors such as brain-derived neurotrophic factor (BDNF).[111] In the D-galactose mouse aging model, supplementation of animals with PCA through the food resulted in a decrease of COX-2 activity and expression, IL-1β, IL-6, TNF-α, and PGE2 in the brain. Protocatechuic acid intake lowered D-galactose-induced elevation in activity, mRNA expression, and protein levels of NF-κB p65.[126] Moreover, PCA was also shown to significantly decrease levels of astro- and microgliosis.[127] VA has shown to have similar effects in mice treated with Aβ by intracerebroventricular injection.[128] Aβ injection caused significant impairments in learning and memory tasks in mice, in addition to enhancing pro-inflammatory markers including iNOS and COX-2 expression and production of IL-1β. NF-κB acetylation and nuclear translocation were also examined and found to be significantly enhanced in this model. Pretreatment with GA, however, potently inhibited these effects, likely through suppression of acetylation of NF-κB p65 subunit and NF-κB activity.[117]

8.2.3 Oxidative Stress

The extensive investigations on the mechanisms by which neurons die have led to the conclusion that the therapeutic use of agents that reduce oxidative stress could be of help in NDDs. Novel therapeutic neuroprotective strategies support the application of ROS effectors, transition metal (*e.g.*, iron and copper) chelators, non-steroidal anti-inflammatory drugs, non-vitamin natural polyphenols, anti-apoptotic drugs such as calcium channel and caspase inhibitors, or as part of an antioxidant cocktail formulation for the treatment of these diseases.[129] There is much evidence showing that naturally occurring compounds such as vitamins, minerals, and (poly)phenolic compounds present in high amounts in fruits like berries, vegetables, and natural products may interact with antioxidant enzymes, enhancing their response in a positive manner. In fact, dietary (poly)phenols and their metabolites have been reported to combat the effect of oxidative stress in several cell and animal models of neuronal oxidative stress (Table 8.3). The treatment of neuronal cell lines with hydrogen peroxide is used as a model of oxidative stress, and as such, it induces a significant degree of cell death.[130] Protocatechuic acid's ability to attenuate neuronal death induced by treatment with various neurotoxins associated with models of PD, such as the mitochondrial complex I inhibitor MPP^+, was evaluated. These studies have indicated that PCA effectively protects PC12 cells from MPP^+-induced toxicity by preventing mitochondrial dysfunction and elevating levels of antioxidant enzymes such as CAT, SOD1, and GSH peroxidase, in addition to augmentation of the levels of GSH and enhancement of the pro-survival protein B-cell lymphoma 2 (Bcl-2) levels.[120,131–134] Similar results were observed when this study was repeated using rotenone, another inhibitor of mitochondrial complex I.[135] The impact of PCA has also been evaluated for cellular models of neurotoxicity observed in AD. Protocatechuic acid mitigates the neurotoxic effects of Aβ in primary rat cortical neurons by attenuating increases in ROS and intracellular calcium levels, in addition to preventing glutamate release caused by this insult.[136] Also, the study by Zhang and colleagues[137] showed that PCA modulates cellular redox status with the upregulated expression of heme oxygenase-1 (HO-1), SOD, and CAT and decreased levels of malondialdehyde, a known lipid peroxidation product. Moreover, PCA also inhibited the activation of NF-κB and the expression of iNOS.[137] Also, pretreatment of SH-SY5Y cells with PCA and cyanidin 3-*O*-glucopyranoside (Cy-3G), aglycon cyanidin (Cy) inhibits H_2O_2-induced ROS formation at different cellular levels: Cy-3G at the membrane level, PCA at cytosolic level, and Cy at both membrane and cytosolic levels. Interestingly, both Cy and PCA, but not Cy-3G, could inhibit H_2O_2-induced apoptotic events, such as mitochondrial functioning loss and DNA fragmentation.[131] Although these results suggest that PCA represents a metabolite with neuroprotective action and may play an important role in brain health promotion, most of the studies use PCA concentrations much higher than it is expected to circulate or even reach the brain. Therefore, mechanistic studies at physiological levels should be performed to confirm its mode of action.

GA presented similar neuroprotective effects to PCA and in few studies for concentrations lower than 5–10 μM, therefore more brain relevant. Indeed, GA treatment significantly attenuated increased calcium levels in addition to mitigating ROS accumulation and preventing the induction of pro-inflammatory enzymes such as COX-2 and p38-MAPK in PC12 cells treated with kainic acid.[138] Similar results were observed in primary cortical neurons challenged with glutamate in which it was shown that GA treatment significantly attenuated cell death by preserving GSH levels and enhancing the activity of SOD and CAT while decreasing expression of pro-inflammatory cytokines.[139] Further study on the mechanism underlying the neuroprotective effects of GA against 6-hydroxydopamine (6-OHDA) in SH-SY5Y cells revealed that GA significantly enhanced expression of anti apoptotic Bcl-2 proteins and antioxidant enzymes that are under the control of nuclear factor erythroid 2-related factor 2 (Nrf2) while reducing expression of pro-apoptotic proteins Bax and caspase-3, and Keap1, which inhibits Nrf2 activity.[140] Moreover, VA, GA, and PCA, all in nutritional relevant circulating concentrations of 5 μM, have been shown to significantly reduce oxidative stress induced by H_2O_2 and preserve cellular viability in the SH-SY5Y neuronal cell line through the upregulation of the expressions of sirtuin-1 (SIRT1) and forkhead box O (FoxO) 3a as well as induce the levels of antioxidant enzymes (SOD 2 and CAT) and Bcl-2 proteins.[134] It has also been shown that 3′,4′-dihydroxy-phenylpropionic acid, DOPAC, and GA prevented neuronal apoptosis *via* attenuation of ROS levels and decreased oxidative stress-induced apoptosis by preventing the caspase-3 activation *via* the mitochondrial apoptotic pathway in SH-SY5Y cells.[141] FA has also been shown to protect PC-12 cells against I/R-induced ROS generation and apoptosis by inhibiting apoptosis, Ca^{2+} influx, superoxide anion (O^{2-}), malondialdehyde,

Table 8.3 Insights on the potential mechanisms of (poly)phenol metabolites on models of oxidative stress.[a]

(Poly)Phenol metabolite	Model	Oxidative stress model	Metabolite dose and time	Mechanistic evidence	Ref.
Benzoic acids					
Vanillic acid/ 4-hydroxy-3-methoxybenzoic acid (VA)	HT22 cells	$A\beta_{1-42}$-induced toxicity	50, 100, and 200 μM for 24 h	↓ ROS levels	122
	SH-SY5Y cells	H_2O_2 induced	1 and 5 μM for 3 h	↑ cell viability ↓ cell death ↓ ROS ↑ Bcl-2	128
	C57BL/6N mice	i.c.v. injection of $A\beta_{1-42}$	30 $mg\,kg^{-1}$ i.p. for 3 weeks	↑ GSH ↓↓ ROS generation ↑ endogenous Nrf2 and HO-1 *via* the activation of Akt and GSK-3β ↓ neuronal apoptosis	122
	Swiss albino mice	Streptozotocin (STZ)-induced neurodegeneration	25, 50, and 100 $mg\,kg^{-1}$ orally for 28 days	Improved spatial learning and memory retention ↓↓ oxidative stress ↓ AChE, corticosterone, TNF-α ↑ SOD, CAT, and GPx levels	142
	Wistar rat	Transient bilateral common carotid artery occlusion and reperfusion	10, 30, 100 $mg\,kg^{-1}$ gavage for 2 weeks	Attenuation of reactive hyperemia and BBB disruption. Improvement of sensory-motor signs and anxiolytic behavior	145
	Wistar rat	Transient bilateral common carotid artery occlusion and reperfusion	100 $mg\,kg^{-1}$ gavage for 2 weeks	Improved locomotion and memory ↓↓ cell death of CA1 neurons	146
	Wistar rat		100 $mg\,kg^{-1}$ gavage for 2 weeks	Restoration of the spatial memory	112

		Transient bilateral common carotid artery occlusion and reperfusion		↓ IL-6, TNF-α, and TUNEL positive cells ↑ IL-10 levels in the hippocampus	
	Sprague-Dawley rat	Middle cerebral artery occlusion	50 and 100 mg kg^{-1} gavage for 14 days	↓ levels of lipid peroxidation ↓ malondialdehyde ↑ SOD and CAT	119
4-Hydroxybenzoic acid (HBA)	PC12 and BV-2 cells	H_2O_2 induced Sodium nitroprusside induced (SND)	50, 100, and 200 μM co-incubation for 24h 10, 30, 100, 300 μM co-incubation for 24h	↓ apoptosis No alteration in apoptosis	114
Protocatechuic acid/ 3,4-dihydroxybenzoic acid (PCA)	PC12 and BV-2 cells	H_2O_2 induced Sodium nitroprusside induced (SND)	50, 100, and 200 μM co-incubation for 24h 10, 30, 100, 300 μM co-incubation for 24h	↓ apoptosis ↓ apoptosis	114
	SH-SY5Y cells	H_2O_2 induced	25, 50, 100 μM for 2h	↓ apoptotic events ↓ ROS formation ↓ mitochondrial functioning loss ↓ DNA fragmentation	125
	PC12 cells	H_2O_2 and sodium nitroprusside induced	60, 300, 600, and 1200 μM for 24 h or 48 h	↓ induced cell death ↑ SOD and CAT activity	126
	PC12 cells	Treatment with MPP^+	330, 650, and 1300 μM for 6 h	↓ cytotoxicity ↓ mitochondrial dysfunction ↓ apoptotic morphology	97
	PC12 cells	H_2O_2 induced	50, 100, 150, and 200 μg mL^{-1} for 1 h	↓ apoptosis ↑ GSH ↑ CAT	127
	PC12 cells	6-OHDA treated	25, 50, 100 μM for 12 h	↑ Nrf2 nuclear translocation ↑ Nrf2 transcriptional activity ↓ malondialdehyde	131
	PC12 cells	H_2O_2 induced	60, 300, 600, and 1200 μM for 24 h or 48 h	↓ apoptosis ↑ GSH ↑ CAT	127

Table 8.3 (*Continued*)

(Poly)Phenol metabolite	Model	Oxidative stress model	Metabolite dose and time	Mechanistic evidence	Ref.
	SH-SY5Y cells	H_2O_2 induced	1 and 5 μM for 3 h	↓ cell death ↓ ROS ↑ Bcl-2	128
	Rat cortical neurons	$A\beta_{25-35}$-induced neurotoxicity	1 and 10 μM for 20 min	Prevention of neuronal cell damage ↓ Ca^{2+} levels Inhibition of glutamate release ↓ generation of ROS ↓ caspase-3 activity.	130
	Rat	Aging	5 mg kg^{-1} i.p. injection daily for 7 days	Improved cognition of aged rats ↓ lipid peroxide ↑ activity of GSH-Px and SOD	140
	Wistar rat	Cadmium-induced neurotoxicity	10 and 20 mg kg^{-1} orally for 21 days	↓ Na+/K + -ATPase activity ↓ acetylcholinesterase, butyrylcholinesterase, and endogenous antioxidant enzymes	139
	Sprague-Dawley rat	Ischemia-induced hippocampal neuronal death	30 mg kg^{-1} per day orally for 7 days	↓ neuronal cell death ↓ oxidative stress ↓ BBB disruption	121
Gallic acid/3,4,5-trihydroxybenzoic acid (GA)	FVB mice PC12 cells	Kainate-induced neuronal injury Kainate-induced neuronal injury	1 mg kg^{-1} by gavage for 3 days 0.1, 1, and 10 μM for 1 or 24 h	↓ lipid peroxidation *in vivo* ↓ in Ca^{2+} release, ROS, lipid peroxidation, expression of COX-2, and p38 MAPK	132
	Primary rat cortex neurons	Glutamate-induced neurotoxicity	147 and 294 μM^b for 2 h	↑ total thiols ↓ malondialdehyde levels ↑↑ SOD, CAT, GSH ↓↓ NO	133
	Wistar rat			↑ levels of GSH, SOD, and CAT	134

		Oxidative stress-induced by 6-OHDA	50, 100, and 200 mg kg^{-1} oral gavage for 10 days	↓ NO concentration ↓ TNF-α, IL-1β, and IL-6.	
	Wistar albino rat	Intracerebroventricular injection of streptozotocin (STZ)	30 mg kg^{-1} oral gavage for 26 days	↑ total thiol content ↑ GSH-Px, SOD, and CAT enzyme activities	141
Syringic acid/3,5-dimethoxy-4-hydroxybenzoic acid	Primary cultures of hippocampal neuronal cells	Oxygen-glucose deprivation followed by reperfusion	0.1, 1, 10, and 20 μM for 24 h	↓ ROS and malondialdehyde ↓ elevations in Ca^{2+} ↑ activity of SOD ↓ Bax and caspase-3 ↑ Bcl-2 ↓ pro-apoptotic signaling through the JNK and p38-MAPK pathways	137
	C57BL/6 mice	MPTP and probenecid (MPTP/p)	20 mg kg^{-1} orally or oral gavage for 35 days	Improvement of motor functions ↑ TH, DAT, and VMAT2 in SN	143
	C57BL/6 mice	Artery occlusion ischemic injury	10 mg kg^{-1} i.p. injection at the time of occlusion	↑ SOD ↑ nuclear respiratory factor 1 levels ↓ caspase-3 and -9 levels	144
Cinnamic acid					
Ferulic acid/4-hydroxy-3-methoxycinnamic acid (FA)	SH-SY5Y cells	H_2O_2 induced	1 and 5 μM for 3 h	↓ cell death ↓ ROS ↑ Bcl-2	128
	Sprague-Dawley rat	Hypoxia/ischemia-induced cell injury	28, 56, and 112 mg kg^{-1} for 5 days[c]	↓ apoptosis ↓ cleaved caspase-3 and Bax ↑ Bcl-2 ↓ ROS generation ↓ Ca^{2+} influx	136

Table 8.3 (*Continued*)

(Poly)Phenol metabolite	Model	Oxidative stress model	Metabolite dose and time	Mechanistic evidence	Ref.
				↓ superoxide anion (O^{2-}), ↓ malondialdehyde content ↑ SOD and GSH-Px activity	
	Wistar rat	Rotenone-induced Parkinson's disease	50 $mg\,kg^{-1}$ i.p. for 4 weeks	↓ cell death ↓ lipid peroxidation ↓ malondialdehyde content ↓ SOD, CAT, and GSH	118
Phenylacetic acid					
3′,4′-Dihydroxyphenyl acetic acid 3′-Hydroxyphenyl acetic acid	SH-SY5Y cells	Neuronal nitrosative stress with SIN-1	0.1, 1, and 10 μM for 18 h	↓ neuronal MAPK p38 and ERK1/2 ↓ caspase-3 activity	138
3′,4′-Dihydroxyphenyl acetic acid 3′-Hydroxyphenyl acetic acid	SH-SY5Y cells	H_2O_2 induced	1, 5, or 10 μM for 6 h	↓ caspase-3 activation *via* the mitochondrial apoptotic pathway ↓ ROS levels ↑ REDOX activity	135

[a]**Abbreviations:** HT22 – mouse hippocampal neuronal cell line; SH-SY5Y – human derived cell line; ROS – reactive oxygen species; Bcl-2 – B-cell lymphoma 2; GSH – glutathione; Nrf – nuclear respiratory factor; HO-1 – heme oxygenase 1; Akt – protein kinase B; GSK-3β – glycogen synthase kinase 3 beta; AChE – acetylcholinesterase; SOD – superoxide dismutase; BBB – blood–brain barrier; CA1 – hippocampal region; CAT – catalase; TNF-α – tumor necrosis factor-alpha; IL1β – interleukin 1 beta; IL-6 – interleukin 6; IL-10 – interleukin 10; GSH-Px – glutathione peroxidase; p38 MAPK – p38 mitogen-activated protein kinase; NO – nitric oxide; Bax – BCL2 associated X; Aβ – amyloid beta; Ca^{2+} – calcium ion; JNK – c-Jun *N*-terminal kinase; TH – tyrosine hydroxylase; DAT – dopamine transporter; VMAT2 – vesicular monoamine transporter 2; SN – substantia nigra; MPTP – 1-methyl-4-phenyl-1,2,3,6-tetrahydropyridine.
[b]Converted from $\mu g\,mL^{-1}$ to μM using the molecular weight of 170.12 $g\,mol^{-1}$.
[c]The exact route for FA treatment was not specified.

and GSH peroxidase (GSH-Px) production in a concentration-dependent manner.[142] Cao and colleagues showed that treatment with syringic acid preserved cellular viability by increasing the levels of antioxidant enzymes, prevented increases in intracellular calcium, and decreased pro-apoptotic signaling through the JNK and p38-MAPK pathways.[143] Another anthocyanin metabolite, 4-hydroxybenzoic acid (HBA), although at levels higher than 10 μM, has also been found to be neuroprotective by reducing apoptosis against H_2O_2-induced oxidative stress and glutamate excitotoxicity.[120] Also, 3′-hydroxy-phenylacetic acid and 3,′4′-hydroxyphenylacetic acid have been shown to be effective at protecting neuroblastoma cells from nitrosative stress.[144]

In a rat model of cadmium-induced heavy metal toxicity, which causes significant disruptions in brain health and cognition, PCA oral administration was found to significantly elevate expression of antioxidant enzymes, and reduce markers of oxidative stress and lipid peroxidation, as well as decrease activity of cholinesterase to preserve cholinergic signaling in the brain.[145] Additionally, intraperitoneally (i.p.) administered PCA significantly improved the cognition of aged rats, reduced the content of lipid peroxide, and increased the activity of GSH-Px and SOD.[146] GA has been proven to have similar neuroprotective effects in the 6-OHDA rat model of PD.[140] GA and VA have also been examined in a model of streptozotocin-induced dementia. In this context, these phenolic acids were shown to significantly improve antioxidant defenses and decrease markers of oxidative stress in the brain, in addition to attenuating the release of pro-inflammatory cytokines.[147,148] It has been shown that IP administration of VA after $A\beta_{1-42}$-injection enhanced GSH levels and abrogated ROS generation accompanied by an induction of the endogenous Nrf2 and HO-1 *via* the activation of Akt and glycogen synthase kinase-3β (GSK-3β) in the mouse brain.[128] Ren and colleagues investigated the protective properties of FA in hypoxia/ischemia-induced cell injury *in vivo* and *in vitro* and revealed that treatment with FA significantly attenuated memory impairment and reduced hippocampal neuronal apoptosis and oxidative stress in a dose-dependent manner.[142] FA treatment also restored the antioxidant enzymes SOD and CAT, prevented depletion of GSH, and inhibited lipid peroxidation.[124] Moreover, orally administered FA strongly inhibited mitochondrial apoptotic signaling molecules, such as Bax, cytochrome c, caspase-3, and poly [ADP-ribose] polymerase 1 (PARP-1), and reversed deregulated synaptic proteins, including postsynaptic density protein 95 (PSD-95), synaptophysin, synaptosomal-associated protein, 25 kDa (SNAP-25) synaptosomal-associated protein, 23 kDa (SNAP-23), and synaptic dysfunction in LPS-treated mice.[134,142] FA has also shown to inactivate the TLR/myeloid differentiation factor 88 (MyD88) pathway.[142] Syringic acid was also shown to improve outcomes in the MPTP mouse model of PD, by preserving dopaminergic neurons through activation of antioxidant enzymes and reductions in pro-inflammatory markers.[149] Also, rats subjected to artery occlusion displayed similar improvements in markers of injury, showing enhanced neuronal survival, increased SOD and CAT activity, and decreased expression of caspase-3 and

9 when treated with syringic acid.[150] Protocatechuic acid was shown to significantly improve cognition in a model of global ischemic injury and was found to enhance neuronal survival in the hippocampus of treated animals.[127] These protective effects were associated with increased levels of GSH and decreased levels of oxidative damage.[127] Several reports have shown similar positive effects on ischemic injury following treatment with VA (Table 8.3). These studies indicate that VA treatment also improves cognition and memory in addition to enhancing neuronal viability, expression of antioxidant enzymes, and long-term potentiation.[118,151,152] In total, the data gathered from animal studies on the action of different metabolites against oxidative stress further validate some targets identified in the mechanistic studies observed in cells. The translation of these effects for an effective dose of ingested dietary parent compounds and ultimately to food consumption remains to be done.

The process of neurodegeneration and neuronal cell death represents the hallmark of all NDDs. It is a multifactorial process, and neurons die as a conjugation of processes of protein aggregation, neuroinflammation, and oxidative stress, among others. These processes are intertwined in all NDDs, and so far, there is no clear image about which of these processes represent a cause and which represent a consequence. What we know now is that all these processes contribute to neurodegeneration, and thus contributing to the disease progression. Since polyphenolic metabolites have been proven to overtly affect all the abovementioned processes, leading to alleviation, or in some cases even a suppression of neuronal cell death, they became very promising pleiotropic neuroprotective agents, and numerous studies were conducted in disease-induced models in order to estimate their neuroprotective capacity. In the studies presented here, although many used concentrations higher than expected in circulation indicate that anthocyanin LMW (poly)phenol metabolites possess significant neuroprotective capacities *in vivo*, which may suggest that they could be effective prophylactic or therapeutic agents for neurodegenerative disorders.

8.3 Limitations and Future Approaches into Mechanistic Studies of Berry Bioactive Compounds

Studies into the efficacy of berry bioactive compounds and in particular anthocyanins and their metabolites for treating neurodegeneration have gained momentum in the past several years, and many reports have emerged describing the use of these compounds both *in vitro* and *in vivo*. Animal and cellular models have provided unparalleled tools for compound screening assays to reveal their mechanisms of action.

Using isolated compounds, these models have helped to understand the possible mechanisms of action against NDDs. Yet, in many cases, these studies use high concentrations or long periods of incubation, reflecting a pharmacological perspective not nutritionally relevant. Furthermore, nutritional studies often reveal only overall effects and most importantly do not identify the relevant metabolites that can comprise the true effectors against neurodegeneration. For a long period of time, research has been conducted analyzing the effect of a defined (poly)phenol or flavonoid present in berries in specific cells/tissues. Thankfully new trends tend to propel the use of more relevant (poly)phenol metabolites using near-physiological concentrations and residence time and therefore nutritionally relevant insights are emerging.

In vitro assays have been crucial to understanding the potential mechanisms of (poly)phenol metabolites underlying potential interactions of (poly)phenols metabolites with other small molecules, proteins, or even metals. These are especially useful when considered experimental conditions close to physiological conditions (concentrations in range 0.1–10 μM). Although the majority of these studies are not nutritionally relevant, since they apply concentrations of (poly)phenol metabolites far beyond those of physiological conditions, such studies can give us clues about the putative mode of action of their (poly)phenol metabolites in a nutritional context. Furthermore, this is an indispensable tool to get a comprehensive picture and tune our understanding for further *in vivo* studies.

In the upcoming years, the study of isolated and relevant (poly)phenols metabolites, at physiological conditions, in relevant neurodegenerative cell models will further elucidate their role, since many LMW (poly)phenol metabolites have never been even evaluated.[73,153,154] Furthermore, we also expect to see *in vitro* studies searching for mechanisms using well-characterized mixtures of compounds to understand their synergetic effects and diversity of (poly)phenol metabolites circulating *in vivo*, under a nutritional approach. It is clear that this synergetic complexity from dietary (poly)phenols may not be completely fulfilled *in vitro* and will require *in vivo* studies.[155] Animal studies will be of utmost importance to solidify this information, searching for the presence of (poly)phenol metabolites and their mechanisms in models of NDDs.

8.4 Conclusion

The increase in the incidence of NDDs due to the increased life span is a big concern. In addition to this problem, current therapies for NDDs act by controlling the disease's symptoms but do not slow the underlying neurodegeneration in the brains of patients. This situation created a critical need to improve our understanding of what are the main mechanisms behind NDDs. Such mechanisms include protein misfolding and aggregation, oxidative stress, and inflammatory stress (Figure 8.2). Such multifactorial causes require the development of new approaches with pleiotropic effects. Berries are suggested to be one of the most interesting bioactive enriched sources already used in our diet and have high amounts of (poly)phenols, in particular anthocyanins that are converted into (poly)phenol metabolites with a large potential of pleiotropic effects (Figure 8.2).

Overall, berry anthocyanins and their metabolites possess significant and varied biological activities that make them uniquely suited for further therapeutic and/or nutrition/nutraceutical development for the prevention and treatment of NDDs, like PD, AD, HD, and ALS. Notwithstanding, future extensive clinical trials are needed to further validate their effects and bring novel therapeutic agents for brain-related diseases.

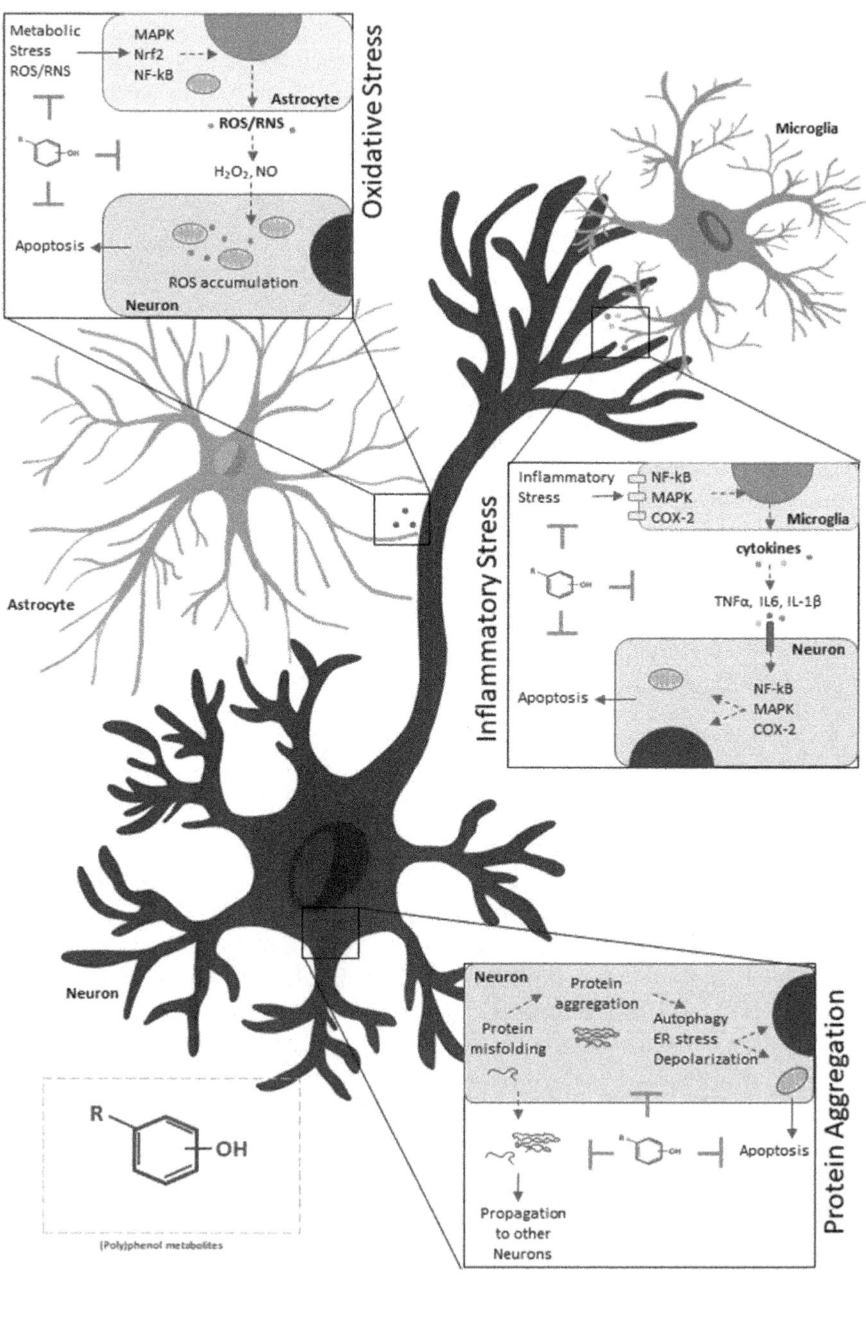

Figure 8.2 Several mechanisms are proposed for the development and propagation of neurodegenerative disease including oxidative and inflammatory stress and protein aggregation. Oxidative stress is mainly generated by metabolic stress from astrocytes, microglia, and neurons. Meanwhile, inflammatory stress arrives from inside or outside the brain, activating inflammatory pathways mainly in microglia and astrocytes. Protein aggregates are known to induce stress and depolarization of the membranes, further culminating in neuronal apoptosis. (Poly)phenol metabolites have pleiotropic effects, modulating these mechanisms reducing the effects of oxidative and inflammatory stress and protein aggregation effectively reducing neuronal death. Nrf2, nuclear factor erythroid 2-related factor 2; NF-κB, nuclear factor kappa B; ROS, reactive oxygen species; RNS, reactive nitrogen species; MAPK, mitogen-activated protein kinases; COX-2, cyclooxygenase-2; TNF-α, tumor necrosis factor-alpha; IL-6, interleukin-6, IL-1β, interleukin-1β; ER, endoplasmic reticulum.

Abbreviations

6-OHDA	6-hydroxydopamine
AD	Alzheimer's disease
Akt	protein kinase B
ALS	amyotrophic lateral sclerosis
Aβ	amyloid β
Bcl-2	B-cell lymphoma 2
BDNF	brain-derived neurotrophic factor
CAT	catalase
CoQ10	coenzyme Q10
COX-2	cyclooxygenase-2
DOPAC	3′,4′-dihydroxyphenylacetic acid
FA	ferulic acid
GA	gallic acid
GFAP	glial fibrillary acidic protein
GSH	glutathione
GSH-Px	glutathione peroxidase
GSK-3β	glycogen synthase kinase-3β
HD	Huntington's disease
HO-1	heme oxygenase 1
Htt	huntingtin
iNOS	inducible nitric oxide synthase
LMW	low molecular weight
LPS	lipopolysaccharides
MAPK	mitogen-activated protein kinase
MD-2	myeloid differentiation factor 2
MPTP	1-methyl-4-phenyl-1,2,3,6-tetrahydropyridine
MyD88	myeloid differentiation factor 88
NF-κB	nuclear factor-κB
NMDA	*N*-methyl-D-aspartate
NOX4	NADPH oxidase 4

Nrf2	erythroid 2-related factor 2
PARP1	poly [ADP-ribose] polymerase 1
PCA	protocatechuic acid
PD	Parkinson's disease
p-JNK	c-Jun N-terminal kinases
PolyQ	polyglutamine
PSD95	postsynaptic density protein 95
SIRT-1	sirtuin-1
SNAP23	synaptosome associated protein 23
SNAP25	synaptosome associated protein 25
SOD	superoxide dismutase
TLR4	Toll-like receptor 4
UPS	ubiquitin/proteasome system
VA	vanillic acid

Acknowledgements

This work has received funding from the European Research Council (ERC) under the European Union's Horizon 2020 research and innovation program under grant agreement No. 804229. iNOVA4Health Research Unit (LISBOA-01-0145-FEDER-007344), which is cofunded by Fundação para a Ciência e Tecnologia (FCT)/Ministério da Ciência e do Ensino Superior, through national funds, and by FEDER under the PT2020 Partnership Agreement, is acknowledged. The authors would like to acknowledge FCT for financial support of D.C (2020.04630.BD).

References

1. T. Wyss-Coray, *Nature*, 2016, **539**, 180–186.
2. C. Martínez-Cué and N. Rueda, Cellular Senescence in Neurodegenerative Diseases, *Front. Cell. Neurosci.*, 2020, **14**, 16.
3. F. U. Hartl and M. Hayer-Hartl, *Nat. Struct. Mol. Biol.*, 2009, **16**, 574–581.
4. C. Soto and S. Pritzkow, *Nat. Neurosci.*, 2018, **21**, 1332–1340.
5. J. Raskin, J. Cummings, J. Hardy, K. Schuh and R. Dean, *Curr. Alzheimer Res.*, 2015, **12**, 712–722.
6. M. G. Spillantini, M. L. Schmidt, V. M. Y. Lee, J. Q. Trojanowski, R. Jakes and M. Goedert, *Nature*, 1997, **388**, 839–840.
7. M. Hashimoto, L. J. Hsu, A. Sisk, Y. Xia, A. Takeda, M. Sundsmo and E. Masliah, *Brain Res.*, 1998, **799**, 301–306.
8. B. I. Giasson, K. Uryu, J. Q. Trojanowski and V. M. Y. Lee, *J. Biol. Chem.*, 1999, **274**, 7619–7622.
9. K. A. Conway, J. D. Harper and P. T. Lansbury, *Biochemistry*, 2000, **39**, 2552–2563.
10. L. C. Serpell, J. Berriman, R. Jakes, M. Goedert and R. A. Crowther, *Proc. Natl. Acad. Sci.*, 2000, **97**, 4897–4902.

11. E. A. Greenbaum, C. L. Graves, A. J. Mishizen-Eberz, M. A. Lupoli, D. R. Lynch, S. W. Englander, P. H. Axelsen and B. I. Giasson, *J. Biol. Chem.*, 2005, **280**, 7800–7807.
12. S. N. Gomperts, *CONTIN. Lifelong Learn. Neurol.*, 2016, **22**, 435–463.
13. M. Arrasate and S. Finkbeiner, *Exp. Neurol.*, 2012, **238**, 1–11.
14. T. Shacham, N. Sharma and G. Z. Lederkremer, Protein Misfolding and ER Stress in Huntington's Disease, *Front. Mol. Biosci.*, 2019, **6**, 20.
15. R. Tapia, Cellular and molecular mechanisms of motor neuron death in amyotrophic lateral sclerosis: a perspective, *Front. Cell. Neurosci.*, 2014, **8**, 241.
16. C. Lagier-Tourenne, M. Polymenidou, K. R. Hutt, A. Q. Vu, M. Baughn, S. C. Huelga, K. M. Clutario, S.-C. Ling, T. Y. Liang, C. Mazur, E. Wancewicz, A. S. Kim, A. Watt, S. Freier, G. G. Hicks, J. P. Donohue, L. Shiue, C. F. Bennett, J. Ravits, D. W. Cleveland and G. W. Yeo, *Nat. Neurosci.*, 2012, **15**, 1488–1497.
17. T. Chitnis and H. L. Weiner, *J. Clin. Invest.*, 2017, **127**, 3577–3587.
18. Z. Zheng, J. Lee and M. Yenari, *Curr. Mol. Med.*, 2003, **3**, 361–372.
19. S. Schaffer, G. P. Eckert, S. Schmitt-Schillig and W. E. Müller, in *Local Mediterranean Food Plants and Nutraceuticals*, 2006, pp. 86–115.
20. M. L. Block and J.-S. Hong, *Prog. Neurobiol.*, 2005, **76**, 77–98.
21. M. R. D'Andrea, G. M. Cole and M. D. Ard, *Neurobiol. Aging*, 2004, **25**, 675–683.
22. P. L. McGeer, S. Itagaki, H. Akiyama and E. G. McGeer, *Ann. Neurol.*, 1988, **24**, 574–576.
23. J. W. Langston, L. S. Forno, J. Tetrud, A. G. Reeves, J. A. Kaplan and D. Karluk, *Ann. Neurol.*, 1999, **46**, 598–605.
24. R. B. Banati, J. Gehrmann, P. Schubert and G. W. Kreutzberg, *Glia*, 1993, **7**, 111–118.
25. P. McGeer and E. McGeer, *Brain Res. Rev.*, 1995, **21**, 195–218.
26. D. Giulian and M. Corpuz, *Adv. Neurol.*, 1993, **59**, 315–320.
27. D. C. Wu, V. Jackson-Lewis, M. Vila, K. Tieu, P. Teismann, C. Vadseth, D.-K. Choi, H. Ischiropoulos and S. Przedborski, *J. Neurosci.*, 2002, **22**, 1763–1771.
28. C. K. Glass, K. Saijo, B. Winner, M. C. Marchetto and F. H. Gage, *Cell*, 2010, **140**, 918–934.
29. I. P. Hargreaves, A. Lane and P. M. A. Sleiman, *Neurosci. Lett.*, 2008, **447**, 17–19.
30. W. M. Johnson, A. L. Wilson-Delfosse and J. J. Mieyal, *Nutrients*, 2012, **4**, 1399–1440.
31. B. Van Houten, V. Woshner and J. H. Santos, *DNA Repair*, 2006, **5**, 145–152.
32. J. Lindsay, *Am. J. Epidemiol.*, 2002, **156**, 445–453.
33. M. Ghosh, V. Aguirre, K. Wai, H. Felfly, W. D. Dietrich and D. D. Pearse, *BioMed Res. Int.*, 2015, **2015**, 1–18.
34. E. S. O'Brien, N. V. Nucci, B. Fuglestad, C. Tommos and A. J. Wand, *J. Biol. Chem.*, 2015, **290**, 30879–30887.

35. B. Uttara, A. Singh, P. Zamboni and R. Mahajan, *Curr. Neuropharmacol.*, 2009, **7**, 65–74.
36. K. Rahman, *Clin. Interventions Aging*, 2007, **2**, 219–236.
37. F. C. Lau, B. Shukitt-Hale and J. A. Joseph, *Neurobiol. Aging*, 2005, **26**, 128–132.
38. E. Esposito, D. Rotilio, V. Dimatteo, C. Digiulio, M. Cacchio and S. Algeri, *Neurobiol. Aging*, 2002, **23**, 719–735.
39. G. H. Kim, J. E. Kim, S. J. Rhie and S. Yoon, *Exp. Neurobiol.*, 2015, **24**, 325–340.
40. E. Lévy, N. El Banna, D. Baïlle, A. Heneman-Masurel, S. Truchet, H. Rezaei, M.-E. Huang, V. Béringue, D. Martin and L. Vernis, Causative Links between Protein Aggregation and Oxidative Stress: A Review, *Int. J. Mol. Sci.*, 2019, **20**, 3896.
41. T. Hussain, B. Tan, Y. Yin, F. Blachier, M. C. B. Tossou and N. Rahu, *Oxid. Med. Cell. Longevity*, 2016, **2016**, 1–9.
42. S. K. Biswas, *Oxid. Med. Cell. Longevity*, 2016, **2016**, 1–9.
43. A. Popa-Wagner, S. Mitran, S. Sivanesan, E. Chang and A.-M. Buga, *Oxid. Med. Cell. Longevity*, 2013, **2013**, 1–14.
44. S. Chen, H. Zhou, G. Zhang, J. Meng, K. Deng, W. Zhou, H. Wang, Z. Wang, N. Hu and Y. Suo, *J. Agric. Food Chem.*, 2019, **67**, 3140–3149.
45. D. E. Bredesen, R. V. Rao and P. Mehlen, *Nature*, 2006, **443**, 796–802.
46. M. J. West, P. D. Coleman, D. G. Flood and J. C. Troncoso, *Lancet*, 1994, **344**, 769–772.
47. J. L. Price, A. I. Ko, M. J. Wade, S. K. Tsou, D. W. McKeel and J. C. Morris, Neuron Number in the Entorhinal Cortex and CA1 in Preclinical Alzheimer Disease, *Arch. Neurol.*, 2001, **58**(9), 1395–1402.
48. T. Gómez-Isla, J. L. Price, D. W. McKeel Jr, J. C. Morris, J. H. Growdon and B. T. Hyman, *J. Neurosci.*, 1996, **16**, 4491–4500.
49. J. W. Olney, D. F. Wozniak and N. B. Farber, *Restor. Neurol. Neurosci.*, 1998, **13**, 75–83.
50. J. J. Miguel-Hidalgo, X. A. Alvarez, R. Cacabelos and G. Quack, *Brain Res.*, 2002, **958**, 210–221.
51. M. D. Ikonomovic, K. Mizukami, D. Warde, R. Sheffield, R. Hamilton, R. J. Wenthold and D. M. Armstrong, *Exp. Neurol.*, 1999, **160**, 194–204.
52. J. Grosch, J. Winkler and Z. Kohl, Early Degeneration of Both Dopaminergic and Serotonergic Axons – A Common Mechanism in Parkinson's Disease, *Front. Cell. Neurosci.*, 2016, **10**, 293.
53. M. J. Farrer, *Nat. Rev. Genet.*, 2006, **7**, 306–318.
54. S. Gandhi and N. W. Wood, *Hum. Mol. Genet.*, 2005, **14**, 2749–2755.
55. S. Przedborski, *Parkinsonism & Related Disorders*, 2005, **11**, S3–S7.
56. D. Sulzer, *Trends Neurosci.*, 2007, **30**, 244–250.
57. S. J. Chia, E. K. Tan and Y. X. Chao, Historical Perspective: Models of Parkinson's Disease, *Int. J. Mol. Sci.*, 2020, **21**(7), 2464. PMID: 32252301; PMCID: PMC7177377.
58. D. Bano, F. Zanetti, Y. Mende and P. Nicotera, *Cell Death Dis.*, 2011, **2**, e228.

59. J. C. Vis, E. Schipper, R. T. de Boer-van Huizen, M. M. Verbeek, R. M. W. de Waal, P. Wesseling, H. J. ten Donkelaar and B. Kremer, *Acta Neuropathol.*, 2005, **109**, 321–328.
60. R. J. Ferrante, T. Kiechle, A. Dedeoglu, J. Kubilus, N. W. Kowall, M. F. Beal, R. M. Friedlander and S. M. Hersch, *NeuroMol. Med.*, 2002, **1**, 183–196.
61. E. G. McGeer and P. L. McGeer, *Nature*, 1976, **263**, 517–519.
62. R. H. Brown, D. L. Longo and A. Al-Chalabi, *N. Engl. J. Med.*, 2017, **377**, 162–172.
63. L. I. Grad, G. A. Rouleau, J. Ravits and N. R. Cashman, Clinical Spectrum of Amyotrophic Lateral Sclerosis (ALS), *Cold Spring Harbor Perspect. Med.*, 2017, 7(8), a024117. PMID: 28003278; PMCID: PMC5538408.
64. S. Mandel and M. B. H. Youdim, *Free Radical Biol. Med.*, 2004, **37**, 304–317.
65. E. E. Devore, J. H. Kang, M. M. B. Breteler and F. Grodstein, *Ann. Neurol.*, 2012, **72**, 135–143.
66. S. Subash, M. M. Essa, S. Al-Adawi, M. A. Memon, T. Manivasagam and M. Akbar, Neuroprotective effects of berry fruits on neurodegenerative diseases, *Neural Regener. Res.*, 2014, **9**(16), 1557–1566. PMID: 25317174; PMCID: PMC4192974.
67. F.-D. Rafael, V.-M. Isidro, A.-M. Héctor-Gabriel, P.-O. Abraham, C.-U. Yolanda, R.-G. Tania, G.-G. Rosa-Isela, P.-C. Lorena and H.-M. Socorro, in *Behavioral Pharmacology - From Basic to Clinical Research*, 2020.
68. D. Del Rio, A. Rodriguez-Mateos, J. P. E. Spencer, M. Tognolini, G. Borges and A. Crozier, *Antioxid. Redox Signaling*, 2013, **18**, 1818–1892.
69. V. Neveu, J. Perez-Jimenez, F. Vos, V. Crespy, L. du Chaffaut, L. Mennen, C. Knox, R. Eisner, J. Cruz, D. Wishart and A. Scalbert, *Database*, 2010, **2010**, bap024.
70. A. N. Winter and P. C. Bickford, Anthocyanins and Their Metabolites as Therapeutic Agents for Neurodegenerative Disease, *Antioxidants*, 2019, **8**(9), 333. PMID: 31443476; PMCID: PMC6770078.
71. S. U. Rehman, S. A. Shah, T. Ali, J. I. Chung and M. O. Kim, *Mol. Neurobiol.*, 2016, **54**, 255–271.
72. S.-C. Chang, A. Cassidy, W. C. Willett, E. B. Rimm, E. J. O'Reilly and O. I. Okereke, *Am. J. Clin. Nutr.*, 2016, **104**, 704–714.
73. D. Carregosa, R. Carecho, I. Figueira and C. N. Santos, *J. Agric. Food Chem.*, 2020, **7**, 1790–1807.
74. C. Czank, A. i. Cassidy, Q. Zhang, D. J. Morrison, T. Preston, P. A. Kroon, N. P. Botting and C. D. Kay, *Am. J. Clin. Nutr.*, 2013, **97**, 995–1003.
75. R. M. de Ferrars, C. Czank, Q. Zhang, N. P. Botting, P. A. Kroon, A. Cassidy and C. D. Kay, *Br. J. Pharmacol.*, 2014, **171**, 3268–3282.
76. C. Manach, G. Williamson, C. Morand, A. Scalbert and C. Rémésy, *Am. J. Clin. Nutr.*, 2005, **81**, 230S–242S.
77. R. Carecho, D. Carregosa and C. N. dos Santos, *Brain Plast.*, 2021, **6**, 193–214.

78. Y. He, Y. Yao, S. E. Tsirka and Y. Cao, *Stroke*, 2014, **45**, 2514–2526.
79. A. Faria, D. Pestana, D. Teixeira, P.-O. Couraud, I. Romero, B. Weksler, V. de Freitas, N. Mateus and C. Calhau, *Food Funct.*, 2011, **2**, 39–44.
80. N. Wang, P. Sun, M. Lv, G. Tong, X. Jin and X. Zhu, *Biomater. Sci.*, 2017, **5**, 1041–1050.
81. S. Talavéra, C. Felgines, O. Texier, C. Besson, A. Gil-Izquierdo, J.-L. Lamaison and C. Rémésy, *J. Agric. Food Chem.*, 2005, **53**, 3902–3908.
82. W. Kalt, J. B. Blumberg, J. E. McDonald, M. R. Vinqvist-Tymchuk, S. A. E. Fillmore, B. A. Graf, J. M. O'Leary and P. E. Milbury, *J. Agric. Food Chem.*, 2008, **56**, 705–712.
83. E. M. Janle, M. A. Lila, M. Grannan, L. Wood, A. Higgins, G. G. Yousef, R. B. Rogers, H. Kim, G. S. Jackson, L. Ho and C. M. Weaver, *J. Med. Food*, 2010, **13**, 926–933.
84. M. Suganuma, *Carcinogenesis*, 1998, **19**, 1771–1776.
85. K. Nakagawa and T. Miyazawa, *J. Nutr. Sci. Vitaminol.*, 1997, **43**, 679–684.
86. H. van Praag, M. J. Lucero, G. W. Yeo, K. Stecker, N. Heivand, C. Zhao, E. Yip, M. Afanador, H. Schroeter, J. Hammerstone and F. H. Gage, *J. Neurosci.*, 2007, **27**, 5869–5878.
87. G. Borges, J. I. Ottaviani, J. J. J. van der Hooft, H. Schroeter and A. Crozier, *Mol. Aspects Med.*, 2018, **61**, 18–30.
88. J. K. Prasain, N. Peng, Y. Dai, R. Moore, A. Arabshahi, L. Wilson, S. Barnes, J. Michael Wyss, H. Kim and R. L. Watts, *Phytomedicine*, 2009, **16**, 233–243.
89. G. Borges, J. J. J. van der Hooft and A. Crozier, *Free Radical Biol. Med.*, 2016, **99**, 128–138.
90. M. A. El Mohsen, J. Marks, G. Kuhnle, K. Moore, E. Debnam, S. K. Srai, C. Rice-Evans and J. P. E. Spencer, *Br. J. Nutr.*, 2007, **95**, 51–58.
91. A. Ishisaka, S. Ichikawa, H. Sakakibara, M. K. Piskula, T. Nakamura, Y. Kato, M. Ito, K.-I. Miyamoto, A. Tsuji, Y. Kawai and J. Terao, *Free Radical Biol. Med.*, 2011, **51**, 1329–1336.
92. J. Bieger, R. Cermak, R. Blank, V. C. J. de Boer, P. C. H. Hollman, J. Kamphues and S. Wolffram, *J. Nutr.*, 2008, **138**, 1417–1420.
93. M. A. El Mohsen, J. Marks, G. Kuhnle, C. Rice-Evans, K. Moore, G. Gibson, E. Debnam and S. K. Srai, *Free Radical Res.*, 2009, **38**, 1329–1340.
94. H. W. Peng, F. C. Cheng, Y. T. Huang, C. F. Chen and T. H. Tsai, *J. Chromatogr. B: Biomed. Sci. Appl.*, 1998, **714**, 369–374.
95. G. Borges, S. Roowi, J.-M. Rouanet, G. G. Duthie, M. E. J. Lean and A. Crozier, *Mol. Nutr. Food Res.*, 2007, **51**, 714–725.
96. C. Andres-Lacueva, B. Shukitt-Hale, R. L. Galli, O. Jauregui, R. M. Lamuela-Raventos and J. A. Joseph, *Nutr. Neurosci.*, 2013, **8**, 111–120.
97. S. Passamonti, U. Vrhovsek, A. Vanzo and F. Mattivi, *J. Agric. Food Chem.*, 2005, **53**, 7029–7034.
98. M. Gasperotti, S. Passamonti, F. Tramer, D. Masuero, G. Guella, F. Mattivi and U. Vrhovsek, *ACS Chem. Neurosci.*, 2015, **6**, 1341–1352.
99. T. K. McGhie and M. C. Walton, *Mol. Nutr. Food Res.*, 2007, **51**, 702–713.

100. I. Figueira, G. Garcia, R. C. Pimpão, A. P. Terrasso, I. Costa, A. F. Almeida, L. Tavares, T. F. Pais, P. Pinto, M. R. Ventura, A. Filipe, G. J. McDougall, D. Stewart, K. S. Kim, I. Palmela, D. Brites, M. A. Brito, C. Brito and C. N. Santos, *Sci. Rep.*, 2017, **7**, 11456.
101. X. Liu, M. Tu, R. S. Kelly, C. Chen and B. J. Smith, *Drug Metab. Dispos.*, 2004, **32**, 132–139.
102. R. Hornedo-Ortega, M. A. Álvarez-Fernández, A. B. Cerezo, T. Richard, A. M. A. Troncoso and M. A. C. Garcia-Parrilla, *J. Agric. Food Chem.*, 2016, **64**, 7722–7732.
103. R. Zhang, X. Hu, H. Khant, S. J. Ludtke, W. Chiu, M. F. Schmid, C. Frieden and J. M. Lee, *Proc. Natl. Acad. Sci.*, 2009, **106**, 4653–4658.
104. H.-N. Zhang, C.-N. An, M. Xu, D.-A. Guo, M. Li and X.-P. Pu, *Biol. Pharm. Bull.*, 2009, **32**, 1866–1869.
105. Y. Liu, J. A. Carver, A. N. Calabrese and T. L. Pukala, *Biochim. Biophys. Acta, Proteins Proteomics*, 2014, **1844**, 1481–1485.
106. M. T. Ardah, K. E. Paleologou, G. Lv, S. B. Abul Khair, A. S. Kazim, S. T. Minhas, T. H. Al-Tel, A. A. Al-Hayani, M. E. Haque, D. Eliezer and O. M. El-Agnaf, Structure activity relationship of phenolic acid inhibitors of α-synuclein fibril formation and toxicity, *Front. Aging Neurosci.*, 2014, **6**, 197. PMID: 25140150; PMCID: PMC4122169.
107. A. N. Khan, M. N. Hassan and R. H. Khan, *J. Mol. Liq.*, 2019, **285**, 27–37.
108. W. Zhou, A. Gallagher, D.-P. Hong, C. Long, A. L. Fink and V. N. Uversky, *J. Mol. Biol.*, 2009, **388**, 597–610.
109. S. Bastianetto, Z.-X. Yao, V. Papadopoulos and R. Quirion, *Eur. J. Neurosci.*, 2006, **23**, 55–64.
110. M. Yu, X. Chen, J. Liu, Q. Ma, Z. Zhuo, H. Chen, L. Zhou, S. Yang, L. Zheng, C. Ning, J. Xu, T. Gao and S.-T. Hou, *Neurobiol. Dis.*, 2019, **124**, 67–80.
111. Y. Song, T. Cui, N. Xie, X. Zhang, Z. Qian and J. Liu, *Int. Immunopharmacol.*, 2014, **20**, 276–281.
112. D. Vauzour, A. Martinsen and S. Layé, *Neurochem. Int.*, 2015, **89**, 63–74.
113. G. Garcia, S. Nanni, I. Figueira, I. Ivanov, G. J. McDougall, D. Stewart, R. B. Ferreira, P. Pinto, R. F. M. Silva, D. Brites and C. N. Santos, *Food Chem.*, 2017, **215**, 274–283.
114. H.-Y. Wang, H. Wang, J.-H. Wang, Q. Wang, Q.-F. Ma and Y.-Y. Chen, *Neurochem. Res.*, 2015, **40**, 1655–1660.
115. S. Thummayot, C. Tocharus, P. Jumnongprakhon, A. Suksamrarn and J. Tocharus, *Acta Pharmacol. Sin.*, 2018, **39**, 1439–1452.
116. J. L. di Gesso, J. S. Kerr, Q. Zhang, S. Raheem, S. K. Yalamanchili, D. O'Hagan, C. D. Kay and M. A. O'Connell, *Mol. Nutr. Food Res.*, 2015, **59**, 1143–1154.
117. M.-J. Kim, A.-R. Seong, J.-Y. Yoo, C.-H. Jin, Y.-H. Lee, Y. J. Kim, J. Lee, W. J. Jun and H.-G. Yoon, *Mol. Nutr. Food Res.*, 2011, **55**, 1798–1808.
118. S. E. Khoshnam, A. Sarkaki, M. Rashno and Y. Farbood, *Life Sci.*, 2018, **211**, 126–132.

119. S. Siddiqui, A. Kamal, F. Khan, K. S. Jamali and Z. S. Saify, *Mol. Biol. Rep.*, 2018, **46**, 997–1011.
120. A. N. Winter, M. C. Brenner, N. Punessen, M. Snodgrass, C. Byars, Y. Arora and D. A. Linseman, *Oxid. Med. Cell. Longevity*, 2017, **2017**, 1–13.
121. M. Koshiguchi, H. Komazaki, S. Hirai and Y. Egashira, *Biosci. Biotechnol. Biochem.*, 2017, **81**, 966–971.
122. S. Chakrabarti, M. Jana, A. Roy and K. Pahan, *Curr. Alzheimer Res.*, 2018, **15**, 894–904.
123. S. U. Rehman, T. Ali, S. I. Alam, R. Ullah, A. Zeb, K. W. Lee, B. P. F. Rutten and M. O. Kim, Ferulic Acid Rescues LPS-Induced Neurotoxicity via Modulation of the TLR4 Receptor in the Mouse Hippocampus, *Mol. Neurobiol.*, 2019, **56**(4), 2774–2790. Epub 2018 Jul 30. PMID: 30058023.
124. S. Ojha, H. Javed, S. Azimullah, S. B. Abul Khair and M. E. Haque, Neuroprotective potential of ferulic acid in the rotenone model of Parkinson's disease, *Drug Des., Dev. Ther.*, 2015, **9**, 5499–5510. PMID: 26504373; PMCID: PMC4603721.
125. J. Wang, Y. Guo and S. Yue Zhang, *Int. J. Pharmacol.*, 2018, **14**, 488–494.
126. S.-J. Tsai and M.-C. Yin, *Food Chem. Toxicol.*, 2012, **50**, 3198–3205.
127. A. R. Kho, B. Y. Choi, S. H. Lee, D. K. Hong, S. H. Lee, J. H. Jeong, K. H. Park, H. K. Song, H. C. Choi and S. W. Suh, Effects of Protocatechuic Acid (PCA) on Global Cerebral Ischemia-Induced Hippocampal Neuronal Death, *Int. J. Mol. Sci.*, 2018, **19**(5), 1420. PMID: 29747437; PMCID: PMC5983751.
128. F. U. Amin, S. A. Shah and M. O. Kim, Vanillic acid attenuates Aβ1-42-induced oxidative stress and cognitive impairment in mice, *Sci. Rep.*, 2017, **7**, 40753. PMID: 28098243; PMCID: PMC5241654.
129. W. Slikkker, M. Youdim, G. C. Palmer, E. Hall, C. Williams and B. Trembly, *Ann. N. Y. Acad. Sci.*, 1999, **890**, 529–533.
130. J. J. P. Gille and H. Joenje, *Mutat. Res., DNAging*, 1992, **275**, 405–414.
131. A. Tarozzi, F. Morroni, S. Hrelia, C. Angeloni, A. Marchesi, G. Cantelli-Forti and P. Hrelia, *Neurosci. Lett.*, 2007, **424**, 36–40.
132. L. J. An, S. Guan, G. F. Shi, Y. M. Bao, Y. L. Duan and B. Jiang, *Food Chem. Toxicol.*, 2006, **44**, 436–443.
133. G. Shui, Y.-M. Bao, J. Bo and L.-J. An, *Eur. J. Pharmacol.*, 2006, **538**, 73–79.
134. N. H. Gay, K. Phopin, W. Suwanjang, N. Songtawee, W. Ruankham, P. Wongchitrat, S. Prachayasittikul and V. Prachayasittikul, *Neurochem. Res.*, 2018, **43**, 619–636.
135. Y.-M. Liu, B. Jiang, Y.-M. Bao and L.-J. An, *Toxicol. In Vitro*, 2008, **22**, 430–437.
136. J. Y. Ban, S. O. Cho, S.-Y. Jeon, K. Bae, K.-S. Song and Y. H. Seong, *Neurosci. Lett.*, 2007, **420**, 184–188.
137. Z. Zhang, G. Li, S. S. W. Szeto, C. M. Chong, Q. Quan, C. Huang, W. Cui, B. Guo, Y. Wang, Y. Han, K. W. Michael Siu, S. M. Yuen Lee and I. K. Chu, *Free Radical Biol. Med.*, 2015, **84**, 331–343.

138. H.-L. Huang, C.-C. Lin, K.-C. G. Jeng, P.-W. Yao, L.-T. Chuang, S.-L. Kuo and C.-W. Hou, *J. Agric. Food Chem.*, 2012, **60**, 2328–2336.
139. S. Maya, T. Prakash and K. Madhu, *Neurochem. Int.*, 2018, **121**, 50–58.
140. M. T. Mansouri, Y. Farbood, M. J. Sameri, A. Sarkaki, B. Naghizadeh and M. Rafeirad, *Food Chem.*, 2013, **138**, 1028–1033.
141. A. González-Sarrías, M. Á. M. Á. Núñez-Sánchez, F. A. Tomás-Barberán, J. C. Espín, A. González-Sarrías, M. í. Á. M. Á. Núñez-Sánchez, F. A. Tomás-Barberán and J. C. Espín, *J. Agric. Food Chem.*, 2016, **65**, 752–758.
142. Z. Ren, R. Zhang, Y. Li, Y. Li, Z. Yang and H. Yang, *Int. J. Mol. Med.*, 2017, **40**, 1444–1456.
143. Y. Cao, L. Zhang, S. Sun, Z. Yi, X. Jiang and D. Jia, *Int. J. Mol. Med.*, 2016, **38**, 567–573.
144. A. Esteban-Fernández, C. Rendeiro, J. P. E. Spencer, D. G. Del Coso, M. D. G. de Llano, B. Bartolomé and M. V. Moreno-Arribas, *Front. Nutr.*, 2017, **4**, 3.
145. S. A. Adefegha, G. Oboh, O. S. Omojokun and O. M. Adefegha, *Biomed. Pharmacother.*, 2016, **83**, 559–568.
146. G.-F. Shi, L.-J. An, B. Jiang, S. Guan and Y.-M. Bao, *Neurosci. Lett.*, 2006, **403**, 206–210.
147. B. Naghizadeh and M. Mansouri, *Drug Res.*, 2014, **65**, 515–520.
148. J. C. H. Singh, R. M. Kakalij, R. P. Kshirsagar, B. H. Kumar, S. S. B. Komakula and P. V. Diwan, *Pharm. Biol.*, 2014, **53**, 630–636.
149. K. R. Rekha, G. P. Selvakumar and R. I. Sivakamasundari, *Biomed. Aging Pathol.*, 2014, **4**, 95–104.
150. M. Güven, A. B. Aras, N. TopaloğLu, A. Özkan, H. M. Şen, Y. Kalkan, A. Okuyucu, A. Akbal, F. Gökmen and M. Coşar, *Turk. J. Med. Sci.*, 2015, **45**, 233–240.
151. S. E. Khoshnam, Y. Farbood, H. Fathi Moghaddam, A. Sarkaki, M. Badavi and L. Khorsandi, *Metab. Brain Dis.*, 2018, **33**, 785–793.
152. S. E. Khoshnam, A. Sarkaki, L. Khorsandi, W. Winlow, M. Badavi, H. F. Moghaddam and Y. Farbood, *Biomed. Pharmacother.*, 2017, **96**, 667–674.
153. P. Mena, L. Bresciani, N. Brindani, I. A. Ludwig, G. Pereira-Caro, D. Angelino, R. Llorach, L. Calani, F. Brighenti, M. N. Clifford, C. I. R. Gill, A. Crozier, C. Curti and D. Del Rio, Phenyl-γ-valerolactones and phenylvaleric acids, the main colonic metabolites of flavan-3-ols: synthesis, analysis, bioavailability, and bioactivity, *Nat. Prod. Rep.*, 2019, **36**(5), 714–752. PMID: 30468210.
154. C. D. Kay, M. N. Clifford, P. Mena, G. J. McDougall, C. Andres-Lacueva, A. Cassidy, D. Del Rio, N. Kuhnert, C. Manach, G. Pereira-Caro, A. Rodriguez-Mateos, A. Scalbert, F. Tomás-Barberán, G. Williamson, D. S. Wishart and A. Crozier, *Am. J. Clin. Nutr.*, 2020, **112**, 1051–1068.
155. R. Verpoorte, Y. H. Choi and H. K. Kim, *J. Ethnopharmacol.*, 2005, **100**, 53–56.

CHAPTER 9

The Role and Mechanisms of Berries and Berry Bioactive Compounds in Cancer

ALEKSANDRA S. KRISTO AND ANGELOS K. SIKALIDIS*

California Polytechnic State University, Food Science and Nutrition Department, 1 Grand Ave., CA 93407, USA
*Email: asikalid@calpoly.edu

9.1 Introduction

Cancer, the uncontrolled growth of cells that can invade and spread to distant sites of the body, is a global health problem with high mortality and disability rates.[1] Cancer is a debilitating, typically terminal disease with a poor prognosis leading to death. Approximately 8.8 million people are dying every year of cancer, amounting to one out of six deaths globally and far exceeding the number of deaths from HIV/AIDS, malaria, and tuberculosis combined.[2,3] The disease burden is greatest in low-income and middle-income countries, where 75% of cancer deaths occur, while the number of cancer cases is rising most rapidly.[2] Importantly, cancer incidence is estimated to double by 2035.[2]

Cancer care costs constitute a significant financial burden to patients, their families, and society as a whole. The combination of increasing population age and cancer prevalence is inducing costs, while the burden of disease is expected to continue to rise.[3] In addition, the development of expensive, targeted treatment strategies that are gradually becoming the standard of care are anticipated to increase cancer costs further. According

Food Chemistry, Function and Analysis No. 33
Berries and Berry Bioactive Compounds in Promoting Health
Edited by Dorothy Klimis-Zacas and Ana Rodriguez-Mateos

Published by the Royal Society of Chemistry, www.rsc.org

to the World Health Organization (WHO), by 2030, global annual cancer deaths will increase to 11.8 million and cancer will be the leading cause of death that year, followed by heart disease and stroke.[2,3] Indicatively, in 2008 alone, the cost of premature death and disability, but not direct medical costs, from cancer was $895 billion worldwide compared to $753 billion for heart disease.[2,3] In the USA, the Agency for Healthcare Research and Quality (AHRQ) estimated that the direct medical costs (total of all health care costs) for cancer in 2015 reached $80.2 billion.[4]

The term cancer is derived from the Greek word "καρκίνος" meaning *crab*. Greek physicians Hippocrates and Galen were the first to note the similarity of crabs (in terms of shape) to some tumors with swollen veins. The word was introduced in English in the modern medical sense approximately at 1600 AD.[5] Prevention along with monitoring offering early and accurate diagnosis when a case is confirmed, constitute a critical approach for optimal outcome, since most therapeutic options do not confer cure but rather a deceleration of cancer progression aiming at life extension and improvement of patients' life quality, albeit with serious and often debilitating side effects. While cancer incidence is expected to rise by 70% by the year 2034, approximately 35% of cancer deaths are attributed to the five leading behavioral and dietary risks: high body mass index (BMI), low fruit and vegetable intake, lack of physical activity, tobacco use (primarily smoking), and harmful alcohol use.[6] In this context, strategies and/or approaches that can prevent or decelerate tumorigenesis can be highly useful when dealing with cancer. A plethora of *in vitro*, *in vivo* and human studies has suggested diet as a critical factor for reducing cancer risk with a particular emphasis on certain foods/food groups, to which specific potential for cancer risk reduction has been attributed.[1]

Thus, exposure is of significant importance in terms of reducing risk for carcinogenesis, with particular emphasis on the food groups of fruits and vegetables. Specifically, diets rich in fruits and vegetables have been associated with a reduced risk of cancer.[7] While in an extensive recent human epidemiological study on single foods items, among single foods, only fruit intake was negatively associated with C-reactive peptide (CRP) levels.[8] This work illustrated that fruit intake reduces inflammation as judged by CRP reduced levels, thus reducing the risk for cancer since inflammation increases the risk for cancer. Moreover, a 2021 systematic review of the evidence investigated the dietary patterns and associations with biomarkers of inflammation in adults and reported that adherence to healthy, Mediterranean, and anti-inflammatory dietary scores appear to be associated with lower inflammatory status cross-sectionally, thus reducing cancer risk.[9] Additionally, an extensive meta-analysis investigating the current dietary recommendations, particularly those on fruits and vegetables, in relation to the burden of disease, supports existing recommendations for fruit and vegetable intakes.[10] According to some analyses, approximately 45% of all cancers are associated with preventable/modifiable risk factors, including tobacco and alcohol use and obesity,[11] while according to WHO, up to one-third of cancer-related deaths could be prevented by modifying risk factors associated with lifestyle, including diet and exercise.[12]

Many non-nutrient plant compounds known as phytochemicals, including numerous phenolic compounds (polyphenols, flavanols, flavones), have been identified and found to exert anti-cancer activity with a significant presence in fruits. A fruit group of high interest in this context is that of berries. Berries are typically rich in carbohydrates but low in calories. They contain minimal amounts of lipid and high content of dietary fiber (cellulose, hemicellulose, pectin); organic acids, such as citric acid, malic acid, tartaric acid, oxalic acid, and fumaric acid; while also certain minerals in trace amounts. With a plethora of bioactive compounds and a variety of profiles, berries have been recently looked at more closely in terms of their notable potential to positively modulate the risk of cancer. Nonetheless, since cancer is a multifactorial disease, understanding how berries and their components directly affect the risk of cancer is somewhat challenging. Several plausible mechanisms *via* which cancer is potentially prevented, or by which the risk of developing a malignant tumor is lowered, have been proposed and identified in the past two decades. Increasing research on berries and cancer has led to the identification of a number of favorable health-protective mechanisms, which are generally based on vitamins, minerals, fiber, and berry bioactive compounds. While a panoply of bioactive compounds naturally found in berries has been shown to extend antioxidative potential, the effective physiological relevance of foods with antioxidants is uncertain, as many investigations are mainly based on *in vitro* assays or *in vivo* work. Therefore, the majority of findings do not necessarily correspond directly with human physiological mechanisms and practical significant risk reduction.

9.2 Cancer Development and Associated Mechanisms

Cancer is defined by the National Cancer Institute as: "class of diseases in which abnormal cells divide without control and are able to invade other tissues".[13] The development of cancer occurs gradually as an aggregation of proliferating cells from tumor(s), which metastasize. The exact mechanism of cancer generation remains elusive. Equally elusive remain the causal events that turn a mammalian healthy cell into a cancerous one. What has been well established, nevertheless, is that there are clearly certain conditions that significantly increase the risk for cancer development. From a biochemical perspective, oxidative stress resulting in the accumulation of reactive oxygen species (ROS) and subsequent inflammatory responses comprise conditions well and strongly associated with tumorigenesis and subsequent cancer progression.[1] Cancer is a stepwise procedure whereby several factors may influence its initiation, development, and progression (the three stages delineating tumorigenesis and development). Oxidative stress specifically facilitates the activation of various transcription factors, including nuclear factor-kappa B/nuclear factor-kappa-light-chain-enhancer of activated B cells (NF-κB), activator protein-1 (AP-1), p53, hypoxia-inducible factor-1α (HIF-1α), peroxisome proliferator-activated receptor gamma (PPAR-γ or PPARG), β-catenin/Wnt, and NF-E2 related factor-2 (Nrf2).[1] Activation of these transcription factors results in the expression of more than 500 different genes, including those encoding for growth factors, inflammatory cytokines, chemokines, cell cycle regulatory molecules, and anti-inflammatory molecules.[1,14] Biological, chemical, and physical factors can all, either in combination or separately, lead to oxidative stress resulting in the production of ROS and/or reactive nitrogen species (RNS).[15] Oxidative stress causes damage to tissues and elicits an immune response that induces inflammation in an attempt of the body to repair the inflicted damage. During the inflammatory response, mast cells and leukocytes are recruited to the injured site, leading to a "respiratory burst" because of an augmented requisite and thus uptake of oxygen, hence inducing release and accumulation of ROS at the site of damage, a development further perpetuating the damage to the already damaged and inflamed site. As the level of inflammatory response is augmented, a surplus of signaling molecules is produced by the inflammatory cells, such as metabolites of arachidonic acid, cytokines, and chemokines, acting to further intensify the inflammatory responses and the continuing production and accumulation of ROS. Simultaneously, the signaling agents produced induce signaling cascades involving NF-κB, signal transducer and activator of transcription 3 (STAT3), HIF1-α, AP-1, nuclear factor of activated T-cells (NFAT), and Nrf2, which mediate immediate cellular stress responses. This includes induction of cyclooxygenase-2 (COX-2), inducible nitric oxide synthase (iNOS), aberrant expression of inflammatory cytokines: tumor necrosis factor-α (TNF-α), interleukin-1 (IL-1), IL-6, and chemokines such as IL-8; CXC chemokine receptor 4 (CXCR4).[15,16] This microenvironment, hence characterized by increasingly high levels of inflammation and oxidative stress, can spread damage to neighboring cells/tissues and, if such a condition is sustained and becomes chronic, it can eventually lead to carcinogenesis. Therefore, chronic inflammation, as mediated by the induction of the immune system due to oxidative stress, can predispose the host to cancer among various chronic illnesses.[16–18] Apart from the oxidative stress/inflammation axis and the related mechanism for induction of carcinogenesis, tumor cells activate autophagy in response to cellular stress and/or increased metabolic demands related to rapid cell proliferation. Autophagy-related stress tolerance can induce cellular pro-survival mechanisms, thus leading to tumor growth and therapeutic resistance. Interestingly, work with preclinical models demonstrated that inhibition of autophagy can restore chemosensitivity and enhance tumor cell death, thus improving response to therapy.[19]

9.3 Berries and Cancer

9.3.1 Berry General Characteristics and Background

According to strict botanical terminology, a "berry" is a simple fruit with seeds and pulp produced from the ovary of a single flower and the pericarp (fruit wall) is fleshy (*i.e.*, the fruit is fleshy from the skin layer inwards and all throughout to its core but the seeds).[20] Under common usage, however, the term berry refers to a small pulpy and often edible fruit. Currently, there is an authentic consensus among researchers, whereby berries are recommended for an individual's health due to their phytochemical composition.[21] Around 400 BC, ancient Greek physician and father of medicine Hippocrates suggested preventing disease by eating nutrient-dense foods.[22] Hippocrates prescribed blackberries in herbal medicinal approaches as the leaves, bark, and roots all contain notable amounts of tannins with astringent properties.[22] The ancient Greeks knew of blackberry's medicinal qualities and considered them an excellent anti-inflammatory remedy for gout,[22] as did the Cherokee, using an infusion of the Texas dewberry for the same ailment.[23]

Research in past recent years has identified that berries contain a number of phytochemicals associated with health benefits. Phytochemicals including anthocyanins, flavonols, flavonoids, and proanthocyanidins, tend to be bioactive and, as such, extend a wide range of biological properties such as anti-carcinogenic, antimutagenic, antioxidants, and cancer preventative. The general health benefits of fruit consumption have been widely established. In 1991, for example, Steinmetz and Potter[24] alluded to the fact that humans have adapted over evolution to a diet that regularly involves the consumption of fruit (including berries) and vegetables.[24] Moreover, it was hypothesized by Steinmetz and Potter that a reduction in fruit and vegetable consumption could potentially lend itself to an increase in the risk of cancer development. It is relatively recent, though, that studies by Neto[25] and Stoner *et al.*[26] have begun to investigate the cancer preventative effects of berry phytochemicals.

9.3.2 Berry Intake and Cancer Outcomes

9.3.2.1 In Vitro *and* In Vivo *Work*

9.3.2.1.1 Colonic Cancer. There have been clear and repeated independent observations illustrating a close association between colon tumorigenesis and chronic inflammation of the intestine in mouse models of inflammation and cancer, with berry dietary supplementation able to attenuate the negative outcomes. More specifically, Shi *et al.* demonstrated that lyophilized strawberry supplementation reduced tumorigenesis and expression of proinflammatory mediators, suppressed nitrosative stress and decreased phosphorylation of phosphatidylinositol-3-kinase, extracellular signal-regulated kinase, and NF-κB in azoxymethane/dextran sodium sulfate-induced colorectal carcinogenesis-induced Crj: CD-1 mice.[27] Tumor incidence of the mice fed with freeze-dried strawberry powder at 2.5%, 5%, and 10% (wt/wt) in the diet was measured at 64%, 75%, and 44%, respectively, *vs.* the 100% of their littermates fed control diet.

Additionally, the protective effects of the mulberry fruit, containing flavonoids and anthocyanins, alkaloids, and carotenoids, were documented through both *in vitro* and *in vivo* studies on colon cancer and intestinal inflammation.[28] Cooke and collaborators used Muc2(−/−) mice, a model of spontaneous chronic intestinal inflammation, to investigate the effects of berry constituent supplementation on intestinal carcinogenesis. A diet enriched with 5% or 10% mulberry extract administered for 3 months, starting at 3–4 weeks of age, produced a reduction in tumorigenesis and intestinal inflammation as estimated by the degree of mucosal damage and lymphocyte infiltration.[28] In the same study, 6- to 8-week-old BALB/c mice supplemented with mulberry extract (5% or 10% wt/wt) for 10 days prior to 3% dextran sulfate sodium (DSS) exposure in drinking water for 9 days, revealed an improvement in DSS-induced acute colitis symptoms such as weight loss, bloody stools, and colorectal histological changes.[28] Moreover, lipopolysaccharide (LPS)-treated RAW264.7 macrophages, whereby mulberry extract was applied, were able to demonstrate attenuated inflammation by reducing the expression of iNOS, COX-2, IL-1β and IL-6, and inhibiting activation of NF-κB/p65 and pERK (phosphorylated Extracellular signal Regulated Kinase) mitogen-activated protein kinase (MAPK) pathways.[28]

In separate experiments using ApcMin mice, an established model of human familial adenomatous polyposis, a condition characterized by multiple intestinal polyps potentially developing to carcinomas, an anthocyanin-rich bilberry extract (40% anthocyanins) fed for 12 weeks at a dietary dose of 0.3%, produced a 30% reduction on adenoma counts in a dose-dependent manner.[29]

Similarly, animal dietary studies with black raspberries reveal chemopreventive capacities of the fruit in models of ulcerative colitis, several rodent models of colon cancer and adenoma. More specifically, Bi and colleagues documented that a 12 week feeding black raspberries regime significantly inhibited intestinal tumor formation, reducing tumor incidence by 45% and tumor multiplicity by 60% in Apc1638+/− mice and tumor incidence and multiplicity by 50% in Muc2−/− mice. Mechanistically, tumor inhibition development in Apc1638+/− mice was achieved by suppressing β-catenin signaling and in Muc2−/− mice by reducing chronic inflammation.[30] Furthermore, Wang *et al.* showed that knockout mice fed 5% black raspberries demonstrated significantly less colonic ulceration compared to knockout mice that consumed the control diet. Researchers investigated the effects of black raspberries on the Wnt pathway and found that the black raspberries-fed knockout mice exhibited a significantly lower level of β-catenin nuclear translocation.[31]

Pan and colleagues showed that a 5% wt/wt black raspberry dietary supplementation to an interleukin-10 knockout mouse, a model of ulcerative colitis-associated colon cancer progression, presented diminished colon ulceration, alongside lower levels of β-catenin nuclear translocation and inhibition of epigenetic events related to abnormal Wnt signaling.[32] The 5% wt/wt black raspberry supplementation for 8 weeks reduced polyp number and size in the Apc(Min/+) mice intestine and colon, inhibiting the development of colonic adenoma. Non-targeted metabolomics revealed that several Apc-related metabolites in the mucosa, liver, and feces were positively modulated by the black raspberry diet, including putrescine and linolenate, which are associated with colorectal cancer in humans.[32]

Similarly, Misikangas *et al.* demonstrated that in DSS-induced ulcerative colitis C57BL/6J mice, a 10% (wt/wt) freeze-dried bilberry, lingonberry, or cloudberry for 10 week supplemented diets, suppressed colon ulceration by restoring epigenetic regulation of systemic inflammation.[33] All three berries significantly inhibited the formation of intestinal adenomas, as indicated by a 15–30% reduction in tumor number ($p < 0.05$). Early changes in gene expression in the normal-appearing mucosa were analyzed by Affymetrix microarrays, which revealed changes in genes implicated in colon carcinogenesis, including the decreased expression of the adenosine deaminase, ecto-5′-nucleotidase, and prostaglandin E_2 (PGE_2) receptor subtype EP4. The authors concluded that berries are potentially able to extend chemopreventive function.[33] In separate *in vivo* experiments on Min mice, Rajakangas *et al.* showed that a 10% white currant dietary supplementation administered for 10 weeks can yield a significant reduction in the number and size of adenomas in the total small intestine, associated with reduced nuclear beta-catenin and NF-κB protein levels in the adenomas.[34] The results of this study argued in favor of a chemopreventive function of white currant berries, at least in the case of Min mice.

Extensive work with Fischer 344 (F344) rats consuming black raspberries at 2.5%, 5%, or 10% (wt/wt) in the diet after azoxymethane injections, revealed that aberrant crypt foci (ACF), total tumor, and adenocarcinoma multiplicity were significantly reduced at varying percentages depending on the berry dose. Moreover, tumor burden demonstrated a non-significant decrease across all berry diet groups. Investigating the dietary dose of black raspberries, the 5% dose was more effective than the 2.5% dose in inhibiting chemically induced adenocarcinoma, while the 10% dose did not produce any greater benefit, suggesting an optimal dose at 5% supplementation. Finally, oxidative stress, as measured by urinary levels of 8-hydroxy-2′-deoxyguanosine (8-OHdG), was reported attenuated significantly in all rats supplemented with black raspberries regardless of the supplementation level.[35]

Many flavonoids have been shown to exert anti-carcinogenic effects in cells and animals. In many of these cases, the compounds are administered in isolation or as part of a whole food, however significantly processed, while usually freeze-dried for practical purposes. Many of the chemopreventive effects observed seem to occur through signaling pathways known to be important in the pathogenesis of cancers. Nonetheless, dietary flavonoid intakes are generally low and their metabolism in humans stands rather complex.[36] Furthermore, the amounts and the optimum mixture of these protective compounds are rather unclear and there is no consensus among researchers in the field as per the profile and doses of bioactives. Certain antioxidants exert undesirable pro-oxidant action when the dosing is high. Therefore, it is important to maintain this perspective when designing feeding regimens and more so when translating a demonstrated proof of concept from *in vivo* to human work and subsequently arguing on possible recommendations.

9.3.2.1.2 Esophageal Cancer. A significant bulk of data, derived from both *in vitro* and *in vivo* work, reveals a series of benefits from berry consumption in terms of both chemoprevention and response to esophageal cancer. Stoner and colleagues described the effects of black raspberries on gene expression in the very early stages of rat esophageal carcinogenesis.[37] The group reported effects on genes that influence carcinogen metabolism. Specifically, out of 2261 dysregulated genes in the esophagi of F344 rats treated for 1 week with nitrosomethylbenzylamine (NMBA), 462 were positively modulated and restored to near-normal levels when a 5% (wt/wt) whole black raspberry powder supplementation to the control AIN-76A diet was introduced.[37] The positively regulated genes were oncogenes, genes involved in oxidative damage, and tumor suppressor genes regulating apoptosis, cell cycling, and angiogenesis.[37] Additionally, in separate feeding experiments involving F344 rats conducted by Wang *et al.*, animals were fed berry-supplemented diets for 2 weeks prior to NMBA treatments and were then maintained on the berry-supplemented diets until the end of the 30 week experimental period, at which point esophageal tumors were assessed. The rats were fed 5% (wt/wt) blackberry powder, an anthocyanin-rich fraction, an organic solvent-soluble extract (each contained approximately 3.8 μmol anthocyanins per gram diet), an organic-insoluble (residue) fraction (containing 0.02 μmol anthocyanins per gram diet), a hexane extract, and a sugar fraction. The experiments indicated that anthocyanins derived from black raspberries were effective in reducing NMBA-induced tumors in the rat esophagus.[38] While several forms of anthocyanins were evaluated, all supplementation forms, *i.e.*, 5% whole raspberries extract, anthocyanin-rich fraction, and organic-solvent soluble extract, produced similar positive results. Notably, the organic insoluble residue fraction also produced comparable results, hence indicating that other compounds from berry anthocyanins may also encompass chemopreventive effects.[38]

Alteration in the immune cell trafficking in esophageal cancer by anthocyanin constituents was demonstrated in another rat feeding study conducted by Peiffer *et al.*[39] Post 5 weeks of NMBA (0.35 $mg\,kg^{-1}$) administration while rats were all on control diet, the animals were switched to their corresponding treatment diets (*i.e.*, 6.1% black raspberries powder, anthocyanin-rich fraction of black raspberries 3.8 $\mu mol\,g^{-1}$, and 500 ppm protocatechuic acid). Esophageal cancer-related inflammatory biomarkers were assessed at three time points (15, 25, and 35 weeks) in the plasma and at the esophagus (tissue). Furthermore, the infiltration of the immune cells in the esophageal tissue was also assessed. Cytokine production was not different among all three different dietary treatments. However, all treatment groups exhibited a decrease in IL-1β and IL-12, while an increase in IL-10 compared to the control, whereas the treatments all produced a significantly lower infiltration of both macrophages and neutrophils to the esophagus proper. These data collectively suggest that anthocyanins can inhibit esophageal tumorigenesis by altering cytokine expression as well as innate immune cell trafficking within the tumors.[39]

The effect of black raspberries (whole food/fruit) on the late stages of rat esophageal carcinogenesis was extensively investigated by Wang *et al.*[40] The impact of a 5% (wt/wt) freeze-dried raspberry-supplemented diet was evaluated on F344 rats in terms of late stages of carcinogenesis for NMBA-induced esophageal cancer. Rats control *vs.* treatment (raspberry supplemented) were evaluated 35 weeks post-NMBA treatment. Results indicated that black raspberries reduced the number of dysplastic lesions while also the number and size of esophageal tumors. Furthermore, black raspberry supplementation was found to exert a positive effect on the expression of a variety of genes associated with preneoplastic esophagus, as well as esophageal papillomas, and was seen to modulate gene expression associated with proliferation, apoptosis, and inflammation in addition to angiogenesis in a positive fashion. The obtained results determine that black raspberries can extend a positive impact regarding molecular signaling of even late stages esophageal carcinogenesis in rats, thus underlining possible benefits from black raspberry consumption in terms of cancer prevention and/or treatment.

Additional evidence pertinent to the positive role of black raspberry consumption was provided by analogous experiments undertaken by Chen and colleagues.[41] More specifically, a 5% (wt/wt) black raspberry diet was demonstrated to reduce both mRNA and protein levels of COX-2, iNOS, and c-Jun as well as the level of PGE_2 in F344 NMBA-treated rats on diets for 25 weeks. These results propose another potential role for black raspberries in esophageal cancer-related molecular signaling. The same group further reported black raspberry angiogenesis-suppression capacity, parallel to the suppression of COX-2 and iNOS, albeit in separate experiments.[42] Reen *et al.* detected similar positive effects of black raspberries in NMBA-induced esophageal cancer in F344 rats, albeit recorded effects were not dose-dependent (5% and 10% wt/wt black raspberry supplementation produced similar effects).[43] Such observations lend support to the concept that fine-tuned dosing is optimizing health benefits, most plausibly, mainly through molecular signaling. Thus, more emphasis needs to be placed on the determination of the "ideal" profile of bioactives' mixture and dosing when considering dietary consultation.

The chemopreventive properties of several berries against esophageal cancer in rat was assessed by Stoner and colleagues.[44] NMBA-treated rats over a period of 5 weeks were provided diets supplemented with 5% (wt/wt) freeze-dried black or red raspberries, strawberries, blueberries, noni, açaí, or wolfberry. The experimental diets were administered for 35 weeks. The study showed that all berry types were roughly equally effective in inhibiting NMBA-induced tumorigenesis in the rat esophagus, although the content of anthocyanins and other comparable phytochemicals varies significantly among these berries. Interestingly, in all cases, the levels of serum cytokines, interleukin 5 (IL-5), and GRO/KC (Growth Regulated Oncogene/Keratinocyte Chemoattractant), the rat homolog for human interleukin-8 (IL-8), were markedly reduced. These findings demonstrate that potential berry chemoprevention may be exerted through a variety of combinations, regarding the type and quantity of bioactive compounds. Therefore, this work underlined not only the importance of quality and synergism but also the importance of overall diet (as a pattern) concerning diet-associated health benefits.

Kresty *et al.* exhibited that cranberry proanthocyanidins inhibit esophageal adenocarcinoma (EAC) both *in vitro* and *in vivo* systems.[45] Particularly, purified cranberry-derived proanthocyanidin extract (C-PAC) was evaluated in its capacity to extend chemoprevention, utilizing acid-sensitive and acid-resistant human EAC cell lines and esophageal tumor xenografts in athymic NU/NU mice.[45] Results showed pleiotropic cell death induction and PI3K/Akt/mTOR axis inactivation upon *in vitro* and *in vivo* exposure to C-PAC, thus indicating a potential mode of action for C-PAC chemoprevention cross-systematically.

9.3.2.1.3 Breast Cancer. Breast cancer risk reduction *via* berry consumption has been exhibited by *in vivo* studies. The preventive and therapeutic potential of highbush blueberry was investigated in (August-Copenhagen-Irish) ACI female rats supplemented with 5% wt/wt blueberry, 2 weeks prior or 12 weeks post-administration of carcinogen 17β-estradiol (E2), correspondingly.[46] The tumor latency for palpable mammary tumors was delayed, while tumor volume and multiplicity were reduced in both intervention schemes. A lower dose of 2.5% blueberry diet dispensed post E2 treatment (*i.e.*, therapeutically) in *in vivo* experiments conducted by another experimental group also demonstrated attenuation of tumor multiplicity.[47] At 5% wt/wt of dietary supplementation, blueberry performed more effectively in reducing tissue proliferation, tumor burden, and downregulation of CYP1A1 expression, while black raspberry delayed tumor latency to a greater extent and downregulated ERα expression.[47]

Notably, a series of phytochemicals abundantly found in berries such as cyanidin, delphinidin, quercetin, kaempferol, ellagic acid, resveratrol, and pterostilbene have been shown by both *in vitro* and *in vivo* studies to interrelate with while also impede, fundamental pathways in breast cancer, as well as stimulate apoptosis and autophagy, therefore attenuating risk for breast cancer development and recurrence.[48]

9.3.2.1.4 Miscellaneous Cancers. According to mounting evidence, berries appear to be able to extend chemoprevention function in cancers beyond those of gastrointestinal (GI) and hormonal ones, which are the most extensively studied. A plethora of studies has exhibited that strawberry consumption in rodents yields chemoprevention properties in several cancers involving oral cavity, breast, lung, and esophageal.[49–53] Potential mechanisms for anti-cancer properties involve suppression of NF-κB, COX-2, and iNOS. Casto *et al.* investigated the potential of lyophilized strawberries to inhibit tumorigenesis on hamsters. Animals were painted three times a week for six weeks with 0.2% 7,12-dimethylbenz[*a*]anthracene

(DMBA) to induce oral cancer.[49] Hamsters were provided diets supplemented with 5% or 10% wt/wt lyophilized strawberries prior to, during, and post, or only post, carcinogen application treatment. At 12 weeks after DMBA initiation, animals were terminated, and numbers of lesions and tumors were counted.[49] Findings demonstrated significantly fewer lesions as well as tumors in the strawberry-supplemented diet groups compared to controls, signifying a potential supportive effect of strawberries in oral cancer. Utilizing the same animal model and carcinogen, Zhu *et al.* demonstrated that tumor incidence, multiplicity, volume, and histological grade of oral precancerous lesions declined significantly in hamsters fed a 5% wt/wt lyophilized strawberry supplemented diet compared to controls.[54] The authors supported that strawberries suppress cell proliferation, angiogenesis, and oncogenic signaling and arachidonic acid metabolism based on their findings.[54] Bishayee *et al.* demonstrated that in rats fed a diet supplemented with pomegranate extract (PE) (1 or 10 $g\,kg^{-1}$) 4 weeks before and 18 weeks following diethylnitrosamine (DENA)-initiated hepatocarcinogenesis, PE dose-dependently suppressed elevated inflammatory markers (COX-2, NF-κB).[55] Research by Balansky *et al.* yielded evidence supporting that berry extracts can extend chemoprevention against lung cancer. Whole-body Swiss ICR mice were exposed to mainstream cigarette smoke at birth and followingly daily for 4 months. Black chokeberry and strawberry aqueous extracts were provided to animals as the only source of drinking water, starting post-weaning and persistent for 7 months. Conclusively, both berry extracts significantly repressed lung adenomas' development.[51]

9.3.2.2 Human Trials on Berries and Cancer

There have been encouraging results obtained from human studies with black raspberry/berries considering the possibility of formulating berry-derived foods with consistent bioactive composition.[56] More recently, the China Kadoorie Biobank Study (462 342 participants and 17 894 deaths) published on fresh fruit intake and multiple causes of death revealing inverse associations with most specific causes of death, including ischemic heart disease; stroke (total, ischemic, and hemorrhagic); other cardiovascular diseases; cancers of the esophagus, stomach, and colorectum; chronic obstructive pulmonary disease; respiratory disease; all other major chronic diseases; and all other causes of death, as well as all causes of death.[57] No association was observed for lung or liver cancer or for transport accidents, and the latter could be considered as a negative control because one would not expect any association with deaths from transport accidents.

Regarding initiation and progression of cancers of the GI tract in humans, it is more plausible that whichever adverse effects of diet are primarily owing to over-consumption of energy, coupled with inadequate intakes of protective compounds, including micronutrients, dietary fiber, and several phytochemicals. Tumors of the esophagus, stomach, and colon, all are, according to epidemiological observations, partially preventable, if diets rich in fruits and vegetables are employed along with other positive lifestyle changes.[58] Therefore, the consumption of berries is sagacious toward risk reduction of colonic cancer. Berry phenolics, despite extensive metabolism and structural changes, seem to conserve their protective effects in relation to colon carcinogenesis.[59]

Knobloch *et al.* demonstrated that black raspberries, when administered as freeze-dried powder supplement in oral torches to oral cancer biopsy-confirmed patients, ameliorate gene expression oral cancer associated profile.[60] The berry supplemented diet group presented significant improvement of transcriptional biomarkers evaluated, specifically marked decline for the proinflammatory genes NFKB1, PTGS2, and pro-survival genes AURKA, BIRC5, and EGFR, all strongly associated with oral cancer risk and progression.[60]

Additionally, research conducted by Mallery and colleagues revealed that in oral intraepithelial neoplasia lesions diagnosed patients, topical application of bio-adhesive gels that contained 10% wt/wt freeze-dried black raspberries for 12 weeks ensued statistically significant diminution in lesion size, histological grades, and loss of heterozygosity events, compared to placebo administered controls.[61]

In a comprehensive review of cancer-focused clinical trials on the use of black raspberries and cancer outcomes, Kresty *et al.* reported that research investigating black raspberries in clinical settings demonstrates positive effects on preneoplastic lesions or cancers of the oral cavity, esophagus, and colon. Black raspberry treatment resulted in histologic regression of oral intraepithelial neoplasia associated with improved histologic grade and significantly reduced loss of heterozygosity at tumor suppressor gene loci, modulated genes linked to RNA processing and growth factor recycling; in the colon, black raspberry inhibited FAP (Fibroblast Activation Protein) associated polyp progression, demethylated tumor suppressor genes, and improved plasma cytokine profiles; in Barrett's patients, black raspberries consumption increased tissue levels of GST-Pi and decreased 8-isoprostane, a marker of lipid peroxidation/oxidative stress.[62] The same group conducted a phase I pilot study evaluating the effects of black raspberries on patients with Barrett's esophagus, aiming to investigate the effect of lyophilized black raspberries (LBR) on urinary metabolites and markers of lipid peroxidation, DNA damage, and tissue markers of cellular proliferation, detoxification, and inflammation. Results after 6 months on LBR treatment (32–45 g per day) showed improvement in urinary lipid peroxidation levels and significantly increased GST-Pi, a marker of detoxification, in BE epithelium.[63] Moreover, Sardo *et al.* reported on a randomized crossover trial of LBR on postprandial inflammation in older overweight males, indicating that consumption of lyophilized BRBs may attenuate postprandial inflammation in overweight or obese males consuming an HFHC meal.[64]

Furthermore, Rossi *et al.* in Italy analyzed a case-control human study of 230 patients and revealed a favorable role of flavonoids and proanthocyanidins for gastric cancer.[65] The same group during the 1991–2008 period conducted a cohort study in Northern Italy, investigating 326 cases of pancreatic cancer patients with respective controls, exploring the relationship between flavonoids and pancreatic cancer risk. Their findings demonstrated that dietary proanthocyanidins (predominantly found in apples, pears, cranberries, and pulses) might deliver a certain level of protection over pancreatic cancer risk.[66]

Overall, as the effects of diet and dietary patterns on cancer have been studied, there is no established specific and defined diet that reduces cancer risk. However, diets rich in fruits and vegetables have been repeatedly found to be beneficial in terms of reducing cancer risk; thus consumption of fruits and vegetables is recommended in this context. Clearly, all studies and recommendations assume a high standard of safety as per contaminants for fruits and vegetables that could potentially be found in foods as a side-effect of agriculture production practices. Interestingly, there has been a notable amount of evidence to suggest that adherence to a Mediterranean diet is associated with reduced risk of overall cancer mortality as well as reduced risk of incidence of several cancers, specifically those of the colorectum, GI tract, breast, stomach, pancreas, prostate, and liver, at least according to observational studies.[67,68] Notably, the Mediterranean diet is a diet rich in fiber, fruits, vegetables, and grains, as well as olive oil and fish, while from a constituents' perspective rich in vitamins, antioxidants, flavonoids, anthocyanins, and other phytochemicals and also ω-3 fatty acids. Moreover, markedly the Mediterranean diet is typically approached as a collection of practices and a lifestyle rather than a sheer diet. According to an interesting study, if fruit and vegetable intake increased to 300 and 400 g per day, respectively, such an increase could confer a significant reduction in gastric cancers, especially in developing countries.[69]

Even though the mechanisms remain not completely clarified, there is convincing evidence from epidemiological and experimental studies that dietary factors are likely to have a major influence on the risk of several types of cancer, particularly of the GI tract.[70–73] Experiments with *in vivo* models using mice showed that diet-induced obesity increases the risk of colonic cancer, while a signaling/modular implication of leptin (both a hormone and a cytokine) was also suggested by the data.[72,73] Finally, and in a more generalist's attitude, berries are demonstrated to extend a positive set of attributes toward health protection in cases of chronic diseases beyond cancer, whereby inflammation and oxidative stress are crucial instigators for disease onset.[74] Therefore, putting greater focus on the dietary constituents, as well as the whole food, while also on dietary patterns that extend health benefits and possible defense *vs.* chronic diseases, emerges as essential.

9.4 Bioactive Compounds in Berries and Cancer with Potential Mechanisms of Action Against Cancer

Polyphenols are widely distributed in plant species, consumption of which has been documented to negatively regulate the NF-κB signaling pathway. They depress the phosphorylation of kinases, inhibit NF-κB translocate into the nucleus, as well as interfere with interactions between NF-κB and DNA.

Through inhibition of NF-κB, polyphenols downregulate inflammatory cascade, induce apoptosis and decrease cell proliferation and metastasis.[13]

Polyphenols regulate immunity by interfering with immune cell regulation, proinflammatory cytokine synthesis, and gene expression. They inactivate NF-κB and modulate MAPK and arachidonic acids pathways. Polyphenolic compounds inhibit phosphatidylinositide 3-kinases/protein kinase B (PI3K/Akt), inhibitor of kappa kinase/c-Jun amino-terminal kinases (IKK/JNK), mammalian target of rapamycin complex 1 (mTORC1), which is a protein complex that controls protein synthesis, and JAK/STAT. They can suppress toll-like receptor (TLR) and proinflammatory gene expression. Their antioxidant activity and ability to inhibit enzymes involved in the production of eicosanoids contribute as well to their anti-inflammation properties. They inhibit certain enzymes involved in ROS production like xanthine oxidase and NADPH oxidase (NOX), while they upregulate other endogenous antioxidant enzymes, including superoxide dismutase (SOD), catalase, and glutathione (GSH) peroxidase (Px). Furthermore, they inhibit phospholipase A2 (PLA2), cyclooxygenase (COX), and lipoxygenase (LOX), leading to a reduction in the production of prostaglandins (PGs) and leukotrienes (LTs) while suppressing inflammation antagonism.

Gu *et al.* evaluated berry phenolic and volatile fractions toward anti-inflammatory effects using LPS-stimulated RAW264.7 macrophage cells by measuring levels of proinflammatory cytokines and the NF-κB signaling pathway.[14] Results showed that LPS-induced excessive production of nitric oxide (NO), PGE_2, COX-2, interleukin-6 (IL-6), and TNF-α, which were inhibited by berry phenolic and volatile extracts. Moreover, berry phenolic and volatile extracts reduced the nuclear translocation of NF-κB, by blocking the phosphorylation of p65 and degradation of IκBα (Inhibitor of kappa light chain gene enhancer in B cells – alpha). These findings showed that berry volatiles from six berries produced comparable anti-inflammatory effects to berry phenolics through the suppression of proinflammatory mediators and cytokines expression *via* NF-κB downregulation, despite being present in the fruit at a lower concentration.[14]

9.4.1 Free-radical Scavenging and Pro-oxidant Activity

The most commonly known and well-established ROS signaling event is *via* the respiratory chain. Respiration involves an electron-transfer chain reaction pathway, whereby the SOD enzymatic complex produces hydrogen peroxide (H_2O_2) in the inner mitochondrial space. Complexes I and III in mitochondrial space are where hydroxyl radical ($^{\bullet}OH$) is generated *via* a Fenton reaction ($H_2O_2 + Fe^{2+} \rightarrow {}^{\bullet}OH + OH^- + Fe^{3+}$). There are other pathways involved in the production of ROS ($O_2^{\bullet -}$), such as *via* an α-ketoglutarate dehydrogenase complex and by several oxidoreductases in the mitochondrion.[75,76] A similar mechanism involving the metabolism of H_2O_2 is also found in peroxisomes and lysosomes. Another pathway stimulating ROS generation is xenobiotic metabolism in the endoplasmic reticulum (ER).

In the ER, oxygen (O_2) initiates a reaction involving lipophilic substrates (such as fatty acids, FH) in the presence of a reducing agent, RH2, following the reaction scheme: $FH + O_2 + RH2 \rightarrow AOH + R + H_2O$, thus producing ROS in the microsomes. In both the ER and plasma membranes, lipid peroxidation occurs by means of the NADPH formation pathways. When NADPH oxidase is activated, considerable amounts of $O_2^{\bullet -}$ and H_2O_2 are generated. ROS produced is shown to induce cell proliferation at least *in vitro*.[77] A conceptual schematic on the potential action of berries to oxidative stress and inflammation is shown in Figure 9.1.

While anthocyanins are extensively metabolized *in vivo*, anthocyanins scavenge free radicals *via* mainly two hypothesized pathways. The first employs the attack of anthocyanin B-ring derived hydroxyl group(s), whereas the second prefigures the attack of oxonium ion on the C-ring.[75,76] While literature supports the fact that the number of hydroxyl groups on the B-ring of the anthocyanin structure positively affects the scavenging activity of the anthocyanin molecules,[78] there is not equal support on if/how the positive charge at the oxygen atom of the C-ring in the anthocyanin structure scavenges free radicals.[76]

Pro-oxidant activity can be initiated by certain conditions such as superoxide leakage. Specifically, the pro-oxidant activities of polyphenols in

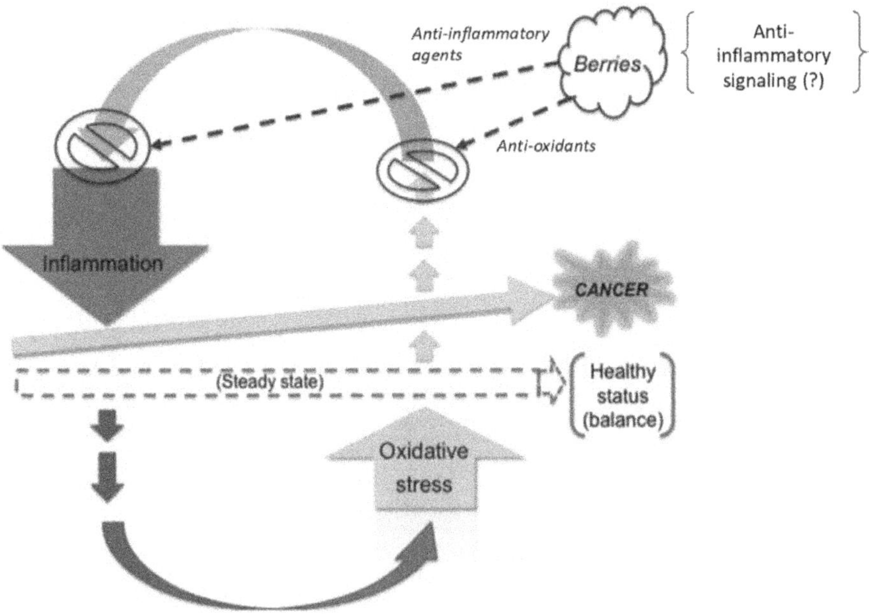

Figure 9.1 Schematic conceptual representation of berry antioxidant and anti-inflammatory action in preventing cancer. Reproduced from ref. 1, https://doi.org/10.3390/antiox5040037, under the terms of the CC BY 4.0 license, https://creativecommons.org/licenses/by/4.0/.

cancer cells can induce apoptosis,[79–81] and cell cycle arrest,[79] while inhibiting proliferation signaling pathways such as epidermal growth factor receptor/mitogen-activated protein kinase, phosphatidylinositide 3-kinases/protein kinase B, as well as NF-κB.[82]

9.4.2 The Cyclooxygenase Pathway

The COX-1, highly expressed in most tissues, enzyme is critical for the formation of thromboxane in blood platelets and for maintaining the integrity of gastrointestinal epithelium.[75] Furthermore, COX-1 is known for cell signaling involvement, endothelial cells–related angiogenesis regulation, and for contribution in maintaining tissue homeostasis. COX-2, also known as prostaglandin-endoperoxide synthase 2, is established as an essential enzyme participating in converting arachidonic acid to prostaglandin H2. COX-2 is overexpressed in benign polyps and adenocarcinomas while it appears to be involved in cell signaling and regulation of cell proliferation and apoptosis.[83]

9.4.3 Mitogen-activated Protein Kinase Pathway

MAPKs constitute protein kinases involved in cell survival, such as cell proliferation, differentiation, migration, and apoptosis. The types of known MAPKs are extracellular signal-regulated kinase 1 and 2 (ERK 1/2), ERK 5, JNK 1–3, and p38 protein isoforms (p38α, β, γ, and δ). Among the MAPKs, p38 MAPKs regulate the expression of many cytokines.[75]

Ho *et al.* showed that peonidin-3-glucoside downregulated ERK 1/2 expression in the H1299 cell line.[84] Afaq and colleagues demonstrated that anthocyanin-rich PE extended a positive inhibitory effect on UVB-mediated phosphorylation of MAPKs pathway in normal human epidermal keratinocytes.[85] Shin *et al.* showed *Vitis coignetiae* Pulliat extract to extend anti-cancer function.[86] The extract contained delphinidin-3,5-diglucoside, cyanidin-3,5-diglucoside, petunidin-3,5-diglucoside, delphinidin-3-glucoside, malvidin-3,5-diglucoside, peonidin-3,5-diglucoside, cyanidin-3-glucoside, petunidin-3-glucoside, peonidin-3-glucoside, and malvidin-3-glucoside and was demonstrated to induce apoptosis of HCT-116 cells by increasing phosphorylation of p38-MAPK and ERK, as well as by suppressing phosphorylation of Akt and JNK.[86] Additionally, recent studies have documented that a phenolic acid wild blueberry extract inhibited human umbilical vein endothelial cell (HUVEC) migration and angiogenesis *via* the PI3K pathway.[87]

9.4.4 Inflammatory Cytokines Signaling

Chronic inflammation is linked to the progression of excessive production of cytokines, inducing alterations in the cellular signaling pattern, in turn causing metabolic dysregulation leading to dysfunction and tumorigenesis. Flavonoids and anthocyanins reduce inflammation through a variety of mechanisms.[88]

In a recent systematic review of the evidence published in 2021 while considering 104 publications, Land Lail *et al.* concluded that the primary mechanisms mediating the anti-inflammatory effects of berries include a reduction in NF-κB signaling possibly secondary to reduced oxidative stress, a downregulation of TLR4 signaling, and an increase in Nrf2, while some studies additionally implicate a downregulation of the NLRP3 inflammasome in adipose tissue macrophages.[89] Berries contain a number of factors that could potentially elicit health benefits, including polysaccharides, fiber, vitamins, minerals, polyphenols, and other phytonutrients. However, since in most of the studies the authors considered utilized freeze-dried berry powders, it is not possible to identify which bioactive compound(s) is/are responsible for the effects of the berries reported in every study. While synergism is most likely an essential requirement, it still becomes challenging to identify potential optimal profiles and doses within mixtures of bioactives. A summary of berry bioactive anti-cancer activity and mode of action is presented in Table 9.1.

Interestingly, there is a number of medications containing polyphenols that are FDA-approved as pharmaceutical medicines. Polyphenon® E, a standardized green tea polyphenol preparation, is an FDA-approved medication to treat genital warts. Another example is the FDA-approved Crofelemer (a medication rich in oligomeric proanthocyanidin) used to manage HIV-associated non-infectious diarrhea.[89] These commercial products and efforts are indicative of the significant traction berry-derived bioactives carry and how the corporate world is taking significant endeavors to capitalize on such potential.

9.5 Dietary Recommendations and Controversies

Increasing evidence, as well as biological plausibility, indicates that bioactives act synergistically[90] while, if further verified, such mode of action could clarify why supplements with one or a few antioxidants appear less effective in extending health benefits regarding chronic disease prevention compared to whole foods. Intervention studies utilizing high doses of fruits and vegetables alone[90] or in combination with additional lifestyle changes[91] have demonstrated that fruit and vegetable intake influences gene expression, in a mode, which fortifies cellular stress-defense[90] and modulation of tumorigenesis, protein metabolism and modification, intracellular protein trafficking, and protein phosphorylation.[91] In screening studies evaluating the total antioxidant content of various foods, it was shown that berries, walnuts, pecans, sunflower seeds, and spices were among those foods with the highest content of antioxidants.[92,93] In meta-analyses of prospective studies, inverse associations were observed between intake of fruits, vegetables, whole grains, and nuts and the risk of coronary artery disease, stroke, cardiovascular disease overall, total cancer, and all-cause mortality.[94] The most prominent decrease in risk was observed with an intake of 800 g per day for fruits and vegetables, 225 g per day for whole grains, and 15–20 g per day for nuts, respectively.[94] It is unclear and highly debatable as to what degree the antioxidant content of different plant foods contributes to the observed health benefits extended.[94]

Table 9.1 Summarized proposed mechanisms of anti-cancer action of berry constituents/bioactives.[a]

Mechanism(s) employed	Mode of action[bc]	Bioactive compound(s)
Antioxidative action	ROS sequestration ↑ GSH	Epigallocatechin-3-gallate (EGCG)
Effects on enzymes of phase I and II	↓ CYP1A1 ↓ LDH	Quercetin, kaempferol
	↑ UDPGT	Ellagic acid
	↑ NQO	Chlorogenic acid
Cell-cycle arrest	↓ cyclin D, E ↓ CDK 1, 2, 4 ↓ PCNA	EGCG
	↑ cyclin E ↓ cyclin A, B1	Ellagic acid
Apoptosis	↑ ROS in cancer cells ↑ caspase-3, -7, -8 ↑ cytochrome c ↑ Blc-2, Blc-X_L	EGCG
	↑ caspase-3, -7, -9 ↑ cytochrome c ↑ PARP cleavage	Quercetin
Anti-proliferation/Anti-survival	↓ GFR/Ras/MAPK & PI3K/Akt ↓ c-fos ↓ erg-1 ↓ PI3K ↓ ERK ↓ Akt phosphorylation ↓ NF-κB	EGCG
Anti-inflammatory action	↓ COX-1 ↓ COX-2	Gallic acid
	↓ TNF-α ↓ COX-2	EGCG
Anti-angiogenesis	↓ VEGF ↓ PDGF ↓ HIF-1α	EGCG, anthocyanin berry extracts
Metastasis inhibition	↓ MMP-9 ↓ mRNA stabilizing factor HuR	EGCG
Cell adhesion and movement inhibition	↓ MRLC phosphorylation ↓ Vimentin phosphorylation	EGCG

[a]**Abbreviations:** Akt/PKB, protein kinase B; CDK, cyclin-dependent kinase; COX, cyclooxygenase; CYP, cytochrome P450; ERK, extracellular regulated kinase; GFR, growth factor receptors; GSH, glutathione; HIF-1α, hypoxia-inducible factor 1α; LDH, lactate dehydrogenase; MAPK, mitogen-activated protein kinase; MMP-9, matrix metallopeptidase 9; MRLC, myosin II regulatory light chain (protein); NF-κB, nuclear factor-kappa (κ) B; NQO, NADPH quinone oxidoreductase; PARP, poly-ADP ribose polymerase; PCNA, proliferating cell nuclear antigen; PDGF, platelet-derived growth factor; PI3K, phosphatidylinositol-3-kinase; ROS, reactive oxygen species; TNF-α, tumor necrosis factor-alpha (α); UDPGT, UDP-glucuronosyl transferase; VEGF, vascular endothelial growth factor.

[b]Upwards vector, (↑): indicates induction/increase.

[c]Downwards vector, (↓): indicates suppression/decrease.

In a comprehensive systematic review and meta-analysis of 95 studies (142 publications), Dagfinn Aune, found 0.97 (95% CI: 0.95, 0.99, I2 = 49%, $n = 12$) for total cancer, and 0.90 (95% CI: 0.87, 0.93, I2 = 83%, $n = 15$) for all-cause mortality per 200 g per day of fruit and vegetable intake (1 serving = 80 g). These findings indicate that there is a positive relationship consistently seen between intake of fruits and vegetables and reduced risk/mortality from cancer.[94] Similar findings pertaining to other chronic metabolic diseases such as cardiovascular disease and type 2 diabetes mellitus were obtained.[94] Analogous associations were still detected for fruits and vegetables even when evaluated separately.[94]

In the EPIC study, inverse associations were observed between fruit intake and mortality from digestive diseases (HR: 0.77; 95% CI: 0.66, 1.00) or unknown causes of death (HR: 0.88; 95% CI: 0.81, 0.97).[94] Other studies such as the PURE study, which is a global cohort study with 135 335 participants and 5796 deaths and with data from 18 low-, middle-, and high-income countries in North America, Europe, South America, the Middle East, South Asia, China, Southeast Asia, and Africa, also recently published on fruit, vegetable, and legume intake and risk of cardiovascular disease and mortality, so not specific data toward cancer-related mortality were produced.[94] The PURE study investigated all-cause mortality, however, whereby cancer is also included, with the lowest risk observed at 3 to <4 servings per day for all-cause mortality (HR: 0.78; 0.69, 0.88) (1 serving was defined as 125 g per day, which is somewhat higher than the 80 g that had been used as a serving size previously).[94] It is, therefore, again challenging to identify a particular dose of fruits and vegetables, let alone berries, regarding optimal anti-cancer defense.

The American Academy of Nutrition and Dietetics recommends including fruit at every meal, which is a qualitative suggestion. Especially for children, the amount of fruit per day children need depends on their age, size, and activity level. According to MyPlate, young kids (2–3 years) need one cup per day of fruit. Older children (4–13 years) need up to $1\frac{1}{2}$ cups per day. Most teenagers also need $1\frac{1}{2}$ cups per day. Adolescent boys may need more—usually approximately two cups of fruit per day.[95] In its 2020–2025 Dietary Guidelines for Americans, the Academy of Nutrition and Dietetics recommends half of the daily plate to be fruits and vegetables.[96] The guidelines also recommend the choice of whole, cut, or pureed fruits – fresh, frozen, dried, or canned in 100% juice.[96] According to the guidelines: "About 80 percent of the U.S. population does not meet fruit recommendations. Most people would benefit from increasing their intake of fruit, mostly as whole fruits in nutrient-dense forms".[96] There are not, however, any specific recommendations currently for any fruit in particular, including berries.

Another important concept as per the effects of berries and their bioactives in regard to chronic diseases, including cancer, lies with the microbiota and its relation to immunity.[97] Emerging research suggests that diets rich in dietary fiber extend significant effects on the microbiota and human health, while plant-based foods specifically constitute notable sources of bioactive compounds and dietary fiber, which are in turn metabolized by microbes to produce different beneficial metabolites.[98] Moreover, dietary anthocyanins undergo complex metabolism following ingestion and interact with endogenous and microbial enzymes, leading to the production of a plethora of circulating and excreted anthocyanin metabolites and catabolic products.[99] Additionally, as the prebiotic-like function of polyphenols on microbiota is becoming more evident, the potential use of polyphenols in preventive/therapeutic strategies for gastrointestinal disorders through a microbiota-related function is recently emerging.[100] While this is still a most novel field, it constitutes an interesting and promising approach, whereby it is postulated that polyphenols can extend health benefits *via* positive microbiome modulation.[101] Interestingly, as the absorption of polyphenols present in food depends mostly on the activity of intestinal microflora,[102] ways to improve the bioavailability and bio-functionality of polyphenols may prove crucial for promoting health,[103] including reducing the risk of cancer. In this context, novel nutrient-delivery products specifically targeting the colon are being designed and developed[104] for gut and subsequent metabolic positive modulation. These compounds may include dietary polyphenols, which have been shown promise as adjuvants in the improvement of colorectal cancer.[105] These approaches present opportunities for creative innovation in the approach of cancer prevention.

9.6 Conclusions

While there is a most significant amount of evidence to support that berry consumption is associated with a lower risk for cancer, there are some aspects of this relationship that are unclear. It remains elusive what are the constituents of berries and specific properties that confer the seemingly positive health benefit. Much emphasis has been placed on a variety of bioactive compounds such as anthocyanins and polyphenols; however, it remains unclear what are the exact compounds, at what doses, and in what combinations that confer the observed and desired results. Another point of uncertainty is the mechanism of action at the molecular and biochemical levels. While ROS scavenging, induction of apoptosis, and cell cycle modulation, are some of the demonstrated mechanisms of action, there appears to be a plethora of potential biochemical mechanisms and pathways involved. Additionally, while there is a fair number of recent studies of *in vitro* and *in vivo* models, there is a significant lack of human trials of berry intake and cancer outcomes. This limitation renders it more challenging to establish potential dietary recommendations for berry consumption, while a commonly accepted evaluation/assessment system regarding berry effectiveness, potency, and efficiency in terms of positively impacting health outcomes is also missing. More research is warranted in order to address some of the aforementioned points, especially when thinking toward establishing strategies and schemes of preventive and/or therapeutic nature.

Abbreviations

AOM	azoxymethane
AP-1	activator protein-1
AHRQ	Agency for Healthcare Research and Quality

BMI	body mass index
CDK	cyclin-dependent kinase
COX-2	cyclooxygenase-2
CRP	C-reactive peptide
CXCR4	CXC chemokine receptor 4
GSH	glutathione
HIV/AIDS	human immunodeficiency virus/acquired immunodeficiency syndrome
IKK/JNK	inhibitor of kappa kinase/c-Jun amino-terminal kinases
IL-1	interleukin-1
iNOS	inducible nitric oxide synthase
LDH	lactate dehydrogenase
LOX	lipoxygenase
LTs	leukotrienes
MAPK	mitogen-activated protein kinase
mTORC1	mammalian target of rapamycin complex 1
NF-κB	nuclear factor kappa B
NMBA	nitrosomethylbenzylamine
NOX	NADPH oxidase
Nrf2	NF-E2 related factor-2
Px	peroxidase
PI3K/Akt	phosphatidylinositide-3-kinases/protein kinase/Akt
PG	prostaglandins
PLA2	phospholipase A2
PPAR-γ	peroxisome proliferator-activated receptor-gamma
ROS	reactive oxygen species
RNS	reactive nitrogen species
SOD	superoxide dismutase
TLR	Toll-like receptor
TNF-α	tumor necrosis factor-alpha
VEGF	vascular endothelial growth factor
WHO	World Health Organization

Acknowledgements

The authors would like to kindly acknowledge the support provided by librarian Ms. Leyla Cabugos (she/her/hers), from California Polytechnic State University, Robert E. Kennedy Library, with distance access and interlibrary loans administered during the SARS-CoV-2/COVID-19 pandemic.

References

1. A. S. Kristo, D. Klimis-Zacas and A. K. Sikalidis, Protective Role of Dietary Berries in Cancer, *Antioxidants*, 2016, **5**(4), 37.

2. G. W. Prager, S. Braga and B. Bystricky, *et al.*, Global cancer control: responding to the growing burden, rising costs and inequalities in access, *ESMO Open.*, 2018, **3**(2), e000285.
3. Ferlay J, Ervik M, Lam F, Colombet M, Mery L, Piñeros M, et al. Global Cancer Observatory: Cancer Today. World Health Organization, Lyon: International Agency for Research on Cancer; 2020. https://gco.iarc.fr/.
4. https://www.cancer.org/cancer/cancer-basics/economic-impact-of-cancer.html.
5. "cancer/ Origin and meaning of cancer by Online Etymology Dictionary" www.etymoline.com.
6. *World Cancer Report*, ed. B. W. Stewart and C. P. Wild, World Health Organization, Geneva, Switzerland, 2014.
7. D. Aune, E. Giovannucci, P. Boffetta, L. T. Fadnes, N. Keum, T. Norat, D. C. Greenwood, E. Riboli, L. J. Vatten and S. Tonstad, Fruit and vegetable intake and the risk of cardiovascular disease, total cancer and all-cause mortality: a systematic review and dose-response meta-analysis of prospective studies, *Int. J. Epidemiol.*, 2017, **46**(3), 1029–1056.
8. E. Piccand, P. Vollenweider, I. Guessous and P. Marques-Vidal, Association between dietary intake and inflammatory markers: results from the CoLaus study, *Public Health Nutr.*, 2019, **22**(3), 498–505.
9. M. J. Hart, S. J. Torres, S. A. McNaughton and C. M. Milte, Dietary patterns and associations with biomarkers of inflammation in adults: a systematic review of observational studies, *Nutr J*, 2021, **20**(1), 24.
10. C. S. C. Yip, W. Chan and R. Fielding, The Associations of Fruit and Vegetable Intakes with Burden of Diseases: A Systematic Review of Meta-Analyses, *J. Acad. Nutr. Diet.*, 2019, **119**(3), 464–481.
11. E. A. Golemis, P. Scheet, T. N. Beck, E. M. Scolnick, D. J. Hunter, E. Hawk and N. Hopkins, Molecular mechanisms of the preventable causes of cancer in the United States, *Genes Dev.*, 2018, **32**(13–14), 868–902.
12. M. Gómez de Cedrón, S. Wagner, M. Reguero, A. Menéndez-Rey and A. R. de Molina, Miracle Berry as a Potential Supplement in the Control of Metabolic Risk Factors in Cancer, *Antioxidants*, 2020, **9**(12), 1282.
13. National Cancer Institute. What is Cancer? Defining Cancer. 2012.
14. M. L. de Castro Barbosa, R. A. da Conceicao, A. G. M. Fraga, B. D. Camarinha, G. C. de Carvalho Silva, A. G. F. Lima, E. A. Cardoso and V. de Oliveira Freitas Lione, NF-κB Signaling Pathway Inhibitors as Anticancer Drug Candidates, *Anti-Cancer Agents Med. Chem.*, 2017, **17**(4), 483–490.
15. L. Zheng, Y. Cai, L. Zhou, P. Huang, X. Ren, A. Zuo, X. Meng, M. Xu and X. Liao, Benzoquinone from Fusarium pigment inhibits the proliferation of estrogen receptor-positive MCF-7 cells through the NF-κB pathway via estrogen receptor signaling, *Int. J. Mol. Med.*, 2017, **39**(1), 39–46.
16. H. Khan, H. Ullah, P. C. M. F. Castilho, A. S. Gomila, G. D'Onofrio, R. Filosa, F. Wang, S. M. Nabavi, M. Daglia, A. S. Silva, K. R. R. Rengasamy, J. Ou, X. Zou, J. Xiao and H. Cao, Targeting NF-κB signaling pathway in

cancer by dietary polyphenols, *Crit. Rev. Food Sci. Nutr.*, 2020, **60**(16), 2790–2800.
17. I. Gu, C. Brownmiller, N. B. Stebbins, A. Mauromoustakos, L. Howard and S. O. Lee, Berry Phenolic and Volatile Extracts Inhibit Pro-Inflammatory Cytokine Secretion in LPS-Stimulated RAW264.7 Cells through Suppression of NF-κB Signaling Pathway, *Antioxidants*, 2020, **9**(9), 871.
18. Z. Zhang, L. Zhou, N. Xie, E. C. Nice, T. Zhang, Y. Cui and C. Huang, Overcoming cancer therapeutic bottleneck by drug repurposing, *Signal Transduction Targeted Ther.*, 2020, **5**(1), 113.
19. Z. J. Yang, C. E. Chee, S. Huang and F. A. Sinicrope, The role of autophagy in cancer: therapeutic implications, *Mol. Cancer Ther.*, 2011, **10**(9), 1533–1541.
20. M. Hickey and C. King, *The Cambridge Illustrated Glossary of Botanical Terms*, 1st edn, 2001, Cambridge University Press, Cambridge, UK.
21. R. A. Barrowclough, The effect of berry consumption on cancer risk, *J. Nutri. Health Food Eng.*, 2015, **2**(1), 1–9.
22. H. King, Let Food Be Thy Medicine, *Hippocrates Now: The 'Father of Medicine' in the Internet Age*, Bloomsbury Academic, London, 2020, pp. 111–132. Bloomsbury Collections.
23. P. B. Hamel and M. Chiltoskey, *Cherokee Plants and Their Uses: A 400-year History*, Herald Publishing Co., Sylva, N.C, 1975.
24. K. A. Steinmetz and J. D. Potter, Vegetables, fruit, and cancer. I. Epidemiology, *Cancer, Causes Control*, 1991, **2**(5), 325–357.
25. C. C. Neto, Cranberry and blueberry: Evidence for protective effects against cancer and vascular diseases, *Mol. Nutr. Food Res.*, 2007, **51**(6), 652–664.
26. G. D. Stoner, L. S. Wang and B. C. Casto, Laboratory and clinical studies of cancer chemoprevention by antioxidants in berries, *Carcinogenesis*, 2008, **29**(9), 1665–1674.
27. N. Shi, S. K. Clinton, Z. Liu, Y. Wang, K. M. Riedl, S. J. Schwartz, X. Zhang, Z. Pan and T. Chen, Strawberry phytochemicals inhibit azoxymethane/dextran sodium sulfate-induced colorectal carcinogenesis in Crj: CD-1 mice, *Nutrients*, 2015, **7**(3), 1696–1715.
28. D. Cooke, M. Schwarz, D. Boocock, P. Winterhalter, W. P. Steward, A. J. Gescher and T. H. Marczylo, Effect of cyanidin-3-glucoside and an anthocyanin mixture from bilberry on adenoma development in the ApcMin mouse model of intestinal carcinogenesis–relationship with tissue anthocyanin levels, *Int. J. Cancer*, 2006, **119**(9), 2213–2220.
29. S. Thomasset, D. P. Berry, H. Cai, K. West, T. H. Marczylo, D. Marsden, K. Brown, A. Dennison, G. Garcea, A. Miller, D. Hemingway, W. P. Steward and A. J. Gescher, Pilot study of oral anthocyanins for colorectal cancer chemoprevention, *Cancer Prev. Res.*, 2009, **2**(7), 625–633.
30. X. Bi, W. Fang, L. S. Wang, G. D. Stoner and W. Yang, Black raspberries inhibit intestinal tumorigenesis in apc1638+/− and Muc2−/− mouse models of colorectal cancer, *Cancer Prev. Res.*, 2010, **3**(11), 1443–1450.

31. L. S. Wang, C. T. Kuo, T. H. Huang, M. Yearsley, K. Oshima, G. D. Stoner, J. Yu, J. F. Lechner and Y. W. Huang, Black raspberries protectively regulate methylation of Wnt pathway genes in precancerous colon tissue, *Cancer Prev. Res.*, 2013, **6**(12), 1317–1327.
32. P. Pan, C. W. Skaer, H. T. Wang, S. M. Stirdivant, M. R. Young, K. Oshima, G. D. Stoner, J. F. Lechner, Y. W. Huang and L. S. Wang, Black raspberries suppress colonic adenoma development in ApcMin/+ mice: relation to metabolite profiles, *Carcinogenesis*, 2015, **36**(10), 1245–1253.
33. M. Misikangas, A. M. Pajari, E. Päivärinta, S. I. Oikarinen, J. Rajakangas, M. Marttinen, H. Tanayama, R. Törrönen and M. Mutanen, Three Nordic berries inhibit intestinal tumorigenesis in multiple intestinal neoplasia/+ mice by modulating beta-catenin signaling in the tumor and transcription in the mucosa, *J. Nutr.*, 2007, **137**(10), 2285–2290.
34. J. Rajakangas, M. Misikangas, E. Päivärinta and M. Mutanen, Chemoprevention by white currant is mediated by the reduction of nuclear beta-catenin and NF-kappaB levels in Min mice adenomas, *Eur. J. Nutr.*, 2008, **47**(3), 115–122.
35. G. K. Harris, A. Gupta, R. G. Nines, L. A. Kresty, S. G. Habib, W. L. Frankel, K. LaPerle, D. D. Gallaher, S. J. Schwartz and G. D. Stoner, Effects of lyophilized black raspberries on azoxymethane-induced colon cancer and 8-hydroxy-2′-deoxyguanosine levels in the Fischer 344 rat, *Nutr. Cancer*, 2001, **40**(2), 125–133.
36. R. Pierini, J. M. Gee, N. J. Belshaw and I. T. Johnson, Flavonoids and intestinal cancers, *Br. J. Nutr.*, 2008, **99**(E Suppl 1), ES53–ES59.
37. G. D. Stoner, A. A. Dombkowski, R. K. Reen, D. Cukovic, S. Salagrama, L. S. Wang and J. F. Lechner, Carcinogen-altered genes in rat esophagus positively modulated to normal levels of expression by both black raspberries and phenylethyl isothiocyanate, *Cancer Res.*, 2008, **68**(15), 6460–6467.
38. L. S. Wang, S. S. Hecht, S. G. Carmella, N. Yu, B. Larue, C. Henry, C. McIntyre, C. Rocha, J. F. Lechner and G. D. Stoner, Anthocyanins in black raspberries prevent esophageal tumors in rats, *Cancer Prev. Res.*, 2009, **2**(1), 84–93.
39. D. S. Peiffer, L. S. Wang, N. P. Zimmerman, B. W. Ransom, S. G. Carmella, C. T. Kuo, J. H. Chen, K. Oshima, Y. W. Huang, S. S. Hecht and G. D. Stoner, Dietary Consumption of Black Raspberries or Their Anthocyanin Constituents Alters Innate Immune Cell Trafficking in Esophageal Cancer, *Cancer Immunol. Res.*, 2016, **4**(1), 72–82.
40. L. S. Wang, A. A. Dombkowski, C. Seguin, C. Rocha, D. Cukovic, A. Mukundan, C. Henry and G. D. Stoner, Mechanistic basis for the chemopreventive effects of black raspberries at a late stage of rat esophageal carcinogenesis, *Mol. Carcinog.*, 2011, **50**(4), 291–300.
41. T. Chen, H. Hwang, M. E. Rose, R. G. Nines and G. D. Stoner, Chemopreventive properties of black raspberries in N-nitrosomethylbenzylamine-induced rat esophageal tumorigenesis: down-regulation of

cyclooxygenase-2, inducible nitric oxide synthase, and c-Jun, *Cancer Res.*, 2006, **66**(5), 2853–2859.
42. T. Chen, M. E. Rose, H. Hwang, R. G. Nines and G. D. Stoner, Black raspberries inhibit N-nitrosomethylbenzylamine (NMBA)-induced angiogenesis in rat esophagus parallel to the suppression of COX-2 and iNOS, *Carcinogenesis*, 2006, **27**(11), 2301–2307.
43. R. K. Reen, R. Nines and G. D. Stoner, Modulation of N-nitrosomethylbenzylamine metabolism by black raspberries in the esophagus and liver of Fischer 344 rats, *Nutr. Cancer*, 2006, **54**(1), 47–57.
44. G. D. Stoner, L. S. Wang, C. Seguin, C. Rocha, K. Stoner, S. Chiu and A. D. Kinghorn, Multiple berry types prevent N-nitrosomethylbenzylamine-induced esophageal cancer in rats, *Pharm. Res.*, 2010, **27**(6), 1138–1145.
45. L. A. Kresty, K. M. Weh, B. Zeyzus-Johns, L. N. Perez and A. B. Howell, Cranberry proanthocyanidins inhibit esophageal adenocarcinoma in vitro and in vivo through pleiotropic cell death induction and PI3K/AKT/mTOR inactivation, *Oncotarget*, 2015, **6**(32), 33438–33455.
46. J. Jeyabalan, F. Aqil, R. Munagala, L. Annamalai, M. V. Vadhanam and R. C. Gupta, Chemopreventive and therapeutic activity of dietary blueberry against estrogen-mediated breast cancer, *J. Agric. Food Chem.*, 2014, **62**(18), 3963–3971.
47. S. Ravoori, M. V. Vadhanam, F. Aqil and R. C. Gupta, Inhibition of estrogen-mediated mammary tumorigenesis by blueberry and black raspberry, *J. Agric. Food Chem.*, 2012, **60**(22), 5547–5555.
48. H. S. Aiyer, A. M. Warri, D. R. Woode, L. Hilakivi-Clarke and R. Clarke, Influence of berry polyphenols on receptor signaling and cell-death pathways: implications for breast cancer prevention, *J. Agric. Food Chem.*, 2012, **60**(23), 5693–5708.
49. B. C. Casto, T. J. Knobloch, R. L. Galioto, Z. Yu, B. T. Accurso and B. M. Warner, Chemoprevention of oral cancer by lyophilized strawberries, *Anticancer Res.*, 2013, **33**(11), 4757–4766.
50. R. R. Somasagara, M. Hegde, K. K. Chiruvella, A. Musini, B. Choudhary and S. C. Raghavan, Extracts of strawberry fruits induce intrinsic pathway of apoptosis in breast cancer cells and inhibits tumor progression in mice, *PLoS One*, 2012, **7**(10), e47021.
51. R. Balansky, G. Ganchev, M. Iltcheva, M. Kratchanova, P. Denev, C. Kratchanov, K. Polasa, F. D'Agostini, V. E. Steele and S. De Flora, Inhibition of lung tumor development by berry extracts in mice exposed to cigarette smoke, *Int. J. Cancer*, 2012, **131**(9), 1991–1997.
52. P. S. Carlton, L. A. Kresty, J. C. Siglin, M. A. Morse, J. Lu, C. Morgan and G. D. Stoner, Inhibition of N-nitrosomethylbenzylamine-induced tumorigenesis in the rat esophagus by dietary freeze-dried strawberries, *Carcinogenesis*, 2001, **22**(3), 441–446.
53. T. Chen, F. Yan, J. Qian, M. Guo, H. Zhang, X. Tang, F. Chen, G. D. Stoner and X. Wang, Randomized phase II trial of lyophilized strawberries in patients with dysplastic precancerous lesions of the esophagus, *Cancer Prev. Res.*, 2012, **5**(1), 41–50.

54. X. Zhu, L. Xiong, X. Zhang and T. Chen, Lyophilized strawberries prevent 7,12-dimethylbenz[α]anthracene (DMBA)-induced oral squamous cell carcinogenesis in hamsters, *J. Funct. Foods*, 2015, **15**, 476–486.
55. A. Bishayee, R. J. Thoppil, A. S. Darvesh, V. Ohanyan, J. G. Meszaros and D. Bhatia, Pomegranate phytoconstituents blunt the inflammatory cascade in a chemically induced rodent model of hepatocellular carcinogenesis, *J. Nutr. Biochem.*, 2013, **24**(1), 178–187.
56. E. M. Brown, G. J. McDougall, D. Stewart, G. Pereira-Caro, R. González-Barrio, P. Allsopp, P. Magee, A. Crozier, I. Rowland and C. I. Gill, Persistence of anticancer activity in berry extracts after simulated gastrointestinal digestion and colonic fermentation, *PLoS One*, 2012, **7**(11), e49740.
57. H. Du, L. Li, D. Bennett, L. Yang, Y. Guo, T. J. Key, Z. Bian, Y. Chen, R. G. Walters, I. Y. Millwood, J. Chen, J. Wang, X. Zhou, L. Fang, Y. Li, X. Li, R. Collins, R. Peto and Z. Chen, China Kadoorie Biobank study. Fresh fruit consumption and all-cause and cause-specific mortality: findings from the China Kadoorie Biobank, *Int. J. Epidemiol.*, 2017, **46**(5), 1444–1455.
58. I. T. Johnson, New approaches to the role of diet in the prevention of cancers of the alimentary tract, *Mutat. Res.*, 2004, **551**(1–2), 9–28.
59. Z. Qian, Z. Wu, L. Huang, H. Qiu, L. Wang, L. Li, L. Yao, K. Kang, J. Qu, Y. Wu, J. Luo, J. J. Liu, Y. Yang, W. Yang and D. Gou, Mulberry fruit prevents LPS-induced NF-κB/pERK/MAPK signals in macrophages and suppresses acute colitis and colorectal tumorigenesis in mice, *Sci. Rep.*, 2015, **5**, 17348.
60. T. J. Knobloch, L. K. Uhrig, D. K. Pearl, B. C. Casto, B. M. Warner, S. K. Clinton, C. L. Sardo-Molmenti, J. M. Ferguson, B. T. Daly, K. Riedl, S. J. Schwartz, Y. Vodovotz, A. J. Buchta, D. E. Schuller, E. Ozer, A. Agrawal and C. M. Weghorst, Suppression of Proinflammatory and Prosurvival Biomarkers in Oral Cancer Patients Consuming a Black Raspberry Phytochemical-Rich Troche, *Cancer Prev. Res.*, 2016, **9**(2), 159–171.
61. S. R. Mallery, M. Tong, B. S. Shumway, A. E. Curran, P. E. Larsen, G. M. Ness, K. S. Kennedy, G. H. Blakey, G. M. Kushner, A. M. Vickers, B. Han, P. Pei and G. D. Stoner, Topical application of a mucoadhesive freeze-dried black raspberry gel induces clinical and histologic regression and reduces loss of heterozygosity events in premalignant oral intraepithelial lesions: results from a multicentered, placebo-controlled clinical trial, *Clin. Cancer Res.*, 2014, **20**(7), 1910–1924.
62. L. A. Kresty, S. R. Mallery and G. D. Stoner, Black raspberries in cancer clinical trials: Past, present and future, *J. Berry Res.*, 2016, **6**(2), 251–261.
63. L. A. Kresty, J. J. Fromkes, W. L. Frankel, C. D. Hammond, N. P. Seeram, M. Baird and G. D. Stoner, A phase I pilot study evaluating the beneficial effects of black raspberries in patients with Barrett's esophagus, *Oncotarget*, 2016, **9**(82), 35356–35372.
64. C. L. Sardo, J. P. Kitzmiller, G. Apseloff, R. B. Harris, D. J. Roe, G. D. Stoner and E. T. Jacobs, An Open-Label Randomized Crossover Trial of Lyophilized Black Raspberries on Postprandial Inflammation in Older Overweight Males: A Pilot Study, *Am. J. Ther.*, 2016, **23**(1), e86–e91.

65. M. Rossi, V. Rosato, C. Bosetti, P. Lagiou, M. Parpinel, P. Bertuccio, E. Negri and C. La Vecchia, Flavonoids, proanthocyanidins, and the risk of stomach cancer, *Cancer, Causes Control*, 2010, **21**(10), 1597–1604.
66. M. Rossi, A. Lugo, P. Lagiou, A. Zucchetto, J. Polesel, D. Serraino, E. Negri, D. Trichopoulos and C. La Vecchia, Proanthocyanidins and other flavonoids in relation to pancreatic cancer: a case-control study in Italy, *Ann. Oncol.*, 2012, **23**(6), 1488–1493.
67. L. Schwingshackl and G. Hoffmann, Does a Mediterranean-Type Diet Reduce Cancer Risk?, *Curr. Nutr. Rep.*, 2016, **5**, 9–17.
68. G. Grosso, S. Buscemi, F. Galvano, A. Mistretta, S. Marventano, V. La Vela, F. Drago, S. Gangi, F. Basile and A. Biondi, Mediterranean diet and cancer: epidemiological evidence and mechanism of selected aspects, *BMC Surg.*, 2013, **13**(Suppl 2), S14.
69. B. Peleteiro, P. Padrão, C. Castro, A. Ferro, S. Morais and N. Lunet, Worldwide burden of gastric cancer in 2012 that could have been prevented by increasing fruit and vegetable intake and predictions for 2025, *Br. J. Nutr.*, 2016, **115**(5), 851–859.
70. A. K. Sikalidis, Amino acids and immune response: a role for cysteine, glutamine, phenylalanine, tryptophan and arginine in T-cell function and cancer?, *Pathol. Oncol. Res.*, 2015, **21**(1), 9–17.
71. A. K. Sikalidis and B. Varamini, Roles of hormones and signaling molecules in describing the relationship between obesity and colon cancer, *Pathol. Oncol. Res.*, 2011, **17**(4), 785–790.
72. A. K. Sikalidis, M. D. Fitch and S. E. Fleming, Risk of colonic cancer is not higher in the obese Lep(ob) mouse model compared to lean littermates, *Pathol. Oncol. Res.*, 2013, **19**(4), 867–874.
73. A. K. Sikalidis, M. D. Fitch and S. E. Fleming, Diet induced obesity increases the risk of colonic tumorigenesis in mice, *Pathol. Oncol. Res.*, 2013, **19**(4), 657–666.
74. A. S. Kristo, N. R. Matthan and A. H. Lichtenstein, Effect of diets differing in glycemic index and glycemic load on cardiovascular risk factors: review of randomized controlled-feeding trials, *Nutrients*, 2013, **5**(4), 1071–1080.
75. B. Baby, P. Antony and R. Vijayan, Antioxidant and anticancer properties of berries, *Crit. Rev. Food Sci. Nutr.*, 2018, **58**(15), 2491–2507.
76. H. E. Khoo, A. Azlan, S. T. Tang and S. M. Lim, Anthocyanidins and anthocyanins: colored pigments as food, pharmaceutical ingredients, and the potential health benefits, *Food Nutr. Res.*, 2017, **61**(1), 1361779.
77. S. Di Meo, T. T. Reed, P. Venditti and V. M. Victor, Role of ROS and RNS Sources in Physiological and Pathological Conditions, *Oxid. Med. Cell. Longevity*, 2016, **2016**, 1245049.
78. T. Kongpichitchoke, J. L. Hsu and T. C. Huang, Number of Hydroxyl Groups on the B-Ring of Flavonoids Affects Their Antioxidant Activity and Interaction with Phorbol Ester Binding Site of PKCδ C1B Domain: In Vitro and in Silico Studies, *J. Agric. Food Chem.*, 2015, **63**(18), 4580–4586.
79. N. Yahfoufi, N. Alsadi, M. Jambi and C. Matar, The Immunomodulatory and Anti-Inflammatory Role of Polyphenols, *Nutrients*, 2018, **10**(11), 1618.

80. T. Nakazato, K. Ito, Y. Ikeda and M. Kizaki, Green tea component, catechin, induces apoptosis of human malignant B cells via production of reactive oxygen species, *Clin. Cancer Res.*, 2005, **11**(16), 6040–6049.
81. L. M. Howells, A. Mitra and M. M. Manson, Comparison of oxaliplatin- and curcumin-mediated antiproliferative effects in colorectal cell lines, *Int. J. Cancer*, 2007, **121**(1), 175–183.
82. S. Balasubramanian, T. Efimova and R. L. Eckert, Green tea polyphenol stimulates a Ras, MEKK1, MEK3, and p38 cascade to increase activator protein 1 factor-dependent involucrin gene expression in normal human keratinocytes, *J. Biol. Chem.*, 2002, **277**(3), 1828–1836.
83. C. S. Williams, M. Mann and R. N. DuBois, The role of cyclooxygenases in inflammation, cancer, and development, *Oncogene*, 1999, **18**(55), 7908–7916.
84. M. L. Ho, P. N. Chen, S. C. Chu, D. Y. Kuo, W. H. Kuo, J. Y. Chen and Y. S. Hsieh, Peonidin 3-glucoside inhibits lung cancer metastasis by downregulation of proteinases activities and MAPK pathway, *Nutr. Cancer*, 2010, **62**(4), 505–516.
85. F. Afaq, A. Malik, D. Syed, D. Maes, M. S. Matsui and H. Mukhtar, Pomegranate fruit extract modulates UV-B-mediated phosphorylation of mitogen-activated protein kinases and activation of nuclear factor kappa B in normal human epidermal keratinocytes paragraph sign, *Photochem. Photobiol.*, 2005, **81**(1), 38–45.
86. D. Y. Shin, W. S. Lee, J. N. Lu, M. H. Kang, C. H. Ryu, G. Y. Kim, H. S. Kang, S. C. Shin and Y. H. Choi, Induction of apoptosis in human colon cancer HCT-116 cells by anthocyanins through suppression of Akt and activation of p38-MAPK, *Int. J. Oncol.*, 2009, **35**(6), 1499–1504.
87. P. Tsakiroglou, J. Weber, S. Ashworth, C. Del Bo' and D. Klimis-Zacas, Angiogenesis is differentially modulated by anthocyanin and phenolic acid extracts from wild blueberry (*V. angustifolium*) through the PI3K Pathway, *J. Med. Food*, 2021, **24**(3), 226–235, DOI: 10.1089/jmf.2020.0066.
88. R. H. Liu, Health benefits of fruit and vegetables are from additive and synergistic combinations of phytochemicals, *Am. J. Clin. Nutr.*, 2003, **78**(3 Suppl), 517S–520SS.
89. H. Land Lail, R. G. Feresin, D. Hicks, B. Stone, E. Price and D. Wanders, Berries as a Treatment for Obesity-Induced Inflammation: Evidence from Preclinical Models, *Nutrients*, 2021, **13**(2), 334.
90. S. K. Bohn, M. C. Myhrstad, M. Thoresen, M. Holden, A. Karlsen, S. H. Tunheim, I. Erlund, M. Svendsen, I. Seljeflot and J. O. Moskaug, *et al.*, Blood cell gene expression associated with cellular stress defense ismodulated by antioxidant-rich food in a randomised controlled clinical trial of male smokers, *BMC Med.*, 2010, **8**, 54.
91. D. Ornish, M. J. Magbanua, G. Weidner, V. Weinberg, C. Kemp, C. Green, M. D. Mattie, R. Marlin, J. Simko and K. Shinohara, *et al.*, Changes in prostate gene expression in men undergoing an intensive nutrition and lifestyle intervention, *Proc. Natl. Acad. Sci. U. S. A.*, 2008, **105**(24), 8369–8374.

92. M. H. Carlsen, B. L. Halvorsen, K. Holte, S. K. Bohn, S. Dragland, L. Sampson, C. Willey, H. Senoo, Y. Umezono and C. Sanada, *et al.*, The total antioxidant content of more than 3100 foods, beverages, spices, herbs and supplements used worldwide, *Nutr. J.*, 2010, **9**, 3.
93. B. L. Halvorsen, K. Holte, M. C. Myhrstad, I. Barikmo, E. Hvattum, S. F. Remberg, A. B. Wold, K. Haffner, H. Baugerod and L. F. Andersen, *et al.*, A systematic screening of total antioxidants in dietary plants, *J. Nutr.*, 2002, **132**(3), 461–471.
94. D. Aune, Plant Foods, Antioxidant Biomarkers, and the Risk of Cardiovascular Disease, Cancer, and Mortality: A Review of the Evidence, *Adv. Nutr.*, 2019, **10**(Suppl_4), S404–S421.
95. Academy of Nutrition and Dietetics: https://www.eatright.org/food/planning-and-prep/snack-and-meal-ideas/a-very-berry-summer.
96. U.S. Department of Agriculture and U.S. Department of Health and Human Services. Dietary Guidelines for Americans, 2020-2025. 9th Edition. December 2020. Available at DietaryGuidelines.gov.
97. A. K. Sikalidis and A. Maykish, The Gut Microbiome and Type 2 Diabetes Mellitus: Discussing a Complex Relationship, *Biomedicines*, 2020, **8**(1), 8.
98. J. Singh, R. Metrani, S. R. Shivanagoudra, G. K. Jayaprakasha and B. S. Patil, Review on Bile Acids: Effects of the Gut Microbiome, Interactions with Dietary Fiber, and Alterations in the Bioaccessibility of Bioactive Compounds, *J. Agric. Food Chem.*, 2019, **67**(33), 9124–9138.
99. L. Tian, Y. Tan, G. Chen, G. Wang, J. Sun, S. Ou, W. Chen and W. Bai, Metabolism of anthocyanins and consequent effects on the gut microbiota, *Crit. Rev. Food Sci. Nutr.*, 2019, **59**(6), 982–991.
100. A. Santino, A. Scarano, S. De Santis, M. De Benedictis, G. Giovinazzo and M. Chieppa, Gut Microbiota Modulation and Anti-Inflammatory Properties of Dietary Polyphenols in IBD: New and Consolidated Perspectives, *Curr. Pharm. Des.*, 2017, **23**(16), 2344–2351.
101. C. G. Fraga, K. D. Croft, D. O. Kennedy and F. A. Tomás-Barberán, The effects of polyphenols and other bioactives on human health, *Food Funct.*, 2019, **10**(2), 514–528.
102. M. Wiciński, J. Gębalski, E. Mazurek, M. Podhorecka, M. Śniegocki, P. Szychta, E. Sawicka and B. Malinowski, The Influence of Polyphenol Compounds on Human Gastrointestinal Tract Microbiota, *Nutrients*, 2020, **12**(2), 350.
103. R. Mahajan, S. Attri, V. Mehta, M. Udayabanu and G. Goel, Microbe-bio-Chemical Insight: Reviewing Interactions between Dietary Polyphenols and Gut Microbiota, *Mini-Rev. Med. Chem.*, 2018, **18**(15), 1253–1264.
104. A. S. Kristo, G. Tzanidaki, A. Lygeros and A. K. Sikalidis, Bile sequestration potential of an edible mineral (clinoptilolite) under simulated digestion of a high-fat meal: an in vitro investigation, *Food Funct.*, 2015, **6**(12), 3818–3827.
105. L. Bracci, A. Fabbri, M. Del Cornò and L. Conti, Dietary Polyphenols: Promising Adjuvants for Colorectal Cancer Therapies, *Cancers*, 2021, **13**(18), 4499.

CHAPTER 10

The Role of Berry Bioactive Compounds in Diabetes Mellitus

DUSTIN W. DAVIS,[a] MACY M. HELM,[a] KENNETH IZUORA[b] AND ARPITA BASU*[a]

[a] Department of Kinesiology and Nutrition Sciences, School of Integrated Health Sciences, University of Nevada, Las Vegas, 4505 S. Maryland Pkwy, Las Vegas, Nevada; [b] Department of Internal Medicine, School of Medicine, University of Nevada, Las Vegas, 4505 S. Maryland Pkwy, Las Vegas, Nevada
*Email: Arpita.basu@unlv.edu

10.1 Diabetes Mellitus: A Global Health Problem

Diabetes mellitus (DM) is a metabolic disease characterized by pancreatic β-cell insufficiency and/or the impaired insulin sensitivity of peripheral body tissues. These two mechanisms disrupt normal glucose metabolism by eliciting prolonged elevations in blood glucose, known as hyperglycemia. Based on pathophysiology, DM is classified into three major types: type 1 DM (T1DM), type 2 DM (T2DM), and gestational DM (GDM).[1] An autoimmune disorder destroys pancreatic β-cells in T1DM, preventing the pancreas from producing sufficient insulin. In T2DM, hyperglycemia is caused by peripheral insulin resistance and inadequate insulin production by the pancreas. Insulin resistance refers to the diminished insulin signaling and glucose uptake by the liver, skeletal muscle, and adipose tissue. Typically, these tissues gradually lose sensitivity to insulin and no longer readily

Food Chemistry, Function and Analysis No. 33
Berries and Berry Bioactive Compounds in Promoting Health
Edited by Dorothy Klimis-Zacas and Ana Rodriguez-Mateos

Published by the Royal Society of Chemistry, www.rsc.org

absorb glucose. Pancreatic β-cells initially respond by producing more insulin, but the cells are damaged by chronic hyperglycemia and progressively produce less insulin until cell death occurs. As suggested by its name, GDM is diabetes that is first detected during pregnancy, typically in the second trimester, when DM was never detected in the mother before pregnancy.[1,2] Further, a mother may not develop GDM during her first pregnancy but may do so in subsequent pregnancies.

Globally, approximately 451 million adults over 18 years of age had DM in 2017, and 49.7% were undiagnosed.[3] Estimates of worldwide DM-related loss of life and financial costs for the same year were over 5 million deaths and \$850 billion.[3] The prevalence of DM varies across nations and is disproportionately burdensome in some countries. In the United States, where 42.4% of adults have obesity,[4] an estimated 88 million and 34 million adults have pre-diabetes and diabetes (34.5 and 13.0%), respectively.[5,6] The deaths of over 250 000 adults in the United States were attributable to DM in 2017, with combined indirect and direct treatment costs exceeding \$300 million.[5] Therefore, DM is a serious public health concern that requires a concerted global effort to reduce the incidence of DM and treat existing and emerging cases.

10.1.1 Diagnosing Diabetes Mellitus

Diagnosing DM in the United States is typically based on criteria delineated by the American Diabetes Association (ADA).[1] A common diagnostic method is measuring fasting plasma glucose (FPG) to determine if the concentration is $\geq$126 mg dL^{-1} or $\geq$7.0 mmol L^{-1} after fasting from all energy intake for at least eight hours. DM is also diagnosed using plasma glucose after the oral glucose load of an oral glucose tolerance test (OGTT). In the OGTT, blood glucose is typically measured 2 hours after a person consumes 75 g of anhydrous glucose dissolved in water. A plasma glucose value $\geq$200 mg dL^{-1} (11.1 mmol L^{-1}) 2 hours after consumption indicates a positive test. Alternatively, DM can be diagnosed based on a random plasma glucose concentration $\geq$200 mg dL^{-1} (11.1 mmol L^{-1}) in patients with classic symptoms of hyperglycemia (polyuria, polydipsia, and polyphagia) or a blood concentration of glycated hemoglobin (HbA_{1c})$\geq$6.5% (48 mmol mol^{-1}).[1]

The diagnostic approach to GDM differs from the approach to T1DM and T2DM. The ADA recommends testing for pre-diabetes or DM in women with risk factors of DM at the initial prenatal visit. If at that visit a woman is not diagnosed with pre-diabetes or DM using their respective non-pregnant diagnostic criteria, one-step or two-step testing for GDM should occur at 24–28 weeks of gestation using pregnancy-specific criteria.[1] One-step testing is conducted as a fasted 75 g OGTT with measurements of fasting, 1 hour, and 2 hour plasma glucose. A positive test (*i.e.* GDM diagnosis) is indicated by any of the following: FPG$\geq$92 mg dL^{-1} (5.1 mmol L^{-1}), 1 hour glucose$\geq$180 mg dL^{-1} (10.0 mmol L^{-1}), or 2 hour glucose$\geq$153 mg dL^{-1} (8.5 mmol L^{-1}).[1] In contrast, two-step testing requires a woman to undergo a non-fasted 50 g glucose load test with measurements of FPG and 1 hour glucose (step one). If the 1 hour glucose is ≥130, 135, or 140 mg dL^{-1} (7.2, 7.5, or 7.8 mmol L^{-1}, respectively), a fasted, 100 g OGTT (step two) is indicated. A diagnosis of GDM is made when at least two of the following measurements are recorded during the 100 g OGTT: FPG ≥ 95 mg dL^{-1} (5.3 mmol L^{-1}), 1 hour glucose ≥ 180 mg dL^{-1}, (10.0 mmol/L) 2 hour glucose ≥ 155 mg dL^{-1} (8.6 mmol L^{-1}), or 3 hour glucose ≥ 140 mg dL^{-1} (7.8 mmol L^{-1}).[1]

10.1.2 Delaying and/or Preventing Diabetes Mellitus

Considering that 90–95% of all DM cases are T2DM,[5] the preventable form, health professionals have focused intensely on identifying and implementing prophylactic lifestyle modifications in people with risk factors. Non-modifiable factors include a family history of DM, advancing age, and race (in the United States: African American/Black, Hispanic/Latino, Native American, Alaska Native, and some Pacific Islanders and Asian Americans).[7] Modifiable factors include poor dietary intake, physical inactivity, overweight, obesity, and pre-diabetes; having had GDM also increases the risk of developing T2DM.[7] In people with risk factors, increasing physical activity and improving the diet and are two critical components of delaying or preventing T2DM.[8–10] Whereas physical inactivity and excessive caloric intake cause overweight, obesity, and T2DM, regular physical activity and a moderate diet delay or prevent these conditions.[11] Specifically, the ADA recommends that health professionals refer at-risk individuals to intensive behavioral lifestyle intervention programs aligned with the Diabetes Prevention Program (DPP). Compared to a placebo treatment, lifestyle intervention in the DPP reduced the progression from pre-diabetes to T2DM by 58%. Metformin only reduced progression by 31%.[12] Long-term follow-up revealed that lifestyle intervention reduced the 15 year progression to T2DM by 27% compared to 18% with metformin alone,[13] indicating that lifestyle can be more effective than pharmacotherapy over time in fending off T2DM.

The two overriding goals of a DPP-based lifestyle intervention for T2DM are to decrease body mass by 7% and maintain the new mass while achieving 150 minutes per week of moderate-intensity physical activity. For many people, a moderate-intensity equates to the intensity of brisk walking, which has reduced T2DM incidence in adults with pre-diabetes.[12] To aid in reducing body mass, the ADA recommends several diets centered on restricting energy intake.[10] The ADA also recommends consuming less dietary fat and processed foods and more foods characteristic of the Mediterranean dietary pattern, including whole grains, legumes, nuts, fruits, and vegetables.[10] While these diets may be self-directed, medical nutrition therapy is a highly valuable alternative that has reduced HbA_{1c}, Diabetes Risk Score, total serum cholesterol, and low-density lipoprotein cholesterol (LDL-C) in individuals with overweight or obesity and pre-diabetes.[10,14] Whether self-administered or administered under professional supervision, diets moderate in caloric energy and dense in nutrients help prevent T2DM.[9]

10.2 A Focus on Fruits and Vegetables

Fruits and vegetables are rich in dietary bioactive compounds, including vitamins, minerals, and fiber, all of which are nutrients and compounds with myriad roles in normal human physiology. At the same time, fruits and vegetables are less dense in macronutrients (*i.e.* carbohydrate, fat, and protein) and hence dietary energy (*i.e.* kilocalories [kcal]) than other food groups. As a result, consuming many fruits and vegetables promotes a healthy nutrition status and satiety while providing a relatively low number of kcal.[9] Additionally, due to their origin from plants, all fruits and vegetables provide non-nutritive (*i.e.* void of kcal), bioactive compounds called phytochemicals.

Many phytochemicals exist, and they vary by chemical structure, physiological activity, and composition in natural foods.[15] Examples of phytochemicals include alkaloids, antioxidant vitamins (*e.g.* β-carotene and vitamins C and E), dietary polyphenols, isothiocyanates, and terpenoids.[16] Many of the phytochemicals contribute to plant development and survival.[16] Once ingested by humans, the compounds exert physiological effects in the body that are postulated to be additive and synergistic; in other words, maximizing health benefits from bioactive phytochemicals may require consuming them as a complex mixture from a diet rich in various whole fruits and vegetables.[15] General health benefits linked to the phytochemicals in fruits and vegetables include reductions in the risk of chronic diseases associated with aging, oxidative stress, and inflammation, including T2DM and cardiovascular disease (CVD).[15,16] In recent years, considerable attention has focused on dietary polyphenols, a major class of phytochemicals that is divided into five subclasses: flavonoids, lignans, phenolic acids, stilbenes, and tannins.[16] They are abundant in colorful fruits and vegetables and are chemically active in living organisms, making them particularly relevant to health and disease.[17] One bountiful and commonly consumed source of dietary polyphenols is berries, the scope of this textbook. The next section briefly describes the nutritional content of a selection of dietary berries and factors that determine their phenolic content.

10.3 Dietary Berries: Consumption and Nutritional Composition

Dietary berries are fleshy fruits that vary widely in size and color, presenting as red, blue, purple, and black. While berries do not contain pits, some contain seeds. Given their widespread availability and sweet taste, many berry variants are popularly consumed fruits worldwide, especially in the United States. For example, in 2018, US adults consumed just over 7 pounds of fresh strawberries and 2 pounds of fresh blueberries per capita compared to approximately 5 pounds and less than 1.5 pounds, respectively, in 2000.[18] Both strawberries and blueberries are low-calorie, low-sodium, and nutrient-dense additions to the diet. One cup of raw, whole strawberries (144 g) contains 46 kcal, 11 g carbohydrate, 0 g fat, and 1 g protein but provides 3 g fiber, 220 mg potassium, 10 μg β-carotene, 1.5 μg vitamin A,

85 mg vitamin C, 0.4 mg vitamin E, various B vitamins, and trace elements such as copper, magnesium, phosphorous, selenium, and zinc.[19] In comparison, one cup of raw, whole blueberries (148 g) contains 84 kcal, 21 g carbohydrate, 0.5 g fat, and 1 g protein but offers 3.5 g fiber, 114 mg potassium, 47 μg β-carotene, 4.5 μg vitamin A, 14 mg vitamin C, and 0.8 mg vitamin E and similarly provides B vitamins and trace elements.[20]

Alongside complex carbohydrates, fiber, and antioxidant vitamins and minerals, dietary berries provide bioactive phytochemicals, the most notable of which are one of the aforementioned subclasses of dietary polyphenols called flavonoids.[21] Flavonoids are subcategorized as anthocyanins, flavan-3-ols, flavanones, flavonols, flavones, and isoflavones.[22] The precise phenolic composition of many dietary berries is becoming increasingly accessible with the development of a comprehensive online database called Phenol-Explorer (https://www.phenol-explorer.eu).[23] The creators of Phenol-Explorer continue to consolidate data on the concentrations of more than 500 distinct polyphenols in over 400 foods determined by 1300 + scientific publications.

Data from Phenol-Explorer suggest that the total polyphenols in commonly consumed berries, such as blackberries, blueberries, cranberries, raspberries, and strawberries, range from 150 to 600 mg per 100 g formula weight (FW) of berry.[24–28] Concentrations of individual anthocyanins are also available. For example, raw blackberries contain 138.7 mg of cyanidin 3-*O*-glucoside per 100 mg FW of berries,[24] and raw highbush blueberries contain 20.5 mg of delphinidin 3-*O*-galactoside per 100 g FW of berries.[25] Raw American cranberries contain 22.0 mg of peonidin 3-*O*-galactoside per 100 mg FW berries,[26] and red raspberries contain 14.9 mg of cyanidin 3-*O*-glucoside per 100 g FW of berries.[27] Finally, raw strawberries contain 47 mg of pelargonidin 3-*O*-glucoside per 100 g FW of berries.[28]

Knowing the phenolic content of dietary berry variants helps identify which are good candidates for increasing the intake of dietary polyphenols. However, it is important to remember that generic names of berries, such as strawberry, represent multiple plant species that produce berries with similar appearances but different genotypes. This means that plants with different genotypes produce visually similar fruits that are classified under the same general name but possess different phenolic contents.[29,30] In addition to genotype, the environment in which berries are cultivated,[31] post-harvest processing (*e.g.* freeze-drying or conversion to juice),[32,33] and storage conditions[33] further modulate the phenolic content of berries by the time they are ready for consumption. Another important consideration is that the concentrations of flavonoids and other dietary polyphenols in raw berries do not equate with the bioavailable concentrations after ingestion. Cooking and baking berries before consumption further modifies blueberry phenolics, for example.[34,35] In addition to the method of preparation, differences in individuals' gut microbiota and the overall food matrix of a meal in which berries are consumed likely affect the bioavailability of polyphenols during digestion and absorption.[36] In summary, there is considerable heterogeneity regarding berry species, cultivation methods, processing, preparation methods, and consumption practices. Collectively, these factors influence the total polyphenol content of berries and the bioavailability and bioactivity of polyphenols inside the human body. As such, it is important to identify which types of berries and berry products provide the most bioavailable and functional dietary polyphenols in the management of DM. In the next section, we describe the process by which we searched for, identified, and evaluated data on this topic.

10.3.1 Methodology

Epidemiological and clinical data included in this chapter were obtained from peer-reviewed articles published online in academic journals between 2000 and 2020. The literature search was conducted in PubMed and Google Scholar using the search terms "berr*," "fruit," "bioactive compounds," "polyphenols," "flavonoids," and "anthocyanins," in combination with "type 2 diabetes," "gestational diabetes," "glucose tolerance," "insulin resistance," and "insulin sensitivity." Epidemiological studies were selected if they evaluated berry consumption as an exposure and the incidence of DM as an outcome. Clinical trials were selected if they intervened with berries in their whole, freeze-dried, and juice forms (but not as extracts) in human participants diagnosed with DM. To keep our focus aligned with the scope of this textbook, we only evaluated studies in people with a diagnosis of DM. Thus, we excluded studies that included people with Metabolic Syndrome or other risks of DM. Clinical trials that implemented widely available and popularly consumed berries were prioritized (*e.g.* blackberries, blueberries, cranberries, pomegranates, raspberries, and strawberries). Additional epidemiological and clinical reports were identified for inclusion by searching the references of reports included after the initial search. Having introduced DM as a major threat to global health, general information on dietary berries and their polyphenols, and the methodology of our literature search and analysis, the next section will discuss how the consumption of dietary berries and their bioactive compounds are associated with the risk of developing DM.

10.3.2 Dietary Berries and Diabetes: Epidemiological Studies

Across the past 20 years, numerous epidemiological studies have investigated the relationship between the intake of fruits and the risk of T2DM.[37–48] In most of the studies, the intake of fruits was combined with the intake of vegetables to represent one exposure: the combined intake of fruits and vegetables. Among the studies cited above, only one found that fruits intake was independently associated with a reduced T2DM risk in Chinese women.[45] Another three studies concluded that higher concentrations of blood biomarkers indicating higher intakes of both fruits and vegetables (*e.g.* β-carotene, total carotenoids, lutein, vitamin C) were associated with a lower risk of T2DM in adults.[39,43,46] Numerous other studies documented that fruit intake did not independently associate with a reduced risk of

T2DM in Chinese men[47] and women,[38] European adults,[40] Japanese adults,[42] adults living in England,[41] Swedish men and women,[48] US women,[37] or older adults (European and US).[44]

While the aforementioned studies suggest a null association between fruit intake and T2DM risk, three *meta*-analyses recently indicated that T2DM incidence tends to be lower in adults eating more fruits,[49–51] and a fourth *meta*-analysis concluded that eating two daily servings of fruits was associated with a lower risk of T2DM than one-half, one, three, or four servings a day.[52] Thus, the totality of the current observational evidence suggests fruits are associated with a lower risk of T2DM. However, a gap in the research exists. Whereas many studies, including *meta*-analyses, have measured associations between the intakes of specific types of vegetables (*e.g.* green leafy, yellow, and cruciferous vegetables) and T2DM risk, relatively few have measured the associations between the intakes of specific fruits and risk. Only one of the *meta*-analyses mentioned above reported summary relative risks (RRs) for specific fruits:[50] 1.02 (95% CI 0.96, 1.09) for citrus, 0.96 (95% CI 0.78, 1.18) for strawberries, and 0.75 (95% CI 0.66, 0.84) for blueberries.[50] These summary RRs indicate that the highest intakes of citrus and strawberries were not associated with a reduced risk of T2DM, but the risk was 25% lower in individuals with the highest intake of blueberries compared to those with the lowest intake.

The gap identified in the previous paragraph was partly addressed in 2016 when a *meta*-analysis examined the association between the intake of dietary berries and anthocyanins and T2DM risk.[53] The analysis included five cohort studies that reported dietary berry intake in approximately 194 000 adults and three cohort studies that reported dietary anthocyanin intake in approximately 200 000 adults. The prominent findings of the *meta*-analysis were curvilinear associations between the intakes of dietary berries and anthocyanins and T2DM risk. Dietary berries and anthocyanins were associated with an 18% (summary, adjusted RR = 0.82; 95% CI 0.76, 0.89) and 15% (summary, adjusted RR = 0.85; 95% CI 0.80, 0.91) lower risk of T2DM, respectively. The summary RRs represent T2DM risk in the highest-intake groups compared to the lowest-intake groups from the cohorts.[53] Data from the individual studies included in the *meta*-analysis are described in chronological order (most recently published to most distantly published) in the text and Table 10.1 below.

10.3.3 Kuopio Ischaemic Heart Disease Risk Factor Study

The Kuopio Ischaemic Heart Disease Risk Factor Study (KIHD) evaluated the relationship between the intakes of fruits, berries, and vegetables (FBV) and the risk of T2DM in Finnish men. Diet was assessed *via* 4 day food records. Hazard ratios (HRs) were calculated using a Cox proportional hazards model to represent the likelihood that participants in the highest quartile of intakes would develop T2DM compared to the reference group, the participants in the lowest quartile of intakes. The HRs were adjusted for age, examination year, BMI, waist-to-hip ratio, smoking status, education level, leisure-time physical activity, family history of DM, and the intake of energy and alcohol. While the intakes of total FBV, fruits, fruit and berry juices, and vegetables were not significantly associated with T2DM risk, the highest intake of berries (> 59.7 g per day) was associated with an adjusted HR of 0.65 (95% CI: 0.49, 0.88, p-trend = 0.003) compared to the lowest intake of berries (<1.3 g per day).[54]

10.3.4 Nurses' Health Study, Nurses' Health Study II, and Health Professionals Follow-up Study

In the Nurses' Health Study (NHS), Nurses' Health Study II (NHS II), and Health Professionals Follow-up Study (HPFS), Muraki *et al.* analyzed the associations between whole fruits and individual fruits consumption and the risk of developing T2DM in female nurses (NHS and NHS II) and male health professionals (HPFS).[55] Pooled HRs were calculated for the three studies combined, adjusting for age, ethnicity, BMI, family history of T2DM, post-menopausal status and hormone replacement, multivitamin use, oral contraceptive use, physical activity, smoking status, fruit juice intake, and the modified alternate healthy eating index score. The pooled HR for weekly whole fruits intake was 0.98 (95% CI: 0.96, 0.99, p-trend = 0.002), comparing ≥ 3 servings per day to <4 servings per week. Fruits were also analyzed individually and in groups: blueberries; prunes; bananas; grapefruit; bananas; strawberries; cantaloupe; grapes and raisins; apples and pears; and peaches, plums, and apricots. These analyses revealed that the highest weekly intake of blueberries, but not strawberries, was independently associated with a 26% reduced rate of T2DM: adjusted HR = 0.74 (95% CI: 0.66, 0.83, p-trend <0.001), comparing the 2–4 servings per week and ≥ 5 servings per week groups (combined due to the number of participants) to the <1 serving per month group.[55]

Using data from the same three studies (NHS, NHS II, and HPFS), Wedick *et al.* reported pooled HRs for anthocyanin intake, comparing the highest quintile to the lowest quintile.[56] The HRs were adjusted for age, ethnicity, BMI, family history of T2DM, multivitamin use, oral contraceptive use, physical activity, post-menopausal status and hormone replacement, smoking status, polyunsaturated:saturated fat ratio, and the intakes of total energy, *trans* fat, alcohol, coffee, fish, high-calorie sodas, red meat, and whole grains. The pooled HR of the fifth quintile of anthocyanin intake (median: 148.4 mg per day) compared to the first quintile of intake (median: 9.0 mg per day) was 0.85 (95% CI: 0.80, 0.91, p-trend <0.001).[56] Additionally, the intake of blueberries, a food rich in anthocyanins, was associated with an adjusted HR of 0.77 (95% CI: 0.68, 0.87, p-trend <0.001), comparing ≥ 2 servings per week to <1 serving per month.[56] Unlike total anthocyanin and blueberry ingestion, the intakes of total flavonoids and other flavonoid subclasses (flavonols, flavones, flavon-3-ols, and flavanones) were not associated with a reduced risk of T2DM.[56]

10.3.5 Finnish Mobile Clinic Health Examination Survey

The Finnish Mobile Clinic Health Examination Survey (FMC) examined whether the intakes of various foods could be associated with T2DM among men and women from 30 communities across Finland.[57] The difference in the daily combined intake of fruits and berries at baseline between participants who developed T2DM (cases) and participants who did not (non-cases) was not significant (mean ± standard deviation: 114 ± 114 g per day *vs.* 103 ± 96.1 g per day, $p = 0.08$). In contrast, the daily intake of berries alone at baseline was significantly lower in those who became cases compared to those who did not (13.7 ± 17.9 g per day *vs.* 16.4 ± 21.1 g per day in non-cases, $p = 0.01$). The highest quartile of fruits and berries intake (> 156 g per day) had an adjusted RR of 0.69 (95% CI: 0.51, 0.92, $p = 0.01$) compared to the lowest quartile (<33 g per day). The adjusted RR for berries intake alone was 0.63 (95% CI: 0.47, 0.85, *p*-trend = 0.002) for the highest quartile of berries intake (> 20 g per day) *vs.* the lowest quartile of intake (<4 g per day).[57]

10.4 Dietary Berries and Gestational Diabetes Mellitus

The epidemiological studies discussed thus far have examined associations between the intakes of dietary berries and anthocyanins and the risk of T2DM. Studies of a similar magnitude that examined the links between similar exposures and GDM exist but are comparatively scarce. In a well-adjusted analysis of over 13 000 women in the NHS II, Chen *et al.* reported that the risk of GDM did not differ between the highest and lowest quintiles of whole fruits intake (median intake of 2.36 *vs.* 0.35 servings per day),[58] nor did GDM risk differ between the highest and lowest quintiles of the combined intake of blueberries and strawberries (median intake of 0.43 *vs.* 0 servings per day).[58] In a separate 2019 analysis of over 1000 women in Iran, the highest quartile of fruits consumption (median intake of 4.2 servings per day) had an adjusted odds ratio (OR) of 0.48 (95% CI: 0.18–0.89, *p*-trend = 0.02) compared to the lowest quartile (median = 0.58 servings per day), but berry consumption was not assessed.[59] Similarly, in an analysis of the Tongji Maternal and Child Health Cohort with Chinese women, Zhou *et al.* reported that whole fruits consumption (median; 25th percentile, 75th percentile) was inversely associated with GDM risk after adjusting for numerous dietary, physiological, and social factors: the adjusted OR in the highest quintile (671; 604, 773 g) was 0.41 (95% CI: 0.27, 0.62, *p*-trend <0.001), compared to the lowest quintile (135; 69, 176 g).[60] While dietary berries were not assessed as an individual category of fruits in this study, the authors also reported low-glycemic index (GI <55) and low-glycemic load fruits to negatively associate with GDM risk.[60] Several berries are low GI. For example, raw blueberries have been shown to have a GI of 53 *vs.* glucose[61] and raw strawberries have a GI of 40 *vs.* glucose.[62] This suggests that there may be value in examining dietary berry intakes in future large-scale epidemiological investigations in GDM.

In addition to the data discussed above that show a lower risk of GDM with greater fruit consumption, other data indicate the opposite. A prospective cohort study with Chinese women reported ORs for fruit intake (median; 25th percentile, 75th percentile) during the second trimester after controlling for numerous confounders. The highest quartile of fruits intake

(710; 602, 870 g) had an adjusted OR of 4.82 (95%: 2.38, 9.76, p-trend <0.001) for GDM compared to the lowest quartile (133; 183, 207 g).[63] Moderate-to-high-GI foods also predicted GDM, as did the intakes of tropical and citrus fruits. The same association was not found for low-GI fruits intake. Berries are low GI, as stated above, and are neither tropical nor citrus fruits. As such, dietary berry consumption during the second trimester may not be associated with an increased risk of GDM, but further research is warranted to test this hypothesis. Both the study by Chen *et al.*[58] and Huang *et al.*[63] were included with eight other observational studies in a recent *meta*-analysis published in 2019.[64] While Pham *et al.* concluded that fruits did not reduce the risk of GDM based on the pooled OR from the two aforementioned studies, the polyphenol-rich Mediterranean dietary pattern was associated with a reduced risk of GDM: pooled OR = 0.57 (95% CI 0.41, 0.79).[64] Considering that dietary berries are rich in polyphenols (especially the flavonoid anthocyanins), they may be a suitable addition to the Mediterranean dietary pattern, which is characterized by high intakes of polyphenol-rich fruits and vegetables. Yet it remains to be elucidated whether such augmentation will benefit the risk of GDM. Flavonoids may not be associated with a lower risk of GDM in all cases. A 2019 cross-sectional study with nearly 800 women in Brazil did not find a significant association between flavonoid or anthocyanin intakes and GDM.[65] It should be noted, however, that the median (25th percentile, 75th percentile) flavonoid intake among the sample was quite low, only 50 (31, 75) mg per day.[65] Such a low intake of flavonoids may be insufficient to associate with the risk and outcomes of GDM.

Dietary patterns other than the Mediterranean pattern have been associated with lower GDM risk in other studies, but these studies have not frequently measured and/or reported on dietary polyphenols. One prospective study evaluated the diet (29 food groups) of 168 women aged 18–40 years between gestational weeks 19 and 24.[66] Only one dietary pattern was associated with a reduced risk of GDM; the pattern was labeled the "prudent dietary pattern" and included fruits and berries. After adjusting for age, energy intake, parity, physical activity, pre-pregnancy weight, and weekly weight gain, the prudent dietary pattern was associated with a reduced risk of GDM, indicated by an adjusted OR of 0.44 (95% CI: 0.21, 0.90). Significantly more participants in the low/medium adherence group developed GDM compared to the highest adherence group (14.3 *vs.* 1.8%, $p = 0.01$). Despite these findings, fruits and berries as a factor of the "prudent dietary pattern" were not independently associated with GDM risk.[66]

In summary, epidemiological data on the intakes of dietary berries and anthocyanins in GDM are presently inconclusive (Table 10.1). On the other hand, both dietary berries and anthocyanins are associated with a reduced risk of T2DM in non-pregnant adults. An important limitation to the epidemiological data presented in this chapter is that the reported associations between berries and anthocyanins and T2DM risk do not prove causation. Consequently, researchers have conducted randomized controlled trials (RCTs) to elucidate the effects of dietary berries on biomarkers related to

Table 10.1 Epidemiological studies on the associations between the intakes of dietary berries and anthocyanins and the risk of type 2 diabetes mellitus.

Authors, Year (Country)	Study	Study Design	Participants	Exposures	Outcomes	Associations
Mursu *et al.*, 2014 (Finland)[54]	KIHD	Prospective, population-based, cohort study	2332 men, aged 42–60 years and without T2DM or IFG at study initiation	Berry intake (g per day)	Incidence of T2DM (questionnaires, FBG, 2 hour OGTT, and reimbursement for medication) across 19 years	↓ HR with highest intake of berries
Muraki *et al.*, 2013 (United States)[55]	NHS NHS II HPFS	Prospective longitudinal cohort studies	66 105 women, aged 30–55 years and without T2DM at study initiation 85 104 women, aged 24–44 years and without T2DM at study initiation 36 173 men, aged 40–75 years and without T2DM at study initiation	Blueberry and strawberry intake (servings per month or week)	Incidence of T2DM (based on self-report and questionnaires) across 394 000 (NHS), 416 000 (NHS II), and 654 000 (HPFS) person years, respectively	↓ HR with highest intake of blueberries; ↔ HR with highest intake of strawberries
Wedick *et al.*, 2012 (United States)[56]	NHS NHS II HPFS	Prospective longitudinal cohort studies	NHS:[a] 70 359 women, mean (range) age of 50 (37–65) years and without T2DM at baseline NHS II:[a] 89 201 women, mean (range) age of 36 (26–45) years without T2DM at baseline HPFS:[a] 41 334 men, mean (range) age of 53 (40–75) years and without T2DM at baseline	Anthocyanin intake (mg per day)	Incidence of T2DM (based on self-report and questionnaires)[b]	↓ HR with highest intake of anthocyanins
Montonen *et al.*, 2005 (Finland)[57]	FMC	Prospective cohort study	4304 men and women, aged 40–69 years without T2DM at study initiation	Berry intake (g per day)	Incidence of T2DM (based on medication use) across 23 years	↓ RR with highest intake of berries

[a]Wedick *et al.* provided age data specifically for the participants they analyzed rather than for the entire NHS, NHS II, and HPFS cohorts.
[b]Person years reported for the NHS, NHS II, and HPFS combined = 3 646 000 years; ↓ lower/decreased; ↔ no effect/did not change; FBG: fasting blood glucose; FMC: Finnish Mobile Clinic Health Examination Survey; HPFS: Health Professionals Follow-up Study; HR: hazard ratio; IFG: impaired fasting glucose; KIHD: Kuopio Ischaemic Heart Disease Risk Factor Study; NHS: Nurses' Health Study; NHS II: Nurses' Health Study II; OGTT: oral glucose tolerance test; RR: relative risk; T2DM: type 2 diabetes mellitus.

glucose, insulin, lipids, and biomarkers of oxidative stress and inflammation in T2DM. Anthocyanins have similarly received attention in RCTs. As recently as 2016–2020, reviews have corroborated the roles of anthocyanins in protecting against oxidative stress and inflammation,[67–69] two physiological processes that modulate insulin sensitivity, glucose tolerance, and the risk of T2DM and CVD. Data on the chronic effects of long-term supplementation and the acute post-prandial effects of short-term supplementation with dietary berries and berry-derived anthocyanins are presented in the next two sections.

10.4.1 Dietary Berries and Diabetes: Long-term Supplementation Studies

In recent years, our group reported a review of human feeding trials showing that dietary berries improve risk factors of chronic diseases such as T2DM.[70] The review illustrated that, in addition to the observational studies discussed above, there is a growing number of RCTs being published to establish causal relationships between the consumption of dietary berries and measures related to DM risks, including body mass, body composition, and blood biomarkers such as glucose, insulin, lipids, and inflammatory cytokines. Overall, the review indicated that in many cases, dietary berries improved post-prandial hyperglycemia and hyperinsulinemia in adults with overweight or obesity and insulin resistance. In adults with the same conditions, long-term supplementation of dietary berries has improved blood glucose, lipids, and the atherosclerotic risk profile. However, at the time of that review, few published studies had examined whether the same benefits of dietary berries occur in T2DM, precluding definitive conclusions about the efficacy of their supplementation in patients with the disease.

To address this gap and determine if new data became available since our last review in 2019 and to offer the latest research to readers of this chapter, we conducted a new literature search as described in our methods section of this chapter. The search revealed over 20 reports of RCTs, many of which tested research-grade berry supplements or fruits that are botanically classified as berries but that are not commonly thought of as berries. We excluded several of these reports to keep the chapter focused on fruits commonly considered to be berries that are accessible to the general public. We only included reports that provided these berries (barberries, blueberries, cranberries, raspberries, strawberries, and whortleberries) in their whole, freeze-dried, and juice forms but not as extracts. Thus, in this section and the next, we present the most pertinent findings of our search and discuss the observed effects of supplementing common dietary berries over long-term (daily for 4–12 weeks) and short-term (post-prandial effects of acute, one-time consumption) periods.

The eight long-term supplementation RCTs identified using our pre-established search criteria have been conducted since 2012 (see Table 10.2).[71–78] One trial was conducted in China,[76] five trials were conducted in Iran,[73–75,77,78] and two trials were conducted in the United States.[71,72] One trial was a triple-blinded, placebo-controlled RCT;[75] four trials were double-blinded, placebo-controlled RCTs;[71,76–78] and three trials were non-blinded RCTs.[72–74] Only one of the trials had a crossover design.[72] The interventions included barberries,[73] blueberries,[71] cranberries,[74,78] strawberries,[77] raspberries,[72] whortleberries,[75] and purified berry anthocyanins.[76] A second RCT that intervened with purified berry anthocyanins was identified but has been excluded from this chapter because the sample included patients with pre-diabetes and early untreated DM, and analyses were conducted using pooled data.[79] In all seven berry trials and the single purified anthocyanin trial, participants had a mean age that exceeded 50 years. Whereas participants had overweight in five trials,[73–75,77,78] participants had obesity in two other trials[71,72] and normal weight in one trial.[76] Participants were required to be taking oral hypoglycemic agents in one trial[73] but not the other seven trials, and one trial required participants to have both T2DM and periodontal disease.[74] Hereafter, we discuss the RCTs in detail.

In men and women with overweight and T2DM who were taking oral hypoglycemic agents, Lazavi *et al.* supplemented 200 mL of barberry juice daily for 8 weeks.[73] Compared to the control, the barberry juice decreased body mass, BMI, systolic blood pressure (SBP), and diastolic blood pressure (DBP). While HbA_{1c} was unchanged, fasting blood glucose (FBG) declined by 7.3% ($p < 0.0001$), and total cholesterol fell by 14.8% ($p = 0.002$). Triacylglycerols (TAGs), LDL-C, high-density lipoprotein cholesterol (HDL-C), and apolipoproteins B and A1 (Apo-B and Apo-A1) were unaffected. Lastly, the concentration of paraoxonase 1 (PON1) increased compared to the control.[73]

In the single RCT that intervened with blueberries, Stote *et al.* gave participants 11 g of freeze-dried blueberries twice daily (22 g per day) for 8 weeks to men with overweight or obesity and T2DM.[71] Blueberries did not significantly decrease body mass, BMI, waist circumference, heart rate, SBP, DBP, FBG, fasting insulin, HbA_{1c}, or fructosamine. However, the blueberry group had lower levels of HbA_{1c} and fructosamine compared to the control group after supplementation (HbA_{1c}: $7.1 \pm 0.5\%$ *vs.* $7.5 \pm 1.0\%$, $p = 0.03$; fructosamine: 6.1% lower, $p = 0.04$). The level of TAGs was also 11.1% lower in the blueberry group ($p = 0.03$), but total cholesterol, LDL-C, HDL-C, and C-reactive protein (CRP) did not differ.[71]

Among two RCTs that provided cranberry juice, one trial provided cranberry juice as the sole intervention,[78] and the other trial provided cranberry juice alone and in combination with omega-3 (n-3) fatty acids.[74] In the former trial from 2012, Shidfar *et al.* had men with T2DM consume 240 mL of cranberry juice daily for 12 weeks.[78] The juice reduced FBG, Apo-B, and Apo-A1 by 14.1%, 14.0%, and 7.8%, respectively, compared to the placebo ($p < 0.05$). The level of lipoprotein(a) was unaffected. Concerning the biomarkers of oxidative stress and inflammation, PON1 activity increased with cranberry juice compared to the placebo.[78] In comparison to this RCT, the latter of the two cranberry trials was conducted by Zare Javid *et al.* in men and women with both T2DM and periodontal disease.[74] In two separate arms, the investigators provided either 200 mL of cranberry juice or 200 mL of cranberry juice with 1 g of *n*-3 fatty acids. Body mass, BMI, waist circumference, and FBG were not affected by either intervention, but HbA_{1c} was decreased by both the cranberry juice ($0.9 \pm 0.9\%$, $p = 0.02$) and juice with *n*-3 fatty acids ($0.4 \pm 0.4\%$, $p = 0.047$). In addition, while neither intervention changed the levels of TAGs, total cholesterol, or LDL-C, only the cranberry juice with *n*-3s increased HDL-C by 21.7% ($p = 0.029$).[74]

Moazen *et al.* supplemented the diets of men and women with overweight and T2DM with 50 g of freeze-dried strawberries daily (25 g in one cup of water twice daily) for 6 weeks.[77] Body mass, body fat percentage, BMI, waist circumference, and FBG did not change, but HbA_{1c} significantly decreased in the strawberry group compared to the control group ($p < 0.01$). Additionally, CRP and malondialdehyde (MDA), two biomarkers of oxidative stress and inflammation, both decreased, and the total antioxidant capacity of blood increased.[77] In a separate RCT, Schell *et al.* reported the effects in men and women with obesity and T2DM who consumed 250 g of red raspberries daily for four weeks.[72] The raspberries did not affect SBP, DBP, small or large artery elasticity, FBG, or TAGs but decreased the levels of interleukin-6 (IL-6) and high-sensitivity tumor necrosis factor-alpha (hsTNF-α). The concentrations of other biomarkers of oxidative stress and inflammation did not decline in the same manner.[72]

In the longest berry supplementation trial that we identified, Mirfeizi *et al.* compared the effects of providing 1 g cinnamon or 1 g of whortleberry to men and women with T2DM for 12 weeks.[75] While the placebo did not change any of the variables, whortleberry significantly decreased FBG (-22.6%, $p = 0.002$) and 2 hour post-prandial glucose (PPG, -26.4%, $p = 0.001$) *vs.* the control. Whortleberries did not change the levels of fasting insulin, HbA_{1c}, homeostatic model assessment of insulin resistance (HOMA-IR), TAGs, total cholesterol, LDL-C, or HDL-C compared to the control.[75]

Unlike the RCTs previously discussed, the final supplementation trial included in this review intervened with anthocyanins purified from bilberries and blackcurrants.[76] Compared to the placebo, 24 weeks of 160 mg of anthocyanins twice daily decreased FBG by 8.5% ($p < 0.05$) and HOMA-IR by 14.2% ($p < 0.05$). Additionally, TAGs declined by 23.0% ($p < 0.01$), as did total cholesterol and LDL-C by 3.7% and 7.9%, respectively ($p < 0.05$). Inversely, HDL-C increased by 19.4% ($p < 0.05$) alongside increases in adiponectin, high-molecular-weight adiponectin, and the high-molecular-weight adiponectin:total adiponectin ratio. Anthocyanins also improved several biomarkers of oxidative stress and inflammation. By week 24, the ferric ion reducing antioxidant power, total radical-trapping antioxidant parameter, and β-hydroxybutyrate level increased. Moreover, the anthocyanins decreased the levels of 8-iso-prostaglandin F2α, 13-hydroxyoctadecadienoic acid, carbonylated protein, IL-6, and TNF-α. Despite the aforementioned changes, anthocyanins did not affect BMI or the levels of percentage of body fat, SBP, DBP, fasting insulin, HbA_{1c}, Apo-A1, Apo-B100, or free fatty acids (FFAs).[76] Collectively, this RCT and the other long-term supplementation studies discussed above (Table 10.2) show that dietary berries and their

Table 10.2 The effects of long-term supplementation of dietary berries and their anthocyanins on anthropometry, body composition, hemodynamics, glucose, insulin, lipids, lipoproteins, adipokines, oxidative stress, and inflammation in type 2 diabetes mellitus.

Authors, Year (Country)	Trial Design	Participants	Intervention (Supplementation)	Anthropometry and Body Composition	Hemodynamics	Glucose and Insulin	Lipids, Lipoproteins, and Adipokines	Oxidative Stress and Inflammation
Stote *et al.*, 2020 (United States)[71]	Double-blind, placebo-controlled, randomized trial	Men with overweight or obesity and T2DM ($n = 52$) Age = 66.9 ± 5.6 years Body mass = 102.6 ± 5.6 kg BMI = 34.1 ± 4.5 kg m^{-2}	8 weeks 22 g freeze-dried blueberries (energy equivalent to 1 cup fresh blueberries; 261.8 mg anthocyanins, 765.6 mg phenolics) or 22 g placebo 11 g twice daily	NS – Body mass, BMI, WC	NS – heart rate, SBP, DBP	↓ HbA_{1c} and fructosamine at 8 weeks NS – FBG, fasting insulin	↓ TAGs at 8 weeks NS – TC, LDL-C, HDL-C	NS – CRP
Schell *et al.*, 2019 (United States)[72]	Randomized, controlled, crossover trial	Men and women with obesity and T2DM ($n = 22$) Age = 54.0 ± 4.2 years Body mass = 104.0 ± 11.0 kg BMI = 35.3 ± 2.0 kg m^{-2}	4 weeks 250 g red raspberries (343 mg polyphenols, 252 mg anthocyanins, 20 mg ellagic acid, 115 mg ellagitannins) or control Once daily	NR	NS – SBP, DBP, LAEI, SAEI	NS – FBG	NS – TAGs	↓ hsTNF-α, IL-6 NS – CRP, IL-1β, PAI-1
Lazavi *et al.*, 2018 (Iran)[73]	Randomized controlled trial	Men and women with T2DM on oral hypoglycemic agents ($n = 42$) Age = 55.4 ± 7.6 years Body mass = 73.7 ± 10.5 kg BMI = 28.5 ± 3.8 kg m^{-2}	8 weeks 200 mL barberry juice (2403 mg gallic acid equivalent) or control Once daily	↓ Body mass, BMI	↓ SBP, DBP	↓ FBG NS – HbA_{1c}	↓ TC NS – TAGs, LDL-C, HDL-C, Apo-A1, Apo-B100	↑ PON1 concentration
Zare Javid *et al.*, 2018 (Iran)[74]	Randomized controlled trial	Men and women with T2DM and periodontal disease ($n = 41$) Age = 55.4 ± 7.1 years	8 weeks T1: 1 g omega-3 FAs T2: 200 mL cranberry juice (195 mg	NS – Body mass, BMI, WC	NR	↓ HbA_{1c} (T1, T3) NS – FBG	↑ HDL-C (T3) NS – TAGs, TC, LDL-C, HDL-C	NR

Table 10.2 (*Continued*)

Authors, Year (Country)	Trial Design	Participants	Intervention (Supplementation)	Anthropometry and Body Composition	Hemodynamics	Glucose and Insulin	Lipids, Lipoproteins, and Adipokines	Oxidative Stress and Inflammation
		Body mass = 72.3 ± 11.2 kg BMI = 27.9 ± 5.0 kg m^{-2}	phenolics, 8 mg anthocyanins, 107 mg proanthocyanidins) T3: 200 mL cranberry juice enriched with 1 g omega-3 FAs (195 mg phenolics, 8 mg anthocyanins, 107 mg proanthocyanidins) T4: control Each treatment twice daily					
Mirfeizi *et al.*, 2016 (Iran)[75]	Triple-blind, placebo-controlled, randomized trial	Men and women with T2DM ($n = 102$) Age = 53.8 ± 11.7 years BMI = 28.7 ± 3.9 kg m^{-2}	12 weeks T1: 1 g cinnamon (91% cinnamaldehyde in essential oil) T2: 1 g whortleberry (9.8 mg anthocyanin) T3: placebo Each treatment once daily	↓ BMI (T1) NR - body mass	NR	↓ FBG (T1, T2), 2 h PPG (T1, T2), fasting insulin (T1, T2), HbA_{1c} (T1, T2), HOMA-IR (T1, T2)	NS – TAGs, TC, LDL-C, HDL-C	NR
Li *et al.*, 2015 (China)[76]	Double-blind, placebo-controlled, randomized trial	Men and women with T2DM ($n = 58$) Age = 57.9 ± 15.5 years BMI = 24.1 ± 17.6 kg m^{-2}	24 weeks 160 mg anthocyanins capsules (80 mg of anthocyanin comprised of 17 anthocyanins purified from bilberries and blackcurrants) or placebo Twice daily	NS – BMI, % body fat	NS – SBP, DBP	↓ FBG, HOMA-IR NS – fasting insulin, HbA_{1c}	↓ TAGs, TC, LDL-C, Apo-B48, Apo-CIII ↑ HDL-C ↑ adiponectin, HMW adiponectin, HMW: total adiponectin ratio NS – Apo-A1, Apo-B100, FFA	↑ β-hydroxybutyrate. FRAP, TRAP ↓ 8-iso-PGF_{2a}, 13-HODE, carbonylated protein, IL-6, TNF-α

Moazen *et al.*, 2013 (Iran)[77]	Double-blind, placebo-controlled, randomized trial	Men and women with T2DM and not receiving lipid-lowering therapy ($n = 36$) Age = 51.6 ± 10.0 years Body mass = 74.5 ± 10.5 kg BMI = 27.9 ± 3.7 kg m^{-2}	6 weeks 25 g FDS in one cup of water (50 g FDS daily = 500 g fresh strawberries; 2006 mg phenolics, 154 mg anthocyanins 50 mg phytosterols) or placebo in one cup of water Twice daily	NS – Body mass, BMI, WC, % body fat	NR	NS – FBG, HbA_{1c}	NR	↓ CRP, MDA ↑ TAC
Shidfar *et al.*, 2012 (Iran)[78]	Double-blind, placebo-controlled, randomized trial	Men with T2DM but not obesity ($n = 58$) Age = 54.8 ± 9.1 years Body mass = 74.0 ± 10.8 kg BMI = 28.8 ± 3.6 kg m^{-2}	12 weeks 240 mL per 1 cup cranberry juice or placebo Once daily	NR	NR	↓ FBG	↑ Apo-A1 ↓ Apo-B NS – Lipoprotein(a)	↑ PON1 activity

anthocyanins are promising interventions for improving metabolic biomarkers related to T2DM.

↓ lower/decreased; ↑ increased; 13-HODE: 13-hydroxyoctadecadienoic acid; 8-iso-$PGF_{2\alpha}$: 8-iso-prostaglandin F2α; Apo-A1: apolipoprotein-A1; Apo-B100: apolipoprotein-B100; Apo-B48: apolipoprotein-B-48; Apo-C-III: apolipoprotein-C-III; BMI: body mass index; CRP: C-reactive protein; DBP: diastolic blood pressure; FAs: fatty acids; FBG: fasting blood glucose; FDS: freeze-dried strawberry; FFAs: free fatty acids; FRAP: ferric ion reducing antioxidant power; HbA_{1c}: glycated hemoglobin; HDL-C: high-density lipoprotein cholesterol; HMW: high molecular weight; HOMA-IR: homeostatic model assessment of insulin resistance; hsTNF-α: high-sensitivity tumor necrosis factor-alpha; IL-1β: interleukin-1-beta; IL-6: interleukin-6; LAEI: large artery elasticity index; LDL-C: low-density lipoprotein cholesterol; MDA: malondialdehyde; *n*: sample size; NR: not reported; NS: non-significant differences between intervention and placebo; PAI-1: plasminogen activator inhibitor-1; PON1: paraoxonase 1; SAEI: small artery elasticity index; SBP: systolic blood pressure; T1-4: treatment 1-4; T2DM: type 2 diabetes mellitus; TAC: total antioxidant capacity; TAGs: triacylglycerols; TC: total cholesterol; TRAP: total radical-trapping antioxidant parameter; WC: waist circumference.

10.5 Dietary Berries and Diabetes: Acute Post-prandial Studies

Three acute post-prandial RCTs published since 2010 met our eligibility criteria. Three other reports were excluded for providing a fruit botanically classified as a dietary berry but not widely considered a berry,[80] providing a berry extract,[81] and overlapping with another included trial.[82] Each of the three included reports was a randomized, controlled, crossover trial conducted in the United States (see Table 10.3).[72,83,84] Due to the nature of the acute feeding trials, none blinded the researchers or participants to the interventions, which included red raspberries[72] and raw and/or dried cranberries.[83,84] Participants were aged in their 50s (years) and had obesity and T2DM in two of the trials.[72,83] In the third trial, participants were aged in their 60s, had obesity, and were on oral hypoglycemic agents.[84] Pertinent findings of the individual RCTs are discussed below.

Most recently, Schell *et al.* provided 250 mg of red raspberries blended in 1 cup of water with a Western-style breakfast to men and women with obesity and T2DM.[72] The raspberries caused the 2 hour and 4 hour PPG levels to decline by 16.5% and 17.6%, respectively, compared to the control phase ($p<0.05$). Additionally, the raspberries decreased post-prandial IL-6 and hsTNF-α but not hemodynamic variables, insulin, lipids, or other biomarkers of oxidative stress and inflammation. Two years before this trial, Schell *et al.* reported a study in which they provided similar participants with 40 g of dried cranberries and an identical Western-style breakfast and reported that the cranberries reduced the 2 hour and 4 hour PPG levels by 15.7% and 13.6%, respectively, compared to the control phase ($p<0.05$).[83] Also, total nitrite was

Table 10.3 The effects of acute supplementation of dietary berries and their anthocyanins on post-prandial hemodynamics, glucose, insulin, lipids, oxidative stress, and inflammation in type 2 diabetes mellitus.[a]

Authors, Year (Country)	Trial Design	Participants	Intervention and Dietary Challenge	Hemodynamics	Glucose and Insulin	Lipids	Oxidative Stress and Inflammation
Schell *et al.*, 2019 (United States)[72]	Randomized, controlled, crossover trial	Men and women with obesity and T2DM ($n = 22$) Age = 54.0 ± 4.2 years Body mass = 104.0 ± 11.0 kg BMI = 35.3 ± 2.0 kg m^{-2}	Intervention: 250 mg red raspberries blended in 1 cup of water (130 kcal, 30 g CHO, 343 mg polyphenols, 225 mg anthocyanins, 20 mg ellagic acid, 115 mg ellagitannins) Dietary challenge: scrambled eggs, butter, hash brown potatoes, buttermilk biscuits, and a sausage patty (974 kcal, 56 g CHO, 31 g PRO, 70 g FAT)	NS – SBP, DBP, LAEI, SAEI	↓ PPG at 2 and 4 h NS – post-prandial insulin	NS – TAGs, TC, LDL-C, HDL-C	↓ IL-6 and hsTNF-α at 4 h NS – CRP, IL-1β, PAI-1
Schell *et al.*, 2017 (United States)[83]	Randomized, controlled, crossover trial	Men and women with obesity and T2DM ($n = 25$) Age = 56.0 ± 6.0 years Body mass = 105.0 ± 13.0 kg BMI = 39.5 ± 32.5 kg m^{-2}	Intervention: 40 g dried cranberries (100 kcal, 30 g CHO, 178 mg polyphenols, 32 mg quercetin-3-rhammnoside, 29 mg quercetin-3-galactoside, 35 mg proanthocyanidins) Dietary challenge: scrambled eggs, butter, hash brown potatoes, buttermilk biscuits, and a sausage patty (974 kcal, 56 g CHO, 31 g PRO, 70 g FAT)	NS – SBP, DBP, SAEI, LAEI	↓ PPG at 2 and 4 h NS – post-prandial insulin, HOMA-IR	NS – TAGs, TC, LDL-C, HDL-C, LDL:HDL ratio	↓ IL-18 at 4 h, HNE and MDA at 4 h ↑ total nitrite at 2 h NS – CRP, IL-6

Table 10.3 (*Continued*)

Authors, Year (Country)	Trial Design	Participants	Intervention and Dietary Challenge	Hemodynamics	Glucose and Insulin	Lipids	Oxidative Stress and Inflammation
Wilson *et al.*, 2010 (United States)[84]	Randomized, controlled, crossover trial	Men and women with T2DM on oral hypoglycemic agents ($n = 13$) Age = 61.6 ± 2.3 years BMI = 33.3 ± 1.2 kg m^{-2}	Intervention: WB (57 g, 160 kcal, 1 g fiber); RC (55 g, 21 kcal, 1 g fiber); SDC (40 g, 138 kcal, 2.1 g fiber); SDC-LS (40 g, 113 kcal, 1.8 g fiber, 10 g polydextrose) Dietary challenge: N/A	NR	↓ PPG with SDC-LS at 1 h *vs.* WB ↓ PPG with SDC and SDC-LS at 2 h *vs.* WB ↓ peak PPG with RC at 30 min *vs.* WB, SDC, and SDC-LS ↓ peak insulin with RC at 30 min *vs.* WB, SDC, and SDC-LS ↓ glucose AUC with RC *vs.* WB, SDC, and SDC-LS ↓ post-prandial insulin with SDC-LS at 1 h *vs.* WB and SDC ↓ insulin AUC in SDC-LS *vs.* WB and SDC ↓ insulin AUC in RC *vs.* WB and SDC	NR	NR

[a] ↓ lower/decreased; ↑ increased; AUC: area under the curve; BMI: body mass index; CHO: carbohydrate; CRP: C-reactive protein; DBP: diastolic blood pressure; FAT: dietary fat HDL-C: high-density lipoprotein cholesterol; HNE: 4-hydroxynonenal; hsTNF-α: high-sensitivity tumor necrosis factor-alpha; IL-18: interleukin-18; IL-1β: interleukin-1-beta; IL-6: interleukin-6; LAEI: large artery elasticity index; LDL-C: low-density lipoprotein cholesterol; MDA: malondialdehyde; *n*: sample size; N/A: not applicable (did not implement a dietary challenge); NR: not reported; NS: non-significant differences between the interventions; PAI-1: plasminogen activator inhibitor-1; PPG: post-prandial glucose; PRO: protein; RC: raw cranberries; SAEI: small artery elasticity index; SBP: systolic blood pressure; SDC: sweetened dried cranberries; SDC-LS: low-sugar sweetened dried cranberries; T2DM: type 2 diabetes mellitus; TAGs: triacylglycerols; TC: total cholesterol; WB: white bread.

elevated at 2 hours post-prandial, and the concentrations of interleukin-18 (IL-18) and hydroxynonenal and MDA combined (hydroxynonenal + MDA) were decreased at 4 hours post-prandial. Similar to the raspberry intervention, the cranberry intervention did not affect hemodynamic variables, lipids, or other biomarkers of oxidative stress and inflammation (*i.e.* CRP and IL-6).[83] In the final acute post-prandial study included in this review, Wilson *et al.* compared the post-prandial effects of white bread, raw cranberries, sweetened dried cranberries, and low-sugar sweetened dried cranberries in men and women with obesity and T2DM who were taking oral hypoglycemic agents.[84] Several findings were reported and are included in Table 10.3. The overall PPG response elicited by the raw cranberries, measured as the area under the curve (AUC), was 22.1%, 19.5%, and 13.7% lower than the responses respectively elicited by the white bread, sweetened dried cranberries, and low-sugar sweetened dried cranberries ($p < 0.05$). Additionally, the cumulative post-prandial insulin response, also measured as AUC, was 30.0% and 29.5% with the low-sugar sweetened dried cranberries compared to the white bread and sweetened dried cranberries, respectively ($p < 0.05$). Finally, the raw cranberries also produced a 53.6% and 53.3% lower post-prandial insulin AUC compared to the white bread and sweetened dried cranberries, respectively ($p < 0.05$). As a group, the three acute studies discussed in this chapter reflect the potential of cranberries and raspberries to beneficially reduce PPG with or without reductions in post-prandial insulin. Additionally, both types of berries may reduce inflammation and oxidative stress after a pro-inflammatory Western-style breakfast. While the data do not currently show that cranberries or raspberries improve post-prandial blood pressure or arterial elasticity, future studies should attempt to verify these findings.

10.6 Dietary Berries and Diabetes: Mechanisms and Animal Models

As illustrated thus far, the consumption of dietary berries is linked to favorable metabolic outcomes in both observational and clinical studies in humans. These studies comprise part of a broader, growing body of literature exploring the purported physiological mechanisms by which dietary berries and their polyphenols, particularly flavonoids, exert anti-diabetic effects in living organisms, including non-human animals and humans. Recent reviews have provided summaries on the topic.[69,85,86]

Three proposed mechanisms of action include reducing oxidative stress and inflammation, upregulating insulin-dependent (phosphoinositide 3-kinase [PI3K]/protein kinase B [Akt/PKB]) and insulin-independent (adenosine-monophosphate-activated protein kinase [AMPK]) metabolic pathways, and inhibiting carbohydrate digestion and absorption.[69] The first two mechanisms are intricately connected because imbalances in the cellular redox state of the pancreas, liver, skeletal muscle, and adipose tissue decrease the activity of insulin-dependent and insulin-independent signaling pathways. The downregulation of these pathways impairs the production, secretion, and action of insulin, thereby promoting insulin resistance subsequent hyperglycemia.[69] Reducing oxidative stress and inflammation to promote metabolic signaling improves preserve β-cell function, insulin resistance, and hyperglycemia. The third prominent, proposed action of berry flavonoids is interfering with carbohydrate digestion and absorption in the small intestine by inhibiting digestive enzymes such as α-amylase and α-glucosidase and the transport of glucose across the intestinal epithelium into the hepatic circulation.[69] Next, we discuss animal studies that have collectively served as the proof of concept for the human feeding trials reviewed in this chapter. The great number of relevant studies and different interventions precludes us from providing a comprehensive summary of the evidence for every dietary berry. Therefore, only a small selection of animal data is discussed here.

A recent animal study provided 0% or 5% freeze-dried raspberries to C57BL/6J mice for 12 weeks to determine if raspberries affected responses to a high-fat diet (HFD). Compared to the 0% control, the 5% group experienced significantly improved insulin sensitivity and browning of white adipose tissue (WAT) as well as decreased WAT hypertrophy, macrophage infiltration, and cytokine expression.[87] Studies have also shown raspberries to attenuate increases in body mass,[88,89] oxidative stress,[89,90] and hepatic lipid accumulation while also normalizing blood glucose and insulin resistance.[89] It is hypothesized that raspberry polyphenols and their metabolites may scavenge some of the excessive free radicals (*i.e.* reactive oxygen species, ROS) in the blood that are overproduced during hyperglycemia (*e.g.* superoxide). This hypothesis was supported by the raspberry group having lower ROS in blood erythrocytes than the control group.[90] Another plausible mechanism by which raspberries reduced oxidative stress is by upregulating antioxidant enzymes, specifically glutathione peroxidase (GPx), which was two times greater in the raspberry group than the control group.[90] By increasing ROS neutralization and attenuating increases in body mass, ROS production, and liver fat, raspberries stave off insulin resistance and glucose intolerance.

Raspberries are just one of many types of commonly consumed berries with demonstrable anti-diabetic effects in animal models. Strawberries have also received great attention, and their effects were reviewed in 2020.[91] In one animal study, Wistar rats with alloxan-induced DM were provided 700 mg of freeze-dried strawberry daily for 45 days. Alloxan exacerbates ROS production in pancreatic β-cells, causing cellular dysfunction, reduced insulin output, and hyperglycemia. The function of β-cells was restored in the strawberry group compared to the control group, as indicated by normalized insulin and glucose levels after the intervention.[92] Additionally, hepatic tissue was afforded protection in the strawberry group *vs.* the control; the serum levels of enzymes indicative of oxidative damage to hepatocellular membrane phospholipids were restored to near-normal levels after the intervention.[92] Similar anti-diabetic effects have also been reported in the case of blueberries[93,94] and cranberries,[93,95] as well as dietary anthocyanins,[96] which are the most common dietary bioactive compounds in berries.

10.7 Recommendations in Diabetes Management

The ADA provides clinical guidelines intended to help health professionals, especially physicians and registered dietitians, and their patients prevent the progression of pre-diabetes to DM and DM to overt CVD.[10] Lifestyle modifications to become more active, improve the diet, and reduce body weight should be implemented as the first therapeutic step upon clinical diagnosis of pre-diabetes or DM.[10,13] While various dietary patterns can delay or prevent DM and improve DM outcomes once diagnosed, moderating the consumption of calories, processed foods, and dietary fat is important. Consuming increasing amounts of whole grains, legumes, nuts, fruits, and vegetables is advisable for increasing the intake of important nutrients such as fiber and antioxidant vitamins and minerals, as well as dietary bioactive compounds.[9,10]

Considering the ADA's recommendations, increasing the intake of fruits by adding whole berries to the diet is a prudent choice. More specifically, dietary berries should be consumed daily in their whole-fruit form over juices and extracts. The observational evidence identified in this chapter suggests that consuming more than five servings of dietary berries weekly,[55] or at least 20 g per day ($\approx$1/6 cup of whole berries for many varieties),[57] may confer protection against T2DM.[55,57] Additionally, consuming $\geq$2 servings of blueberries weekly ($\approx$ two cups of whole berries),[56] or approximately 60 g ($\approx$1/2 cup of whole berries) daily,[54] was associated with a reduced risk. Presumably, people at risk for T2DM could reach the weekly consumption of dietary berries associated with reduced risks by eating a serving of blueberries, cranberries, raspberries, or strawberries as a snack or part of a meal once per day. As a result of industrialization and globalized trade, each of these berry variants is available year-round in several forms (dried, juiced, and whole raw or frozen) in many grocery stores across the United States and, presumably, in other countries.

Our recommendation to consume dietary berries daily can be extended from people at risk of developing T2DM to people diagnosed with the disease. The long-term supplementation studies included in this review showed that, overall, the daily intake of dietary berries often improved hyperglycemia. While lipids were not as consistently improved by berry intake, in some cases, there were marked reductions in TAGs and augmentations of the HDL-C concentration shown to improve cardiovascular risks linked to T2DM. Changes in glycemia, lipidemia, and biomarkers related to oxidative stress and inflammation were observed with whole, freeze-dried, and juiced dietary berries, demonstrating that a variety of berry fruits and forms are conducive to improving metabolism in T2DM. Regarding the timing of dietary berry consumption, the acute post-prandial studies in this review suggest that eating the berries simultaneously with a Western-style meal (processed and fat-laden) attenuates post-prandial hyperglycemia and inflammation. Given the common consumption of this type of meal by many people in Western societies, particularly the United States, it may be advantageous to consume the berries with a meal rather than alone. However, more data will be required to support this recommendation.

10.8 Conclusion and Future Directions

This chapter began with an overview of the etiology and diagnosis of DM, a chronic metabolic disease with a burgeoning global prevalence. Then, the chapter provided a brief explanation of the bioactive compounds in fruits and vegetables that, when ingested, elicit favorable physiological effects in humans, especially in reducing risks of chronic metabolic diseases such as DM and CVD. A special focus was placed on dietary berries due to their popular worldwide consumption and high concentrations of complex carbohydrates, fiber, antioxidant vitamins and minerals, and other bioactive phytochemicals by weight and caloric value. After describing the methodology of our literature search and selection process, we introduced readers to large-scale observational studies that associated dietary berries with a reduced risk of T2DM. Then, readers were presented with experimental data from RCTs and mechanisms from animal studies that revealed several promising effects of consuming dietary berries over the long-term and acutely in T2DM, such as ameliorating fasting and post-prandial hyperglycemia, lipidemia, oxidative stress, and inflammation. We concluded the chapter by recommending that berries should be incorporated into current dietary practices intended to treat and manage T2DM.

In addition to sharing our view that dietary berries may reduce the risk of initially developing T2DM and improve outcomes in overt T2DM, we highlight the need for future studies to examine the effects of dietary berries in GDM. Observational studies offer conflicting data on the associations between intakes of dietary fruits and berries and GDM risk. Moreover, to our knowledge, no RCT has been published on the topic of dietary berries and GDM. While the pathophysiology of GDM is similar to that of T2DM, GDM is considered a condition that is unique to pregnancy and has its own environmental and genetic influences and health ramifications. Therefore, we are hopeful that new investigations will examine supplementation of dietary berries in GDM-affected pregnancies to elucidate their efficacy in improving maternal and fetal health. We also wish to highlight the need for more acute post-prandial studies with dietary berries, as we identified only a handful during our search. Lastly, we recommend that investigations with dietary berries in both GDM and T2DM primarily intervene with berries in their whole, dried, juiced, or powdered forms that could be commercially produced on a large scale and made widely available beyond the laboratory. The current prevalence of DM, over 450 million cases globally, is rapidly growing and comes with enormous personal, societal, and economic costs. The identification of affordable and accessible foods that improve T2DM risk and outcomes, such as dietary berries, will be critical to alleviating this public health burden.

Abbreviations

13-HODE	13-hydroxyoctadecadienoic acid
8-iso-$PGF_{2\alpha}$	8-iso-prostaglandin F2α
ADA	American Diabetes Association
Akt	protein kinase B
AMPK	adenosine-monophosphate-activated protein kinase
Apo-A1	apolipoprotein A1
Apo-B	apolipoprotein B
Apo-B48	apolipoprotein-B-48
Apo-C-III	apolipoprotein-C-III
AUC	area under the curve
BMI	body mass index
CHO	carbohydrates
CI	confidence interval
CRP	C-reactive protein
CVD	cardiovascular disease
DBP	diastolic blood pressure
DM	diabetes mellitus
DPP	Diabetes Prevention Program
FA	fatty acids
FBG	fasting blood glucose
FBV	fruits, berries, and vegetables
FDS	freeze-dried strawberries
FFA	free fatty acid
FMC	Finnish Mobile Clinic Health Examination Survey
FPG	fasting plasma glucose
FRAP	ferric ion reducing antioxidant power
FW	formula weight
GDM	gestational diabetes mellitus
GI	glycemic index
GPx	glutathione peroxidase
HbA_{1c}	blood concentration of glycated hemoglobin
HDL-C	high-density lipoprotein cholesterol
HFD	high-fat diet
HMW	high molecular weight
HNE	4-hydroxynonenal
HOMA-IR	homeostatic model assessment of insulin resistance
HPFS	Health Professionals Follow-up Study
HR	hazard ratio
hsTNF-α	high-sensitivity tumor necrosis factor-alpha
IFG	impaired fasting glucose
IL-18	interleukin-18
IL-1β	interleukin-1-beta
IL-6	interleukin-6
KIHD	Kuopio Ischaemic Heart Disease Risk Factor Study

LAEI	large artery elasticity index
LDL-C	low-density lipoprotein cholesterol
MDA	malondialdehyde
n-3	omega-3
n	sample size
N/A	not applicable
NHS	Nurses' Health Study
NHS II	Nurses' Health Study II
NR	not reported
NS	non-significant differences between interventions and placebo or between interventions
OGTT	oral glucose tolerance test
OR	odds ratio
p	probability
PAI-1	plasminogen activator inhibitor-1
PI3K	phosphoinositide 3-kinase
PKB	protein kinase B
PON1	paraoxonase 1
PPG	post-prandial glucose
PRO	protein
RC	raw cranberries
RCT	randomized controlled trial
ROS	reactive oxygen species
RR	relative risk
SAEI	small artery elasticity index
SBP	systolic blood pressure
SDC-LS	low-sugar sweetened dried cranberries
SDC	sweetened dried cranberries
T1–4	treatment 1–4
T1DM	type 1 diabetes mellitus
T2DM	type 2 diabetes mellitus
TAC	total antioxidant capacity
TAG	triacylglycerols
TC	total cholesterol
TRAP	total radical-trapping antioxidant parameter
WAT	white adipose tissue
WB	white bread
WC	waist circumference

Acknowledgements

We acknowledge the support of the California Strawberry Commission, the Cranberry Institute, and the US Highbush Blueberry Council for supporting our studies.

References

1. American Diabetes Association, *Diabetes Care*, 2020, **43**, S14.
2. H. D. McIntyre, P. Catalano, C. Zhang, G. Desoye, E. R. Mathiesen and P. Damm, *Nat. Rev. Dis. Primers*, 2019, **5**, 47.
3. N. H. Cho, J. E. Shaw, S. Karuranga, Y. Huang, J. D. da Rocha Fernandes, A. W. Ohlrogge and B. Malanda, *Diabetes Res. Clin. Pract.*, 2018, **138**, 271.
4. National Center of Health Statistics, https://www.cdc.gov/nchs/data/databriefs/db360-h.pdf, (accessed January 2020).
5. Centers for Disease Control and Prevention, https://www.cdc.gov/diabetes/data/statistics-report/index.html, (accessed January 2020).
6. American Diabetes Association, https://www.diabetes.org/resources/statistics/statistics-about-diabetes, (accessed January 2020).
7. Centers for Disease Control and Prevention, https://www.cdc.gov/diabetes/basics/risk-factors.html, (accessed October 2020).
8. Office of Disease Prevention and Health Promotion, https://health.gov/paguidelines/second-edition/pdf/Physical_Activity_Guidelines_2nd_edition.pdf, (accessed January 2020).
9. Office of Disease Prevention and Health Promotion, https://health.gov/our-work/food-nutrition/2015-2020-dietary-guidelines/guidelines/, (accessed January 2020).
10. American Diabetes Association, *Diabetes Care*, 2020, **43**, S32.
11. M. Asif, *J. Educ. Health Promot.*, 2014, **3**, 54.
12. W. C. Knowler, E. Barrett-Connor, S. E. Fowler, R. F. Hamman, J. M. Lachin, E. A. Walker and D. M. Nathan, *N. Engl. J. Med.*, 2002, **346**, 393.
13. D. M. Nathan, E. Barrett-Connor, J. P. Crandall, S. L. Edelstein, R. B. Goldberg, E. S. Horton, W. C. Knowler, K. J. Mather, T. J. Orchard, X. Pi-Sunyer, D. Schade and M. Temprosa, *Lancet Diabetes Endocrinol.*, 2015, **3**, 866.
14. A. R. Parker, L. Byham-Gray, R. Denmark and P. J. Winkle, *J. Acad. Nutr. Diet.*, 2014, **114**, 1739.
15. R. H. Liu, *Am. J. Clin. Nutr.*, 2003, **78**, 517S.
16. C. Forni, F. Facchiano, M. Bartoli, S. Pieretti, A. Facchiano, D. D'Arcangelo, S. Norelli, G. Valle, R. Nisini, S. Beninati, C. Tabolacci and R. N. Jadeja, *BioMed Res. Int.*, 2019, **2019**, 8748253.
17. R. Tsao, *Nutrients*, 2010, **2**, 1231.
18. U.S. Department of Agriculture, https://www.ers.usda.gov/data-products/fruit-and-tree-nuts-data/fruit-and-tree-nuts-yearbook-tables/, (accessed August 2020).
19. U.S. Department of Agriculture, https://fdc.nal.usda.gov/fdc-app.html#/food-details/747448/nutrients, (accessed August 2020).
20. U.S. Department of Agriculture, https://fdc.nal.usda.gov/fdc-app.html#/food-details/171711/nutrients, (accessed August 2020).
21. D. Del Rio, G. Borges and A. Crozier, *Br. J. Nutr.*, 2010, **104**, S67.
22. C. Vetrani, G. Costabile, M. Vitale and R. Giacco, *J. Funct. Foods*, 2020, **71**.
23. Phenol-Explorer, http://phenol-explorer.eu/, (accessed August 2020).

24. Phenol-Explorer, http://phenol-explorer.eu/foods/79, (accessed June 2020).
25. Phenol-Explorer, http://phenol-explorer.eu/foods/95, (accessed June 2020).
26. Phenol-Explorer, http://phenol-explorer.eu/contents/food/74, (accessed August 2020).
27. Phenol-Explorer, http://phenol-explorer.eu/foods/67, (accessed June 2020).
28. Phenol-Explorer, http://phenol-explorer.eu/contents/food/69, January 2020).
29. J. Scalzo, A. Politi, N. Pellegrini, B. Mezzetti and M. Battino, *Nutrition*, 2005, **21**, 207.
30. S. Tulipani, B. Mezzetti, F. Capocasa, S. Bompadre, J. Beekwilder, C. H. Ric De Vos, E. Capanoglu, A. Bovy and M. Battino, *J. Agric. Food Chem.*, 2008, **56**, 696.
31. K. Šavikin, M. Mikulič-Petkovšek, B. Djordjević, G. Zdunić, T. Janković, D. Djurović and R. Veberič, *Sci. Hortic.*, 2013, **160**, 20.
32. A. Wojdyło, A. Figiel and J. Oszmiański, *J. Agric. Food Chem.*, 2009, **57**, 1337.
33. L. R. Howard, R. L. Prior, R. Liyanage and J. O. Lay, *J. Agric. Food Chem.*, 2012, **60**, 6678.
34. A. Rodriguez-Mateos, T. Cifuentes-Gomez, T. W. George and J. P. E. Spencer, *J. Agric. Food Chem.*, 2014, **62**, 3979.
35. Y. Zhao, X. Wu, L. Yu and P. Chen, *J. Food Compos. Anal.*, 2017, **56**, 55.
36. T. Bohn, *Nutr. Rev.*, 2014, **72**, 429.
37. S. Liu, M. Serdula, S. J. Janket, N. R. Cook, H. D. Sesso, W. C. Willett, J. E. Manson and J. E. Buring, *Diabetes Care*, 2004, **27**, 2993.
38. R. Villegas, X. O. Shu, Y. T. Gao, G. Yang, T. Elasy, H. Li and W. Zheng, *J. Nutr.*, 2008, **138**, 574.
39. A. H. Harding, N. J. Wareham, S. A. Bingham, K. T. Khaw, R. Luben, A. Welch and N. G. Forouhi, *Arch. Intern. Med.*, 2008, **168**, 1493.
40. A. J. Cooper, N. G. Forouhi, Z. Ye, B. Buijsse, L. Arriola, B. Balkau, A. Barricarte, J. W. J. Beulens, H. Boeing, F. L. Büchner, C. C. Dahm, B. de Lauzon-Guillain, G. Fagherazzi, P. W. Franks, C. Gonzalez, S. Grioni, R. Kaaks, T. J. Key, G. Masala, C. Navarro, P. Nilsson, K. Overvad, S. Panico, J. Ramón Quirós, O. Rolandsson, N. Roswall, C. Sacerdote, M. J. Sánchez, N. Slimani, I. Slujis, A. M. W. Spijkerman, B. Teucher, A. Tjonneland, R. Tumino, S. J. Sharp, C. Langenberg, E. J. M. Feskens, E. Riboli, N. J. Wareham and The InterAct Consortium, *Eur. J. Clin. Nutr.*, 2012, **66**, 1082.
41. A. J. Cooper, S. J. Sharp, M. A. H. Lentjes, R. N. Luben, K. Khaw, N. J. Wareham and N. G. Farouhi, *Diabetes Care*, 2012, **35**, 1293.
42. K. Kurotani, A. Nanri, A. Goto, T. Mizoue, M. Noda, M. Kato, M. Inoue and S. Tsugane, *Br. J. Nutr.*, 2013, **109**, 709.
43. A. J. M. Cooper, S. J. Sharp, R. N. Luben, K. T. Khaw, N. J. Wareham and N. G. Forouhi, *Eur. J. Clin. Nutr.*, 2015, **69**, 449.

44. L. Mamluk, M. G. O'Doherty, P. Orfanos, G. Saitakis, J. V. Woodside, L. M. Liao, R. Sinha, P. Boffetta, A. Trichopoulou and F. Kee, *Eur. J. Clin. Nutr.*, 2017, **71**, 83.
45. J. Liang, Y. Zhang, A. Xue, J. Sun, X. Song, B. Xue, F. Ji, W. Gao, L. He, Z. Pang, Q. Qiao and F. Ning, *Asia Pac. J. Clin. Nutr.*, 2017, **26**, 255.
46. J. S. Zheng, S. J. Sharp, F. Imamura, R. Chowdhury, T. E. Gundersen, M. Steur, I. Sluijs, Y. T. van der Schouw, A. Agudo, D. Aune, A. Barricarte, H. Boeing, M. D. Chirlaque, M. Dorronsoro, H. Freisling, D. El-Fatouhi, P. W. Franks, G. Fagherazzi, S. Grioni, M. J. Gunter, C. Kyrø, V. Katzke, T. Kühn, K. T. Khaw, N. Laouali, G. Masala, P. M. Nilsson, K. Overvad, S. Panico, K. Papier, J. R. Quirós, O. Rolandsson, D. Redondo-Sánchez, F. Ricceri, M. B. Schulze, A. M. W. Spijkerman, A. Tjønneland, T. Y. N. Tong, R. Tumino, E. Weiderpass, J. Danesh, A. S. Butterworth, E. Riboli, N. G. Forouhi and N. J. Wareham, *BMJ*, 2020, **370**, DOI: 10.1136bmj.m2194.
47. C. Wu, P. Liu and Z. Yuan, *Int. J. Food Sci. Nutr.*, 2020, **72**, 208.
48. A. Ahmed, A. Lager, P. Fredlund and L. S. Elinde, *J. Nutr. Sci.*, 2020, **9**, 1.
49. M. Li, Y. Fan, X. Zhang, W. Hou and Z. Tang, *BMJ Open*, 2014, **4**, e005497.
50. P. Y. Wang, J. C. Fang, Z. H. Gao, C. Zhang and S. Y. Xie, *J. Diabetes Invest.*, 2016, **7**, 56.
51. L. Schwingshackl, G. Hoffmann, A. M. Lampousi, S. Knüppel, K. Iqbal, C. Schwedhelm, A. Bechthold, S. Schlesinger and H. Boeing, *Eur. J. Epidemiol.*, 2017, **32**, 363.
52. Y. Wu, D. Zhang, X. Jiang and W. Jiang, *Nutr., Metab. Cardiovasc. Dis.*, 2015, **25**, 140.
53. X. Guo, B. Yang, J. Tan, J. Jiang and D. Li, *Eur. J. Clin. Nutr.*, 2016, **70**, 1360.
54. J. Mursu, J. K. Virtanen, T. P. Tuomainen, T. Nurmi and S. Voutilainen, *Am. J. Clin. Nutr.*, 2014, **99**, 328.
55. I. Muraki, F. Imamura, J. E. Manson, F. B. Hu, W. C. Willett, R. M. van Dam and Q. Sun, *BMJ*, 2013, **347**, f5001.
56. N. M. Wedick, A. Pan, A. Cassidy, E. B. Rimm, L. Sampson, B. Rosner, W. Willett, F. B. Hu, Q. Sun and R. M. van Dam, *Am. J. Clin. Nutr.*, 2012, **95**, 925.
57. J. Montonen, R. Järvinen, M. Heliövaara, A. Reunanen, A. Aromaa and P. Knekt, *Eur. J. Clin. Nutr.*, 2005, **59**, 441.
58. L. Chen, F. B. Hu, E. Yeung, D. K. Tobias, W. C. Willett and C. Zhang, *Diabetes Care*, 2012, **35**, 1079.
59. P. Mirmiran, S. Hosseinpour-Niazi, L. Moghaddam-Banaem, M. Lamyian, A. Goshtasebi and F. Aziz, *Int. J. Vitam. Nutr. Res.*, 2019, **89**, 37.
60. X. Zhou, R. Chen, C. Zhong, J. Wu, X. Li, Q. Li, W. Cui, N. Yi, M. Xiao, H. Yin, G. Xiong, W. Han, L. Hao, X. Yang and N. Yang, *Nutrition*, 2019, **60**, 129.
61. The University of Sydney, https://glycemicindex.com/gi-search/&?food_name=wild, (accessed August 2020).
62. The University of Sydney, https://glycemicindex.com/gi-search/&?food_name=strawberries, (accessed August 2020).

63. W. Q. Huang, Y. Lu, M. Xu, J. Huang, Y. X. Su and C. X. Zhang, *Sci. Rep.*, 2017, **7**, 43620.
64. N. M. Pham, V. Van Do and A. H. Lee, *Eur. J. Clin. Nutr.*, 2019, **73**, 647.
65. M. de Andrade Balbi, L. C. Crivellenti, D. C. C. Zuccolotto, L. J. Franco and D. S. Sartorelli, *Arch. Endocrinol. Metab.*, 2019, **63**, 241.
66. E. A. Tryggvadottir, H. Medek, B. E. Birgisdottir, R. T. Geirsson and I. Gunnarsdottir, *Eur. J. Clin. Nutr.*, 2016, **70**, 237.
67. J. F. Reis, V. V. S. Monteiro, R. de Souza Gomes, M. M. do Carmo, G. V. da Costa, P. C. Ribera and M. C. Monteiro, *J. Transl. Med.*, 2016, **14**, 315.
68. V. A. do Rosario, J. Spencer, K. Weston-Green and K. Charlton, *Curr. Nutr. Rep.*, 2020, **9**, 381.
69. B. Burton-Freeman, M. Brzeziński, E. Park, A. Sandhu, D. Xiao and I. Edirisinghe, *Nutrients*, 2019, **11**, 841.
70. A. Calvano, K. Izuora, E. C. Oh, J. L. Ebersole, T. J. Lyons and A. Basu, *Food Funct.*, 2019, **10**, 6227.
71. K. S. Stote, M. M. Wilson, D. Hallenbeck, K. Thomas, J. M. Rourke, M. I. Sweeney, K. T. Gottschall-Pass and A. R. Gosmanov, *Curr. Dev. Nutr.*, 2020, **4**, nzaa030.
72. J. Schell, N. M. Betts, T. J. Lyons and A. Basu, *Ann. Nutr. Metab.*, 2019, **74**, 165.
73. F. Lazavi, P. Mirmiran, G. Sohrab, O. Nikpayam, P. Angoorani and M. Hedayati, *Complement. Ther. Clin. Pract.*, 2018, **31**, 170.
74. A. Zare Javid, L. Maghsoumi-Norouzabad, E. Ashrafzadeh, H. A. Yousefimanesh, M. Zakerkish, K. A. Angali, M. Ravanbakhsh and H. Babaei, *J. Am. Coll. Nutr.*, 2018, **37**, 71.
75. M. Mirfeizi, Z. Mehdizadeh Tourzani, S. Z. Mirfeizi, M. Asghari Jafarabadi, H. R. Rezvani and M. Afzali, *J. Diabetes*, 2016, **8**, 647.
76. D. Li, Y. Zhang, Y. Liu, R. Sun and M. Xia, *J. Nutr.*, 2015, **145**, 742.
77. S. Moazen, R. Amani, A. H. Rad, H. Shahbazian, K. Ahmadi and M. T. Jalali, *Ann. Nutr. Metab.*, 2013, **63**, 256.
78. F. Shidfar, I. Heydari, S. J. Hajimiresmaiel, S. Hosseini, S. Shidfar and F. Amiri, *J. Res. Med. Sci.*, 2012, **17**, 355.
79. L. Yang, W. Ling, Y. Yang, Y. Chen, Z. Tian, Z. Du, J. Chen, Y. Xie, Z. Liu and L. Yang, *Nutrients*, 2017, **9**, 1104. DOI: 10.3390/nu9101104.
80. S. A. Banihani, S. M. Makahleh, Z. El-Akawi, R. A. Al-Fashtaki, O. F. Khabour, M. Y. Gharibeh, N. A. Saadah, F. H. Al-Hashimi and N. J. Al-Khasieb, *Nutr. Res.*, 2014, **34**, 862.
81. N. Hoggard, M. Cruickshank, K. M. Moar, C. Bestwick, J. J. Holst, W. Russell and G. Horgan, *J. Nutr. Sci.*, 2013, **2**, e22. DOI: 10.1017/jns.2013.16.
82. T. Wilson, S. L. Meyers, A. P. Singh, P. J. Limburg and N. Vorsa, *J. Food Sci.*, 2008, **73**, H241.
83. J. Schell, N. M. Betts, M. Foster, R. H. Scofield and A. Basu, *Food Funct.*, 2017, **8**, 3083.

84. T. Wilson, J. L. Luebke, E. F. Morcomb, E. J. Carrell, M. C. Leveranz, L. Kobs, T. P. Schmidt, P. J. Limburg, N. Vorsa and A. P. Singh, *J. Food Sci.*, 2010, **75**, H218.
85. I. Edirisinghe and B. Burton-Freeman, *J. Berry Res.*, 2016, **6**, 237.
86. A. Hameed, M. Galli, E. Adamska-Patruno, A. Krętowski and M. Ciborowski, *Nutrients*, 2020, **12**, 2538.
87. T. Xing, Y. Kang, X. Xu, B. Wang, M. Du and M. J. Zhu, *Mol. Nutr. Food Res.*, 2018, **62**, 1035.
88. T. Luo, O. Miranda-Garcia, A. Adamson, G. Sasaki and N. F. Shay, *J. Berry Res.*, 2016, **6**, 213.
89. M. J. Zhu, Y. Kang, Y. Xue, X. Liang, M. P. G. García, D. Rodgers, D. R. Kagel and M. Du, *J. Nutr. Biochem.*, 2018, **53**, 96.
90. G. D. Noratto, B. P. Chew and L. M. Atienza, *Food Chem.*, 2017, **227**, 305.
91. M. D. Putri, B. Wiboworini and P. Dirgahayu, *Int. J. Nutr. Sci.*, 2020, **5**, 1.
92. S. S. Abdulazeez, *J. Taibah Univ. Medical Sci.*, 2014, **9**, 268.
93. D. M. Rocha, A. P. Caldas, B. P. da Silva, H. H. Hermsdorff and R. D. Alfenas, *Crit. Rev. Food Sci. Nutr.*, 2019, **59**, 1816.
94. M. Shi, H. Loftus, A. J. McAinch and X. Q. Su, *J. Funct. Foods*, 2017, **30**, 16.
95. K. Kowalska and A. Olejnik, *J. Funct. Foods*, 2016, **20**, 171.
96. R. A. S. Sancho and G. M. Pastore, *Food Res. Int.*, 2012, **46**, 378–386.

CHAPTER 11

The Role of Berries and Their Bioactive Compounds on Obesity-induced Inflammation

C. DEL BO',*[a] S. VENDRAME,[b] D. MARTINI,[a] P. RISO[a] AND D. KLIMIS-ZACAS*[b]

[a] Department of Food, Environmental and Nutritional Sciences, Division of Human Nutrition, Università degli Studi di Milano, 20133, Italy; [b] School of Food and Agriculture, University of Maine, Orono ME 04469, USA
*Emails: cristian.delbo@unimi.it; dorothea@maine.edu

11.1 Introduction

11.1.1 The Obesity Epidemic

Although largely preventable and reversible by means of simple lifestyle interventions, such as diet and physical activity, the incidence and prevalence of overweight (defined as having a BMI > 25) and obesity (BMI > 30) has been following a constantly growing trend over the last half-century.[1]

According to the WHO, worldwide obesity has almost tripled from 1975 to 2016, with 39% of all adults being overweight and 13% obese. Even more alarmingly, the prevalence of overweight and obesity in the 5–19 years age group has increased more than fourfold since 1975, with 18% of all children and adolescents worldwide being overweight or obese. As a result of the epidemic proportions to which the problem has grown, over 4 million people die each year as a direct consequence of being obese or overweight.[2]

Food Chemistry, Function and Analysis No. 33
Berries and Berry Bioactive Compounds in Promoting Health
Edited by Dorothy Klimis-Zacas and Ana Rodriguez-Mateos

Published by the Royal Society of Chemistry, www.rsc.org

The recent lifestyle changes associated with the COVID-19 pandemic, including stay-at-home orders, closure of gyms, restriction of social activities, and disruption of daily routines, have exacerbated the obesogenic environment of hundreds of millions of households worldwide, further reducing physical activity, access to fresh healthy food and increasing stress-related inflammation. It is projected that its impact on obesity rates will be particularly severe in the pediatric population, where the fastest-growing rate of severe obesity has been registered in recent years.[3] Thus, it will become ever more crucial to develop simple and effective dietary and lifestyle strategies to attenuate the spread of obesity and its health-related complications in the general population.

11.1.2 Adiposity and Inflammation

Overweight and obesity, particularly if associated with a sizable proportion of visceral adiposity, is the most widely recognized cause of a chronically sustained, subclinical, systemic inflammatory status, which is a major risk factor contributing or leading to a plethora of chronic conditions, including cardiovascular disease, type 2 diabetes mellitus (DMT2), neurodegenerative disease and some form of cancers.[4,5] By causing unnecessary tissue damage, chronic inflammation leads to endothelial dysfunction and impaired vascular reactivity.[6] Inflammation and oxidative stress are strongly interlinked, so inflammation increases the production of reactive oxygen species (ROS) and this, in turn, sustains a pro-inflammatory environment.[7] Inflammation also leads to impairments in glucose and insulin metabolism, which in turn exacerbate inflammation and tissue damage. Such tendency of risk factors to cluster results in a high-risk condition called Metabolic Syndrome (MetS).[8]

The reason underlying the association between visceral adiposity and chronic inflammation is the inherent ability of adipocytes to secrete inflammatory mediators, in particular adipokines (adipocyte-derived cytokines) and hormones, such as leptin.[9] This innate pro-inflammatory activity of adipocytes is unclear, but it is likely an evolutionary adaptation meant to enhance the disease-fighting ability of the body when extra energy is available.[10] However, when fat accumulation is permanent and uncontrolled, the resulting constant release of pro-inflammatory mediators becomes highly detrimental. In fact, fat accumulation has been shown to recruit and convert macrophages to a pro-inflammatory phenotype, which leads to the production of pro-inflammatory proteins such as tumor necrosis factor-alpha (TNF-α), interleukin-6, (IL-6), and monocyte chemoattractant protein-1 (MCP-1), contributing further to the pro-inflammatory state.[6] This could explain the critical link between obesity and the development of insulin resistance, type 2 diabetes, and CVD. Furthermore, the local tissue damage induced by inflammation and oxidative stress leads to the migration of monocytes from the bloodstream to the adipocytes. Here, adipokines activate monocytes to macrophages, which in turn release a wide range of pro-inflammatory mediators, further sustaining inflammation and contributing to the progression from localized inflammation to a systemic inflammatory state.[11]

Alterations in gut microbial species and related dysfunctions in intestinal permeability have also been suggested as a potentially relevant link between obesity and inflammation, both as a result and as triggers of inflammatory processes.[12]

The pro-inflammatory state associated with visceral adiposity is also a likely causative factor leading to the impaired immune function associated with obesity. A constant and chronic release of pro-inflammatory adipokines and leptin may in effect impair the ability of the body to effectively distinguish between the chronic production of such molecules and when they signal a situation actually requiring immunocompetence.[13] For this combination of reasons, the chronic pro-inflammatory state induced by visceral adiposity, also called lipoinflammation, represents a significant risk factor for MetS and contributes to the pathogenesis of a number of obesity-associated comorbidities.[14]

Although the term 'berry' has a precise and distinct botanical definition, this name is commonly used to refer to a variety of small edible fruits, growing on bushes or low plants, but not trees, native to woodlands in cold and temperate climates, and is characterized by red, blue and purple colors.

11.1.3 The Potential of Berry Bioactives to Attenuate Obesity-induced Inflammation

The most common 'berries' under this non-botanical meaning are from the genera *Rubus* (blackberries and raspberries), *Vaccinium* (blueberries, bilberries, and cranberries), *Ribes* (black, white, and red currants), *Aronia* (chokeberries), and Fragaria (strawberries).[15,16]

From a nutritional standpoint, these fruits are a rich source of multiple bioactive compounds known for their antioxidant and anti-inflammatory potential. In particular, they contain significant amounts of phenolic compounds, including flavonoids (anthocyanins [ACNs], flavonols, flavanols), phenolic acids (hydroxybenzoic and hydroxycinnamic acids, chlorogenic acid), condensed tannins (proanthocyanidins), hydrolyzable tannins (ellagitannins and gallotannins), stilbenes (resveratrol), and lignans.[15,16] They are also a good source of dietary fiber, vitamins such as vitamin C, and minerals such as manganese, all potentially contributing to antioxidant, anti-inflammatory, and vascular protective roles.[16]

Thus, several berries have been widely investigated with increasing interest in recent years as a potentially useful dietary strategy to promote health and prevent chronic diseases such as diabetes, cardiovascular diseases, neurodegenerative diseases, and cancer development;[17–22] however, the results are not always univocal.[23] In particular, clinical research studies have focused on the role of whole berries or their components, in particular polyphenols, in the modulation of oxidative stress, inflammation, vascular function, blood pressure, lipid metabolism, and glucose metabolism.

Among polyphenols, ACNs are the most abundant polyphenolic components in berries.[24] They are particularly concentrated in the outermost layers of the pericarp. ACNs are glycosylated forms of basic flavonoids with a characteristic structure called anthocyanidins: the predominant anthocyanidins are pelargonidin, cyanidin, delphinidin, petunidin, peonidin, and malvidin.[25] ACNs have been shown to exert a remarkable anti-inflammatory effect by modulating several enzymes and transcription factors involved in the mediation of the inflammatory response. Indeed, their ability to attenuate inflammation and improve markers of different inflammatory diseases has been confirmed repeatedly in both *in vitro* and *in vivo* studies.[26,27] These molecules exert their anti-inflammatory effect, both directly and indirectly through their antioxidant power, *via* the modulation of nuclear factor-κB and mitogen-activated protein kinase signaling cascades.[28] Among flavonoids, another important class is represented by flavonols, such as quercetin, kaempferol, and myricetin. They are also of particular interest for their potential benefits in the modulation of inflammatory responses.[29,30]

Among berries, chokeberries, blackcurrants, blueberries, and raspberries are particularly rich in ACNs (over 100 mg per 100 g), followed by raspberries, blackberries, and cherries.[24] Cranberries have a lower ACNs content but have a high flavonol content, with an abundance of quercetin.[31]

Another important class of polyphenols is represented by phenolic acids.[32] They belong to two main classes, hydroxycinnamic acids (including cinnamic acid, ferulic acid, caffeic acid, *p*-coumaric acid, and sinapic acid) and hydroxybenzoic acids (such as benzoic and *p*-hydroxybenzoic acids, ellagic acid, gallic acid, salicylic acid, and vanillic acid).[32] Both cinnamic and benzoic acid derivatives are mainly present in bound forms as esters with sugars (glycosides) or organic acids, using both the phenolic group and the carboxyl group as bonding sites. Chlorogenic acid, a powerful antioxidant molecule, is an ester of caffeic acid with quinic acid. Esters of gallic and ellagic acid form hydrolyzable tannins. Hydroxybenzoic acids can also be involved as components in the complex structure of lignins.

Chlorogenic acid is abundant in blackberries and blueberries, ferulic acid in blueberries and cranberries, and *p*-coumaric acid in bilberries and blackcurrants, while gallic acid is abundant in blueberries and blackberries. Ellagic acid is the main phenolic acid, generally esterified to glucose in hydrolyzable ellagitannins, strawberries, and raspberries.[15,16]

A growing body of evidence both from animal and clinical studies reports the anti-inflammatory activity of phenolic acids, with major evidence for specific compounds such as gallic acid, caffeic acid, and chlorogenic acid.[33] The ability of such molecules to reverse a number of metabolic dysfunctions associated with MetS, including insulin resistance, hyperglycemia, dyslipidemia, inflammation, and oxidative stress, is also promising.[34]

Another important class of polyphenols is represented by tannins. They are a large and diverse group of complex molecules, mainly divided into two classes, based on their chemical structure and reactivity to hydrolyzing agents: condensed tannins, also known as proanthocyanidins, and hydrolyzable tannins.[35]

Condensed tannins are polymers of a group of flavonoids called catechins (flavan-3-ols), but upon hydrolysis in an acidic environment, they release anthocyanidins, hence their name.[35] Hydrolyzable tannins are highly polymerized molecules with a basic structure consisting of esters of gallic and ellagic acids, often bound to glucose and other sugars.[35]

Among berries, proanthocyanidins are abundant in chokeberries, blackberries, and blueberries, while hydrolyzable tannins, particularly ellagitannins, are abundant in raspberries and strawberries.[36] Besides their unique role in contributing to sensory properties such as astringence and color, they also play antiseptic and antioxidant roles in foods. Ellagitannins have been studied for their antioxidant capacity and vasorelaxation properties, while the anti-inflammatory potential associated with their presence is likely due to their metabolites obtained after hydrolysis to phenolic acids such as gallic acid and ellagic acid.[37] Proanthocyanidins have a strong antioxidant activity and a direct anti-inflammatory effect that has been repeatedly investigated in intestinal epithelial cells.[38–40]

Berry bioactives are recognized to exert a plethora of biological activities, including antioxidant, antiaging, anti-obesity, anticarcinogen, antiatherosclerosis, antihypertensive, and anti-inflammation.[36,41,42] In this regard, we have demonstrated the capacity of a blueberry-rich extract and its single components/metabolites to counteract TNF-α-induced inflammatory response and related pro-inflammatory cytokines and to inhibit NF-κB pathway activation driven by interleukin-1-beta (IL-1β).[43–46] Numerous other studies reported similar findings by inhibiting the production of nitric oxide, as well as IL-1β and TNF-α, iNOS and COX-2 in different cell models by using different berry extracts.[47–49]

Similar findings were also documented *in vivo* using animal models of obesity. In obese Zucker rats, a wild blueberry–enriched diet significantly reduced plasma concentrations of TNF-α, IL-6, and CRP, increased plasma adiponectin concentrations, and down-regulated the expression of NF-kB, TNF-α, and IL-6 both in the liver and the abdominal adipose tissue.[50] Wild blueberry consumption also attenuated local inflammation in the aortic perivascular adipose tissue, as evidenced by reduced concentrations of IL-8 and MCP-1 and increased adiponectin levels.[51]

Given the promising anti-inflammatory potential of berry bioactives also in obesity conditions, the aim of the present work was to review the available clinical evidence on the role of berry fruit consumption in attenuating chronic inflammation in overweight and obese individuals.

In the following paragraph reported is the methodology used for the bibliographic research and the inclusion and exclusion criteria adopted for the studies selection.

11.2 Methodology

A search of the literature was carried out by using PubMed[52] and Embase[53] databases (updated 21 November 2020) with the addition of other scientific papers of relevance found in web sources or in previously published reviews. The searches used the following terms and text words: "metabolic syndrome" AND "overweight" AND "obese" AND "inflammation" AND "randomized human intervention study" AND "berries" OR "blueberry" OR "cranberry" OR "strawberry" OR "bilberry" OR "raspberry" OR "chokeberry" OR "Whortleberry" OR "blackcurrant". In the research, a filter for clinical trial and randomized controlled trial was applied. The search was limited to the last 20 years of publication and English-language articles. Two independent reviewers conducted the literature search in the scientific databases and assessed the eligibility of the studies based on the title and abstract. Disagreement between reviewers was resolved by a third reviewer. The full text was analyzed and data extraction was performed by two independent reviewers. The following information was collected: (i) the type of berry; (ii) the duration of the treatment; (iii) the study design; (iv) the number and the subjects' characteristics; (v) the treatment in terms of dose/amount of berry administered; (vi) the main findings on inflammatory markers; and (vii) first author name and year of publication. A total of 523 records were identified from the database search and other sources. After the removal of duplicate papers and those not meeting the inclusion criteria or not of interest/pertinent, a total of 44 papers were included and analyzed. A summary of included studies and their results related to inflammation is reported in Table 11.1.

11.3 Effect of Different Berries on Inflammatory Markers

11.3.1 Blueberries

The role of blueberry in the modulation of inflammation and MetS parameters to date has been investigated in eight studies. Du and colleagues[54] documented that a 4 month intervention with 40 g per day of freeze-dried whole blueberry powder (equivalent to 2 cups of fresh blueberries) blunted inflammation in 63 obese subjects with symptomatic knee osteoarthritis. Nair *et al.*[55] reported that a 6 week intervention with yogurt and skim milk–based smoothie with 45 g of freeze-dried blueberry powder (equivalent to approximately 2 cups of fresh blueberries) exerted an immunomodulatory effect and attenuated oxidative stress and inflammation in 15 adults with MetS.

Conversely, Stote *et al.*[56] documented that 7-day consumption of 240 mL wild blueberry juice failed to induce significant changes in cardiometabolic, oxidative stress, vascular function, and inflammatory markers in overweight and obese subjects at risk of type 2 diabetes. Johnson and coworkers[57,58] reported that the consumption of 480 mL blueberry drink (22 g freeze-dried blueberry powder corresponding to 1 fresh cup blueberries) for 4 and 8 weeks did not reduce marker of oxidative stress, inflammation, and antioxidant defense, while they showed an increase in NO plasma levels and

Table 11.1 A summary of dietary intervention studies relating berry consumption to inflammatory markers.

Berry	Duration	Study design	Participants[a]	Treatment	Effect on inflammation[b]	Reference
Blueberry	4 months	Double-blind, placebo-controlled, parallel arms	Sixty-three with knee osteoarthritis (age 56.5 ± 1.2 years; BMI 31.9 ± 0.9 kg m^{-2})	40 g freeze-dried whole blueberry powder daily	= TNF-α, IL-1β, IL-6, IL-10, IL-13, MMP-3, MMP-13, MCP-1 (trends for ↑ IL-13 and ↓ MCP-1)	Du *et al.*[54]
	6 weeks	Double-blind, placebo-controlled, parallel arms	Twenty-seven with MetS (age 56.5 ± 2.5 years; BMI 34.7 ± 1.1 kg m^{-2})	Twice daily a 12 oz (~356 mL) yogurt and skim milk–based smoothie with 22.5 g of freeze-driedblueberry powder (total 45 g per day, equivalent to approximately 2 cups of fresh berries)	↓ expression of TNF-α, TLR4, IL-6 = IFN-α, IL12p70, FLT3L	Nair *et al.*[55]
	1 week	Single-blind, placebo-controlled, crossover	Nineteen with type 2 diabetes risk (W; age 39–64 years; BMI 27-37 kg m^{-2})	240 mL wild blueberry juice	= TNF-α, CRP, IL-6, IL-10, serum amyloid A, sICAM, sVCAM↑ plasma NO	Stote *et al.*[56]
	8 weeks	Double-blind, placebo-controlled, parallel arms	Forty-eight postmenopausal women with pre- and stage 1-hypertension (age 58.5 ± 4.7 years; BMI 31.4 ± 6.4 kg m^{-2})	480 mL blueberry drink (22 g freeze-dried blueberry powder corresponding to 1 cup fresh berries)	↓ TNF-α (time effect), 8-OHdG (treatment effect)↑ NO = CRP, SOD, GR, GPx, oxLDL, 8-iso, TBARS	Johnson *et al.*[57,58]
	6 weeks	Double-blinded, placebo-controlled, parallel arms	Twenty-seven obese, insulin-resistant subjects (age >20 years; BMI 32–45 kg m^{-2})	Smoothie prepared with 45 g of blueberry powder (equivalent to 2 cups of fresh blueberries)	= CRP, TNF-α, MCP1	Stull *et al.*[59]
	8 weeks	Single-blind, controlled, parallel arms	Forty-eight with MetS (age 50.0 ± 3.0 years; BMI 37.8 ± 2.3 kg m^{-2})	480 mL blueberry drink from 50 g freeze-dried blueberries (corresponding to 350 g fresh berries)	↓ plasma ox-LDL, MDA, and HNE levels = CRP, adiponectin, IL-6, sICAM-1, sVCAM-1	Basu *et al.*[60]

	Single dose	Controlled, crossover	Five with MetS (age 26–61 years; BMI 28–40 kg m^{-2})	150 g fresh blueberries following a high-fat/ high-glycemic-load meal	↓ IL-6 expression ↑ TGF-β expression = IL-1β, TNF-α, IL-10, IL-4	Sobolev *et al.*[61]
Bilberry	8 weeks	Controlled, parallel arms	Twenty-seven with MetS (Bilberry group:age 53 ± 6 years; BMI 31.4 ± 4.7 kg m^{-2}; Control group: age 50 ± 7 years; 32.9 ± 3.4 kg m^{-2})	200 g ofbilberry purée + 40 g of dried bilberries (equivalent to 400 g fresh bilberries)	↓ serum hs-CRP, IL-6, IL-12, and inflammation score ↓ expression of MMD and CCR2 transcripts associated with monocyte and macrophage function associated genes	Kolehmainen *et al.*[62]
	5 weeks	Controlled, crossover	Eighty overweight or obese (W; age 44.2 ± 6.2 years; BMI 29.6 ± 2.1 kg m^{-2})	100 g fresh bilberries	↓ VCAM-1, TNF-α, adiponectin = serum IL-6	Lehtonen *et al.*[63]
	4 weeks	Controlled, parallel arms	Sixty-two with CVD risk factors (age 30–68 years; BMI 18–32 kg m^{-2})	330 mL bilberryjuice per day (diluted to 1 L with water)	↓ CRP, IL-6, IL-15, TNF-α, IFN-gamma	Karlsen *et al.*[64]
Blackcurrant	Single dose	Single-blind, placebo-controlled, crossover	Eleven (M; age 37 ± 2 years; BMI 32 ± 1 kg m^{-2})	250 g blackcurrant juice providing 617 total phenolics, following high-energy meal	= PON, TNF-α, IL-6, IL-1β	Huebbe *et al.*[65]
Cranberry	12 weeks	Pre-post	Thirty with abdominal obesity (M; age 51 ± 10 years; BMI 27.8 ± 3.2 kg m^{-2})	125 mL per day cranberry juice (4 weeks) + 250 mL per day (4 weeks) + 500 mL per day (4 weeks)	↓ s-VCAM, ICAM = E-selectin plasma levels	Ruel *et al.*[66]
	12 weeks	Placebo-controlled, double-blind, parallel arms	Thirty with type 2 diabetes (age 65.5 ± 2 years; BMI 26.1 ± 0.9 kg m^{-2})	500 mg per capsule cranberry powder extract	= CRP	Lee *et al.*[67]
	8 weeks	Double-blind, placebo-controlled, parallel arms	Thirty-one with MetS (W; age 52.0 ± 8.0 years; BMI 40.0 ± 7.7 kg m^{-2})	480 mL per day cranberry juice	= CRP, IL-6	Basu *et al.*[68]

Table 11.1 (*Continued*)

Berry	Duration	Study design	Participants[a]	Treatment	Effect on inflammation[b]	Reference
	Single dose	Pre-post	Fifteen with coronary artery disease(age 62 ± 8 years; BMI not reported but overweight or obese)	Cranberry group:480 mL cranberry juice following meal	= CRP, ICAM-1	Dohadwala *et al.*[69]
	4 weeks	Placebo-controlled, crossover	Forty-four with coronary artery disease(age 62 ± 10 years; BMI 29.5 ± 4.5 kg m^{-2})	480 mL per day cranberry juice	= CRP, ICAM-1	Dohadwala *et al.*[69]
	4 weeks	Controlled, parallel arms	Fifty-six with MetS (median age 50 years; median BMI 31 kg m^{-2})	700 mL per day reduced-energy cranberry juice	↓serum homocysteine = CRP, TNF-α, IL-1, IL-6	Simao *et al.*[70]
	4 weeks	Double-blind, placebo-controlled, crossover	Thirty-five with abdominal obesity, with or without MetS (M, age 45 ± 10 years, BMI 28.3 ± 2.4 kg m^{-2})	500 mL per day of low-calorie cranberry juice (27% juice)	= ICAM-1, VCAM-1, E-selectin	Ruel *et al.*[71]
	90 days	Controlled, parallel arms	Forty-one with rheumatoid arthritis (W; age 53 [40–65] years; BMI 28 [22–32] kg m^{-2})	500 mL cranberry juice	= CRP	Thimoteo *et al.*[72]
	8 weeks	Placebo-controlled, parallel arms	Fifty-six (age 50 ± 11 years; BMI 28 ± 4 kg m^{-2})	Twice daily cranberry juice (amount not reported) providing 173 or 62 mg total phenolics	↓ CRP = ICAM-1, VCAM-1, IL-6, Il-10, IL-1β, TNF-α	Novotny *et al.*[73]
	Single dose	Controlled, crossover	Forty obese with type 2 diabetes (age 56 ± 6 years; BMI 40 ± 7 kg m^{-2})	40 g dried cranberry powder following high-fat meal, test after 1 h, 2 h, and 4 h	↓ IL-18 at 4 h = CRP, IL-6	Schell *et al.*[74]
	8 weeks	Double-blind, placebo-controlled, parallel arms	Seventy-eight obese or overweight (age 43 ± 1 years; BMI 31 ± 1 kg m^{-2})	450 mL high phenolics cranberry extract	↓ CRP = IL-6, IL-10, IL-23, TNF-α, IFN-γ	Chew *et al.*[75]

	Single dose	Double-blind, placebo-controlled, parallel arms	Seventy-eight obese or overweight (age 43 ± 1 years; BMI 31 ± 1 kg m^{-2})	450 mL high phenolics cranberry extract	↓endothelin-1 ↑NO, IFN-g =CRP, IL-6, IL-10, IL-23, TNF-α	Chew *et al.*[75]
Raspberry	12 weeks	Controlled, parallel arms	Seventy-seven with MetS (age 60.0 ± 9.4 years; BMI 25.7 ± 4.2 kg m^{-2})	750 mg per day of black raspberry powder as capsules	↓ IL-6, TNF-α ↑ adiponectin =CRP, ICAM, VCAM	Jeong *et al.*[76]
	12 weeks	Double-blind, placebo-controlled, parallel arms	Fifty-one with MetS (age 59 ± 10 years, BMI 25 ± 4 kg m^{-2})	750 mg daily dried unripe black raspberry powder	↓ TNF-α, IL-6, sICAM-1, CD34/KDR cells, ↑adiponectin, CD34/CD133, CD34/CD117 cells, =CRP, VCAM-1	Jeong *et al.*[77]
	8 weeks	Double-blind, placebo-controlled, parallel arms	Forty-five with pre-hypertension (age 57 ± 12 years, BMI 25 ± 3 g m^{-2})	1500 mg or 2500 mg daily dried unripe black raspberry powder	↓ IL-6 and TNF-α with highest dose =CRP, IL-18, ICAM-1, VCAM-1, PAI-1	Jeong *et al.*[78]
	14 days	Controlled, crossover	Ten overweight or obese (M; age 64.7 ± 6.9 years; BMI 31.4 ± 2.7 kg m^{-2})	45 g per day of lyophilized black raspberry powder for 4 days following high-fat meal + 6 days wash-out + 4 days intervention	↓ IL-6 =TNF-α, CRP	Sardo *et al.*[79]
	2 weeks	Double-blind, placebo-controlled, parallel arms	Sixty-seven overweight or obese(age 44 ± 8 years; BMI 25 ± 2 kg m^{-2})	30 g daily freeze-dried Korean black raspberry powder (equivalent to 100 g fresh fruit), providing 900 mg total phenolics. Test performed before and	↓ IL-6 =TNF-α	Kim *et al.*[80]

Table 11.1 (*Continued*)

Berry	Duration	Study design	Participants[a]	Treatment	Effect on inflammation[b]	Reference
				after exercise challenge		
	4 weeks	Controlled, crossover	Twenty-two obese (age 54 ± 4 years; BMI 33 ± 2 kg m^{-2})	Daily raspberry supplement providing 225 mg ACN and 343 mg total phenolics	↓ IL-6, TNF-α = CRP, IL-1β, PAI-1	Schell *et al.*[81]
	Single dose	Controlled, crossover	Twenty-five obese with type 2 diabetes(age 54 ± 4 years; BMI 35 ± 2 kg m^{-2})	Post-prandialassessment with or without raspberry supplement providing 225 mg ACN and 343 mg total phenolics	↓ IL-6 and ↓TNF-α at 4 h = CRP, IL-1β, PAI-1	Schell *et al.*[81]
	12 weeks	Double-blind, placebo-controlled, parallel arms	Forty-four with pre-diabetes(age 59 ± 8 years; BMI 25 ± 2 kg m^{-2})	Daily 900 mg or 1800 mg raspberry extract	↓ MCP-1	An *et al.*[82]
Chokeberry	18 weeks	Pre-post	Fifty-eight with mild hypercholesterolemia (M; age 54.1 ± 5.6 years; BMI 27.7 ± 2.9 kg m^{-2})	150 mL per day chokeberry juice (6 weeks intervention + 6 weeks wash-out + 6 weeks intervention)	= hs-CRP	Skoczyñska *et al.*[83]
Strawberry	12 weeks	Double-blind placebo-controlled, crossover	Seventeen with osteoarthritis(age 57 ± 3 years; BMI 39.1 ± 1.5 kg m^{-2})	50 g per day Control group: 50 g per day matched for calories and fiber	↓ hs-TNF-a; sTNF-R2; = APRIL/TNFSF13, BAFF/TNFSF13B, sSC30/TNFRSF8, sTNF-R1, IL-19, TWEAK/TNFSF12, sCD163, CHI3L1, Osteocalcin, pentraxin-3	Basu *et al.*[84]
	12 weeks	Double-blind, placebo-controlled, crossover	Seventeen with knee osteoarthritis (age 57 ± 7 years; BMI 39.1 ± 1.5 kg m^{-2})	50 g freeze-dried strawberry powder (equivalent to approximately 500 g of fresh strawberries)	↓ IL-6, IL-1β, MMP3 = hs-CRP, MMP8	Schell *et al.*[85]

			reconstituted in water (twice a day)		
6 weeks	Double-blind placebo-controlled, parallel arms	Thirty-six with type 2 diabetes(age, 51.6 ± 11.1 years; BMI 28.0 ± 3.8 kg m^{-2})	50 g per day freeze-dried strawberry (equivalent to 500 g fresh strawberries)	↓ hs-CRP	Moazen *et al.*[86]
8 weeks	Controlled, parallel arms	Twenty-seven with MetS (age 47.0 ± 3.0 years; BMI 37.5 ± 2.15 kg m^{-2})	Four cups daily strawberry drink with 25 g per cup of freeze-dried strawberry powder	↓ VCAM-1 = ICAM-1	Basu *et al.*[87]
12 weeks	Controlled, parallel arms	Sixty with CVD risk factors (age 48.8 ± 10 years; BMI 36.1 ± 5.3 kg m^{-2})	Low dose (25 g) or high dose (50 g) freeze-dried strawberries powder in 2 cups (474 mL per day) of water daily (corresponding to 250 or 500 g of fresh strawberries)	= VCAM-1, ICAM-1, hs-CRP	Basu *et al.*[88]
4 weeks	Pre-post	Sixteen with MetS (W; age 39-71 years; BMI 38.6 ± 2.3 kg m^{-2})	Two cups daily strawberry drink with 25 g per cup of freeze-dried strawberry powder	= hs-CRP, adiponectin	Basu *et al.*[89]
7 weeks	Double-blind, controlled, crossover	Twenty obese(age 20–50 years; BMI 30–40 kg m^{-2})	Two servings of strawberry powder equivalent to 320 g per day of frozen strawberries (mixed as a milkshake, in yogurt, cream cheese, or water)	= hs-CRP, IL-1b, IL-6, IL-8, TNF-α, sICAM-1, sVCAM-1	Zunino *et al.*[90]
4 weeks	Controlled, crossover	Twenty-eight with hyperlipidemia (age 38–75	454 g per day strawberries	= CRP	Jenkins *et al.*[91]

Table 11.1 (*Continued*)

Berry	Duration	Study design	Participants[a]	Treatment	Effect on inflammation[b]	Reference
			years; BMI 19.8–32.3 kg m^{-2})			
	6 weeks	Single-blind, placebo-controlled, parallel arms	Twenty-four overweight and obese (age 50.9 ± 15.0 years; BMI 29.2 ± 2.3 kg m^{-2})	10 g per day strawberry powder in 305 mL water following high carbohydrates moderate-fat meal	= hs-CRP, IL-6, PAI-1, IL-1β, TNF-α after 6-week intervention ↓ PAI-1, IL-1β following 6 week intervention + post-prandial intervention = hs-CRP, IL-6, TNF-α after 6 week intervention + post-prandial intervention	Ellis *et al.*[92]
	Single dose	Single-blind, placebo-controlled, crossover	Twenty-six overweight (age 50.9 ± 15.0 years; BMI 29.2 ± 2.3 kg m^{-2})	10 g strawberry powder in water following high-carbohydrate, moderate-fat meal	↓ hs-CRP, IL-6 = PAI-1, IL-1β, TNF-α	Ederisinghe *et al.*[93]
	Single dose	Controlled, crossover	Twenty-one with abdominal obesity and insulin resistance(age 39.8 ± 13.8 years; BMI 40.2 ± 7.2 kg m^{-2})	0, 10, 20, or 40 g freeze-dried whole strawberry powder following meal	= IL-6	Park *et al.*[94]
	Single dose	Placebo-controlled, crossover	Thirty overweight and obese (age 28.2 ± 2.4 years; BMI 31.4 ± 0.7 kg m^{-2})	40 g freeze-dried strawberry powder (equivalent to 450 g fresh strawberries)	↓ CRP	Richter *et al.*[95]

				following a high-fat meal		
Whortleberry	4 weeks	Double-blind, placebo-controlled, parallel arms	Fifty with hyperlipidemia (age 47.2 ± 16.5 years; BMI 25.3 ± 1.9 kg m^{-2})	2 capsules per day whortleberry	= hs-CRP	Soltani *et al.*[96]
Berry mix	6 weeks	Double-blind, controlled, parallel arms	Forty-one overweight or obese with insulin resistance (age 59 ± 1 years; BMI 31 ± 1 kg m^{-2})	Daily strawberry and cranberry beverage (333 mg)	= CRP, adiponectin, PAI-1, IL-6, TNF-α	Paquette *et al.*[97]
	20 weeks	Controlled, parallel arms	Sixty-one female with MetS (age 35–52 years, BMI 29.4 ± 2.0 kg m^{-2})	163 g per day mix of northern berries (lingonberry, sea buckthorn berry, bilberry, and blackcurrant)	= TNF-α, adiponectin, sICAM, sVCAM, E-selectin, hs-CRP	Lehtonen *et al.*[98]

[a]Unless otherwise specified, study participants are both men (M) and women (W).
[b]From each study, only results related to inflammation are selected and reported here.

superoxide dismutase activity in overweight and obese postmenopausal women. Stull *et al.*[59] documented an increase in insulin sensitivity after 6 weeks of a smoothie prepared with 45 g of blueberry powder, while no effect was found on markers of inflammation, lipid profile, and blood pressure in 15 obese subjects with insulin resistance. Finally, Basu and coworker[60] did not observe an effect on body weight, lipid profile, and inflammation after an 8 week intervention with 480 mL blueberry drink (50 g freeze-dried blueberries corresponding to 350 g fresh berries) in subjects with MetS, while they documented a positive effect on blood pressure and oxidative stress markers.

Only one study reported the short-term effects of blueberry on inflammation. Sobolev *et al.*[61] showed a reduction in mRNA gene expression of IL-6 and an increase in transforming growth factor-β (TGF-β) and pro- and anti-inflammatory ion cytokines, respectively, following post-prandial high-fat/high-glycemic-load meal enriched with blueberry (150 g) in overweight and obese subjects, indicating a positive impact of blueberry ingestion in the reduction of risk of inflammation.

Overall, the available studies show only a weak association between blueberry consumption and inflammation in overweight or obese humans, but they suggest a trend toward an anti-inflammatory effect.

11.3.2 Bilberries

The role of bilberry in the modulation of MetS and related markers has been investigated in three studies conforming to our selection criteria. In a randomized controlled trial, Kolehmainen *et al.*[62] documented that daily consumption of a diet rich in bilberries (equivalent dose of 400 g fresh fruits) for 8 weeks reduced low-grade inflammation in terms of high sensitivity C-reactive protein, IL -6 and IL -12, and inflammation score, while no effect was reported for body weight, lipid profile, and glycemia in subjects with MetS. In another trial, Lehtonen *et al.*[63] reported that a 5 week intervention with 100 g of fresh bilberry reduced TNF-α, ICAM-1, and VCAM-1, but not IL-6, serum levels in overweight and obese women. Finally, Karlsen *et al.*[64] showed that 4 weeks of dietary intervention with 330 mL bilberry juice per day modulate NF-κB related inflammatory markers in subjects with CVD features while no effect was found on oxidative stress markers.

In conclusion, all of the three available long-term studies report a significant effect of bilberry consumption on markers of inflammation in overweight or obese humans.

11.3.3 Blackcurrants

The impact of blackcurrant on inflammatory markers was reported only in one study that failed to show an effect after an oral administration of 250 g of blackcurrant-based juice plus a high-energy meal in overweight/obese subjects with hyperlipidemia.[65]

11.3.4 Cranberries

The effect of cranberry consumption was analyzed in 12 dietary intervention studies. Ruel and colleagues[66] tested the effect of 250 mL and 500 mL cranberry juice consumption on 30 middle-aged abdominally obese men, 9 of which with MetS. The results obtained documented a reduction in plasma ICAM-1 and VCAM-1 concentrations but only at the highest dose with a stronger effect in the subset of subjects with MetS. No significant effect was observed at the lowest dose, and no effect other than cardiometabolic markers including plasma E-selectin was observed at any dose.[66] In a 12 week, placebo-controlled, randomized intervention, 30 older overweight subjects with T2D received a daily capsule of 500 mg cranberry powder extract, resulting in a lack of an effect on CRP levels, while a reduction was documented for total and low-density lipoprotein (LDL) cholesterol, and total:high-density lipoprotein (HDL) cholesterol ratio.[67] In a placebo-controlled, randomized intervention, 31 obese women with MetS were given 480 mL per day of cranberry juice for 8 weeks, resulting in reduced circulating levels of lipid peroxidation markers, while blood pressure, blood lipid and glucose levels, and inflammatory markers did not change.[68] When a serving of 480 mL cranberry juice was administered to 44 subjects with coronary artery disease in a 4 week, placebo-controlled, crossover intervention, no changes were reported for markers of inflammation such as CRP and ICAM-1.[69] In 56 overweight or obese subjects with MetS were given 700 mL per day of low-calorie cranberry juice for four weeks, inflammatory markers CRP, TNF-α, IL-1, and IL-6 did not change.[70] Thirty-five abdominally obese men with or without MetS were given 500 mL per day of low-calorie 27% cranberry juice for four weeks in a placebo-controlled, crossover intervention. The results have shown no significant effect on ICAM-1, VCAM-1, and E-selectin as inflammatory markers, as well as oxidized LDLs and blood pressure, while global endothelial function and arterial stiffness significantly decreased following cranberry consumption compared to baseline, but not compared to post-placebo.[71] Thimóteo *et al.*[72] showed that 3 months of 500 mL of cranberry juice per day did not reduce CRP levels as well as other cardiometabolic markers in women with rheumatoid arthritis. A similar effect was observed by Novotny and colleagues[73] after 8 weeks of cranberry juice (2 portions per day), who showed a reduction in the markers of inflammation in subjects with a mild overweight. Different findings were also documented by Schell *et al.*,[74] who observed showed a reduction in IL-18 while no effect was observed on other inflammatory and cardiometabolic markers following a single dose of cranberries (40 g dried cranberry) and high-fat meal in obese subjects with T2D. Finally, Chew and coworkers[75] evaluated the effect of acute (one portion) and chronic (8 weeks) consumption of 450 mL cranberry juice on numerous markers of inflammation in obese and overweight subjects, finding a significant reduction in CRP after 8 weeks (but not with a single dose) and endothelin-1 after a single-dose, as well as a significant increase in NO and IFN-γ after a single-dose. However, no changes were observed for IL-6, IL-10, IL-23, and TNF-α in both populations.[75]

Overall, the results are mixed: out of the nine long-term studies investigating cranberry consumption in obese or overweight humans, five studies reported no effect on inflammation, and the remaining four reported a positive outcome only for one of the several inflammatory markers that were analyzed. Of the three single-dose studies, one reported no effect, and the other two reported a positive effect for at least one inflammatory marker.

11.3.5 Raspberries

The impact of raspberry on inflammation has been evaluated in eight studies. In a 12 week, randomized, controlled intervention, a group of 77 MetS patients received 750 mg per day of black raspberry powder in the form of capsules or a fiber- and sugar-matched control.[76] Raspberry intake resulted in decreased total serum cholesterol and total:HDL cholesterol ratio, with no other changes in serum lipid profile. Inflammatory markers IL-6 and TNF-α significantly decreased, and anti-inflammatory adiponectin significantly increased, while CRP, ICAM-1, and VCAM-1 were unaffected. Brachial artery flow-mediated dilation significantly improved.[76] More recently, the same authors found that the administration of 750 mg per day of a black raspberry extract for 12 weeks reduced serum TNF-α, IL-6, sICAM-1, and CD34/KDR cells, while it increased serum levels of adiponectin and CD34/CD133 and CD34/CD117 cells in 26 overweight subjects with MetS. No effect on CRP and sVCAM-1 levels was observed.[77] Interestingly, using a similar experimental design, the authors found that the administration of higher doses of black raspberry extract (1500 or 2500 mg per day) for a shorter time (8 weeks) to a group of pre-hypertensive subjects did not significantly affect CRP, ICAM-1, VCAM-1, PAI-1, or IL-18 and resulted in a reduction of serum IL-6 and TNF-α only with the highest dose.[78]

In a controlled, crossover intervention on a group of 10 older overweight or obese men, the short-term effect of relatively large amounts of lyophilized black raspberry powder (45 g per day for 4 days) was evaluated in countering the post-prandial inflammatory effects of a high-fat meal. Serum IL-6 concentrations significantly decreased, but not TNF-α and CRP concentrations.[79] In a 2 week, double-blind intervention, Kim and colleagues[80] showed that 30 g daily freeze-dried Korean black raspberry powder reduced IL-6 and oxidative stress markers while no effect was observed on TNF-α. Recently, Schell *et al.*[81] reported a significant effect on IL-6 and TNF-α levels in obese subjects with T2D both following post-prandial and 4-week consumption of raspberry supplement (providing 225 mg ACN and 343 mg total phenolics).

In a double-blind, placebo-controlled study on 44 pre-diabetic subjects, An *et al.* found a significant decrease in MCP-1 concentrations following daily consumption of either 900 mg or 1800 mg raspberry extract for 12 weeks.[82]

In conclusion, the available evidence shows a consistent association between raspberry consumption and an anti-inflammatory effect in obese or overweight humans. All of the eight available studies report a positive outcome for at least one of the inflammatory markers that were analyzed.

11.3.6 Chokeberries

Only one study investigated the effect of chokeberries on inflammation. A long-term intervention (18 weeks) on overweight and obese subjects with mild hypercholesterolemia receiving 150 mL per day of chokeberry failed to affect hs-CRP protein levels, while a significant decrease in serum total and LDL cholesterol, triglycerides, glucose, homocysteine, and fibrinogen levels, as well as blood pressure, was observed. An increase in serum HDL-2 cholesterol was also found.[83]

11.3.7 Strawberries

Twelve studies investigated the effect of strawberry consumption on inflammation. Eight studies were chronic interventions, three studies were acute/post-prandial studies, while one study performed both post-prandial and chronic intervention. In a 12 week, randomized, crossover, double-blind, placebo-controlled trial, Basu *et al.*[84] showed that 50 g per day of a lyophilized strawberry powder decreased circulating levels of hs-TNF-α and sTNF-R2 in 17 obese adults with knee osteoarthritis, while no effect was observed on a plethora of other inflammatory markers (*e.g.* APRIL/TNFSF13, BAFF/TNFSF13B, sSC30/TNFRSF8, sTNF-R1, IL-19, TWEAK/TNFSF12, sCD163, CHI3L1). In a 12 week, randomized, double-blind, crossover intervention, Schell *et al.*[85] documented that the administration of 50 g of freeze-dried strawberry powder (equivalent to approximately 500 g of fresh strawberries) reconstituted in water (twice a day) reduced IL-6, IL-1β, and MMP3 circulating levels but not hs-CRP, MMP8, and other metabolic markers in 17 obese adults with knee osteoarthritis. In a 6 week, randomized double-blind controlled trial, Moazen and colleagues[86] reported a significant reduction in hs-CRP serum levels after intervention with 50 g per day of a strawberry powder (equivalent to 500 g fresh strawberries) in a group of 19 overweight subjects with type 2 diabetes. In an 8 week intervention study by Basu *et al.*,[87] the consumption of 4 cups per day of strawberry drink (each cup containing 25 g of freeze-dried strawberry powder) reduced circulating levels of VCAM-1 in a group of obese subjects with MetS. Conversely, the same authors[88] have recently reported a lack of an effect on inflammatory markers (VACM-1, ICAM-1, hs-CRP) following a 12 week, randomized, controlled, parallel intervention study with a low and high dose of a freeze-dried strawberry powder (25 g *vs.* 50 g; equivalent to 250 g *vs.* 500 g of raw strawberries per day) in 30 obese subjects with cardiovascular risk factors. Yet in another study from the same authors, no effect on hs-CRP and adiponectin serum concentrations was found following a 4 week intervention with two cups of strawberry drink per day (each cup containing 25 g of freeze-dried strawberry powder) in 16 obese female subjects with MetS.[89] Similar findings were also observed by Zunino *et al.*[90] in a 7 week, double-blind, randomized, crossover intervention following the administration of two servings of strawberry powder (amount not reported; equivalent to 320 g per day of frozen strawberries) mixed in a milkshake, yogurt, cream cheese, or water-based sweetened beverage, that showed no effect on several inflammatory markers such as CRP, IL-1β, IL-6, IL-8, TNF-α, sICAM-1, and sVCAM-1. Jenkins and colleagues[91] in a 4 week, randomized, controlled, crossover intervention with 454 g per day strawberries performed in 28 overweight/obese hyperlipidemic subjects reported no significant effect on CRP concentrations. Finally, Ellis *et al.*[92] reported that a 6 week intervention (randomized, single-blind, placebo-controlled, parallel trial) with 10 g per day of strawberry powder dissolved in water failed to reduce hs-CRP, IL-6, PAI-1, IL-1β, and TNF-α serum levels in overweight and obese subjects. On the contrary, the same authors showed a reduction in PAI-1 and IL-1β serum concentrations in the same group of subjects when the 6 week intervention was followed by a post-prandial trial, in which a high-fat meal was administered.[92] Regarding the effect of acute intervention trials, Ederisinghe and colleagues[93] reported a reduction in serum hs-CRP and IL-6 but not PAI-1, IL-1β, and TNF-α in 26 overweight subjects following the intake of a high-carbohydrate, moderate-fat meal plus strawberry drink (10 g strawberry powder). Accordingly, Park *et al.*[94] showed no effect on circulating levels of VACM-1, ICAM-1, and hs-CRP following the consumption of single different doses of whole strawberry powder (10 *vs.* 20 *vs.* 40 g) in 20 subjects with abdominal obesity and insulin resistance. Conversely, Richter *et al.* reported a significant decrease in CRP concentrations following a single dose of 40 g freeze-dried strawberry powder administered following a high-fat meal to a group of 30 overweight or obese subjects.[95]

In conclusion, of the nine long-term studies investigating the association between strawberry consumption and inflammation on obese or overweight humans, four studies reported a consistent anti-inflammatory effect, one study reported a weak association, and the remaining four studies did not report any association, while two out of the three single-dose interventions reported a significant anti-inflammatory effect. Thus, the evidence suggests an important, although inconsistent, anti-inflammatory effect associated with strawberry consumption.

11.3.8 Whortleberry

The effect of whortleberry on inflammation has been evaluated only in one study. Soltani *et al.*[96] reported a lack of effect on hs-CRP levels following 4-week supplementation with 2 capsules per day (500 mg each) in a group of 25 overweight hyperlipidemic subjects.

11.3.9 Mixed Berries

The impact of mixed berries on inflammation has been analyzed in two studies. In a 6 week randomized, double-blinded, controlled, parallel intervention, Paquette *et al.*[97] documented no significant effect of a beverage formulated with a mix of strawberry and cranberry polyphenols (333 mg) on circulating levels of CRP, adiponectin, PAI-1, IL-6, and TNF-α in a group of overweight/obese individuals with insulin resistance. Lehtonen *et al.*[98] reported no effect of a 20 week, randomized, controlled, parallel intervention with 163 g per day mix of northern berries (lingonberry, sea buckthorn berry, bilberry and blackcurrant) on a panel of inflammatory markers (TNF-α, adiponectin, sICAM, sVCAM, E-selectin and hs-CRP) in a group of 31 overweight subjects with MetS.

11.4 Conclusions and Future Directions

This chapter provided an overview of the link between overweight/obesity and inflammation and how obesity-associated inflammation drives the risk of numerous chronic diseases. A specific focus of the chapter was devoted to the identification of the evidence of the potential health benefits associated with berry consumption. In particular, a brief explanation of the main polyphenol bioactive compounds in berries potentially involved in the favorable anti-inflammatory effects was reported, followed by the main evidence deriving from RCTs, including both long-term and post-prandial intervention studies. Specifically, this review took into consideration the effect of nine different berries and the effect of mixed berries.

Overall, the evidence seems to suggest a potential role of berries in mitigating inflammation in overweight and obese subjects; however, there are discrepancies in the results obtained due to differences in study designs including food matrix (*e.g.* whole fruit, juice, powder, berry mix), dosage, duration, subject characteristics including the metabolic status of subjects (*e.g.* MetS, diabetes, arthritis), existing comorbidities, and different ethnicities/countries with different food habits. It is to be noted, however, that all of the reviewed studies either found an anti-inflammatory effect or failed to find an effect on inflammation, while no study reported a negative effect of berry consumption on inflammation.

The studies already available performed with raspberries, strawberries, and bilberries seem to support the attenuation or reduction of inflammation. Conversely, those carried out with blueberries and cranberries report conflicting results, while very few studies investigated the effects of chokeberries, whortleberries, sea buckthorn berries, and blackcurrants, making it difficult to draw conclusions. Thus, based on recent results, evidence from human intervention studies deserves more substantiation. In particular, well-designed human intervention studies are highly recommended to recruit and select carefully the participants through thorough nutritional assessment and define dose efficacy, frequency of intake, and optimal duration. In addition, a full nutritional characterization of the berries, also in terms of bioactive compounds, must be provided as composition changes are based on climatic conditions, agronomic practices including soil condition, year-to-year bioactive compound variations, and storage length and conditions. Future studies should be designed in order to fill research gaps and provide more evidence also corroborating findings already obtained. Finally, studies ascertaining the potential mechanisms through which berry bioactive compounds, in particular polyphenols (but not only limited to this group), may affect inflammation are encouraged.

Abbreviations

ACNs	anthocyanins
APRIL/TNFSF13	a proliferation-inducing ligand/tumor necrosis factor ligand superfamily member 13
BAFF/TNFSF13	B-cell activating factor
BMI	body mass index
CCR	chemokine receptor
CHI3L1	chitinase-3-like protein 1
CRP	C-reactive protein
FLT3L	FMS-related tyrosine kinase 3 ligand
GPX	glutathione peroxidase
HNE	hydroxynonenal
ICAM	intercellular adhesion molecules
IFN-γ	interferon-gamma
IL	interleukin
iNOS	inducible nitric oxide synthase
MCP-1	monocyte chemoattractant protein
MDA	malondialdehyde
MetS	metabolic syndrome
IFN-γ	gamma interferon
MMD	monocyte to macrophage differentiation-associated
MMP	matrix metalloproteinase
NF-κB	nuclear factor kappa-B
NO	nitric oxide
Ox-LDL	oxidized low-density lipoproteins
PAI-1	plasminogen activator inhibitor 1
PON	paraoxonase
sCD163	cluster of differentiation 163
sSC30/TNFRSF8	TNF receptor superfamily member 8
SOD	superoxide dismutase
sTNF-R1	soluble tumor necrosis factor receptors
TBARS	thiobarbituric acid reactive substances
TNF	tumor necrosis factor
TGF-β	transforming growth factor-β
T2D	type 2 diabetes
TWEAK/TNFSF12	TNF-related weak inducer of apoptosis
VCAM-1	vascular cell adhesion molecule 1
8-OXdG	8-oxo-7,8-dihydro-2-deoxyguanosine

References

1. K. B. Smith and M. S. Smith, Obesity Statistics, *Prim. Care*, 2016, **43**, 121–135.
2. https://www.who.int/health-topics/obesity.
3. N. T. Browne, J. A. Snethen, C. S. Greenberg, M. Frenn, J. F. Kilanowski, B. Gance-Cleveland, P. J. Burke and L. Lewandowski, When pandemics

collide: The impact of COVID-19 on childhood obesity, *J. Pediatr. Nurs.*, 2021, **56**, 90–98.
4. T. Christen, S. Trompet, P. C. N. Rensen, K. Willems van Dijk, H. J. Lamb, J. W. Jukema, F. R. Rosendaal, S. le Cessie and R. de Mutsert, The role of inflammation in the association between overall and visceral adiposity and subclinical atherosclerosis, *Nutr., Metab. Cardiovasc. Dis.*, 2019, **29**, 728–735.
5. M. S. Burhans, D. K. Hagman, J. N. Kuzma, K. A. Schmidt and M. Kratz, Contribution of Adipose Tissue Inflammation to the Development of Type 2 Diabetes Mellitus, *Compr. Physiol.*, 2018, **13**, 1–58.
6. V. Z. Rocha and P. Libby, Obesity, inflammation, and atherosclerosis, *Nat. Rev. Cardiol.*, 2009, **6**, 399–409.
7. O. Izaola, D. de Luis, I. Sajoux, J. C. Domingo and M. Vidal, Inflammation and obesity (lipoinflammation), *Nutr. Hosp.*, 2015, **31**, 2352–2358.
8. S. M. Grundy, H. B. Brewer Jr, J. I. Cleeman, S. C. Smith Jr and C. Lenfant, National Heart, Lung, and Blood Institute; American Heart Association. Definition of metabolic syndrome: report of the National Heart, Lung, and Blood Institute/American Heart Association conference on scientific issues related to definition, *Arterioscler. Thromb. Vasc. Biol.*, 2004, **24**, e13–18.
9. A. Tchernof and J. P. Després, Pathophysiology of human visceral obesity: an update, *Physiol. Rev.*, 2013, **93**, 359–404.
10. E. Stolarczyk, Adipose tissue inflammation in obesity: a metabolic or immune response?, *Curr. Opin. Pharmacol.*, 2017, **37**, 35–40.
11. A. B. Engin, Adipocyte-Macrophage Cross-Talk in Obesity, *Adv. Exp. Med. Biol.*, 2017, **960**, 327–343.
12. A. J. Cox, N. P. West and A. W. Cripps, Obesity, inflammation, and the gut microbiota, *Lancet Diabetes Endocrinol.*, 2015, **3**, 207–215.
13. F. P. de Heredia, S. Gómez-Martínez and A. Marcos, Obesity, inflammation and the immune system, *Proc. Nutr. Soc.*, 2012, **71**, 332–338.
14. R. Monteiro and I. Azevedo, Chronic inflammation in obesity and the metabolic syndrome, *Mediators Inflammation*, 2010, **2010**, 289645.
15. A. Basu, R. Michael and L. J. Timothy, Berries: emerging impact on cardiovascular health, *Nutr. Rev.*, 2010, **68**, 168–177.
16. A. Rodriguez-Mateos, C. Heiss, G. Borges and A. Crozier, Berry (poly)-phenols and cardiovascular health, *J. Agric. Food Chem.*, 2014, **62**, 3842–3851.
17. M. Marino, C. Del Bo', D. Martini, M. Porrini and P. Riso, A Review of Registered Clinical Trials on Dietary (Poly)Phenols: Past Efforts and Possible Future Directions, *Foods*, 2020, **9**, 1606.
18. D. Martini, M. Marino, D. Angelino, C. Del Bo', D. Del Rio, P. Riso and M. Porrini, Role of berries in vascular function: a systematic review of human intervention studies, *Nutr. Rev.*, 2020, **78**, 189–206.
19. B. Baby, P. Antony and R. Vijayan, Antioxidant and anticancer properties of berries, *Crit. Rev. Food Sci. Nutr.*, 2018, **58**, 2491–2507.

20. C. Del, Bo', D. Martini, M. Porrini, D. Klimis-Zacas and P. Riso, Berries and oxidative stress markers: an overview of human intervention studies, *Food Funct.*, 2015, **6**, 2890–2917.
21. C. Govers, M. Berkel Kasikci, A. A. van der Sluis and J. J. Mes, Review of the health effects of berries and their phytochemicals on the digestive and immune systems, *Nutr. Rev.*, 2018, **76**, 29–46.
22. K. Miller, W. Feucht and M. Schmid, Bioactive Compounds of Strawberry and Blueberry and Their Potential Health Effects Based on Human Intervention Studies: A Brief Overview, *Nutrients*, 2019, **11**, 1510.
23. S. V. Joseph, I. Edirisinghe and B. M. Burton-Freeman, Fruit Polyphenols: A Review of Anti-inflammatory Effects in Humans, *Crit. Rev. Food Sci. Nutr.*, 2016, **56**, 419–444.
24. S. Skrovankova, D. Sumczynski, J. Mlcek, T. Jurikova and J. Sochor, Bioactive Compounds and Antioxidant Activity in Different Types of Berries, *Int. J. Mol. Sci.*, 2015, **16**, 24673–24706.
25. H. Speer, N. M. D'Cunha, N. I. Alexopoulos, A. J. McKune and N. Naumovski, Anthocyanins and Human Health-A Focus on Oxidative Stress, Inflammation and Disease, *Antioxidants*, 2020, **9**, 366.
26. S. J. Maleki, J. F. Crespo and B. Cabanillas, Anti-inflammatory effects of flavonoids, *Food Chem.*, 2019, **299**, 125124.
27. S. Vendrame, C. Del Bo', S. Ciappellano, P. Riso and D. Klimis-Zacas, Berry Fruit Consumption and Metabolic Syndrome, *Antioxidants*, 2016, **5**, 34.
28. S. Vendrame and D. Klimis-Zacas, Anti-inflammatory effect of anthocyanins via modulation of nuclear factor-κB and mitogen-activated protein kinase signaling cascades, *Nutr. Rev.*, 2015, **73**, 348–358.
29. Y. Li, J. Yao, C. Han, J. Yang, M. T. Chaudhry, S. Wang, H. Liu and Y. Quercetin, Inflammation and Immunity, *Nutrients*, 2016, **8**, 167.
30. K. P. Devi, D. S. Malar, S. F. Nabavi, A. Sureda, J. Xiao, S. M. Nabavi and M. Daglia, Kaempferol and inflammation: From chemistry to medicine, *Pharmacol. Res.*, 2015, **99**, 1–10.
31. S. H. Häkkinen, S. O. Kärenlampi, I. M. Heinonen, H. M. Mykkänen and A. R. Törrönen, Content of the flavonols quercetin, myricetin, and kaempferol in 25 edible berries,*J. Agric. Food Chem.*, 1999, **47**, 2274–2279.
32. R. Zadernowski, M. Naczk and J. Nesterowicz, Phenolic acid profiles in some small berries, *J. Agric. Food Chem.*, 2005, **53**, 2118–2124.
33. S. S. Ali, W. A. N. W. Ahmad, S. B. Budin and S. Zainalabidin, Implication of dietary phenolic acids on inflammation in cardiovascular disease, *Rev. Cardiovasc. Med.*, 2020, **21**, 225–240.
34. O. B. Ibitoye and T. O. Ajiboye, Dietary phenolic acids reverse insulin resistance, hyperglycaemia, dyslipidaemia, inflammation and oxidative stress in high-fructose diet-induced metabolic syndrome rats, *Arch. Physiol. Biochem.*, 2018, **124**, 410–417.
35. A. Smeriglio, D. Barreca, E. Bellocco and D. Trombetta, Proanthocyanidins and hydrolysable tannins: occurrence, dietary intake and pharmacological effects, *Br. J. Pharmacol.*, 2017, **174**, 1244–1262.

36. S. H. Nile and S. W. Park, Edible berries: Bioactive components and their effect on human health, *Nutrition*, 2014, **30**, 134–144.
37. W. Mullen, J. McGinn, M. E. Lean, M. R. MacLean, P. Gardner, G. G. Duthie, T. Yokota and A. Crozier, Ellagitannins, flavonoids, and other phenolics in red raspberries and their contribution to antioxidant capacity and vasorelaxation properties, *J. Agric. Food Chem.*, 2002, **50**, 5191–5196.
38. L. Yang, D. Xian, X. Xiong, R. Lai, J. Song and J. Zhong, Proanthocyanidins against Oxidative Stress: From Molecular Mechanisms to Clinical Applications, *BioMed Res. Int.*, 2018, **2018**, 8584136.
39. K. Gil-Cardoso, R. Comitato, I. Ginés, A. Ardévol, M. Pinent, F. Virgili, X. Terra and M. Blay, Protective Effect of Proanthocyanidins in a Rat Model of Mild Intestinal Inflammation and Impaired Intestinal Permeability Induced by LPS, *Mol. Nutr. Food Res.*, 2019, **63**, e1800720.
40. S. L. Coleman and O. M. Shaw, Progress in the understanding of the pathology of allergic asthma and the potential of fruit proanthocyanidins as modulators of airway inflammation, *Food Funct.*, 2017, **8**, 4315–4324.
41. C. Del Bo', A. S. Kristo, A. Z. Kalea, S. Ciappellano, P. Riso, M. Porrini and D. Klimis-Zacas, The temporal effect of a wild blueberry (Vaccinium angustifolium)-enriched diet on vasomotor tone in the Sprague-Dawley rat, *Nutr., Metab. Cardiovasc. Dis.*, 2012, **22**, 127–132.
42. C. Del Bo', D. Martini, S. Vendrame, P. Riso, S. Ciappellano, D. Klimis-Zacas and M. Porrini, Improvement of lymphocyte resistance against H(2)O(2)-induced DNA damage in Sprague-Dawley rats after eight weeks of a wild blueberry (Vaccinium angustifolium)-enriched diet, *Mutat. Res.*, 2010, **703**, 158–162.
43. C. Del Bo', M. Roursgaard, M. Porrini, S. Loft, P. Møller and P. Riso, Different effects of anthocyanins and phenolic acids from wild blueberry (Vaccinium angustifolium) on monocytes adhesion to endothelial cells in a TNF-α stimulated proinflammatory environment, *Mol. Nutr. Food Res.*, 2016, **60**, 2355–2366.
44. C. Del Bo', M. Marino, P. Riso, P. Møller and M. Porrini, Anthocyanins and metabolites resolve TNF-α-mediated production of E-selectin and adhesion of monocytes to endothelial cells, *Chem.-Biol. Interact.*, 2019, **300**, 49–55.
45. M. Marino, C. Del Bo', M. Tucci, D. Klimis-Zacas, P. Riso and M. Porrini, Modulation of Adhesion Process, E-Selectin and VEGF Production by Anthocyanins and Their Metabolites in an in vitro Model of Atherosclerosis, *Nutrients*, 2020, **12**, 655.
46. V. Taverniti, D. Fracassetti, C. Del Bo', C. Lanti, M. Minuzzo, D. Klimis-Zacas, P. Riso and S. Guglielmetti, Immunomodulatory effect of a wild blueberry anthocyanin-rich extract in human Caco-2 intestinal cells, *J. Agric. Food Chem.*, 2014, **62**, 8346–8351.
47. H. Land Lail, R. G. Feresin, D. Hicks, B. Stone, E. Price and D. Wanders, Berries as a Treatment for Obesity-Induced Inflammation: Evidence from Preclinical Models, *Nutrients*, 2021, **13**, 334.

48. F. C. Laua, J. A. Josepha, J. E. McDonald and W. Kalt, Attenuation of iNOS and COX2 by blueberry polyphenols is mediated through the suppression of NF-jB activation, *J. Funct. Foods*, 2009, **1**, 274–283.
49. C. Yahfoufi, N. Alsadi, M. Jambi and C. Matar, The Immunomodulatory and Anti-Inflammatory Role of Polyphenols, *Nutrients*, 2018, **10**, 1618.
50. S. Vendrame, A. Daugherty, A. S. Kristo, P. Riso and D. Klimis-Zacas, Wild blueberry (Vaccinium angustifolium) consumption improves inflammatory status in the obese Zucker rat model of the metabolic syndrome, *J. Nutr. Biochem.*, 2013, **24**, 1508–1512.
51. S. Vendrame, P. Tsakiroglou, A. S. Kristo, D. A. Schuschke and D. Klimis-Zacas, Wild blueberry consumption attenuates local inflammation in the perivascular adipose tissue of obese Zucker rats, *Appl. Physiol., Nutr., Metab.*, 2016, **41**, 1045–1051.
52. http://www.ncbi.nlm.nih.gov/pubmed.
53. http://www.embase.com.
54. C. Du, A. Smith, M. Avalos, S. South, K. Crabtree, W. Wang, Y. H. Kwon, P. Vijayagopal and S. Juma, Blueberries Improve Pain, Gait Performance, and Inflammation in Individuals with Symptomatic Knee Osteoarthritis, *Nutrients*, 2019, **11**, 290.
55. A. R. Nair, N. Mariappan, A. J. Stull and J. Francis, Blueberry supplementation attenuates oxidative stress within monocytes and modulates immune cell levels in adults with metabolic syndrome: a randomized, double-blind, placebo-controlled trial, *Food Funct.*, 2017, **8**, 4118–4128.
56. K. S. Stote, M. I. Sweeney, T. Kean, D. J. Baer, J. A. Novotny, N. L. Shakerley, A. Chandrasekaran, P. M. Carrico, J. A. Melendez and K. T. Gottschall-Pass, The effects of 100% wild blueberry (Vaccinium angustifolium) juice consumption on cardiometablic biomarkers: a randomized, placebo-controlled, crossover trial in adults with increased risk for type 2 diabetes, *BMC Nutr.*, 2017, **3**, 45.
57. S. A. Johnson, R. G. Feresin, N. Navaei, A. Figueroa, M. L. Elam, N. S. Akhavan, S. Hooshmand, S. Pourafshar, M. E. Payton and B. H. Arjmandi, Effects of daily blueberry consumption on circulating biomarkers of oxidative stress, inflammation, and antioxidant defense in postmenopausal women with pre- and stage 1-hypertension: a randomized controlled trial, *Food Funct.*, 2017, **8**, 372–380.
58. S. A. Johnson, A. Figueroa, N. Navaei, A. Wong, R. Kalfon, L. T. Ormsbee, R. G. Feresin, M. L. Elam, S. Hooshmand, M. E. Payton and B. H. Arjmandi, Daily blueberry consumption improves blood pressure and arterial stiffness in postmenopausal women with pre- and stage 1-hypertension: a randomized, double-blind, placebo-controlled clinical trial, *J. Acad. Nutr. Diet.*, 2015, **115**, 369–377.
59. A. J. Stull, K. C. Cash, W. D. Johnson, C. M. Champagne and W. T. Cefalu, Bioactives in blueberries improve insulin sensitivity in obese, insulin-resistant men and women, *J. Nutr.*, 2010, **140**, 1764–1768.

60. A. Basu, M. Du, M. J. Leyva, K. Sanchez, N. M. Betts, M. Wu, C. E. Aston and T. J. Lyons, Blueberries decrease cardiovascular risk factors in obese men and women with metabolic syndrome, *J. Nutr.*, 2010, **140**, 1582–1587.
61. A. P. Sobolev, A. Ciampa, C. Ingallina, L. Mannina, D. Capitani, I. Ernesti, E. Maggi, R. Businaro, M. Del Ben, P. Engel, A. M. Giusti, L. M. Donini and A. Pinto, Blueberry-Based Meals for Obese Patients with Metabolic Syndrome: A Multidisciplinary Metabolomic Pilot Study, *Metabolites*, 2019, **9**, 138.
62. M. Kolehmainen, O. Mykkänen, P. V. Kirjavainen, T. Leppänen, E. Moilanen, M. Adriaens, D. E. Laaksonen, M. Hallikainen, R. Puupponen-Pimiä, L. Pulkkinen, H. Mykkänen, H. Gylling, K. Poutanen and R. Törrönen, Bilberries reduce low-grade inflammation in individuals with features of metabolic syndrome, *Mol. Nutr. Food Res.*, 2012, **56**, 1501–1510.
63. H. M. Lehtonen, J. P. Suomela, R. Tahvonen, B. Yang, M. Venojärvi, J. Viikari and H. Kallio, Different berries and berry fractions have various but slightly positive effects on the associated variables of metabolic diseases on overweight and obese women, *Eur. J. Clin. Nutr.*, 2011, **65**, 394–401.
64. A. Karlsen, I. Paur, S. K. Bøhn, A. K. Sakhi, G. I. Borge, M. Serafini, I. Erlund, P. Laake, S. Tonstad and R. Blomhoff, Bilberry juice modulates plasma concentration of NF-kappaB related inflammatory markers in subjects at increased risk of CVD, *Eur. J. Nutr.*, 2010, **49**, 345–355.
65. P. Huebbe, K. Giller, S. de Pascual-Teresa, A. Arkenau, B. Adolphi, S. Portius, C. N. Arkenau and G. Rimbach, Effects of blackcurrant-based juice on atherosclerosis-related biomarkers in cultured macrophages and in human subjects after consumption of a high-energy meal, *Br. J. Nutr.*, 2012, **108**, 234–244.
66. G. Ruel, S. Pomerleau, P. Couture, S. Lemieux, B. Lamarche and C. Couillard, Low-calorie cranberry juice supplementation reduces plasma oxidized LDL and cell adhesion molecule concentrations in men, *Br. J. Nutr.*, 2008, **99**, 352–359.
67. I. T. Lee, Y. C. Chan, C. W. CW, W. J. Lee and W. H. Sheu, Effect of cranberry extracts on lipid profiles in subjects with Type 2 diabetes, *Diabet. Med.*, 2008, **25**, 1473.
68. A. Basu, N. M. Betts, J. Ortiz, B. Simmons, M. Wu and T. J. Lyons, Low-energy cranberry juice decreases lipid oxidation and increases plasma antioxidant capacity in women with metabolic syndrome, *Nutr. Res.*, 2011, **31**, 190–196.
69. M. M. Dohadwala, M. Holbrook, N. M. Hamburg, S. M. Shenouda, W. B. Chung, M. Titas, M. A. Kluge, N. Wang, J. Palmisano, P. E. Milbury, J. B. Blumberg and J. A. Vita, Effects of cranberry juice consumption on vascular function in patients with coronary artery disease, *Am. J. Clin. Nutr.*, 2011, **93**, 934–940.
70. T. N. Simão, M. A. Lozovoy, A. N. Simão, S. R. Oliveira, D. Venturini, H. K. Morimoto, L. H. Miglioranza and I. Dichi, Reduced-energy cranberry juice increases folic acid and adiponectin and reduces homocysteine and

oxidative stress in patients with the metabolic syndrome, *Br. J. Nutr.*, 2013, **110**, 1885–1894.

71. G. Ruel, A. Lapointe, S. Pomerleau, P. Couture, S. Lemieux, B. Lamarche and C. Couillard, Evidence that cranberry juice may improve augmentation index in overweight men, *Nutr. Res.*, 2013, **33**, 41–49.
72. N. S. B. Thimóteo, T. M. V. Iryioda, D. F. Alfieri, B. E. F. Rego, B. M. Scavuzzi, E. Fatel, M. A. B. Lozovoy, A. N. C. Simão and I. Dichi, Cranberry juice decreases disease activity in women with rheumatoid arthritis, *Nutrition*, 2019, **60**, 112–117.
73. J. A. Novotny, D. J. Baer, C. Khoo, S. K. Gebauer and C. S. Charron, Cranberry juice consumption lowers markers of cardiometabolic risk, including blood pressure and circulating C-reactive protein, triglyceride, and glucose concentrations in adults, *J. Nutr.*, 2015, **145**, 1185–1193.
74. J. Schell, N. M. Betts, M. Foster, R. H. Scofield and A. Basu, Cranberries improve postprandial glucose excursions in type 2 diabetes, *Food Funct.*, 2017, **8**, 3083–3090.
75. B. Chew, B. Mathison, L. Kimble, D. McKay, K. Kaspar, C. Khoo, C. O. Chen and J. Blumberg, Chronic consumption of a low calorie, high polyphenol cranberry beverage attenuates inflammation and improves glucoregulation and HDL cholesterol in healthy overweight humans: a randomized controlled trial, *Eur. J. Nutr.*, 2019, **58**, 1223–1235.
76. H. S. Jeong, S. J. Hong, T. B. Lee, J. W. Kwon, J. T. Jeong, H. J. Joo, J. H. Park, C. M. Ahn, C. W. Yu and D. S. Lim, Effects of black raspberry on lipid profiles and vascular endothelial function in patients with metabolic syndrome, *Phytother. Res.*, 2014, **28**, 1492–1498.
77. H. S. Jeong, S. J. Hong, J. Y. Cho, T. B. Lee, J. W. Kwon, H. J. Joo, J. H. Park, C. W. Yu and D. S. Lim, Effects of Rubus occidentalis extract on blood pressure in patients with prehypertension: Randomized, double-blinded, placebo-controlled clinical trial, *Nutrition*, 2016b, **32**, 461–467.
78. H. S. Jeong, S. Kim, S. J. Hong, S. C. Choi, J. H. Choi, J. H. Kim, C. Y. Park, J. Y. Cho, T. B. Lee, J. W. Kwon, H. J. Joo, J. H. Park, C. W. Yu and D. S. Lim, Black Raspberry Extract Increased Circulating Endothelial Progenitor Cells and Improved Arterial Stiffness in Patients with Metabolic Syndrome: A Randomized Controlled Trial, *J. Med. Food*, 2016a, **19**, 346–352.
79. C. L. Sardo, J. P. Kitzmiller, G. Apseloff, R. B. Harris, D. J. Roe, G. D. Stoner and E. T. Jacobs, An Open-Label Randomized Crossover Trial of Lyophilized Black Raspberries on Postprandial Inflammation in Older Overweight Males: A Pilot Study, *Am. J. Ther.*, 2016, **23**, e86–91.
80. Y. J. Kim, I. Huh, J. Y. Kim, S. Park, S. H. Ryu, K. B. Kim, S. Kim, T. Park and O. Kwon, Integration of Traditional and Metabolomics Biomarkers Identifies Prognostic Metabolites for Predicting Responsiveness to Nutritional Intervention against Oxidative Stress and Inflammation, *Nutrients*, 2017, **9**, 233.

81. J. Schell, N. M. Betts, T. J. Lyons and A. Basu, Raspberries Improve Postprandial Glucose and Acute and Chronic Inflammation in Adults with Type 2 Diabetes, *Ann. Nutr. Metab.*, 2019, **74**, 165–174.
82. J. H. An, D. L. Kim, T. B. Lee, K. J. Kim, S. H. Kim, H. K. Kim, H. Y. Kim, D. S. Choi and S. G. Kim, Effect of Rubus Occidentalis Extract on Metabolic Parameters in Subjects with Prediabetes: A Proof-of-concept, Randomized, Double-blind, Placebo-controlled Clinical, *Trial. Phytother. Res.*, 2016, **30**, 1634–1640.
83. A. Skoczyñska, I. Jêdrychowska, R. Porêba, A. AffelskaJercha, B. Turczyn, A. Wojakowska and R. Andrzejak, Influence of chokeberry juice on arterial blood pressure and lipid parameters in men with mild hypercholesterolemia, *Pharmacol. Rep.*, 2007, **59**, 177–182.
84. A. Basu, B. T. Kurien, H. Tran, B. Byrd, J. Maher, J. Schell, E. Masek, J. R. Barrett, T. J. Lyons, N. M. Betts and R. Hal Scofield, Strawberries decrease circulating levels of tumor necrosis factor and lipid peroxides in obese adults with knee osteoarthritis, *Food Funct.*, 2018, **9**, 6218–6226.
85. J. Schell, R. H. Scofield, J. R. Barrett, B. T. Kurien, N. Betts, T. J. Lyons, Y. D. Zhao and A. Basu, Strawberries Improve Pain and Inflammation in Obese Adults with Radiographic Evidence of Knee Osteoarthritis, *Nutrients*, 2017, **9**, 949.
86. S. Moazen, R. Amani, A. Homayouni Rad, H. Shahbazian, K. Ahmadi and M. Taha Jalali, Effects of freeze-dried strawberry supplementation on metabolic biomarkers of atherosclerosis in subjects with type 2 diabetes: a randomized double-blind controlled trial, *Ann. Nutr. Metab.*, 2013, **63**, 256–264.
87. A. Basu, D. X. Fu, M. Wilkinson, B. Simmons, M. Wu, N. M. Betts, M. Du and T. J. Lyons, Strawberries decrease atherosclerotic markers in subjects with metabolic syndrome, *Nutr. Res.*, 2010, **30**, 462–469.
88. A. Basu, N. M. Betts, A. Nguyen, E. D. Newman, D. Fu and T. J. Lyons, Freeze-dried strawberries lower serum cholesterol and lipid peroxidation in adults with abdominal adiposity and elevated serum lipids, *J. Nutr.*, 2014, **144**, 830–837.
89. A. Basu, M. Wilkinson, K. Penugonda, B. Simmons, N. M. Betts and T. J. Lyons, Freeze-dried strawberry powder improves lipid profile and lipid peroxidation in women with metabolic syndrome: baseline and post intervention effects, *Nutr. J.*, 2009, **8**, 43.
90. S. J. Zunino, M. A. Parelman, T. L. Freytag, C. B. Stephensen, D. S. Kelley, B. E. Mackey, L. R. Woodhouse and E. L. Bonnel, Effects of dietary strawberry powder on blood lipids and inflammatory markers in obese human subjects, *Br. J. Nutr.*, 2012, **108**, 900–909.
91. D. J. Jenkins, T. H. Nguyen, C. W. Kendall, D. A. Faulkner, B. Bashyam, I. J. Kim, C. Ireland, D. Patel, E. Vidgen, A. R. Josse, H. D. Sesso, B. Burton-Freeman, R. G. Josse, L. A. Leiter and W. Singer, The effect of strawberries in a cholesterol-lowering dietary portfolio, *Metabolism*, 2008, **57**, 1636–1644.

92. C. L. Ellis, I. Edirisinghe, T. Kappagoda and B. Burton-Freeman, Attenuation of meal-induced inflammatory and thrombotic responses in overweight men and women after 6-week daily strawberry (Fragaria) intake. A randomized placebo-controlled trial, *J. Atheroscler. Thromb.*, 2011, **18**, 318–327.
93. I. Edirisinghe, K. Banaszewski, J. Cappozzo, K. Sandhya, C. L. Ellis, R. Tadapaneni, C. T. Kappagoda and B. M. Burton-Freeman, Strawberry anthocyanin and its association with postprandial inflammation and insulin, *Br. J. Nutr.*, 2011, **106**, 913–922.
94. E. Park, I. Edirisinghe, H. Wei, L. P. Vijayakumar, K. Banaszewski, J. C. Cappozzo and B. Burton-Freeman, A dose-response evaluation of freeze-dried strawberries independent of fiber content on metabolic indices in abdominally obese individuals with insulin resistance in a randomized, single-blinded, diet-controlled crossover trial, *Mol. Nutr. Food Res.*, 2016, **60**, 1099–1109.
95. C. K. Richter, A. C. Skulas-Ray, T. L. Gaugler, J. D. Lambert, D. N. Proctor and P. M. Kris-Etherton, Incorporating freeze-dried strawberry powder into a high-fat meal does not alter postprandial vascular function or blood markers of cardiovascular disease risk: a randomized controlled trial, *Am. J. Clin. Nutr.*, 2017, **105**, 313–322.
96. R. Soltani, M. Hakimi, S. Asgary, S. M. Ghanadian, M. Keshvari and N. Sarrafzadegan, Evaluation of the Effects of Vaccinium arctostaphylos L. Fruit Extract on Serum Lipids and hs-CRP Levels and Oxidative Stress in Adult Patients with Hyperlipidemia: A Randomized, Double-Blind, Placebo-Controlled Clinical Trial, *Evidence-Based Complementary Altern. Med.*, 2014, **2014**, 217451.
97. M. Paquette, A. S. Medina Larqué, S. J. Weisnagel, Y. Desjardins, J. Marois, G. Pilon, S. Dudonné, A. Marette and H. Jacques, Strawberry and cranberry polyphenols improve insulin sensitivity in insulin-resistant, non-diabetic adults: a parallel, double-blind, controlled and randomised clinical trial, *Br. J. Nutr.*, 2017, **117**, 519–531.
98. H. M. Lehtonen, J. P. Suomela, R. Tahvonen, J. Vaarno, M. Venojärvi, J. Viikari and H. Kallio, Berry meals and risk factors associated with metabolic syndrome, *Eur. J. Clin. Nutr.*, 2010, **64**, 614–621.

CHAPTER 12

Mechanisms of Action of Berry Bioactives on Obesity-induced Inflammation

F. GIAMPIERI,[a,b] DANILA CIANCIOSI,[c] J. ANSARY,[c] T. Y. FORBES-HERNANDEZ,[d] J. M. ALVAREZ-SUAREZ[e,f,g] AND M. BATTINO*[c,h]

[a] Research Group on Food, Nutritional Biochemistry and Health, Universidad Europea del Atlántico, Santander, Spain; [b] Department of Biochemistry, Faculty of Sciences, King Abdulaziz University, Jeddah, Saudi Arabia; [c] Department of Clinical Sciences, Polytechnic University of Marche, Ancona 60131, Ancona, Italy; [d] Department of Physiology, Institute of Nutrition and Food Technology "José Mataix", Biomedical Research Centre, University of Granada, Granada, 1800, Spain; [e] Departamento de Ingeniería en Alimentos. Colegio de Ciencias e Ingenierías. Universidad San Francisco de Quito, Quito, Ecuador, 170157, Ecuador; [f] Instituto de Investigaciones en Biomedicina iBioMed, Universidad San Francisco de Quito, Quito, Ecuador; [g] King Fahd Medical Research Center, King Abdulaziz University, 21589, Jeddah, Saudi Arabia; [h] International Joint Research Laboratory of Intelligent Agriculture and Agri-products Processing, Jiangsu University, Zhenjiang, China
*Email: m.a.battino@univpm.it

12.1 Introduction

Metabolic syndrome (MetS) and obesity represent two of the major public health challenges in the world. MetS can be described as the association of

Food Chemistry, Function and Analysis No. 33
Berries and Berry Bioactive Compounds in Promoting Health
Edited by Dorothy Klimis-Zacas and Ana Rodriguez-Mateos

Published by the Royal Society of Chemistry, www.rsc.org

different factors, including dyslipidaemia (increased low-density lipoprotein [LDL], triglycerides and decreased high-density lipoprotein [HDL] levels), elevated blood glucose, insulin resistance, hypertension and central obesity.[1] Its prevalence depends on the age of the population, the ethnicity and the definition applied: indeed, the two most used definitions are those proposed by the International Diabetes Federation and the American Heart Association/ National Heart, Lung, and Blood Institute (based on the previous definition of the National Cholesterol Education Program Adult Treatment Panel III [ATP III]). Specifically, the first definition considers the waist circumference, and thus the visceral obesity, as the mandatory diagnostic criterion and takes into account that different nationalities, ethnicities and populations may present diverse distributions of norms for waist circumference and body weight; the latter considers three of the five following possible components: levels of fasting blood sugar higher than 100 $mg\,dL^{-1}$, levels of triglyceride higher than 150 $mg\,dL^{-2}$, levels of HDL cholesterol less than 40 and 50 $mg\,dL^{-1}$ for men and women, respectively, blood pressure higher than 130/85 mm Hg and waist circumference over 40 and 35 inches for men and women, respectively.[2–4] MetS is associated with a higher risk of developing different diseases, including type 2 diabetes, cardiovascular diseases, coronary heart diseases and various types of cancer in both sexes.[5,6]

On the other hand, overweight and obesity are metabolic disorders with pandemic proportion, with 650 million adults obese and more than 1.9 billion overweight in 2016 worldwide, according to the World Health Organization.[7] Obesity can be defined as a long-term energy imbalance between calories consumed and calories expended, where the energy excess is continuously deposited as fats into the adipose tissue. This unbalance may depend on different genetic, social, psychological and environmental factors,[7] such as hormone problems, sedentary lifestyle, high-glycaemic diet, stress, poor sleep, oversized food portions, food marketing and the use of some drugs, like psychotropic ones.[7,8] Like MetS, obesity is strictly correlated with a higher risk of developing some comorbidities, including type 2 diabetes, cardiovascular diseases, hypertension and certain kinds of cancer.[9]

Being overweight and obese may evolve into MetS through several pathological mechanisms, of which the chronic low-grade inflammation and excess of adipose tissue play a key role, being implicated both on its onset and on its pathological consequences.[10]

12.2 Inflammation

Inflammation is the biological response of the immune system against different harmful agents, which can be of different natures (Figure 12.1).[11]

Besides the classic stimuli, inflammation can also be induced by body fat accumulation and poor and unhealthy diet: the overproduction of free radicals, especially at the mitochondrial level, stimulated by an excess of energy intake, may lead to a cellular redox imbalance, which in turn activates redox-sensitive signalling molecules, promoting the increase of oxidative stress

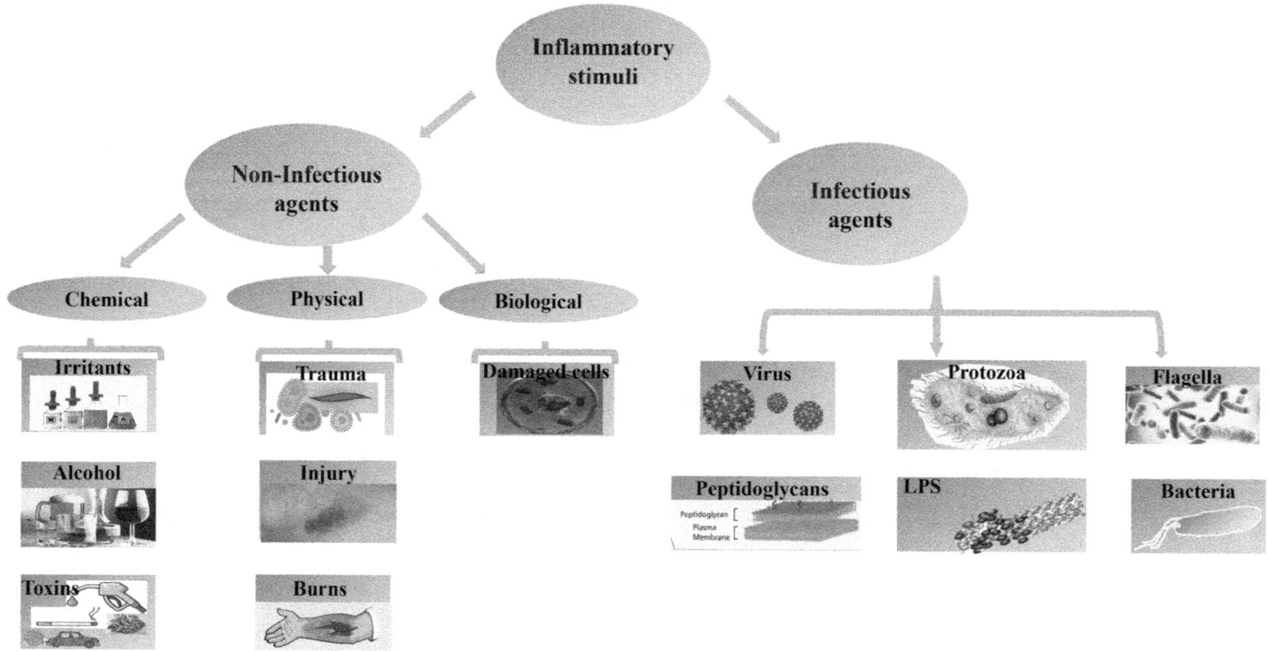

Figure 12.1 Principal harmful agents that can promote inflammation. They can be non-infectious or infectious agents: the former comprise chemical (irritants, alcohol, toxins, *etc.*), physical (trauma, physical injury, burns, *etc.*) and biological (damaged cells) factors; the latter include bacteria, viruses, fungi, protozoa and their fragments (microbial DNA, lipopolysaccharide [LPS], lipopeptides, peptidoglycans, flagellin, *etc.*).

markers and inflammatory genes expression.[12,13] In other words, inflammation and oxidative stress, defined as an imbalance between the production and accumulation of reactive oxygen species (ROS) and the antioxidant system, are strongly correlated, so inflammation induces the production of ROS and *vice versa* and both events play a key role in the onset of the main chronic diseases, such as type 2 diabetes, obesity and MetS.

Inflammation can be acute or chronic. During acute inflammation, which usually lasts a few days, microcirculation is altered to promote the passage of plasma proteins from blood to tissues and leucocyte migration from the microcirculation to the injured tissue, where they are activated and eliminate the harmful agent by producing several inflammatory molecules.[14,15] Therefore, typical signs of acute inflammation are swelling, redness, pain, heat and loss of tissue function, which are essentially due to the local vascular, inflammatory and immune responses to the injury. When the inflammation is resolved through the activation of some negative feedback mechanisms, including the secretion of anti-inflammatory cytokines, the activation of regulatory cells, the secretion of pro-resolving lipid mediators and the reduction of the membrane receptors expression, a process named resolution, the repair of the damaged tissue, is initiated.[14,15] In contrast, during chronic inflammation, which can last for several months or even years and which has lower intensities compared to the acute inflammatory status, a continuous change in inflammatory cells and in tissue destruction/repair occurs due to the loss of regulatory processes, resulting in an increase of activated inflammatory cells and biomarkers in the bloodstream, as demonstrated for example in subjects affected by MetS or type 2 diabetes.[16,17]

12.2.1 Inflammation in Metabolic Syndrome and Obesity

Obesity and metabolic diseases are closely linked to a state of chronic or low-grade inflammation, characterized by the abnormal production of inflammatory cytokines and acute phase mediators and by the continued activation of inflammatory pathways that are responsible for the development and the complications of these pathologies.[18,19]

In humans, adipose tissue can be categorized into brown adipose tissue (BAT) and white adipose tissue (WAT): the first is essentially involved in the regulation of body temperature, and the latter basically stores excess energy as triglycerides. In people with obesity, adipose tissue is generally increased at the thorax and abdomen sites (*i.e.*, abdominal, central or visceral obesity): the prolonged state of positive energy balance, together with the incessant accumulation of fats, generally leads to adypocyte hyperthropy that facilitates cell rupture, creating new sites of inflammation: additionally, fat may deposit in organs different from the adipose tissue, such as the liver.[20,21] Moreover, in adipose tissue, a condition of hypoxia commonly exists, due essentially to the progressive adipocyte enlargement and to the increase of the intra-abdominal pressure, which can reduce blood supply and lead to the necrosis of adipocytes. Of note, hypoxia is considered one of the most important factors for the infiltration of macrophages, eosinophils and neutrophils, as well as for the overproduction of inflammatory mediators, which contributes to the spread of the overall systemic inflammation.[19,22,23] Therefore, besides its function in fat accumulation, the adipose tissue is a highly active crucial site for the generation and secretion of several inflammatory mediators, mainly cytokines, chemokines, acute-phase proteins and hormones, among others. In brief, the most common cytokines are essentially the classic peptide mediators of inflammation, *i.e.*, tumor necrosis factor (TNF)-α; interleukin (IL)-1 and its receptor antagonist; IL-6, IL-7, IL-10, IL-18 and macrophage migration inhibitory factor, while the main chemokines include IL-8; regulated on activation, normal T expressed and secreted (RANTES); monocyte chemoattractant protein (MCP-1); macrophage inflammatory protein-1α and macrophage inflammatory protein-1β and metallothioneins. Among acute-phase proteins, there are C-reactive protein (CRP), haptoglobin, plasminogen activator inhibitor 1 (PAI-1), lipocalin and pentraxin-3, while the main hormones produced that may affect the immune system and insulin sensitivity are leptin, with pro-inflammatory activities, and adiponectin, with anti-inflammatory properties. Finally, adipose tissue also produces and secretes some components of the alternative complement system (factors C2, C3, C4, C7, B and D) and growth factors, such as transforming growth factor (TGF)-β.[22–24]

Recently, it has been shown that, besides adipose tissue, other organs, mainly liver, skeletal muscle, hypothalamus and pancreas, are also involved in the production of these pro-inflammatory mediators, boosting the state of chronic inflammation typical of obesity and MetS.[25,26]

12.2.1.1 Molecular Pathways Involved in Inflammation

Several molecular pathways are responsible for nutrient intake and are involved in the modulation of the low-grade inflammation in obesity and metabolic disease conditions (Figure 12.2).

In general, the inflammatory response initiates when the germline-encoded pattern-recognition receptors (PRRs), located in the plasma membrane of both non-immune and immune cells, recognize harmful foreign molecules, the so-called pathogen-associated molecular patterns, or endogenous ligands typical of the obese state.[25,26] The most studied PRRs are the Toll-like receptors (TLRs), highly conserved in mammalians, that comprise more than 110 members, are present in the adipose tissue and participate in the activation of the inflammatory response. In particular, concerning obesity, scientific attention has been focused on TLR4, because it can be activated by free fatty acids (FAs) and can consequently induce the inflammatory response, as demonstrated in TRL4-deficient mice, where the absence of TLR4 has been correlated with the decrease of inflammation and insulin resistance induced by lipid perfusion,[26,27] and on TLR5, since it can sense and modulate gut

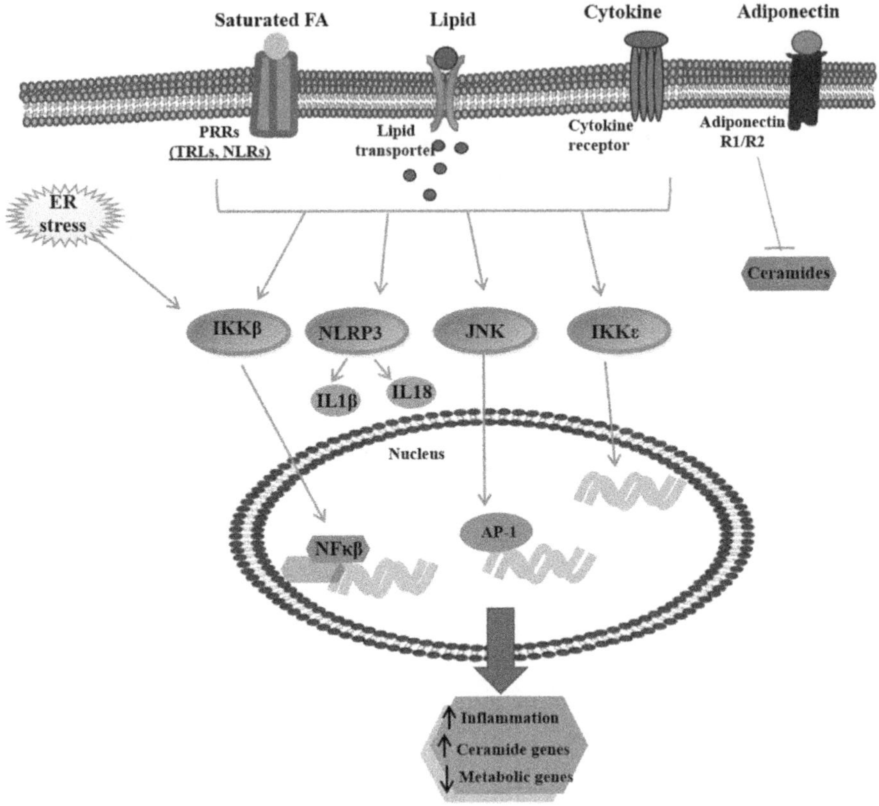

Figure 12.2 Schematic representation of the main molecular pathways involved in inflammation.

microbiota, contributing to the regulation of nutrient supply and energy expenditure and storage, as shown in TLR5$^{-/-}$ mouse.[25,26,28]

Another important PRR family is the NOD-like receptors (NLRs), which, after being activated by several obesity-induced signals, such as those from stressed or dying cells, mobilize leukocytes and macrophages to repair tissue damage. In particular, in macrophages, NLRs stimulate the NLR family pyrin domain containing 3 (NLRP3) inflammasome, inducing the production of IL-1β and IL-18 through the activity of caspase-1;[26] of note is that an obesity-induced diet stimulates the production of IL-1β and caspase-1 also at the adipose tissue level.[29] In addition, the activation of these pathways leads to the death of pancreatic β-cell due to chronic hyperglycaemia, affecting diabetes progression.[30]

After activation, TLRs stimulate an intracellular signalling cascade that can involve the adaptor protein MyD88, as occurs in hypothalamic inflammation in obesity,[31] and that eventually leads to the induction of IκB kinase-β (IKKβ).[26] The ultimate target of TLR/IKKβ signals is the nuclear factor kappa-light-chain-enhancer of activated B cells (NF-κB), a transcription factor that promotes the transcription of inflammatory genes, such as those responsible for the production of cytokines and other inflammatory mediators, such as cell adhesion molecules, chemokines, acute phase proteins and growth factors.[26,32] In addition, in adipose tissue, under high-fat diet (HFD) conditions, NF-κB may stimulate the transcription of the protein kinase IKKε, which contributes to suppressing thermogenesis, thus contributing to maintaining body weight and insulin resistance.[33]

In insulin-responsive tissues, especially in response to obesity biomarker signals, including insulin, FAs, inflammatory cytokines and hyperglycaemia, IKKβ may also activate (JNK) pathway, belonging to the mitogen-activated protein kinase family;[34] once activated, JNK leads to the phosphorylation and activation of different transcription factors, including NF-κB and activator protein-1 (AP-1), inducing the expression of pro-inflammatory mediators.[19] In particular, JNK1 plays a key role in protecting against diet-induced obesity, since its inactivation has been correlated with a marked decrease in inflammation and with the inhibition of insulin resistance development.[26,35,36]

Besides NF-κB, ceramide is another downstream target of TLR4, especially when it is activated by FAs, as occurs in several metabolic tissues, such as muscle and hypothalamus, where it can inhibit glucose uptake by hindering Akt/PKB activation, thus impairing insulin signalling.[37] Additionally, ceramide production is regulated by adiponectin, an adipokine that is able to activate ceramidase and modulate the ceramides/sphingosine-1-phosphate ratio, thus decreasing inflammation and promoting cell survival.[38]

Finally, in obesity, endoplasmic reticulum (ER) stress and inflammation are closely related: on the one hand, in obese subjects, an increase of the main ER stress components (*i.e.*, inositol-requiring enzyme 1 and activating transcription factor 6) has been found and the pharmacological inhibition of ER stress has been linked to the improvement of the metabolic dysfunction;[25,39] on the other hand, ER stress molecules may activate both JNK1 and IKK/NF-κB pathways, especially in metabolic tissues, such as adipose tissue and hypothalamus, thus increasing the pro-inflammatory status.[40]

12.2.1.2 *Inflammatory and Insulin Signalling Pathways*

In obesity, chronic inflammation is a key factor in the development of insulin resistance. In brief, insulin and insulin growth factor receptors are members of the receptor tyrosine kinase family that, once activated, phosphorylates itself and several intracellular substrates, such as the insulin receptor substrate (IRS) proteins (IRS-1 to IRS-6), thus starting the downstream signal events.[18,41] The inhibition of the insulin-receptor pathway is the main mechanism by which inflammation promotes insulin resistance: high levels of TNF-α and FAs can induce the phosphorylation of serine residues of IRS-1, reducing its capacity to interact with the insulin-receptor signalling and thus affecting insulin action.[18,19,42] (Figure 12.3).

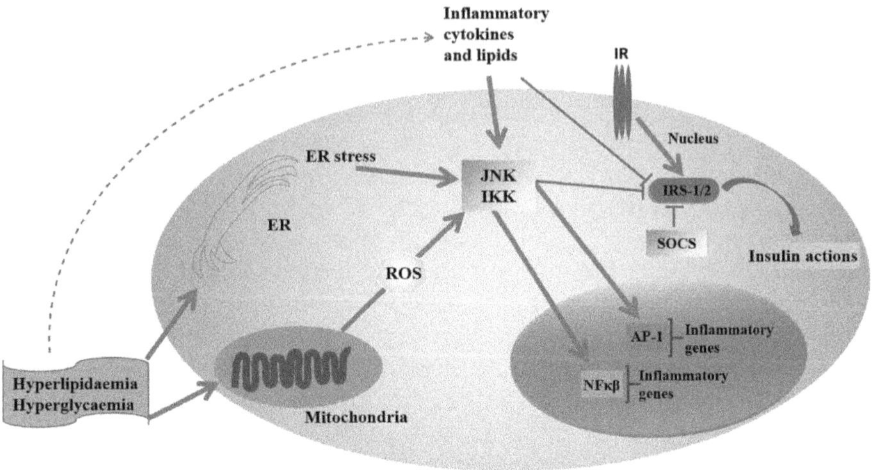

Figure 12.3 Simplified representation of inflammatory and insulin signalling pathways. Pro-inflammatory mediators, dietary lipids and overnutrition contribute to chronic inflammation and insulin resistance in obesity and metabolic syndrome through different mechanisms.

In addition, also ER stress and increased levels of mitochondrial ROS, typical of obesity and metabolic disorders, may activate inflammatory signalling pathways, thus contributing to insulin resistance.[18]

Moreover, many kinases that are activated by inflammatory signals and that contribute to perpetuating chronic inflammation are also involved in the development of insulin resistance: IKKβ, JNK and conventional protein kinase C (PKC) are able to inhibit insulin action by the inhibitory phosphorylation of IRS-1.[19,42,43] For example, in experimental animal models and humans, the administration of high-dose salicylates, an inhibitor of IKKβ, restored insulin action,[19,44] while in mice, genetic JNK1 deficiency, preventing the phosphorylation of IRS-1, hindered the development of insulin resistance, diabetes and fatty liver.[19,45] Protein kinase C can also be activated by some fatty acid metabolites, including diacylglycerides and fatty acyl-coenzyme A (in particular PKC-θ in muscles and PKC-δ in the liver); its inactivation has been correlated with protection against fatty-acid-induced insulin resistance *in vivo*.[19,46,47] Interestingly, some interactions occur between IKK, JNK and PKC, even if further studies are necessary to identify these interactions in the context of metabolic disorders. The inhibition of these kinases may represent a promising tool to decrease the low-grade inflammation and, at the same time, improve insulin sensitivity and glucose homeostasis in obesity and metabolic disease.[19]

Besides kinases, other molecules involved in insulin resistance are the members of the suppressor of cytokine signalling (SOCS) family, especially SOCS1, -3 and -6, which inhibit insulin signalling by interfering with the phosphorylation of IRS-1 and IRS-2 or by targeting IRS-1 and IRS-2 for proteasomal degradation,[48,49] and of nitric oxide (NO), whose overproduction by the inducible nitric oxide synthase (iNOS) seems to be correlated with the impairment of insulin activity on muscle and β-cells in obesity condition.[50,51]

12.3 Inflammation and Diet

Many risk factors are associated with the development of chronic inflammation: some of these are not modifiable, such as age (mostly due to the increase of free radicals and mitochondrial dysfunction over time) or hormone imbalance (testosterone and estrogen usually regulate the production and secretion of several pro-inflammatory cytokines), while other risk factors are strictly linked to our lifestyles, such as smoking, stress and unhealthy diet (enriched with refined sugar, *trans*-fats or saturated fats), and they should be modified in order to decrease risk; a fruitful strategy to decrease a chronic state of inflammation should be adopting a healthy diet, stopping smoking and being involved in regular physical activity.[52]

Among dietary sources, berries are common fruits, spread all over the world, with high levels of well-known antioxidant and anti-inflammatory bioactive compounds, including micronutrients, such as vitamins and minerals, and phytochemicals, especially polyphenols. The variety and the amount of these bioactive compounds exert synergistic and cumulative effects on the promotion of human health and the prevention of many common diseases, especially those related to inflammation, such as obesity and metabolic diseases. In the following chapters, the main *in vitro* and *in vivo* studies describing the effects of the main commonly consumed berries on chronic inflammation in obesity and MetS are shown (Tables 12.1–12.4). However, regarding *in vitro* models, studies concerning the inflammation in a broad sense, *i.e.*, studies performed exclusively on macrophages stressed with lipopolysaccharides (LPSs) (without any connection with metabolism), such as those of Gasparrini *et al.*[53] or Perez *et al.*,[54] have not been taken into consideration, because, even if they are very interesting, they go beyond the topic of this chapter; for these reasons, most of the studies presented here have been performed on animal models or humans, providing findings more pertinent to obesity and MetS.

12.3.1 Blueberry

Blueberries (*Vaccinium corymbosum L.*, family: Arecaceae) are a rich source of polyphenols, such as anthocyanins, flavonols, phenolic acids and tannins, which are mainly known for their biological activities, including their anti-inflammatory capacity.[55] In the past few years, numerous studies have evaluated the role of blueberry extract or its phenolic fractions or isolated compounds in inflammation in obesity and MetS (Table 12.1).

For example, in macrophages treated with conditioned media from fully differentiated 3T3-L1 adipocytes or co-cultured with 3T3-L1 adipocytes, blueberry extract prevented nitric oxide release and IL-10 production and secretion;[56] similar results were obtained for maqui extracts, a typical Chilean berry,

Table 12.1 Effects of blueberry on inflammation associated with obesity and metabolic syndrome.

Extract/Fraction/ Compounds	Experimental Model	Dosage and duration	Effects	Reference
In vitro studies				
Blueberry extract Maqui extract	RAW264.7 macrophages treated with conditioned media from fully differentiated 3T3-L1 adipocytes or co-cultured with 3T3-L1 adipocytes	100 μM for 24 h	– Prevention of NO release – Reduction of IL-10 production and secretion – Reduction of iNOS and TNF-α expression and secretion	56
Anthocyanin-enriched fractions from blackberry–blueberry beverages	3T3-L1 adipocytes treated with TNF-α or with LPS-induced RAW264.7 macrophage-conditioned media	100 μM for 24 h	– Prevention of glycerol release – Reduction of MCP-1 expression – Restoration of insulin-induced glucose uptake	57
Piceatannol	Cocultured 3T3-L1 adipocyte and RAW264.7 macrophage system	5 and 10 μM for 24 h	– Prevention of TNF-α and MCP-1 release – Reduction of IL-6 expression – Improvement of the malfunction of insulin-stimulated glucose uptake – Inhibition of NF-κB and JNK pathways	58
In vivo studies				
Blueberry powder	C57Bl/6J mice fed an HFD	4% for 8 weeks	– Downregulation of TNF-α, IL-6, iNOS and MCP-1 – Decrease of CD11c+ adipose tissue macrophages – Reduction of oxidative stress and adipocyte death	59

Freeze-dried blueberry	C57BL mice fed an HFD	5% or 10% (w/w) for 3 months	– Inhibition of body weight gain – Improvement of type 1 pro-inflammatory responsiveness – Prevention of systolic blood pressure increase	60
Blueberry anthocyanins Blackberry purified anthocyanins	C57BL mice fed an HFD	200 mg kg^{-1} food for 12 weeks	– Decrease of serum and liver lipids – Increase of hepatic antioxidant enzyme activities – Reduction of liver TNF-α, IL-6 and NF-κB expression – Improvement of glucose metabolic pathways	61
Blueberry powder	Wistar rats fed an HFD	10% (w/w) for 8 weeks	– Alteration of microbiota composition – Decrease of TNF-α and IL-1β levels in visceral fat – Improvement of insulin sensitivity	62
Freeze-dried strawberry–blueberry (5 : 1) powder	Wistar rats fed an HFD	10% (w/w) for 8 weeks	– Reduction of body weight gain and food efficiency – Decrease of visceral adiposity accumulation – Reduction of serum insulin and MCP-1 levels	63
Freeze-dried blueberry powder	Obese Zucker rats with MetS	12% (w/w) for 15 weeks	– Improvement of glucose tolerance renal function parameters – Decrease of oxidative stress – Reduction of TLR4, NF-kB, IL-1β, IL-18 and TNF-α in renal tissue	64

Table 12.1 (*Continued*)

Extract/Fraction/ Compounds	Experimental Model	Dosage and duration	Effects	Reference
Blueberry powder	Obese Zucker rats with MetS	8% (w/w) for 8 weeks	– Reduction of TNF-α, IL-6 and CRP in plasma, liver and adipose tissue	65
Blueberry powder	Obese Zucker rats with MetS	8% (w/w) for 90 days	– Improvement of plasma lipid profile and insulin resistance – Reduction of abdominal fat mass – Decrease of TNF-α, IL-6 and CRP in plasma, liver and adipose tissue	66
Blueberry powder	Obese Zucker rats with MetS	8% (w/w) for 8 weeks	– Improvement of maximal constrictor force in aortic rings – Decrease of TNF-α, MCP-1 and IL-8 in perivascular adipose tissue	67
Blueberry pomace Blueberry pomace ethanol extract Blueberry residue from ethanol extract	Syrian Golden hamsters fed an HFD	8% (w/w) blueberry pomace/6% blueberry pomace ethanol extract/2% blueberry residue from ethanol extract for 3 weeks	– Decrease of plasma total cholesterol levels – Increase of bile acid and cholesterol synthesis – No change in the anti-inflammatory markers	68

that was also effective in reducing the expression and secretion of iNOS and TNF-α.[56] Additionally, in adipocytes treated with TNF-α or with LPS-induced macrophage-conditioned media, anthocyanin-enriched fractions from blackberry–blueberry beverages reinstated adiponectin-blunted gene expression, reduced glycerol release and MCP-1 expression and restored insulin-induced glucose uptake.[57] Moreover, piceatannol, a bioactive compound present in blueberry and grape, decreased the release of TNF-α and MCP-1 in a co-cultured adipocyte and macrophage system, as well as reduced the expression of IL-6 and improved the malfunction of insulin-stimulated glucose uptake by blocking NF-κB and JNK activation, suggesting its capability in improving chronic inflammation and insulin sensitivity in adipose tissues.[58]

Regarding *in vivo* studies, in mice fed an HFD, blueberry powder protected against adipose tissue inflammation and insulin resistance by down-regulating TNF-α, IL-6, iNOS and MCP-1, decreasing pro-inflammatory CD11c+ adipose tissue macrophages, reducing oxidative stress and adipocyte death,[59] together with improving type 1 pro-inflammatory responsiveness, reducing the prevalence of interferon-gamma-producing T-cells and preventing systolic blood pressure increase.[60] Similarly, blueberry and blackberry purified anthocyanins inhibited body weight gain, decreased lipid levels in serum and liver, increased the activity of hepatic antioxidant enzymes and reduced the liver expression of TNF-α, IL-6 and NF-κB, as well as improved glucose metabolic pathways.[61]

Similar results were found in rats fed an HFD, where blueberry, alone or in combination with strawberry, decreased the levels of TNF-α and IL-1β in visceral fat, ameliorated markers of insulin sensitivity, by reducing the phosphorylation of IRS-1 and modified microbiota composition, as well as decreased body weight gain, visceral adiposity accumulation, serum insulin and MCP-1 levels.[62,63] Moreover, in obese rats with MetS, a blueberry-enriched diet improved glucose tolerance and renal function parameters and decreased oxidative stress and gene and protein expression of TLR4, NF-kB, IL-1β, IL-18 and TNF-α in renal tissue;[64] similar results were found in plasma, liver and abdominal adipose tissue of an obese Zucker rat model of MetS fed a wild blueberry–enriched diet.[65–67] However, in Syrian Golden hamsters fed an HFD, supplementation with blueberry peels or peel extracts did not exert anti-inflammatory effects on liver and adipose tissue but lowered the plasma levels of very-low-density lipoprotein cholesterol and total cholesterol concentrations and increased the synthesis of bile acid and cholesterol.[68]

In conclusion, the main mechanisms through which blueberry exerts anti-inflammatory effects in obesity and MetS are (i) the inhibition of NF-κB and JNK pathways; (ii) the reduction of TLR4 activation; (iii) the reduction of the expression and secretion of pro-inflammatory cytokines, including TNF-α, IL-1β, IL-6, IL-10 and IL-18; (iv) the decrease of the expression of the main chemokines, such as MCP-1 and (v) the diminution of the expression and activities of iNOS and NO release.

Table 12.2 Effects of raspberry on inflammation associated with obesity and metabolic syndrome.

Extract/Fraction/ Compounds	Experimental model	Dosage and duration	Effects	Reference
In vitro studies				
Raspberry polyphenolic extract	Hepatoblastoma (HepG2) HB-8965® cell line treated with the plasma of Wistar rats fed an HFD	0.5 nmol L^{-1} for 3 h	– Increase of angiopoietin-like 4, signal transducer and activator of transcription 1 expression – Increase of CD44 expression – Reduction of COX-2 and of Ahr expression – No effect on TLR4 expression	70
Raspberry fruit extract	Hypertrophied 3T3-L1 adipocytes	5, 10 and 20 mg mL^{-1} for 24 h	– Down-regulation of IL-6, TNF-α, IL-1β, MCP-1 and leptin – Inhibition of the decrease of adiponectin and IL-10 expression – Reduction of intracellular ROS – Increase of antioxidant enzymes and inhibition of NADPH oxidase 4 – Decrease of lipid accumulation	71
Ellagic acid	3T3-L1 adipocytes stressed with LPS	25 and 50 μM for 24 h	– Inhibition of TNF-α, iNOS, IL6 and MCP-1 expression – No effects on COX-2 – Attenuation of lipid accumulation	72
In vivo studies				
Freeze-dried raspberry	C57BL/6J mice fed an HFD	5% for 12 weeks	– Decrease of NF-kβ; IL-1β, IL-6, IL-18; MCP-1 and CD14 and CD68 – Improvement of insulin sensitivity – Reduction of macrophage infiltration and WAT hypertrophy – Improvement of WAT browning by modulating AMP-activated protein kinase pathway	73

Raspberry fruit, seed and pulp	C57BL/6J mice fed an HFD	0.4% for 16 weeks	– Increase of energy expenditure – Reduction of obesity, insulin resistance and adipocyte hypertrophy – Decrease of adipose tissue macrophages recruitment and NLRP3 inflammasome	74
Raspberry fruit extract	C57BL/6J mice fed an HFD	1.67 g kg^{-1} diet for 14 weeks	– Decrease of IL-1β, TNF-α, iNOS in serum and aortic tissue – Reduction of lipids and lipid peroxides in serum and aortic tissues	75
Freeze-dried raspberry	C57BL/6J mice fed an HFD	5% for 12 weeks	– Reduction of TNF-α, IL-1β, IL-6, p65 and triglycerides in skeletal muscle – Improvement of insulin sensitivity by decreasing the phosphorylation of IRS-1 and PKC-θ – Stimulation of mitochondrial biogenesis by activating the AMPK pathway	76
Freeze-dried raspberry	$R26^{Cre}/AMPK\alpha 1^{fl/fl}$ mice fed an HFD	5% for 10 weeks	– Reduction of TNF-α, IL-1β, IL-6, IL18 and lipids in skeletal muscle – Improvement of insulin sensitivity by decreasing the phosphorylation of IRS-1 and PKC-θ – Stimulation of mitochondrial biogenesis by activating the AMPK pathway	77
Freeze-dried raspberry	C57BL/6J mice fed an HFD	5% for 12 weeks	– Improvement of blood glucose level, glucose intolerance and insulin resistance – Reduction of hepatic lipid accumulation and NLRP3 inflammasome activation – Decrease of IL-1β and IL-18 production in lever – Normalization of liver mitochondria biogenesis	78
Freeze-dried raspberry	Obese diabetic (db/db) mice ($Lepr^{db/db}$)	0.8 g day^{-1} for 8 weeks	– Decrease in oxidative and inflammatory stress by reducing the expression of NAD(P)H dehydrogenase, glutathione *S*-transferase, heat shock protein 10, guanine nucleotide-binding protein beta 2 in heart tissue	79

Table 12.2 (*Continued*)

Extract/Fraction/ Compounds	Experimental model	Dosage and duration	Effects	Reference
Raspberry seed flour	C57BL/6 mice fed a high-sucrose diet	0.03% of ellagic acid for 12 weeks	– Improvement of glucose homeostasis – Decrease of ER stress and ROS levels in liver tissue – Reduction of adipocyte hypertrophy, IL-6, IL-8, TNF-α and MCP-1 expression and macrophage infiltration in adipose tissue	80
Raspberry fruit extract	Sprague-Dawley rats fed an HFD	0.6% for 8 weeks	– Decrease of NF-kβ, IL-1, IL-6 and COX-2 expression in liver tissue – Improvement of serum lipid profile	81
Ellagic acid	Wistar rats fed a high-fat high carbohydrate diet	0.8 $g\,kg^{-1}$ food for 16 weeks	– Decrease of NF-kβ and Nrf2 expression in heart and liver tissue – Improvement of ventricular function and glucose tolerance	82
Raspberry ketone in combination with Garcinia cambogia	Wistar rats fed a high-fructose diet	55 $mg\,kg^{-1}$ for 8 weeks	– Improvement of glucose homeostasis and lipid profile – Increase of Nrf2, glutathione expression in adipose tissue – Increase of insulin pathway in adipose tissue	83

12.3.2 Raspberry

Raspberry (*Rubus* sp., family: Rosaceae) has recently received worldwide attention for its health benefits, mainly due to the presence and high amount of ellagic acid.[55,69] Considering the anti-inflammatory effect of this berry in obesity and MetS, interesting results have been obtained recently (Table 12.2).

For instance, in hepatocytes treated with plasma collected from rats fed an HFD, raspberry polyphenolic extract led to an increase of some factors involved in fat metabolism and immune response, including angiopoietin-like 4, signal transducer and activator of transcription 1 and CD44, as well as a reduction of cyclooxygenase (COX)-2, which modulates the synthesis of prostaglandins from arachidonic acid and fatty acids, and aryl hydrocarbon receptor (AhR), which is involved in the signalling cascade of "Toll-like" receptor, even if no effect was found on TLR4 expression.[70] Moreover, in hypertrophic adipocytes, raspberry fruit extract downregulated the expression of pro-inflammatory factors (IL-6, TNF-α, IL-1β, MCP-1 and leptin), counteracted the decrease of anti-inflammatory adiponectin and IL-10 expression, reduced intracellular ROS accumulation by promoting the expression of antioxidant enzymes and inhibiting NADPH oxidase 4 and decreased lipid accumulation.[71] Similar results were found for ellagic acid, the main polyphenol present in raspberry, which in adipocytes stressed with LPS inhibited the expression of TNF-α, iNOS, IL6, MCP-1 and attenuated lipid accumulation by regulating the gene expression of glucose transporter type 4.[72]

Regarding *in vivo* findings, most studies have been performed on mice and rats (Table 12.2). For instance, in mice fed an HFD, raspberry supplementation suppressed the expression of pro-inflammatory cytokines, including NF-kβ, IL-1β, IL-6 and IL-18, MCP-1 and Cd14 and Cd68 in WAT, ameliorated insulin sensitivity, decreased macrophage infiltration and the hypertrophy of WAT and improved its browning by modulating the AMP-activated protein kinase (AMPK) pathway.[73,74] Similar results were found in (i) serum and aortic tissues, where raspberry supplementation decreased the levels of IL-1β, TNF-α, iNOS, lipids and lipid peroxides;[75] (ii) skeletal muscle, where raspberry-enriched diet alleviated inflammation by reducing the expression of TNF-α, IL-1β, IL-6 and p65, decreased the levels of triglyceride, improved insulin sensitivity by decreasing the phosphorylation of IRS-1 and PKC-θ and stimulated mitochondrial biogenesis by activating the AMPK pathway;[76,77] (iii) hepatic tissue, where raspberry consumption improved blood glucose level, glucose intolerance and insulin resistance, decreased lipid accumulation, NLRP3 inflammasome activation and IL-1β and IL-18 production and normalized the biogenesis of mitochondria.[78] Similarly, in obese mice, freeze-dried raspberry attenuated the levels of oxidative and inflammatory stress by decreasing the expression of NADPH dehydrogenase, glutathione *S*-transferase, heat shock protein 10, guanine nucleotide-binding protein beta 2 in heart tissue, while in diabetic mice, raspberry seed flour restored glucose homeostasis and reduced ER stress and ROS levels in liver tissue as well as decreased adipocyte hypertrophy, pro-inflammatory gene expression (IL-6, IL-8, TNF-α and MCP-1) and macrophage infiltration in adipose tissue.[79,80]

In addition, in rats fed an HFD, raspberry extract and ellagic acid decreased the expression of NF-kβ, IL-1, IL-6, COX-2 and nuclear factor erythroid 2-related factor 2 (Nrf2) in liver and heart tissues and improved serum lipid profile,[81,82] while in rats fed a high-fructose diet, raspberry ketone, one of the major aromatic compounds of this berry, in combination with *Garcinia cambogia*, improved serum glucose homeostasis and lipid profile as well as increased the expression of the Nrf2, glutathione and insulin pathway in adipose tissue.[83]

To conclude, raspberries have been shown to decrease obesity-related inflammation by inhibiting the pathway of NF-κB and by activating those of AMPK and Nfr2, with the consequent (i) reduction of the expression and secretion of pro-inflammatory cytokines, including TNF-α, IL-1β, IL-6, IL-10, IL-18 and COX-2; (ii) decrease of the expression of the main chemokines, such as MCP-1, Cd14 and Cd68; (iii) reduction of the expression of adiponectin; (iv) inhibition of NADPH oxidase 4; (v) diminution of the expression and activities of iNOS and PKC-θ; (vi) decrease of IRS-1 phosphorylation and (vii) increase of the expression and activities of the main antioxidant enzymes.

12.3.3 Strawberry

Strawberry (*Fragaria X ananassa Duch.*; family: Rosaceae) is a remarkable source of micronutrients, phytochemicals, especially anthocyanins (*i.e.*, pelargonidin and cyanidin derivates) and ellagitannins and fibers.[55] Several studies have highlighted the potential role of strawberries against oxidative stress and inflammation.[53,84,85] To the best of our knowledge, studies evaluating the effects of strawberry on inflammation related to obesity and MetS have been performed exclusively *in vivo* (Table 12.3).

In diabetic mice, strawberry consumption decreased blood pressure, monocyte binding to aortic vessels and endothelial-dependent vasorelaxation and reduced levels of IKKβ and NADPH oxidase 2,[86] while in obese mice, strawberry-enriched diet lowered the blood levels of glucose and CRP but did not affect other inflammatory cytokines.[87] In the same way, in rats with MetS or with inflammatory bowel disease, pelargonidin-3-glucoside-enriched strawberry lowered body weight gain and abdominal fat, ameliorated ventricular stiffness, blood pressure and liver and cardiac structure, as well as decreased inflammatory cell infiltration,[88] while in rats fed an HFD, strawberry ellagitannin-rich extract decreased body weight gain and hepatic fat and blood levels of lipids, biomarkers of oxidative stress and TNF-α, IL-1 and IL-6.[89] Moreover, in diabetic rats, different types of strawberry extract improved serum glucose and lipid profile; increased the liver antioxidant status and reduced the levels of NF-kβ, IL6, TN-Fα and sterol regulatory element-binding transcription factor, while they upregulated liver expression of peroxisome proliferator–activated receptor-γ (PPAR-γ), thus improving hepatic fatty acid metabolism;[90] similar results were found in the hippocampus

Table 12.3 Effects of strawberry on inflammation associated with obesity and metabolic syndrome.

Extract/Fraction/ Compounds	Experimental model	Dosage and duration	Effects	Reference
Freeze-dried strawberry	db/db diabetic mice	2.35% for 10 weeks	– Decrease of blood pressure, monocytes binding to aortic vessels and the endothelial-dependent vasorelaxation – Reduction of IKKβ and NADPH oxidase 2	86
Freeze-dried strawberry	C57BL/6J mice fed an HFD	2.6% for 10 weeks	– Reduction of blood glucose and CRP levels	87
Strawberry powder	Wistar rats fed a high carbohydrate, HFD	10.3 g kg^{-1} for 8 weeks	– Reduction of body weight gain and abdominal fat – Improvement of ventricular stiffness, blood pressure and liver and cardiac structure – Decrease of inflammatory cell infiltration	88
Strawberry ellagitannin-rich extract	Wistar rats fed an HFD	0.24–0.72% for 4 weeks	– Reduction of body weight, hepatic fat and blood levels of lipids, biomarkers of oxidative stress – Decrease of blood TNF-α, IL-1 and IL-6	89
Aqueous, hydroalcoholic, alcoholic strawberry extracts	Wistar rats intraperitoneally injected with streptozotocin	2 g kg^{-1} bodyweight for 4 weeks	– Improvement of serum glucose, liver lipid profile – Increase of liver antioxidant status – Reduction of liver NF-kβ, IL6, TN-Fα and sterol regulatory element-binding transcription factor – Upregulation of liver PPAR-γ	90
Strawberry leaf extract	Sprague-Dawley rats intraperitoneally injected with streptozotocin	200 mg kg^{-1} for 4 weeks	– Reduction of hippocampus ROS, lipid peroxidation – Increase of hippocampus antioxidant enzymes and Nrf2 – Decrease of hippocampus TN-Fα, IL6, caspase-3 and caspase-9	91

of diabetic rats, where strawberry leaf extract reduced the level of ROS and lipid peroxidation, increased the activity of antioxidant enzymes and Nrf2 and lowered the expression of TN-Fα, IL6, caspase-3 and caspase-9.[91]

From the above data, it emerges that strawberry exerts anti-inflammatory effects on obesity and MetS through the modulation of multiple molecular pathways. Indeed, it decreases inflammation and oxidative stress by (i) blocking NADPH oxidase 2, with a consequent decrease of intracellular ROS and lipid peroxidation; (ii) downregulating the NF-κB pathway through the inhibition of IKKβ, leading to a decrease of the expression of the main pro-inflammatory cytokines, including TNF-α, IL-1 and IL-6; (iii) activating the Nfr2 pathway and increasing the expression and activities of the principal antioxidant enzymes and (iv) upregulating PPAR-γ levels.

12.3.4 Blackcurrant and Blackberry

Blackcurrant (*Ribes nigrum* L.) belongs to the family Grossulariaceae and is a rich source of phytonutrients, with antioxidant and anti-inflammatory activities.[55] Up to date, few investigations have been carried out on obesity and MetS (Table 12.4).

For example, in mouse and human macrophages, blackcurrant extract suppressed M1 polarization, typical of obesity-associated inflammation, while, in mouse macrophages treated with the serum from obese mice fed blackcurrant, a reduction of the expression of pro-inflammatory genes, including IL-1β, IL-6 and TNF-α, was found.[92] Concerning *in vivo* findings, in mice fed an HFD, blackcurrant consumption lowered body weight and fat mass; reduced the number of crown-like structures (a morphological parameter for macrophage infiltration and inflammation), the expression of IKKε in epididymal adipose tissue and the plasma levels of inflammatory PAI-1 and increased mitochondrial biogenesis and energy metabolism in skeletal muscle.[93–95] Similar results were found in mice fed a high-fat/high-sucrose diet,[96] highlighting the ability of blackcurrant to prevent obesity-induced inflammation and insulin resistance.

Blackberry (*Rubus fruticosus* L.) belongs to the family of Rosaceae and is known especially for its antioxidant capacity due to the high presence of tannins, ellagitannins, quercetin and anthocyanins.[55] In the last decade, three *in vivo* studies have investigated its effect on inflammation in obesity or MetS (Tables 12.1 and 12.4). The first two have been already discussed in a previous section (3.1),[57,61] while, regarding the third one, in ovariectomized rats, cyanidin 3-O-b-D-glucoside-rich blackberries exerted anti-obesity effects by lowering bodyweight and decreasing the expression of NF-kβ, COX-2 and PPAR-γ in liver samples.[97]

In summary, also blackcurrants and blackberries modulate the expression of the NF-kβ pathway, decrease the gene expression of some pro-inflammatory interleukins, such as IL-1β, IL-6 and TNF-α and reduce the expression of COX-2 and PPAR-γ, leading to a general improvement of inflammation and oxidative stress associated with obesity.

Table 12.4 Effects of blackcurrant and blackberry on inflammation associated with obesity and metabolic syndrome.

Extract/Fraction/ Compounds	Experimental model	Dosage and duration	Effects	Reference
Blackcurrant				
In vitro studies				
Blackcurrant extract	Marrow-derived macrophages and human THP-1 macrophages Marrow-derived macrophages	50 μg mL^{-1} of blackcurrant extract for 24 h 15% serum from obese mice fed blackcurrant for 18 h	– Suppression of M1 polarization – Reduction of pro-inflammatory genes, including IL-1β, IL-6 and TNF-α	92
In vivo studies				
Polyphenol-rich blackcurrant Extract	C57BL/6J mice fed an HFD	0.1% for 12 weeks	– Decrease of body weight and fat mass – Reduction of crown-like structures and IKKε in epididymal adipose tissue – Increase of mitochondrial biogenesis and energy metabolism in skeletal muscle	93
Blackcurrant powder	C57BL/6J mice fed an HFD	5.7% for 12 weeks	– Reduction of crown-like structures in the epididymal fat pad	94
Freeze-dried blackcurrant	C57BL/6J mice fed an HFD	20% w/w for 13 weeks	– Reduction of body weight gain and fasting insulin levels – Decrease of body fat content, hepatic lipid accumulation – Reduction of plasma levels ofPAI-1 – Improvement of glucose homeostasis	95
Freeze-dried blackcurrant	C57BL/6J mice fed a high-fat/high-sucrose diet	6% for 24 weeks	– Decrease of liver weights and triglyceride – Reduction of liver macrophages infiltration and percentages of monocytes of total splenocytes – Decrease of IL-1β and TNF-α in splenocytes	96
Blackberry				
Freeze-dried blackberries	Sprague-Dawley ovariectomized rats	5–10% for 100 days	– Decrease of body weight – Reduction of liver NF-kβ, COX-2 and PPAR-γ expression	97

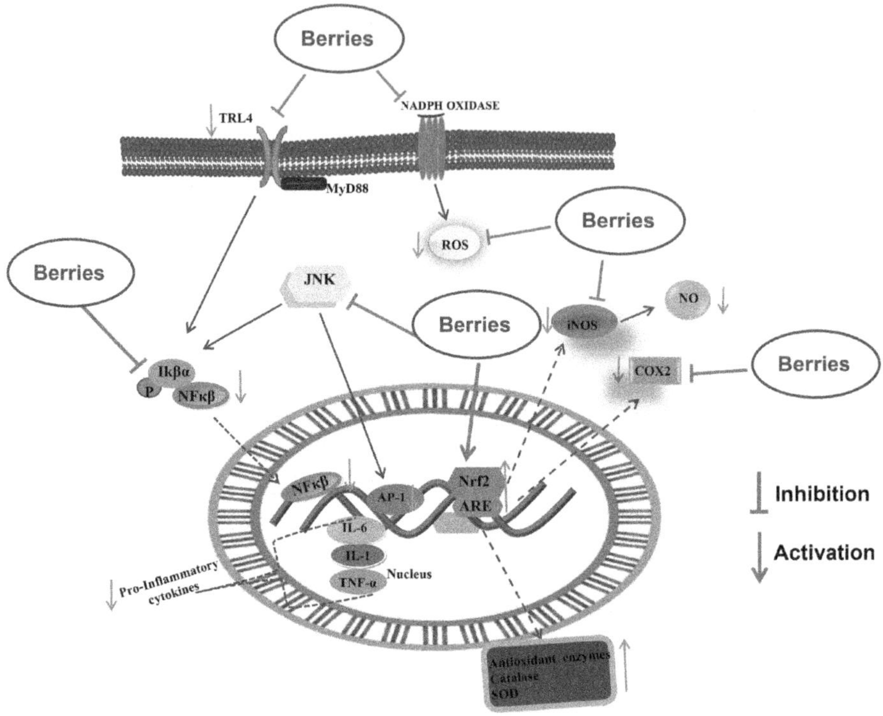

Figure 12.4 Simplified scheme of the anti-inflammatory effect of berries in obesity. Berries attenuated the inflammatory and oxidative stress through the downregulation of different factors, including TLR4, NADPH oxidase, JNK, NF-κB, iNOS and COX-2 and the concomitant upregulation of antioxidant transcription factors, such as Nrf2.

12.4 Conclusion

Berries are rich sources of nutrients and bioactive compounds with multiple health-promoting activities. Several *in vitro* and *in vivo* studies have demonstrated the efficacy of berry extracts or their fractions against chronic inflammation associated with metabolic disorders. The main mechanisms involved in their anti-inflammatory activities comprise, on the one hand, the inhibition of NADPH oxidase, TLR4, NF-kB and JNK activation, with the consequent reduction of pro-inflammatory cytokines and adiponectins, including IL-1, IL-6 and TNF-α, and, on the other hand, the induction of the Nrf2 pathway, which leads to the increase of the expression and activities of the main cellular antioxidant enzymes and to a reduction of intracellular ROS and NO levels (Figure 12.4). Altogether, these processes may represent an important tool for the control and management of obesity and metabolic diseases.

Even if the role of berries in counteracting inflammation is widely accepted, several aspects need to be still elucidated, because too many differences exist in published studies, such as, for example, the methods used for obtaining berry extracts, the dose and the duration of each treatment, the amount of polyphenols applied in cell cultures or the experimental models used. In addition, particular attention should also be paid to berry metabolites that are formed after *in vivo* digestion and that exert the actual anti-inflammatory effects.

Abbreviations

AhR	aryl hydrocarbon receptor
AMPK	AMP-activated protein kinase
AP-1	activator protein-1
BAT	brown adipose tissue
COX	cyclooxygenase
CRP	C-reactive protein
ER	endoplasmic reticulum
FAs	fatty acids
HDL	high-density lipoprotein
HFD	high-fat diet
IKK	IκB kinase
IL	interleukin
iNOS	inducible nitric oxide synthase
IRS	insulin receptor substrate
JNK	c-Jun N-terminal kinases
LDL	low-density lipoprotein
LPS	lipopolysaccharide
MCP-1	monocyte chemoattractant protein
MetS	metabolic syndrome
NF-κB	nuclear factor kappa-light-chain-enhancer of activated B cells
NLRP3	NLR family pyrin domain containing 3
NLRs	Nod-like receptors
NO	nitric oxide
Nrf2	nuclear factor erythroid 2-related factor 2
PAI-1	plasminogen activator inhibitor 1
PKC	protein kinase C
PPAR- γ	peroxisome proliferator–activated receptor-γ
PRRs	germline-encoded pattern-recognition receptors
RANTES	regulated on activation, normal T expressed and secreted
ROS	oxygen reactive species
SOCS	suppressor of cytokine signalling
TGF	transforming growth factor
TLR	Toll-like receptor
TNF	tumor necrosis factor
WAT	white adipose tissue

Acknowledgements

This work was supported by "Progetto Strategico di Ateneo 2017–2020", funded by the Polytechnic University of Marche.

References

1. R. H. Eckel, S. M. Grundy and P. Z. Zimmet, The metabolic syndrome, *Lancet*, 2005, **365**, 1415–1428.
2. K. G. M. M. Alberti, P. Zimmet and J. Shaw, The metabolic syndrome—a new worldwide definition, *Lancet*, 2005, **366**, 1059–1062.
3. S. M. Grundy, J. I. Cleeman, S. R. Daniels, K. A. Donato, R. H. Eckel, B. A. Franklin, D. J. Gordon, R. M. Krauss, P. J. Savage, S. C. Smith, J. A. Spertus, F. Costa and American Heart Association, National Heart Lung Blood Institute Diagnosis and management of the metabolic syndrome: an American Heart Association/National Heart, Lung, and Blood Institute scientific statement, *Circulation*, 2005, **112**, 2735–2752.
4. P. L. Huang, A comprehensive definition for metabolic syndrome, *Dis. Models Mech.*, 2009, **2**, 231–237.
5. M. P. Stern, K. Williams, C. Gonźalez-Villalpando, K. J. Hunt and S. M. Haffner, Does the metabolic-syndrome improve identification of individuals at risk of type 2 diabetes and/or cardiovascular disease?, *Diabetes Care*, 2007, **27**, 2676–2681.
6. P. Pothiwala, S. K. Jain and S. Yaturu, Metabolic syndrome and cancer, *Metab. Syndr. Relat. Disord.*, 2009, **7**, 279–288.
7. https://www.who.int/news-room/fact-sheets/detail/obesity-and-overweight.
8. https://www.nichd.nih.gov/health/topics/obesity/conditioninfo/cause.
9. A. Hruby, J. E. Manson, L. Qi, V. S. Malik, E. B. Rimm, Q. Sun, W. C. Willett and F. B. Hu, Determinants and Consequences of Obesity, *Am. J. Public Health*, 2016, **106**, 1656–1662.
10. M. Qatanani and M. A. Lazar, Mechanisms of obesity associated insulin resistance: many choices on the menu, *Genes Dev.*, 2007, **21**, 1443–1455.
11. R. Medzhitov, Inflammation 2010: new adventures of an old flame, *Cell*, 2010, **140**, 771–776.
12. S. V. Joseph, I. Edirisinghe and B. M. Burton-Freeman, Berries: anti-inflammatory effects in humans,*J. Agric. Food Chem.*, 2014, **62**, 3886–3903.
13. F. Giampieri, T. Y. Forbes-Hernandez, M. Gasparrini, J. M. Alvarez-Suarez, S. Afrin, S. Bompadre, J. L. Quiles, B. Mezzetti and M. Battino, Strawberry as a health promoter: an evidence based review, *Food Funct.*, 2015, **6**, 1386–1398.
14. A. G. Stewart and P. M. Beart, Inflammation: maladies, models, mechanisms and molecules, *Br. J. Pharmacol.*, 2016, **173**, 631–634.
15. M. M. Rogero and P. C. Calder, Obesity, Inflammation, Toll-Like Receptor 4 and Fatty Acids, *Nutrients*, 2018, **10**, 432.
16. H. Kolb and T. Mandrup-Poulsen, The global diabetes epidemic as a consequence of lifestyle-induced low-grade inflammation, *Diabetologia*, 2010, **53**, 10–20.
17. J. M. Fernandez-Real and J. C. Pickup, Innate immunity, insulin resistance and type 2 diabetes, *Trends Endocrinol. Metab.*, 2008, **19**, 10–16.
18. K. E. Wellen and G. S. Hotamisligil, Inflammation, stress, and diabetes, *J. Clin. Invest.*, 2005, **115**, 1111–1119.

19. G. S. Hotamisligil, Inflammation and metabolic disorders, *Nature*, 2006, **444**, 860–867.
20. R. Monteiro, E. Keating, P. Castro and I. Azevedo, Abdominal cavity compliance: a participant more in the building up of visceral obesity, *Obesity*, 2009, **17**, 937.
21. J. K. Sethi and A. J. Vidal-Puig, Thematic review series: adipocyte biology. Adipose tissue function and plasticity orchestrate nutritional adaptation, *J. Lipid Res.*, 2007, **48**, 1253–1262.
22. F. Emanuela, M. Grazia, M. de Robertis, M. P. Luconi, G. Furlani and M. Boscaro, Inflammation as a Link between Obesity and Metabolic Syndrome, *J. Nutr. Metab.*, 2012, **2012**, 476380.
23. R. Monteiro and I. Azevedo, Chronic inflammation in obesity and the metabolic syndrome, *Mediators Inflammation*, 2010, **2010**, 289645.
24. P. C. Calder, N. Ahluwalia, F. Brouns, T. Buetler, K. Clement, K. Cunningham, K. Esposito, L. S. Jönsson, H. Kolb, M. Lansink, A. Marcos, A. Margioris, N. Matusheski, H. Nordmann, J. O'Brien, G. Pugliese, S. Rizkalla, C. Schalkwijk, J. Tuomilehto, J. Wärnberg, B. Watzl and B. M. Winklhofer-Roob, Dietary factors and low-grade inflammation in relation to overweight and obesity, *Br. J. Nutr.*, 2011, **106**, S5–S78.
25. M. M. Rogero and P. C. Calder, Obesity, Inflammation, Toll-Like Receptor 4 and Fatty Acids, *Nutrients*, 2018, **10**, 432.
26. C. N. Lumeng and A. R. Saltiel, Inflammatory links between obesity and metabolic disease, *J. Clin. Invest.*, 2011, **121**, 2111–2117.
27. H. Shi, M. V. Kokoeva, K. Inouye, I. Tzameli, H. Yin and J. S. Flier, TLR4 links innate immunity and fatty acid-induced insulin resistance, *J. Clin. Invest.*, 2006, **116**, 3015–3025.
28. M. Vijay-Kumar, J. D. Aitken, F. A. Carvalho, T. C. Cullender, S. Mwangi, S. Srinivasan, S. V. Sitaraman, R. Knight, R. E. Ley and A. T. Gewirtz, Metabolic syndrome and altered gut microbiota in mice lacking Toll-like receptor 5, *Science*, 2010, **328**, 228–231.
29. B. Vandanmagsar, Y. H. Youm, A. Ravussin, J. E. Galgani, K. Stadler, R. L. Mynatt, E. Ravussin, J. M. Stephens and V. D. Dixit, The NLRP3 inflammasome instigates obesity-induced inflammation and insulin resistance, *Nat. Med.*, 2011, **17**, 179–188.
30. R. Zhou, A. Tardivel, B. Thorens, I. Choi and J. Tschopp, Thioredoxin-interacting protein links oxidative stress to inflammasome activation, *Nat. Immunol.*, 2010, **11**, 136–140.
31. A. Kleinridders, D. Schenten, A. C. Könner, B. F. Belgardt, J. Mauer, T. Okamura, F. T. Wunderlich, R. Medzhitov and J. C. Brüning, MyD88 signaling in the CNS is required for development of fatty acid-induced leptin resistance and diet-induced obesity, *Cell Metab.*, 2009, **10**, 249–259.
32. R. G. Baker, M. S. Hayden and S. Ghosh, NF-kappaB, inflammation, and metabolic disease, *Cell Metab.*, 2011, **13**, 11–22.
33. S. H. Chiang, M. Bazuine, C. N. Lumeng, L. M. Geletka, J. Mowers, N. M. White, J. T. Ma, J. Zhou, N. Qi, D. Westcott, J. B. Delproposto,

T. S. Blackwell, F. E. Yull and A. R. Saltiel, The protein kinase IKKepsilon regulates energy balance in obese mice, *Cell*, 2009, **138**, 961–975.
34. G. Solinas and M. Karin, JNK1 and IKKbeta: molecular links between obesity and metabolic dysfunction, *FASEB J.*, 2010, **24**, 2596–2611.
35. G. Solinas, C. Vilcu, J. G. Neels, G. K. Bandyopadhyay, J. L. Luo, W. Naugler, S. Grivennikov, A. Wynshaw-Boris, M. Scadeng, J. M. Olefsky and M. Karin, JNK1 in hematopoietically derived cells contributes to diet-induced inflammation and insulin resistance without affecting obesity, *Cell Metab.*, 2007, **6**(5), 386–397.
36. S. N. Vallerie, M. Furuhashi, R. Fucho and G. S. Hotamisligil, A predominant role for parenchymal c-Jun amino terminal kinase (JNK) in the regulation of systemic insulin sensitivity, *PLoS One*, 2008, **3**, e3151.
37. W. L. Holland, B. T. Bikman, L. P. Wang, G. Yuguang, K. M. Sargent, S. Bulchand, T. A. Knotts, G. Shui, D. J. Clegg, M. R. Wenk, M. J. Pagliassotti, P. E. Scherer and S. A. Summers, Lipid-induced insulin resistance mediated by the proinflammatory receptor TLR4 requires saturated fatty acid-induced ceramide biosynthesis in mice, *J. Clin. Invest.*, 2011, **121**, 1858–1870.
38. W. L. Holland, R. A. Miller, Z. V. Wang, K. Sun, B. M. Barth, H. H. Bui, K. E. Davis, B. T. Bikman, N. Halberg, J. M. Rutkowski, M. R. Wade, V. M. Tenorio, M. S. Kuo, J. T. Brozinick, B. B. Zhang, M. J. Birnbaum, S. A. Summers and P. E. Scherer, Receptor-mediated activation of ceramidase activity initiates the pleiotropic actions of adiponectin, *Nat. Med.*, 2011, **17**, 55–63.
39. U. Ozcan, E. Yilmaz, L. Ozcan, M. Furuhashi, E. Vaillancourt, R. O. Smith, C. Z. Görgün and G. S. Hotamisligil, Chemical chaperones reduce ER stress and restore glucose homeostasis in a mouse model of type 2 diabetes, *Science*, 2006, **313**, 1137–1140.
40. G. S. Hotamisligil, Endoplasmic reticulum stress and the inflammatory basis of metabolic disease, *Cell*, 2010, **140**, 900–917.
41. G. S. Hotamisligil, P. Peraldi, A. Budavari, R. Ellis, M. F. White and B. M. Spiegelman, IRS-1-mediated inhibition of insulin receptor tyrosine kinase activity in TNF-alpha- and obesity-induced insulin resistance, *Science*, 1996, **271**, 665–668.
42. V. Aguirre, T. Uchida, L. Yenush, R. Davis and M. F. White, The c-Jun NH2-terminal kinase promotes insulin resistance during association with insulin receptor substrate-1 and phosphorylation of Ser307, *J. Biol. Chem.*, 2000, **275**, 9047–9054.
43. Z. Gao, D. Hwang, F. Bataille, M. Lefevre, D. York, M. J. Quon and J. Ye, Serine phosphorylation of insulin receptor substrate 1 by inhibitor kappa B kinase complex, *J. Biol. Chem.*, 2002, **277**, 48115–48121.
44. M. Yuan, N. Konstantopoulos, J. Lee, L. Hansen, Z. W. Li, M. Karin and S. E. Shoelson, Reversal of obesity- and diet-induced insulin resistance with salicylates or targeted disruption of Ikkbeta, *Science*, 2001, **293**, 1673–1677.

45. J. Hirosumi, G. Tuncman, L. Chang, C. Z. Görgün, K. T. Uysal, K. Maeda, M. Karin and G. S. Hotamisligil, A central role for JNK in obesity and insulin resistance, *Nature*, 2002, **420**, 333–336.
46. J. K. Kim, J. J. Fillmore, M. J. Sunshine, B. Albrecht, T. Higashimori, D. W. Kim, Z. X. Liu, T. J. Soos, G. W. Cline, W. R. O'Brien, D. R. Littman and G. I. Shulman, PKC-theta knockout mice are protected from fat-induced insulin resistance, *J. Clin. Invest.*, 2004, **114**, 823–827.
47. G. Boden, P. She, M. Mozzoli, P. Cheung, K. Gumireddy, P. Reddy, X. Xiang, Z. Luo and N. Ruderman, Free fatty acids produce insulin resistance and activate the proinflammatory nuclear factor-kappaB pathway in rat liver, *Diabetes*, 2005, **54**, 3458–3465.
48. L. Rui, M. Yuan, D. Frantz, S. Shoelson and M. F. White, SOCS-1 and SOCS-3 block insulin signaling by ubiquitin-mediated degradation of IRS1 and IRS2, *J. Biol. Chem.*, 2002, **277**, 42394–42398.
49. K. Ueki, T. Kondo and C. R. Kahn, Suppressor of cytokine signaling 1 (SOCS-1) and SOCS-3 cause insulin resistance through inhibition of tyrosine phosphorylation of insulin receptor substrate proteins by discrete mechanisms, *Mol. Cell. Biol.*, 2004, **24**, 5434–5446.
50. M. Shimabukuro, M. Ohneda, Y. Lee and R. H. Unger, Role of nitric oxide in obesity-induced beta cell disease, *J. Clin. Invest.*, 1997, **100**, 290–295.
51. M. Perreault and A. Marette, Targeted disruption of inducible nitric oxide synthase protects against obesity-linked insulin resistance in muscle, *Nat. Med.*, 2001, **7**, 1138–1143.
52. D. Furman, J. Campisi, E. Verdin, P. Carrera-Bastos, S. Targ, C. Franceschi, L. Ferrucci, D. W. Gilroy, A. Fasano, G. W. Miller, A. H. Miller, A. Mantovani, C. M. Weyand, N. Barzilai, J. J. Goronzy, T. A. Rando, R. B. Effros, A. Lucia, N. Kleinstreuer and G. M. Slavich, Chronic inflammation in the etiology of disease across the life span, *Nat. Med.*, 2019, **25**, 1822–1832.
53. M. Gasparrini, T. Y. Forbes-Hernandez, F. Giampieri, S. Afrin, J. M. Alvarez-Suarez, L. Mazzoni, B. Mezzetti, J. L. Quiles and M. Battino, Anti-inflammatory effect of strawberry extract against LPS-induced stress in RAW 264.7 macrophages, *Food Chem. Toxicol.*, 2017, **102**, 1–10.
54. C. Pérez, M. L. Ruiz del Castillo, C. Gil, G. P. Blanch and G. Flores, Supercritical fluid extraction of grape seeds: extract chemical composition, antioxidant activity and inhibition of nitrite production in LPS-stimulated Raw 264.7 cells, *Food Funct.*, 2015, **6**, 2607–2613.
55. S. Afrin, F. Giampieri, M. Gasparrini, T. Y. Forbes-Hernandez, A. Varela-López, J. L. Quiles, B. Mezzetti and M. Battino, Chemopreventive and therapeutic effects of edible berries: a focus on colon cancer prevention and treatment, *Molecules*, 2016, **21**, 169.
56. M. Reyes-Farias, K. Vasquez, A. Ovalle-Marin, F. Fuentes, C. Parra, V. Quitral, P. Jimenez and D. F. Garcia-Diaz, Chilean native fruit extracts inhibit inflammation linked to the pathogenic interaction between adipocytes and macrophages, *J. Med. Food*, 2015, **18**, 601–608.

57. D. F. Garcia-Diaz, M. H. Johnson and E. G. de Mejia, Anthocyanins from fermented berry beverages inhibit inflammation-related adiposity response in vitro, *J. Med. Food*, 2015, **18**, 489–496.
58. Y. Li, P. Yang, Q. Chang, J. Wang, J. Liu, Y. Lv, T. Wang, B. Gao, Y. Zhang and L. L. Yu, Inhibitory effect of piceatannol on TNF-α-mediated inflammation and insulin resistance in 3T3-L1 adipocytes, *J. Agric. Food Chem.*, 2017, **65**, 4634–4641.
59. J. DeFuria, G. Bennett, K. J. Strissel, J. W. Perfield, P. E. Milbury, A. S. Greenberg and M. S. Obin, Dietary blueberry attenuates whole-body insulin resistance in high fat-fed mice by reducing adipocyte death and its inflammatory sequelae, *J. Nutr.*, 2009, **139**, 1510–1516.
60. O. T. Mykkänen, A. Huotari, K. H. Herzig, T. W. Dunlop, H. Mykkänen and P. V. Kirjavainen, Wild blueberries (Vaccinium myrtillus) alleviate inflammation and hypertension associated with developing obesity in mice fed with a high-fat diet, *PLoS One*, 2014, **9**, e114790.
61. T. Wu, Y. Gao, X. Guo, M. Zhang and L. Gong, Blackberry and blueberry anthocyanin supplementation counteract high-fat-diet-induced obesity by alleviating oxidative stress and inflammation and accelerating energy expenditure, *Oxid. Med. Cell. Longevity*, 2018, **2018**, 4051232.
62. S. Lee, K. I. Keirsey, R. Kirkland, Z. I. Grunewald, J. G. Fischer and C. B. de La Serre, Blueberry supplementation influences the gut microbiota, inflammation, and insulin resistance in high-fat-diet-fed rats, *J. Nutr.*, 2018, **148**, 209–219.
63. P. Aranaz, A. Romo-Hualde, M. Zabala, D. Navarro-Herrera, M. Ruiz de Galarreta, A. G. Gil, J. A. Martinez, F. I. Milagro and C. J. González-Navarro, Freeze-dried strawberry and blueberry attenuates diet-induced obesity and insulin resistance in rats by inhibiting adipogenesis and lipogenesis, *Food Funct.*, 2017, **8**, 3999–4013.
64. A. R. Nair, C. M. Elks, J. Vila, F. Del Piero, D. B. Paulsen and J. Francis, A blueberry-enriched diet improves renal function and reduces oxidative stress in metabolic syndrome animals: potential mechanism of TLR4-MAPK signaling pathway, *PLoS One*, 2014, **9**, e111976.
65. S. Vendrame, A. Daugherty, A. S. Kristo, P. Riso and P. D. Klimis-Zacas, Wild blueberry (Vaccinium angustifolium) consumption improves inflammatory status in the obese Zucker rat model of the metabolic syndrome, *J. Nutr. Biochem.*, 2013, **24**, 1508–1512.
66. E. M Seymour, I. I. Tanone, D. E. Urcuyo-Llanes, S. K. Lewis, A. Kirakosyan, M. G. Kondoleon, P. B. Kaufman and S. F. Bolling, Blueberry intake alters skeletal muscle and adipose tissue peroxisome proliferator-activated receptor activity and reduces insulin resistance in obese rats, *J. Med. Food*, 2011, **14**, 1511–1518.
67. S. Vendrame, P. Tsakiroglou, A. S. Kristo, D. A. Schuschke and D. Klimis-Zacas, Wild blueberry consumption attenuates local inflammation in the perivascular adipose tissue of obese Zucker rats, *Appl. Physiol., Nutr., Metab.*, 2016, **41**, 1045–1051.

68. H. Kim, G. E. Bartley, A. M. Rimando and W. Yokoyama, Hepatic gene expression related to lower plasma cholesterol in hamsters fed high-fat diets supplemented with blueberry peels and peel extract, *J. Agric. Food Chem.*, 2010, **58**, 3984–3991.
69. X. Wu, J. Sun, A. Jaspreet, D. B. Haytowitz, P. Chen, B. Burton-Freeman and P. R. Pehrsson, Anthocyanins in processed red raspberries on the US market, *J. Berry Res.*, 2019, **9**, 603–613.
70. B. Fotschki, J. M. Laparra and M. Sójka, Raspberry polyphenolic extract regulates obesogenic signals in hepatocytes, *Molecules*, 2018, **23**, 2103.
71. K. Kowalska, A. Olejnik, J. Zielińska-Wasielica and M. A Olkowicz, Raspberry (*Rubus idaeus* L.) fruit extract decreases oxidation markers, improves lipid metabolism and reduces adipose tissue inflammation in hypertrophied 3T3-L1 adipocytes, *J. Funct. Foods*, 2019, **62**, 103568.
72. L. Cisneros-Zevallos, W. Y. Bang and C. Delgadillo-Puga, Ellagic Acid and Urolithins A and B differentially regulate fat accumulation and inflammation in 3T3-L1 adipocytes while not affecting adipogenesis and insulin sensitivity, *Int. J. Mol. Sci.*, 2020, **21**, 2086.
73. T. Xing, Y. Kang, X. Xu, B. Wang, M. Du and M. J. Zhu, Raspberry supplementation improves insulin signaling and promotes brown-like adipocyte development in white adipose tissue of obese mice, *Mol. Nutr. Food Res.*, 2018, **62**, 1701035.
74. R. Fan, M. You, A. M. Toney, J. Kim, D. Giraud, Y. Xian, F. Ye, L. Gu, A. E. Ramer-Tait and S. Chung, Red raspberry polyphenols attenuate high-fat diet-driven activation of NLRP3 inflammasome and its paracrine suppression of adipogenesis via histone modifications, *Mol. Nutr. Food Res.*, 2020, **64**, e1900995.
75. S. Kim, C. K. Kim, K. S. Lee, J. H. Kim, H. Hwang, D. Jeoung, J. Choe, M. H. Won, H. Lee, K. S. Ha, Y. G. Kwon and Y. M. Kim, Aqueous extract of unripe Rubus coreanus fruit attenuates atherosclerosis by improving blood lipid profile and inhibiting NF-κB activation via phase II gene expression, *J. Ethnopharmacol.*, 2013, **146**, 515–524.
76. T. Zou, Y. Kang, B. Wang, J. M. de Avilab, J. You, M. J Zhu and M. Du, Raspberry supplementation reduces lipid accumulation and improves insulin sensitivity in skeletal muscle of mice fed a high-fat diet, *J. Funct. Foods*, 2019, **63**, 103572.
77. L. Zhao, T. Zou, N. A. Gomez, B. Wang, M. J. Zhu and M. Du, Raspberry alleviates obesity-induced inflammation and insulin resistance in skeletal muscle through activation of AMP-activated protein kinase (AMPK) α1, *Nutr. Diabetes*, 2018, **8**, 39.
78. M. J. Zhu, Y. Kang, Y. Xue, X. Liang, M. García, D. Rodgers, D. R. Kagel and M. Du, Red raspberries suppress NLRP3 inflammasome and attenuate metabolic abnormalities in diet-induced obese mice, *J. Nutr. Biochem.*, 2018, **53**, 96–103.
79. G. Noratto, B. P. Chew and I. Ivanov, Red raspberry decreases heart biomarkers of cardiac remodeling associated with oxidative and

inflammatory stress in obese diabetic db/db mice, *Food Funct.*, 2016, **7**, 4944–4955.
80. I. Kang, J. C. Espín, T. P. Carr, F. A. Tomás-Barberán and S. Chung, Raspberry seed flour attenuates high-sucrose diet-mediated hepatic stress and adipose tissue inflammation, *J. Nutr. Biochem.*, 2016, **32**, 64–72.
81. T. Lim, J. Ryu, K. Lee, S. Y. Park and K. T. Hwang, Protective effects of black raspberry (*Rubus occidentalis*) extract against hypercholesterolemia and hepatic inflammation in rats fed high-fat and high-choline diets, *Nutrients*, 2020, **12**, 2448.
82. S. K. Panchal, L. Ward and L. Brown, Ellagic acid attenuates high-carbohydrate, high-fat diet-induced metabolic syndrome in rats, *Eur. J. Nutr.*, 2013, **52**, 559–568.
83. R. T. Attia, Y. Abdel-Mottaleb, D. M. Abdallah, H. S. El-Abhar and N. N. El-Maraghy, Raspberry ketone and Garcinia Cambogia rebalanced disrupted insulin resistance and leptin signaling in rats fed high fat fructose diet, *Biomed. Pharmacother.*, 2019, **110**, 500–509.
84. F. Giampieri, J. M. Alvarez-Suarez, M. D. Cordero, M. Gasparrini, T. Y. Forbes-Hernández, S. Afrin, C. Sanrtos-Buelga, A. M. Gonzales-Paramas, D. P. Astolfi, C. Rubini, A. Zizzi, S. Tulipani, J. L. Quiles, B. Mezzetti and M. Battino, Strawberry consumption improves aging-associated impairments, mitochondrial biogenesis and functionality through the AMP-Activated Protein Kinase signaling cascade, *Food Chem.*, 2017, **234**, 464–471.
85. T. Y. Forbes-Hernández, S. Afrin, D. Cianciosi, P. P Manna, J. J. Zhang, M. Gasparrini and P. Reboredo-Rodriguez, Strawberry extract attenuates oxidative stress in 3T3-L1 cells, *J. Berry Res.*, 2018, **8**, 193–203.
86. C. Petersen, D. Bharat, B. R. Cutler, S. Gholami, C. Denetso, J. E. Mueller, J. M. Cho, J. S. Kim, J. D. Symons and P. V. Anandh Babu, Circulating metabolites of strawberry mediate reductions in vascular inflammation and endothelial dysfunction in db/db mice, *Int. J. Cardiol.*, 2018, **263**, 111–117.
87. M. A. Parelman, D. H. Storms, C. P. Kirschke, L. Huang and S. J. Zunino, Dietary strawberry powder reduces blood glucose concentrations in obese and lean C57BL/6 mice, and selectively lowers plasma C-reactive protein in lean mice, *Br. J. Nutr.*, 2012, **108**, 1789–1799.
88. N. K. Ghattamaneni, A. Sharma, S. K. Panchal and L. Brown, Pelargonidin 3-glucoside-enriched strawberry attenuates symptoms of DSS-induced inflammatory bowel disease and diet-induced metabolic syndrome in rats, *Eur. J. Nutr.*, 2020, **59**, 2905–2918.
89. E. Żary-Sikorska, B. Fotschki, A. Jurgoński, M. Kosmala, J. Milala, K. Kołodziejczyk, M. Majewski, K. Ognik and J. Juśkiewicz, Protective effects of a strawberry ellagitannin-rich extract against pro-oxidative and pro-inflammatory dysfunctions induced by a high-fat diet in a rat model, *Molecules*, 2020, **25**, 5874.
90. P. Mandave, S. Khadke, M. Karandikar, V. Pandit, P. Ranjekar, A. Kuvalekar and N. Mantri, antidiabetic, lipid normalizing, and

nephroprotective actions of the strawberry: a potent supplementary fruit, *Int. J. Mol. Sci.*, 2017, **18**, 124.
91. L. Zhang, Q. Ma and Y. Zhou, Strawberry leaf extract treatment alleviates cognitive impairment by activating Nrf2/HO-1 signaling in rats with streptozotocin-induced diabetes, *Front. Aging Neurosci.*, 2020, **12**, 201.
92. Y. Lee and J. Y. Lee, Blackcurrant (Ribes nigrum) extract exerts an anti-inflammatory action by modulating macrophage phenotypes, *Nutrients*, 2019, **11**, 975.
93. T. Benn, B. Kim, Y. K. Park, C. J. Wegner, E. Harness, T. G. Nam, D. O. Kim, J. S. Lee and J. Y. Lee, Polyphenol-rich blackcurrant extract prevents inflammation in diet-induced obese mice, *J. Nutr. Biochem.*, 2014, **25**, 1019–1025.
94. B. Kim, S. G. Lee, Y. K. Park, C. S. Ku, T. X. Pham, C. J. Wegner, Y. Yang, S. I. Koo, O. K. Chun and J. Y. Lee, Blueberry, blackberry, and black-currant differentially affect plasma lipids and pro-inflammatory markers in diet-induced obesity mice, *Nutr. Res. Pract.*, 2016, **10**, 494–500.
95. L. Heyman, U. Axling, N. Blanco, O. Sterner, C. Holm and K. Berger, Evaluation of Beneficial Metabolic Effects of Berries in High-Fat Fed C57BL/6J Mice, *J. Nutr. Metab.*, 2014, **2014**, 403041.
96. Y. Lee, T. X. Pham, M. Bae, S. Hu, E. O'Neill, O. K. Chun, M. J. Han, S. I. Koo, Y. K. Park and J. Y. Lee, Blackcurrant (Ribes nigrum) prevents obesity-induced nonalcoholic steatohepatitis in mice, *Obesity*, 2019, **27**, 112–120.
97. L. Kaume, W. C. Gilbert, C. Brownmiller, L. R. Howard and L. Devareddy, Cyanidin 3-O-b-D-glucoside-rich blackberries modulate hepatic gene expression, and anti-obesity effects in ovariectomized rats, *J. Funct. Foods*, 2012, **4**, 480–488.

CHAPTER 13

The Role of Berries as Potential Wound Healing Agents

RUPALI S. PRASAD,[a] DAMIKI LALOO,[b] ASHISH P. BHARNE[c] AND SATYENDRA K. PRASAD*[d]

[a] Department of Pharmaceutical Sciences, Rashtrasant Tukadoji Maharaj Nagpur University, Nagpur, Maharashtra, India; [b] Girijananda Chowdhury Institute of Pharmaceutical Sciences, Guwahati, Assam, India; [c] Indegene Pvt. Ltd., Bangalore, India; [d] Department of Pharmaceutical Sciences, Rashtrasant Tukadoji Maharaj Nagpur University, Nagpur, Maharashtra, India
*Emails: tiwari.rup01@gmail.com; damiki.laloo@gmail.com; ashishbharne2@gmail.com; skprasad.itbhu@gmail.com

13.1 Introduction

Wound is a type of injury that can be open (when skin is torn, cut, or punctured) or closed (a trauma causes a contusion inside the body). Wound healing is a physiological, however complex, and dynamic process that replaces necrotic and degrading cellular structures and tissue layers.[1] It occurs in different stages, *i.e.* hemostasis, inflammation, proliferation and maturation or remodeling coordinated by various intracellular factors.[2] There are several conditions that lead to impairment of the wound healing process when wound closure is delayed at the different stages.[3] The most common factors affecting wound healing in chronic wounds are age, infection, poor nutrition, lack of hydration, edema and chronic disease condition such as diabetes. There are several changes in the body of patients with diabetes such as hyperglycemia, hypoxia, impaired angiogenesis and neovascularization, oxidative stress, formation of advanced glycation end-products and

Food Chemistry, Function and Analysis No. 33
Berries and Berry Bioactive Compounds in Promoting Health
Edited by Dorothy Klimis-Zacas and Ana Rodriguez-Mateos

Published by the Royal Society of Chemistry, www.rsc.org

neuropathy. Due to this multifactorial impairment of wound healing in diabetes, it is a challenge to treat it. Because of their wide spectrum of activity and safety margin, medicinal plants have been widely used in the treatment of various ailments including wounds and also diabetic wounds. There are several medicinal plants that have been proven to have a potential wound healing activity and have also shown promising results in treating diabetic wounds.[3,4] Thus, keeping the above view into consideration, the present chapter has been presented to put focus on the wound healing potential of berries.

13.2 Physiological Process of Wound Healing

Physiologically wound healing is a continuous process; however, it occurs in phases coordinated by a subtle interplay of humoral and cellular factors. Traditionally, this is divided into hemostasis, inflammation, proliferation, and maturation or remodeling (Figure 13.1). Hemostasis occurs within an hour after injury with the leaking of blood out of the body due to injury and is characterized by vasoconstriction and subsequent clotting.[5] It is initiated within seconds of the rupture of a blood vessel's epithelial wall by adherence of platelets to the sub-endothelial surface. Platelets' role is not only to initiate the series of clotting events but also to facilitate the secretion of growth factors and cytokines, which initiate the process of healing.[6]

At the end of hemostasis or overlapping, an inflammation phase begins that can continue for approximately up to 7 days. In normal skin wound healing, this phase ends with the removal of the harmful stimuli, though the immune reactions remain throughout the wound healing process.[7] In the early inflammatory response, in parallel with hemostasis, mobilization of local and systemic defense responses at the site of the wound are observed, facilitated through neutrophil granulocytes to prevent bacterial contamination and cleanse the wound from cell debris. This phase begins when the injured blood vessels leak transudate (made of water, salt, and protein) after the injury and results in localized swelling.[8] Physiologically, this helps to control bleeding and prevents infection. The latter effect is mediated by the differentiation of monocytes into the wound macrophages. These monocytes get attracted to the wound by chemotactic factors, while macrophages help to remove bacteria and non-viable tissue *via* phagocytosis. Various growth factors required to stimulate fibroplasia and angiogenesis are also released.[9]

Some days post-injury, the proliferation phase is initiated and lasts up to 20 days, depending on the severity of the injury. This phase is largely characterized by tissue granulation and angiogenesis (formation of new blood vessels) and re-epithelialization (covering the wound surface).[10] The angiogenic process is mediated by growth factors such as platelet-derived growth factor, macrophage angiogenesis factor and angiotensin. Concomitant epithelialization is then commenced to cover the granulation tissue with a cellular barrier.[11] The last phase of wound healing is tissue remodeling, which depends on the severity of the injury, and may last from

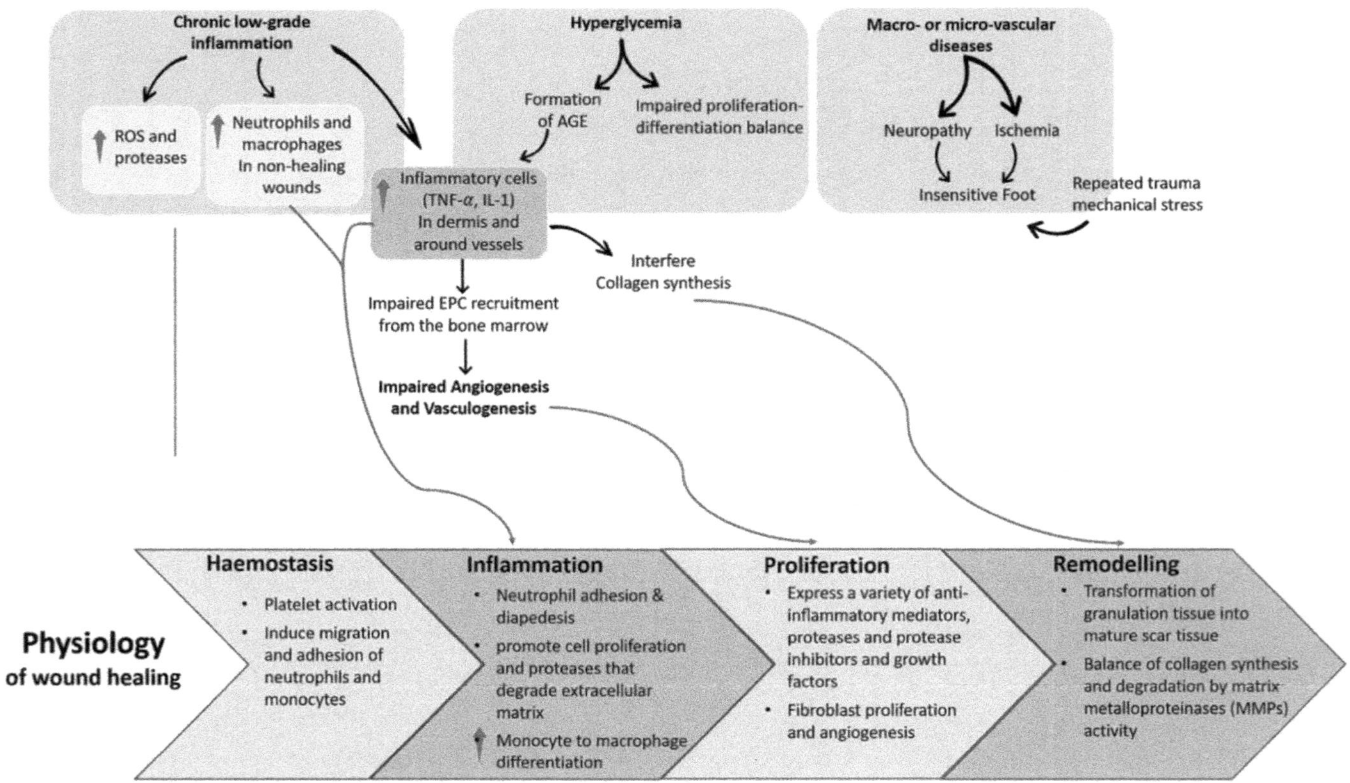

Figure 13.1 Illustration of the different factors responsible for the impairment of wound healing process in patients living with diabetes. Chronic low-grade inflammation, hyperglycemia and micro- and macro-vascular diseases impair healing by altering various endogenous factors delaying hemostasis, inflammation, proliferation and remodeling.

1 week up to 6 months post-injury. During this phase, the interim wound matrix is replaced with proteoglycan and collagen molecules, which readily become organized into thicker bundles resulting in stronger but more rigid scar tissue.[12]

13.3 Etiology of Diabetic Foot Ulceration

The etiology of diabetic foot ulcers generally has many components with a series of micro- and macro-vascular changes. Though diabetic wounds start as acute wounds, it is a medical challenge to treat due to interrupted and stalled healing process. This interruption occurs at different phases of wound healing mediated by different internal as well as external factors.[13] These factors broadly involve peripheral sensory neuropathy, trauma, deformity, peripheral vascular disease, immune suppression, ischemia, callus formation and edema. Infection is also a main factor in diabetic wound healing interruption, although infection is rarely implicated in the etiology of diabetic foot ulcers. This is due to the fact that ulcers are susceptible to infection once the wound is present. Many of the risk factors for foot ulcers are also predisposing factors for amputation, because ulcers are the primary causes leading to amputation.[14] The factors that delay wound healing in diabetes are multiple and relate both to the impaired glucose metabolism and to the effect of neurovascular complications.

13.3.1 Hyperglycemia

Hyperglycemia itself has a deleterious effect on wound healing *via* its damaging effect on many body systems in people with diabetes.[15] High blood glucose gradually leads to blood vessel narrowing, stiffening of the arteries and damage to neurons called diabetic neuropathy. These changes increase the risk of wound development as well as complications in the healing process. Hyperglycemia-induced advanced glycation end-product (AGE) formation in turn increases the production of inflammatory molecules (tumor necrosis factor-alpha [TNF-α], interleukin-1 [IL-1]) and interferes with collagen synthesis. Another hyperglycemia-mediated plausible mechanism is the imbalance in proliferation and differentiation of keratinocytes.[16–18]

13.3.2 Neuropathy

Loss of sensation due to neuropathy is considered a major cause of injury in patients living with diabetes and one which prevents healing.[19] Diabetes is associated with irreversible nerve destruction that leaves diabetics less sensitive to pain in their limbs, thereby preventing blisters, infections and wound healing. In addition, a reduced density of nerves, also called polyneuropathy, in the skin and a reduced level of nerve growth factor impair wound healing.[20]

13.3.3 Narrowed Blood Vessels: Impaired Circulation

Predominant manifestations of macrovascular disease in diabetes are damage to the coronary arteries, cerebrovascular and peripheral vessels.[21] Physiological wound healing depends on good blood circulation at the wound site. Narrowed blood vessels due to diabetes carry less oxygen to the wound and the tissues do not heal as quickly.[22] Furthermore, hyperglycemia reduces the functioning of red blood cells in terms of carrying nutrients to the injured area and limits the effectiveness of white blood cells against infections.

13.3.4 Altered Immune Function

High blood sugar levels also lower the body's ability to fight infections by decreased chemotaxis, phagocytosis, bacterial killing and reduced heat shock protein expression. There is altered leukocyte infiltration and an increased level of pro-inflammatory IL-6 in the wound milieu, which hampers wound healing in diabetes.[4] In addition, growth factors also play an important role in wound healing and increased degradation of growth factors in the wound fluid of diabetic subjects contributes to the impaired wound healing.[23] Due to the limited functioning of immune cells, the risk of bacterial infections also increases in diabetic wounds. The slower movement of white blood cells in the wound site concomitant with nominal immune responses prolongs the inflammation stage of wounds and ultimately delays the healing process.[24] Collectively, it is suggested that enzymes and hormones generated following chronic blood glucose levels can attenuate the effectiveness of the immune system.[24]

13.3.5 Chronic Inflammation

The chronic low-grade inflammatory state is a characteristic of diabetes. Inflammation is essential as well as one of the natural stages of wound healing; however, the prolonged inflammatory response after injury is detrimental to wound healing. In chronic wounds, the balance between producing and degrading collagen is lost and wounds do not heal. The prolonged inflammatory phase is characterized by sustained expression and increased levels of pro-inflammatory cytokines. This includes TNF-α, IL-1 and IL-6.[25] Detailed factors responsible for the impairment of the wound healing process in patients living with diabetes are graphically represented in Figure 13.1.

13.4 Natural Products and Their Role on Wound Healing

In the life of an organism, wounds are perhaps unavoidable events, and at times they are extremely dangerous and may be life threatening.

The strategic management and treatment of wounds still remains a major global health challenge. Prolonging the healing process of wounds or failure to heal may result in increased financial strain, social distress and decreased quality of life in patients. Over the years, there has been an increased incidence of the occurrence of both chronic and acute wounds in most countries, in which geriatric patients are greatly affected with diabetic foot ulcers, venous stasis ulcers and pressure ulcers.[26] Chronic wounds are generally characterized by various factors such as they are prone to infection, cause intense pain, result in lack of mobility and, in serious cases, may lead to amputations (as seen in diabetic foot ulcer) and even death. With a drastic increase in the prevalence of chronic wound cases and the emerging high cost of synthetic medicines, most people, especially those belonging to under-developed and developing countries, prefer alternative medicines from herbal sources as their main choice of treatment, and this has been attributed to the fact that herbal medicines are significantly effective, readily available and bear a low cost. In this context, there has been an urgent need to notify prospective users and influencers of the safety, efficacy and likelihood of lack of toxicity of naturally derived wound healing products for treating a variety of diseases and disorders.[27,28]

Since ancient times, nature has always been an alternative source for the supply of plant-based medicine, which continues to play a vital role in primary health care around the world in the management of disease including skin wound care.[29,30] Many potential medicinal plants and their active components have been investigated for their efficacy in treating wounds and such agents have shown promising results in advancing the healing process and regeneration of tissue through several connected mechanisms.[31] The mechanisms of wound healing properties of natural products are attributed to their anti-bacterial, antioxidant, anti-inflammatory and pro-collagen synthesis actions. As per literature, it was observed that natural products with wound healing properties are distributed among various chemical families including saponins, phenolics, tannins, flavonoids, alkaloids, essential oils, terpenoids and vitamins.[32] Each bioactive agent may have a definite function and specific mechanism of action in healing the various stages of wounds. For instance, phenolic and flavonoid compounds have significant anti-septic, anti-bacterial and antioxidant activities, whereas saponins promote the healing of wounds by enhancing the synthesis of pro-collagen.[33] Tannin components, on the other hand, are astringents and are highly responsible for wound contraction and wound closure with an increased rate of epithelialization at the granulation formation and scar remodeling phases.[34,35] Vitamins such as retinol (vitamin A), niacinamide (vitamin B_3), L-ascorbic acid (vitamin C) and α-tocopherol (vitamin E) have also been reported to heal wounds very effectively with probable mechanisms of action including anti-inflammatory and antioxidant effects.[36] Taking into account the wound healing activity of volatile oils, the mechanisms underlying the healing effects include anti-microbial, antioxidant and anti-inflammatory properties.[37]

The extraction, isolation, purification and structural characterization of compounds from natural sources have brought new perspectives in the drug discovery process, along with the therapeutic potential contributing toward the pharmacological aspects. Natural products have mainly contributed toward the development of front-line drugs for the treatment of many infectious and non-infectious diseases.[38] The search for wound healing agents is one of the oldest challenges in the field of medicine, because the tissue repair mechanism is still not fully understood. There are a number of pure isolated natural compounds that have been pre-clinically and clinically evaluated for their effectiveness in curing wounds.[39] Some of the commonly studied isolated components with wound healing activity include curcumin from *Curcuma longa* (turmeric), asiaticoside from *Centella asiatica* (pennywort), catechins from *Camellia sinensis* (tea), aloe-emodin, glucomannan and vitamins A, B, C, E from *Aloe vera* juice, silymarin from *Silybum marianum*, anthocyanidin from *Alchemilla vulgaris* and many others.[39] With respect to berries, because of their rich composition in polyphenolics and proanthocyanins, they tend to play a pivotal role in healing the various stages of wounds. For instance, anthocyanins (ACNs) isolated from strawberry and blackberry revealed their wound healing effects attributed to their anti-inflammatory and antioxidant activity.[40]

13.5 Medicinal Role and Therapeutic Effects of Berries

As per the dietary guidelines, it has been recommended worldwide that a healthy diet containing fruits (including berries) and vegetables is enriched with nutrients and essential phytochemicals that can help prevent various ranges of human diseases and disorders.[41] Most berries make up the largest category of fruits which are edible and are widely consumed either in fresh or frozen preserved forms in the human diet for maintaining health benefits.[42,43] There are varieties of edible berries that are processed as derived products, including dried and canned fruits, beverages, yogurts, jellies and jams.[44] As an essential part of a healthy diet, berries are a rich source of an essential phytochemical class comprising of polyphenolics and other components including minerals, dietary fibers, fatty acids and vitamins. In addition, berries also contain a high amount of tannin and ACN (plant pigments) content, resulting in the coloration and therapeutic activity of the plant.[45–47] Worldwide based on research, there is growing evidence that consumption of berries is an important part of a healthy diet. Some of the well-known colorful berries that are widely used as functional foods and dietary supplements in the human diet, either fresh or in processed forms, include black raspberry, strawberry, cranberry, blackberry, blueberry and red raspberry.[41,42] It has been widely reported that berries possess biological properties that are mainly attributed to the presence of proanthocyanins (found in cranberries and blueberries) and ellagitannins (commonly distributed in strawberries, blackberries, red and black raspberries). For instance, cranberry appears to be distinctive among all the berry fruits with significant anti-bacterial anti-adhesive properties.[48] Such activity is attributed to its oligomeric proanthocyanidins (PACs), which chemically consist of an A-type linkage in its structure. On the contrary, tannin-rich extract of blueberries (a PAC-rich fruit) and strawberries (an ellagitannin-rich fruit) showed a distinct biological role in neuronal brain function and behavior activity in aging animals.[49,50] Berries rich in ACNs have a wide range of pharmacological activities, including antioxidant, anti-inflammatory and anti-cancer properties. Eventually, the chief constituents, which are polyphenolics and vitamins, found in berries are potential antioxidant agents, which help protect the body against various destructive properties generated by free radicals (reactive oxygen species), resulting in chronic diseases that are associated with aging. Apart from polyphenolics, vitamins C and E present in berries are also a contributing factor toward antioxidation efficacy.[51,52] Raspberries, like blueberries, strawberries and cranberries, comprise an abundant quantity of vitamin C, which is helpful for the development of strong connective tissues, thereby providing significant protection to the immune system from attacks of various pathogenic microorganisms. Several studies have reported that potential constituents of berries also help to control and maintain many physiological functions of the body, including regulation of the metabolizing enzyme activities, repair DNA oxidative damage and also modulate nuclear receptors, gene expression and subcellular signaling pathways.[53,54] The literature revealed that the chief composition from the fruits and leaves of berries acts as front-line agents to treat both minor as well as major complications including sore mouths, sore throats (soothing), nausea, aphtha, stomatitis, diabetes, diarrhea, cancer, inflammation and dysentery.[55] In addition, apart from the medicinal and therapeutic uses, berry phenolics also serve many diverse biological functions, including roles in plant growth, development and defense. They provide pigmentation, anti-microbial and anti-fungal functions, insect-feeding deterrence, ultraviolet radiation protection, chelation of toxic heavy metals and antioxidant quenching of free radicals generated during photosynthesis.[56,57]

13.6 Berries with Wound Healing Potential

13.6.1 Acai Berry (*Euterpe oleracea*)

Kang *et al.* evaluated the skin wound healing effect of acai berry (*E. oleracea* Mart., Arecaceae) water extracts (ABWEs) using HS68 fibroblast cells followed by the determination of mRNA expression of pro-collagen, fibronectin and MMP-1. The study also included the determination of mRNA expression levels of wound-related genes. The results showed that ABWE increased the migration of HS68 fibroblast cells and mRNA expression levels of fibronectin; however, there was a significant decrease in the mRNA expression levels of MMP-1. Further, the *in vivo* wound healing potential of ABWE was also confirmed through macroscopical and histopathological studies and expression studies of wound-related genes.[58]

In another study, ABWE was subjected to *in vitro* antioxidant study along with an estimation of its total phenolics content. The wound healing potential of the ABWE on Sprague–Dawley rats was evaluated using 50% acetic acid, which was applied to the rat's oral mucosa. The oral treatment of the ABWE at 50 µL per day was continued for 6 days, where the extract was found to have potential wound healing activity based on microscopical and histopathological observations, which may be attributed to the presence of the high amount of phenolics compounds.[59]

13.6.2 Barbados Gooseberry (*Pereskia aculeate*)

Pinto *et al.* evaluated the topical wound healing of a macerated methanolic extract of the leaves from the plant *P. aculeate* Miller (Cactaceae), commonly known as Barbados gooseberry, subjected to fractionation using hexane as solvent. The extract and its hexane fraction were then tested for topical wound healing activity using the excision wound healing model in mice. The wounds were treated with 30 µl of the gel containing 5% of each methanolic extract and its hexane fraction for 14 days. Both the extract and fraction showed significant wound healing activity; however, from the laser Doppler perfusion imaging, the hexane fraction showed increased blood flow rate and better collagen deposition at the wound site. The wound healing potential of both the extract and fraction was also confirmed from histological studies showing very less fibroplasia and inflammatory cells with more collagen deposition as compared to vehicle control animals. Thus, the study scientifically validated the traditional use of the leaves of the plant *P. aculeate* in treating wounds.[60]

13.6.3 Barberry (*Berberis vulgaris*)

The anti-microbial, anti-itch and blood glucose-lowering properties of *B. vulgaris* L., Berberidaceae (barberry), were utilized to evaluate its wound healing potential in streptozotocin (STZ)-induced diabetic rats. Wounds were created on the dorsal lumbar portion of the Wistar rats and the impact of hydroalcoholic extract of *B. vulgaris* on healing was checked for a period of 21 days.[61] The results demonstrated the potential wound healing effect of the extract, which was evidenced through a significant reduction in the lesion length in treated rats as compared to control animals, which may be attributed to its anti-diabetic and anti-microbial effect. Thus, the study confirmed the wound healing potential of barberry in diabetic rats; however, their mechanism for the above effect needs to be investigated further.[61]

13.6.4 Blackberry (*Rubus fruticosus*)

Blackberry (*R. fruticosus* L., Rosaceae) has been found to be very rich in polyphenolic content. In a study conducted by Velde *et al.*,[40] polyphenolic profiling of the blackberry was analyzed using high-performance liquid chromatography–time-of-flight-mass spectrometry (HPLC–TOF-MS) and its anthocyanin-enriched fractions (AEFs) and PAC-enriched fractions (PEFs) were studied for *in vitro* anti-inflammatory and wound healing activity. The inflammatory study was performed using lipopolysaccharide (LPS)-stimulated RAW 264.7 macrophage assay, while the wound healing study was performed using skin fibroblast migration and proliferation study. The results revealed a significant ROS reduction capacity of blackberry crude extracts and also documented reduced *COX2* gene expression. PEF was found to be more prominent in attenuating gene expression of interleukins as compared to AEF, which also reduced the expression of iNOS. The results demonstrated the potential anti-inflammatory property of PACs, where ACNs were more effective as wound healing agents.

13.6.5 Blackcurrant (*Ribes nigrum*)

A study in Turkey was performed to evaluate the wound healing potential of different *Ribes* species, *viz. R. anatolica* Behçet, *R. alpinum* L., *R. multiflorum* Kit. ex Romer & Schultes, *R. nigrum* L., *R. petraeum* Wulfen, *R. orientale* Desf., *R. rubrum* L. and *R. uva-crispa* L. using excision and incision wound model in rats and mice.[62] Wounds were created on the dorsal portion of the rats and mice and ointments of the methanolic extracts prepared from the leaves of the plants were applied once daily. The results demonstrated a significant wound-breaking strength of *R. nigrum* and *R. multiflorum* and also showed maximum reduction in the wound area along with the highest antioxidative potential as compared to the other tested *Ribes* species. From the overall study, the ethyl acetate sub-extract of *R. nigrum* methanolic extract showed the best wound healing potential (1.67 ± 0.79 mm^2 wound area on the 15th day compared to control having a wound area of 10.38 ± 0.56 mm^2), which was quite comparable with the standard ointment, FITO Krem®.[62] Thus, from the above investigation, we may presume that the ethyl acetate sub-extract of *R. nigrum* can be utilized for the preparation of a potential wound healing agent.

13.6.6 Bog blueberry (*Vaccinium uliginosum*)

The traditional use of one of the Alaskan berries, *Vaccinium uliginosum* L., Ericaceae, in the treatment of skin wounds was utilized to evaluate its wound healing potential using a cell migration assay with human dermal fibroblasts. The results revealed that the berry extract promoted the migration of fibroblast. The PEF of the extract was found to be most effective, showing suppression of inflammatory mediators such as NO, prostaglandin-endoperoxide synthase 2, iNOS and ROS. The bioactive molecules from the berry, such as B2 dimer and epicatechin, at 10 μM each were found to be mainly responsible in wound repair and tissue regeneration showing the highest impacts on early stages of wound healing *via* the stimulation of mitochondrial bioenergetics and up-regulation of important extracellular matrix; proteins; integrin-β1 and collagen type I α2 chain.[63] The findings of the study suggested the use of PAC and bioactives such as epicatechins from bog blueberry contributed to regenerative wound healing and connective tissue repair.

13.6.7 Cranberry (*Vaccinium macrocarpon*)

The oil of cranberry (*Vaccinium macrocarpon* Aiton., Ericaceae) was subjected to evaluation of its wound healing activity using excision wound models in rats. The oil at a dose level of 100 mg kg^{-1} was administered to male Sprague Dawley rats for a period of 13 days and wound contraction was examined as compared to that of control and standard drug mupirocin ointment. The results showed a significant wound healing potential of the cranberry oil with an 88.1% reduction in the wound area, which was found to be significantly higher than with the control (74.1%) and the standard drug-treated rats (78.4%). The results were also confirmed with a concomitant increase in the content of hydroxyproline in the treated animals as compared to control animals.[64] The authors concluded that the observed wound healing activity of the oil from cranberry could be attributed to the anti-microbial, anti-inflammatory and antioxidant potential of its active phytoconstituents in combination. However, further studies of isolated molecules from the oil could help us in identifying the exact phytoconstituents responsible for wound healing activity with their specific mode of action.

13.6.8 Crowberry (*Empetrum nigrum*)

The wound healing potential of crowberry (*Empetrum nigrum* L., Ericaceae) was evaluated on the cell migration assay using human dermal fibroblasts. The study revealed a wound healing divergent effect related to compositional disparities, where the extract and its PAC fraction showed a significant decline in the LPS-stimulated inflammatory variables. The PAC fractions were found to exert significantly higher scratch-wound closure as compared to control and were most effective in the early stages of wound healing by stimulating mitochondrial bioenergetics such as ATP production, basal respiration and maximum respiratory capacity and up-regulation of extracellular matrix. Thus, the study confirmed the potential of crowberry in wound repair and wound closure.[63]

13.6.9 Elderberry (*Sambucus ebulus*)

Leaves of the elderberry plant (*Sambucus ebulus* L., Adoxaceae) were subjected to extraction using different solvents such as methanol, diethyl ether, ethyl acetate and *n*-hexane and were subjected to incision and excision wound models in rats and mice. The results demonstrated the methanolic extract to be most effective, which was further fractionated using the successive chromatographic technique, and each of the fractions was evaluated biologically using bioactivity-guided fractionation techniques. The ointment prepared from methanolic extract at 1% concentration was found to be most effective with 84.3% wound closure compared to the negative control, which was quite close to standard Madecassol. The incision wound model also depicted the methanolic extract to be most effective, showing a 43.7% increase in tensile strength. The flavonoid quercetin 3-*O*-glucoside was isolated from the sub-fraction and identified to be responsible for the observed activity of the elderberry.[65]

Ghabaee *et al.*[66] also reported the wound healing potential of elderberry using a dorsal full-thickness wound model in male Wistar albino rats. The extract in the form of 5% ointment was applied to the wound area daily for a period of 12 days. The study reported significant wound contraction and a systemic organization of granulation tissue as observed by histopathology. The study also revealed a marked increase in the collagen deposition and showed an increased healing rate by 95% upon treatment with extract as compared to the control group animal showing 84% healing of wounds, thus confirming a potential wound healing activity of elderberry.

13.6.10 Golden Himalayan Raspberry (*Rubus ellipticus*)

Different leaf extracts of the plant *Rubus ellipticus* Smith. (Rosaceae), also known as golden evergreen raspberry or golden Himalayan raspberry, were initially subjected to *in vitro* antioxidant activity, where the methanolic extract was found to be most active. Therefore, methanolic extract was further processed for *in vivo* antioxidant, anti-tumor and wound healing activity. The extract was found to be safe up to 2 $g\,kg^{-1}$, as confirmed through acute oral and dermal toxicity studies, and also showed potential wound healing properties *via* incision, excision and *Staphylococcus aureus*-induced infected wound models showing complete epithelialization between the 13th and 19th day. The extract at 250 $mg\,kg^{-1}$ also exhibited a prolongation of the life span of mice by 46.76% in Ehrlich ascites carcinoma and also reduced the volume of Dalton's lymphoma ascites solid tumors, confirming its role as a potential anti-tumor agent.[67]

13.6.11 Holy Bramble (*Rubus sanctus*)

Four different extracts of the aerial parts of the plant *Rubus sanctus* Schreber (Rosaceae) were prepared using methanol, chloroform, *n*-hexane and ethyl acetate and their wound healing effect was studied using incision and excision wound models in mice. From all the tested extracts, the ointment with 1% methanol extract showed the best wound healing potential with 56.5% wound contraction on day 6 and 80.6% wound contraction on day 12, which was also supported with the histopathological examination and was quite comparable with that of standard drug Madecassol.[68] Therefore, the traditional claim of the plant *R. sanctus* was scientifically justified through this study; however, the exact mechanism of wound healing still needs to be investigated.

13.6.12 Indian Gooseberry (*Phyllanthus emblica*)

The fruits of *P. emblica* L., Phyllanthaceae (also known as *Emblica officinalis,* amla or Indian gooseberry), are said to be a major component in various traditional medicines described in the traditional systems of medicines such as Ayurveda, Unani and Siddha. The fruits are used in the treatment of various ailments, where vitamin C is found to be the major contributory factor in the therapeutic value of amla. In the present investigation, an attempt was made to determine the wound healing pattern of amla in the Wistar albino rat using histological and biochemical investigations. The *in vivo* study confirmed a 95% regeneration of dermis and epidermis in treated rats within a period of 16 days, where the granulation tissue showed greater amounts of newly synthesized components of extracellular matrix as compared to control group rats. The treated rats also showed a higher amount of type 1 collagen with better maturation and cross-linking along with enhanced aldehyde content of type 1 collagen, thus confirming the role of amla in wound healing and its management.[69]

In another study, five traditional anti-aging formulations based on Ayurveda were prepared using cow ghee, flaxseed oil, *P. emblica* fruits, *Shorea robusta* resin and Yashada bhasma as ingredients and their wound healing potential was determined using excision and incision wound models in Swiss Wistar rats. The results demonstrated a better wound healing potential of the formulation containing Yashada bhasma along with *S. robusta* resin and flaxseed oil, showing higher wound contraction, higher collagen content and better skin breaking strength.[70]

The effect of *E. officinalis* on wound healing was studied on full-thickness excision wound models, where the topical application of *Emblica* showed a significant rate of wound contraction and exhibited increased cellular proliferative activity, which also increased cross-linking of collagen at the site of wounds.[59] This was also confirmed through an increase in extracellular signal-regulated kinase (ERK1/2) activity and also documented significantly higher levels of tensile strength, DNA, type III collagen, acid-soluble collagen and aldehyde content along with an enhanced *in vivo* antioxidant status. The study confirmed the potential wound healing activity of Emblica, which may be attributed to the presence of its major biomarkers such as ascorbic acid, emblicanin A and emblicanin B.[71]

13.6.13 Lingonberry (*Vaccinium vitis-idaea*)

A cell migration assay with human dermal fibroblasts (isolated from adult skin [HDFa, Invitrogen C-013-5C]) was used to determine the wound healing potential of *V. vitis-idaea,* L. (Ericaceae), which demonstrated a potential healing effect of the berry at the concentration of 50 $\mu g\,mL^{-1}$. The study also revealed the suppression of LPS-stimulated inflammatory variables such as ROS, NO production, COX2 and iNOS. The extract and its PEFs were found to be most active in the early wound healing stage by stimulating mitochondrial bioenergetics and by the up-regulation of the expression of important extracellular matrix.[63] Thus, the overall study confirmed the role of lingonberry and its PAC fraction in wound repair, which may contribute to aberrant extracellular matrix (ECM) remodeling and collagen fragmentation mediated through immunomodulation and timely inflammatory solutions.

13.6.14 Poison Gooseberry (*Withania somnifera*)

The hydroalcoholic root extract of the plant *Withania somnifera* L. Dunal (Solanaceae), also known as poison gooseberry, was subjected to wound healing activity using the excision wound model. Eucerin was used as ointment base and different concentrations of extract ointment were prepared using 20, 40 and 60% w/w of extract. The respective ointments were applied to the open wound of the rats for 14 days. The group treated with 60% ointment showed maximum wound healing potential from the 8th day of treatment, as evident through the significant reduction in the wound area compared to negative control group rats.[72]

13.6.15 Star Gooseberry (*Sauropus androgynus*)

The 5% cream prepared from the leaf extract of the plant *Sauropus androgynus* L. Merr. (Phyllanthaceae) was studied for its potential in wound healing activity using incision and excision wound models in Wistar rats. The results demonstrated significant wound healing activity of the extract ointment confirmed through increased rate of wound contraction showing complete 100% healing as compared to $30.5 \pm 2.4\%$ healing of control group animals on the 17th day. The study also showed an increased tensile strength of 162.86 ± 6.3 g in the case of animals treated with ointment as compared to control group animals showing a breaking strength of 95 ± 5 g and demonstrated higher re-epithelialization rate. The histopathological studies also confirmed the wound healing activity showing high collagen formation, fibroblast and low macrophage numbers in treated animals as compared to control animals.[73] Studies related to the identification of the exact mechanism of wound healing and the responsible factor from the plant are still required to understand the exact role of star gooseberry in healing.

13.6.16 Strawberry (*Fragaria ananassa*)

AEFs and PEFs of strawberry (*Fragaria ananassa* Duchesne., Rosaceae) were subjected to *in vitro* anti-inflammatory and wound healing activity using an LPS-stimulated RAW 264.7 macrophage model and a skin fibroblast migration and proliferation assay. The results showed that AEF and PEF significantly reduced the gene expression of IL-1β, IL-6 and iNOS. From the overall observations, the PEF was found to be more promising as an anti-inflammatory agent, while AEF was found to be effective as potential wound healing agent.[40]

13.6.17 Wild Blueberry (*Vaccinium angustifolium*)

The present investigation deals with evaluating the potential of ACN, phenolic acid (PA) fractions and their combination (at different concentrations *i.e.* 0.002, 8, 15, 60, and 120 $\mu g\,mL^{-1}$) extracted from the wild blueberry (*V. angustifolium* Aiton and Benth., Ericaceae) on endothelial cell migration, gene expression and protein levels of *RAC1* and *RHOA*. The observation revealed a significant ($P \leq 0.05$) inhibition in the endothelial cell migration on treatment with ACN at 60 $\mu g\,mL^{-1}$, whereas PA significantly increased the endothelial cell migration at different concentrations *i.e.* 0.002 $\mu g\,mL^{-1}$ ($P \leq 0.0001$), 60 $\mu g\,mL^{-1}$ ($P \leq 0.0001$), and 120 $\mu g\,mL^{-1}$ ($P \leq 0.01$). Further, ACN downregulated the gene expression of *AKT1* and endothelial nitric oxide synthase (eNOS) and PA upregulated the gene expression of *AKT1* and vascular endothelial growth factor (VEGF).[74] In other studies, Human umbilical vein endothelial cells (HUVEC) treatment with the ACN extract decreased overall endothelial cell tube formation inhibiting angiogenesis, while treatment with PAs increased overall endothelial cell tube formation compared to control promoting angiogenesis.[75] Thus, the present investigations confirmed the role of ACNs from wild blueberry in inhibiting endothelial cell migration and angiogenesis, while the phenolic acid fraction increased migration of endothelial cells and angiogenesis in a concentration-dependent manner.[74,75] Therefore, we may presume that the type of extract from wild blueberry modulates the angiogenesis process, pointing to a wound healing potential of the wild blueberry.

13.7 Conclusion

In conclusion, we have made an attempt to review the role of different types of berries and medicinal plants bearing berries as a potential source of wound healing agents. We have discussed wound healing in brief, including its physiological process and etiology and complications hampering wound healing such as diabetes, diabetic neuropathy, diabetic foot ulcer, narrowed blood vessels and chronic inflammation. It also includes a section detailing natural products in wound healing and the therapeutic benefits of consuming different berries. The chapter presents the details of different wound healing studies performed with various berries using *in vivo* or *in vitro* models. Some of the major berries possessing potential wound healing activity include acai berry, barberry, gooseberry, blackberry, bog blueberry cranberry, crowberry, elderberry, golden Himalayan raspberry, Indian gooseberry, star gooseberry, strawberry and wild blueberry. The role of these berries in inhibiting certain biochemical pathways/signaling molecules hampering wound healing, like interleukins, nitric oxide, iNOS, ROS and other inflammatory mediators, have also been presented. Berries have also been shown to significantly increase cross-linking of collagen in tissues, increase epithelialization rate drastically and promote cell migration and angiogenesis, which scientifically justifies their traditional use as potential wound healing agents. Thus, the present overview on berries as wound healing agents could prove beneficial for researchers as well as pharmaceutical and naturopathic companies for further exploring their medicinal values.

Abbreviations

ABWE	Acai berry water extracts
ACN	anthocyanin
AEF	Anthocyanin-enrich fraction
AGE	Advanced glycation end-product
ATP	Adenosine triphosphate
COX2	Cyclooxygenase-2
HPLC–TOF-MS	High-performance liquid chromatography–time-of-flight-mass spectrometry
IL-1β	Interleukin 1-beta
IL-6	Interleukin 6
iNOS	Inducible nitric oxide
LPS	Lipopolysaccharide
MMP-1	Matrix metalloproteinase-1
PA	phenolic acid
PAC	Proanthocyanidin
PEF	Proanthocyanidin-enriched fraction
STZ	Streptozotocin
TNF-α	Tumor necrosis factor-alpha

References

1. H. Sorg, D. J. Tilkorn, S. Hager, J. Hauser and U. Mirastschijski, *Eur. Surg. Res.*, 2017, **58**, 81.
2. A. Gosain and L. A. DiPietr, *World J. Surg.*, 2004, **28**, 321.
3. A. Joshi, V. K. Joshi, D. Pandey and S. Hemalatha, *J. Ethnopharmacol.*, 2016, **191**, 95.
4. K. M. Parmar, P. R. Shende, N. Katare, M. Dhobi and S. K. Prasad, *J. Pharm. Pharmacol.*, 2018, **70**, 1389.
5. G. S. Schultz, *Acute and Chronic Wounds: Nursing Management*, Mosby (Elsevier), Philadelphia, 2nd edn, 2000.
6. E. M. Golebiewska and A. W. Poole, *Blood Rev.*, 2015, **29**, 153.
7. N. Strbo, N. Yin and O. Stojadinovic, *Adv. Wound Care*, 2014, **3**, 492.
8. S. A. Eming, M. Hammerschmidt, T. Krieg and A. Roers, *Semin. Cell Dev. Biol.*, 2009, **20**, 517.
9. J. E. Park and A. Barbul, *Am. J. Surg.*, 2004, **187**, 11S.
10. K. Lau, R. Paus, S. Tiede, P. Day and A. Bayat, *Exp. Dermatol.*, 2009, **18**, 921.
11. I. Pastar, O. Stojadinovic, N. C. Yin, H. Ramirez, A. G. Nusbaum, A. Sawaya, S. B. Patel, L. Khalid, R. R. Isseroff and M. Tomic-Canic, *Adv. Wound Care*, 2014, **3**, 445.

12. M. Xue and C. J. Jackson, *Adv. Wound Care*, 2015, **4**, 119.
13. R. Blakytny and E. Jude, *Diabetic Med.*, 2006, **23**, 594.
14. A. J. Boulton, *Med. Clin. North Am.*, 2013, **97**, 775.
15. V. P. Singh, A. Bali, N. Singh and A. S. Jaggi, *Physiol. Pharmacol.*, 2014, **18**, 1.
16. V. Falanga, *Lancet*, 2005, **366**, 1736.
17. N. Spravchikov, G. Sizyakov, M. Gartsbein, D. Accili, T. Tennenbaum and E. Wertheimer, *Diabetes*, 2001, **50**, 1627.
18. M. L. Usui, J. N. Mansbridge, W. G. Carter, M. Fujita and J. E. Olerud, *J. Histochem. Cytochem.*, 2008, **56**, 687.
19. M. Volmer-Thole and R. Lobmann, *Int. J. Mol. Sci.*, 2016, **17**, 917.
20. A. R. Barker, G. D. Rosson and A. L. Dellon, *Ann. Plast. Surg.*, 2006, **57**, 339.
21. S. Pendsey, *Diabetic Foot: A Clinical Atlas*, Jaypee Brothers Medical Publishers, India, 2003.
22. C. K. Sen, *Wound Repair Regen.*, 2009, **17**, 1.
23. G. Crovetti, G. Martinelli, M. Issi, M. Barone, M. Guizzardi, B. Campanati, M. Moroni and A. Carabelli, *Transfus. Apher. Sci.*, 2004, **30**, 145.
24. Wound Source, *How Diabetes Impacts Wound Healing*. December 4, 2017, https://www.woundsource.com/blog/how-diabetes-impacts-wound-healing.
25. T. Dinh, F. Tecilazich, A. Kafanas, J. Doupis, C. Gnardellis, E. Leal, A. Tellechea, L. Pradhan, T. E. Lyons, J. M. Giurini and A. Veves, *Diabetes*, 2012, **61**, 937.
26. J. Caporusso, R. Abdo, J. Karr, M. Smith and A. Anaim, *Wounds*, 2019, **31**, S19.
27. G. Y. Zhang, E. A. Langan, N. T. Meier, W. Funk, F. Siemers and R. Paus, *PLoS One*, 2019, **14**, e0212659.
28. I. Pastar, L. L. Wong, A. N. Egger and M. Tomic-Canic, *Exp. Dermatol.*, 2018, **27**, 551.
29. N. Pazyar, R. Yaghoobi, E. Rafiee, A. Mehrabian and A. Feily, *Skin Pharmacol. Physiol.*, 2014, **27**, 303.
30. R. F. Pereira and P. J. Bártolo, *Adv. Wound Care*, 2016, **5**, 208.
31. T. Maver, U. Maver, K. S. Kleinschek, D. M. Smrke and S. Kreft, *Int. J. Dermatol.*, 2015, **54**, 740.
32. R. Thakur, N. Jain, R. Pathak and S. S. Sandhu, *Evidence-Based Complementary Altern. Med.*, 2011, 438056.
33. N. I. Ibrahim, S. K. Wong, I. N. Mohamed, N. Mohamed, K. Y. Chin, S. Ima-Nirwana and A. N. Shuid, *Int. J. Environ. Res. Public Health*, 2018, **15**, 2360.
34. M. Chaudhari and S. Mengi, *Phytother. Res.*, 2006, **20**, 799.
35. J. Li, J. Chen and R. Kirsner, *Clin. Dermatol.*, 2007, **25**, 9.
36. M. Barchitta, A. Maugeri, G. Favara, R. M. S. Lio, G. Evola, A. Agodi and G. Basile, *Int. J. Mol. Sci.*, 2019, **20**, 1119.
37. M. Georgescu, O. Marinas, M. Popa, T. Stan, V. Lazar, S. V. Bertesteanu and M. C. Chifiriuc, *Worldwide Wound Healing-Innovation in Natural and Conventional Methods*, Intech Open, London, 2016.

38. A. de Fatima, L. V. Modolo, A. C. C. Sanches and R. R. Porto, *Mini-Rev. Med. Chem.*, 2008, **8**, 879.
39. A. Shedoeva, D. Leavesley, Z. Upton and C. Fan, *Evidence-Based Complementary Altern. Med.*, 2019, 2684108.
40. F. V. Velde, D. Esposito, M. H. Grace, M. E. Pirovani and M. A. Lila, *Food Res. Int.*, 2019, **121**, 453.
41. J. L. Slavin and B. Lloyd, *Adv. Nutr.*, 2012, **3**, 506.
42. L. A. Bazzano, *World Health Organization and Joint FAO/WHO Workshop on Fruit and Vegetables for Health*, World Health Organization, Japan, 2005.
43. S. Y. Wang and H. Jiao, *J. Agric. Food Chem.*, 2000, **48**, 5677.
44. N. P. Seeram, *Berries*, Academic Press, London (United Kingdom), 2nd edn, 2006.
45. F. L. Meyskens and E. Szabo, *Cancer Epidemiol., Biomarkers Prev.*, 2005, **14**, 1366.
46. N. P. Seeram, *J. Agric. Food Chem.*, 2008, **56**, 627.
47. S. H. Nile and S. W. Park, *Nutrition*, 2014, **30**, 134.
48. H. Xue, R. M. Aziz, N. Sun, J. M. Cassady, L. M. Kamendulis, Y. Xu, G. D. Stoner and J. E. Klaunig, *Carcinogenesis*, 2002, **22**, 351.
49. B. S. Hale, A. N. Carey, D. Jenkins, B. M. Rabin and J. A. Joseph, *Neurobiol. Aging*, 2007, **28**, 1187.
50. A. Szajdek and E. J. Borowska, *Plant Foods Hum. Nutr.*, 2008, **63**, 147.
51. D. Ferreira, G. G. Gross, H. Kolodziej and T. Yoshida, *Phytochem.*, 2005, **66**, 1969.
52. A. B. Howell, *Mol. Nutr. Food Res.*, 2007, **51**, 732.
53. B. T. F. Kraft, M. Dey, R. B. Rogers, D. M. Ribnicky, D. M. Gipp, W. T. Cefalu, I. Raskin and M. A. Lila, *J. Agric. Food Chem.*, 2008, **56**, 654.
54. T. Nurmi, J. Mursu, M. Heinonen, A. Nurmi, R. Hiltunen and S. Voutilainen, *J. Agric. Food Chem.*, 2009, **57**, 2274.
55. A. Schieber, F. C. Stintzing and R. Carle, *Trends Food Sci. Technol.*, 2001, **12**, 401.
56. C. Beer, R. A. Myers, J. H. Sorenson and L. R. Bucci, *Curr. Top. Nutraceutical Res.*, 2004, **2**, 227.
57. J. Parry, L. Su, M. Luther, K. Zhou, M. P. Yurawecz, P. Whittaker and L. Yu, *J. Agric. Food Chem.*, 2005, **53**, 566.
58. M. H. Kang, S. Choi and B. Kim, *Toxicol. Res.*, 2017, **33**, 149.
59. N. H. Kang and B. Kim, *Toxicol. Res.*, 2018, **34**, 97.
60. N. C. C. Pinto, P. C. Vieira, E. M. Souza-Fagundes, L. S. Barcelos, M. C. M. N. Castañon and E. Scio, *J. Ethnopharmacol.*, 2016, **194**, 131.
61. M. Pashaee, A. Shiravi and V. Hojati, *J. Chem. Health Risks*, 2016, **6**, 319.
62. G. Kendir, I. Süntar, A. O. Çeribaşı and A. Köroğlu, *J. Ethnopharmacol.*, 2019, **237**, 141.
63. D. Esposito, J. Overall, M. H. Grace, S. Komarnytsky and M. A. Lila, *Front. Pharmacol.*, 2019, **10**, 1058.
64. B. S. Nayak, D. D. Ramdath, J. R. Marshall, G. Isitor, S. Xue and J. Shi, *Phytother. Res.*, 2011, **25**, 1201.

65. I. P. Süntar, E. K. Akkol, F. N. Yalçin, U. Koca, H. Keleş and E. Yesilada, *J. Ethnopharmacol.*, 2010, **129**, 106.
66. D. N. Z. Ghabaee, M. A. Ebrahimzadeh, J. Akbari and F. T. Amiri, *Int. J. Pharm. Sci. Res.*, 2017, **8**, 132.
67. B. P. George, T. Parimelazhagan, Y. T. Kumar and T. Sajeesh, *J. Acupunct. Meridian Stud.*, 2015, **8**, 134.
68. I. Süntar, U. Koca, H. Keleş and E. K. Akkol, *Evidence-Based Complementary Altern. Med.*, 2011, 816156.
69. S. Kirubanandan and S. Renganathan, *Int. J. MediPharm Res.*, 2016, **2**, 71.
70. H. S. Datta, S. K. Mitra and B. Patwardhan, *Evidence-Based Complementary Altern. Med.*, 2011, 134378.
71. M. Sumitra, P. Manikandan, V. S. Gayathri, P. Mahendran and L. Suguna, *Wound Repair Regen.*, 2009, **17**, 99.
72. K. Roshanai and N. Ajand, *J. Shahid Sadoughi Univ. Med. Sci.*, 2015, **23**, 900.
73. A. Bhaskar, K. V. Ramesh and S. Rajeshwari, *J. Nat. Rem.*, 2009, **9**, 159.
74. P. Tsakiroglou, J. Webber, S. Ashworth, C. D. Bo and D. Klimis-Zacas, *J. Cell. Biochem.*, 2019, **120**, 11056.
75. P. Tsakiroglou, J. Webber, S. Ashworth, C. D. Bo and D. Klimis-Zacas, *J. Med. Food*, 2021, **24**, 226.

Subject Index